SYSTEMS APPROACH
TO AIR POLLUTION CONTROL

SYSTEMS APPROACH TO AIR POLLUTION CONTROL

ROBERT J. BIBBERO, P.E.
IRVING G. YOUNG, Ph.D.
Advanced Technology Staff
Honeywell Inc.
Process Control Division

A WILEY-INTERSCIENCE PUBLICATION
JOHN WILEY & SONS

NEW YORK   LONDON   SYDNEY   TORONTO

Copyright © 1974, by John Wiley & Sons, Inc.

All rights reserved. Published simultaneously in Canada.

No part of this book may be reproduced by any means, nor transmitted, nor translated into a machine language without the written permission of the publisher.

Library of Congress Cataloging in Publication Data:
Bibbero, Robert J
   Systems approach to air pollution control.

  "A Wiley-Interscience publication."
  Includes bibliographical references.
   1. Air-Pollution. 2. Air quality management.
I. Young, Irving G., joint author. II. Title.
[DNLM: 1. Air pollution—Prevention and control.

WA754 B581s]
TD883.1.B52   363.6   74-8905
ISBN 0-471-07205-2

Printed in the United States of America

10 9 8 7 6 5 4 3 2 1

For Herbert and Ira

and

For Jeffrey and Arlene

Preface

The authors, like many of their technologically oriented contemporaries, gave little thought to ecological matters prior to the last decade. Our generation of scientists and engineers has been oriented toward concepts of industrial growth and expansion, increased efficiency in production, and the exploitation of resources. Conventional economic criteria and concepts of optimum design give little room for concern about the long-range effects of waste accumulation or of possible harmful side effects of products.

Technology, in partnership with science, has, within the last century, given man the capability to alter his environment profoundly, whereas in the first million years of his existence man was merely one more element in the vast interlinked chain of nature. In the twentieth century we have begun to realize the full measure of the disruptive influence of urban and rural industrialization and have paid the first installment of its toll on ourselves and our fellow species. Air pollution, although not new, has increased to intolerable proportions in many localities, and the irreversible pollution of rivers, lakes, and entire seas has become a melancholy commonplace.

When we realize that these unpleasant and frightening effects are linked to our growth in numbers and consumption per capita, made possible by the gifts of technology, our first, unthinking reaction is to call a halt to expansion and return to a simpler mode of life. On reflection, it becomes clear that this is both undesirable and impossible to achieve, short of an accompanying cataclysm. The answer is not to turn our back on science, which makes it possible to inhabit the earth in such numbers and to exert such power over our environment, but to use its gifts more wisely. Science is the only means by which we can learn how we can fit our urban industrialization into the complex ecology of the earth without destroying ourselves in the process.

Our need for the means by which we can live at peace with nature, as well as with ourselves, is our greatest threat and challenge. The search for truth—science—is the only way we can attain this goal. Technology, then, will enable us to cancel or alleviate, at minimum penalty, the unwanted side effects of our exploitation of the earth.

Our utilization of nature's resources cannot continue to grow forever. The fossil fuels (coal and oil) and the minerals that become the metal skeleton of our civilization are being removed from the earth at rates vastly greater than their formation and replacement. Ultimately, we must learn to replace or reuse all that we have. Our production, then, will become a closed cycle; we will re-form, but not alter, the sum of our material wealth. In nature there is no waste, so that when we learn to blend with nature, our industrial activities will not be the cause of pollution. The economic basis of our production efficiency will then change so as to reflect the true costs of re-forming materials or of converting to our use the energy that must ultimately stem from the sun.

The minimization of these true costs and the organization of an efficient production structure truly representing them is the goal of the systems design to which we have reference in this book. We do not underestimate the complexity of the problem. As a systems problem, the elimination of pollution, that of the air included, encompasses all the technological disciplines—life sciences as well as the harder varieties. Furthermore, these disciplines or the lessons they teach us must enter into the framework of the political power structure in order to have the necessary effect. Without underestimating the enormously long and arduous path we must tread to compensate for our errant wanderings since Galileo's day, we are optimistic that science, people, and politics are equal to the challenge. Realization of ignorance is the first step toward knowledge; we have that realization and have faith that the solution will be forthcoming in the next generation.

ACKNOWLEDGMENTS

We wish to express appreciation to Honeywell, Inc., for permission to utilize portions of our unpublished internal reports and studies and to our associates in the Process Control Division for their generous support, consultation, and use of facilities. In particular, we thank James E. Myers, James R. Berrett, and Robert L. Wilson for encouragement and numerous helpful discussions; Ethlyn Thomson for invaluable library support; and Margie Nelis, who typed the manuscript with skill and patience. We are grateful to James M. Lufkin of Honeywell's Corporate Staff for smoothing the way toward the entire project. We acknowledge with thanks permission of the *IEEE Spectrum* to use, in Chapters 5–9, material adapted from a three-part article, "Systems Approach Toward Nationwide Air-Pollution Control," published in the October, November, and December 1971 issues of the *Spectrum*.

Fort Washington, Pennsylvania *Robert J. Bibbero*
February 1974 *Irving G. Young*

Contents

	1 INTRODUCTION	1
1.1	Historical Perspective	1
1.2	Air Pollution as a Societal Systems Problem	6
1.3	Societal Viewpoints	11
1.4	Definitions and Scientific Fundamentals	15
	References	36

	2 AIR POLLUTANTS: SOURCES, SINKS, RESIDENCE TIMES	41
2.1	What Is a Pollutant? Background Concentrations in the Atmosphere	41
2.2	The Sources of Materials Presently Considered Pollutants	43
2.3	Reactions in the Atmosphere and Sinks for Pollutants	65
2.4	Residence Times	73
2.5	Summary	76
	References	78

	3 SOCIETAL AND ECONOMIC COSTS OF AIR POLLUTION	81
3.1	Evaluation of Air Environmental Insults	81
3.2	Global Effects	82
3.3	Urban Effects and Public Health	89
3.4	Economic Costs of Air Pollution	104
3.5	Aesthetic Degradation from Air Pollution	120
3.6	Growth of Air Pollution	128
	References	134

	4 LEGISLATIVE BASIS OF AIR-POLLUTION CONTROL SYSTEMS	139
4.1	Evolution of Federal Legislation	140
4.2	The Clean Air Amendments of 1970	145

4.3	State Legislation and Regulations	163
4.4	Worldwide Air-Pollution Regulations	184
	References	184

5 CONTROL STRATEGIES AND SYSTEMS DESIGN — 187

5.1	Overall Strategies	181
5.2	Overview of Existing U.S. Control System	190
5.3	Constraints and Cost Effectiveness	199
5.4	System Objectives	204
5.5	Second-Generation System	207
5.6	Summary	209
	References	210

6 DATA ACQUISITION AND MONITORING — 213

6.1	Characteristics of Air Pollution Data	215
6.2	Network Design for Data Acquisition and Handling	233
6.3	Summary	250
	References	250
	Appendix	253

7 APPLICATIONS OF AIR POLLUTION DATA — 267

7.1	Indexes	268
7.2	Forecasting and Trend Prediction	281
7.3	Meteorological Measurements	297
	References	299

8 MATHEMATICAL MODELS OF AIR POLLUTION — 301

8.1	Quantitative Basis of Air-Pollution Models	305
8.2	Application of Models	316
8.3	Experimental Verification of Plume Models	331
8.4	Photochemical Models	334
8.5	Implementation of Urban Models	338
	References	346

9 COMMUNITY AIR-QUALITY MANAGEMENT: CONTROL, MONITORING, AND ABATEMENT — 349

9.1	Prior Controls	350
9.2	Posterior Control of Stationary Sources	351

9.3	Current Ambient Monitoring Systems	356
9.4	Short-Term and Episode Control	360
	References	363

10 GAS- AND VAPOR-MONITORING INSTRUMENTS 365

10.1	System Requirements of Pollutant Monitors	365
10.2	Instruments and Methods Requiring Separation Before Measurement	388
10.3	Instruments and Techniques That Measure the Sample Directly	401
10.4	Calibration and Standardization	439
	References	443

11 PARTICULATE-MONITORING INSTRUMENTS 451

11.1	Nomenclature and Definitions	451
11.2	Sources and Distribution	451
11.3	Mechanisms of Aerosol Formation	453
11.4	Size, Shape, and Mass Parameters	456
11.5	Composition, Reactions, Fate	457
11.6	Measurement Problems of Particulate Matter	460
11.7	Integrated Information Measurements	465
11.8	Methods Adapted to Continuous Monitoring of Particulates	477
11.9	Evaluation of Particulate Monitors for System Purposes	484
11.10	Calibration of Measurements on Particulates	489
	References	490

12 THE FUTURE OF AIR-POLLUTION CONTROL SYSTEMS 495

12.1	New Data and Revised Objectives	495
12.2	Interface with Other Societal Problems	498
12.3	Preservation of "Clean" Areas	500
12.4	Cost–Effective and Cost–Benefit Analysis	506
	References	511

Author Index	513
Subject Index	517

1

Introduction

1.1. HISTORICAL PERSPECTIVE

If by *air pollution* we mean substantial changes in the composition of the atmosphere sufficient to alter the ecological balance of the earth's life forms, then the origin of air pollution was concurrent with the origin of life itself.

The primitive, abiogenic atmosphere, some 4 billion years ago, contained no free oxygen; it consisted largely of hydrogen and its compounds, including water vapor, ammonia, methane, and some hydrogen sulfide. What little oxygen was produced by the action of the short ultraviolet radiation in sunlight on water vapor was quickly absorbed by the unsaturated rocks. Free oxygen did not appear on earth before green plants flourished.[1]

Under the primitive earth conditions—reducing atmosphere, warm oceans made alkaline by dissolved ammonia, energy from lightning and ultraviolet radiation—life was initiated. The steps in this process have been demonstrated recently by research workers in the United States. Simulating the primordial atmospheric environment, they showed the spontaneous production of amino acids; polypeptides (long chains of amino acids that resemble proteins); and nucleotide phosphates, the precursors of nucleic acids and of ATP, an essential element of the photosynthesis process.[2] The lack of free oxygen was an essential element in the genesis of these protoliving materials.

From the origin of life to the first evidence of photosynthesis, fossil algae and molecular fragments of chlorophyll in sediments 2.7 billion years old, encompassed a span of 1 to 1.5 billion years. Yet another 1.5 billion years elapsed before the emergence of an oxygen atmosphere on earth resembling our present air. The original living organisms had over 2.5 billion years to adapt to it, but they could never have come into being or even survived in the oxidizing atmosphere of today. Oxygen, to the earliest life-form, would have been a poisonous gas. It is suggested, parenthetically, that earth will have only this one chance to generate life as we recognize it—a significant warning. But the more pertinent lesson is that life did adapt, thrive, and evolve, despite the drastic alteration of the atmosphere, only because of the unimaginable stretch of time, measured in billions of years, which accommodated the change.

What is happening to our urban and global atmosphere today because of the technical activities of our species is insignificant compared to these ancient geological transformations, but the time scale of current changes is also miniscule, so that the rate of change may be tremendously greater. The combustion of fossil fuels, coal and petroleum, is the principle source of our present air changes. But coal only came into use in the thirteenth century and petroleum, in the twentieth. Yet the chief product of combustion of these substances, carbon dioxide, has increased by more than 10% since 1890 and is currently growing at the rate of 0.27% per year. We doubt if man can adapt to changes of this speed and magnitude by processes of evolution. Hence, the problem of air pollution as a threat to the welfare or survival of mankind looms today, in our accelerating technological growth, for the first time in our existence.

Although the unpleasant, and even lethal, effects of air pollution must have plagued men since fire was first introduced into caves, the ability to contaminate large areas of urban air space awaited the dawn of the industrial age, in the form of the capacity to mine quantities of cheap coal and to transport it by sea. So immediate was the concern for the effect on health of this smoky, bituminous "sea" coal that its entry into London and other English cities was followed by the first smoke-abatement act, in 1273.[3] In 1306, a royal proclamation was issued, forbidding the burning of coal during sessions of Parliament. Artisans and manufacturers of the period were forced to utilize the more expensive wood fuel, and opposition was such that one man was hanged for violating the ordinance. (In the New World, during the same early period, a similar story is told of the populous and highly developed Aztec civilization in Mexico City. There, an inhabitant was allegedly executed for polluting the thin high altitude air of the city with charcoal smoke.)

By the seventeenth century the edict of 1306 had apparently been forgotten, and the quality of London's air had fallen so low as to be strongly castigated in pamphlets, notably one by John Evelyn, which was published in 1661 and again in 1772.[4] But this and other similar public protests, decrees, and investigating commissions failed to solve the problem. Its continual exacerbation in the fast-growing city culminated in modern times in the disaster of 1952. In December of that year a "smog" (heavy fog and smoke) developed, and nearly 4000 people, most of them elderly, died of effects attributed to the air pollution during one week and another 8000 over

1.1. Historical Perspective

the next three months. The Beaver commission, appointed to recommend means to prevent recurrence of the London "episode," resulted in Great Britain's Clean Air Act of 1956. Although strict enforcement of this act has no doubt improved the deteriorating situation in London, at least one other serious episode occurred in December 1962, which is alleged to have caused 750 deaths.[4]

Other parts of the world have also suffered serious losses from the combination of coal smoke or other industrial fumes and unfavorable atmospheric conditions that prevent its dissipation, but none have so aroused world opinion and action. Over 1000 persons died from the effects of air pollution in Glasgow during 1909. The 1936 tragedy of the Meuse Valley of Belgium, in which 63 lives were lost, will be recalled as having first been thought to be the result of the escape of war gases, so drastic was the effect of the polluted air. But later it was proven to have been caused by emissions from industrial plants concentrated in the region.

Air pollution in the United States has followed a similar course, although we have fortunately been spared, to date, the magnitude of London's misfortune. In 1948 an industrial episode similar to that in the Meuse occurred in Donora, Pennsylvania, located in a confined valley along the Monongahela River, causing the death of 20 and illness to thousands. The combination of a narrow valley; poor atmospheric ventilation; and concentrated industrial emissions from steel mills, smelters, coke, and other plants; was blamed. Local protest movements sprang up to demand protection of the public from smoke and dirt from factory stacks. Ultimately, these led to pioneer smoke-abatement ordinances, notably in the cities of Chicago, Pittsburgh, Cincinnati, and St. Louis. These experiments in prevention proved that high smoke densities could be avoided, although at high cost, by the employment of less smoky fuels and by enforcing the utilization of more efficient combustion processes.

The first specific air-pollution control law in the United States was adopted by the city of Chicago in 1881. It prohibited the emission of "dense smoke" and provided for fines of $5 to $50 for offenders. Later amendments gave the city control over new and modified furnace plants. This law was followed by other local control ordinances.

The efficacy of Pittsburgh's smoke-control ordinances has offered an encouraging example. The city is situated on a triangular plot of land in a deep valley formed by the confluence of the Monongahela and Allegheny rivers. Its reputation and its appellation as the "Smoky City" derived from the seven railroads that served it and from its concentrated industry—over 20 million tons annual capacity for steel alone.

Since the turn of the century this city's commercial leaders have sponsored smoke-abatement committees which, in 1906, effected the passage of one of the earliest American smoke-control laws. The ordinance relied chiefly on self-regulation, and its success was predicated on educational campaigns, waged by the civic groups, to improve combustion equipment and methods. As in early London, moderate measures, protests, and education proved to be of little avail against the powerful forces of urban growth, and air quality worsened. In 1941 the aroused city passed a new and stringent law, successively stiffened by amendments over the next 10 years. These regulations were jointly administered by city and county officials, the resulting cooperation and lack of administrative conflict adding to the effectiveness of enforcement. (Monitoring methods currently used by the enforcing agency will be described in more detail in a later chapter).

Ordinance and amendments controlled both industry and domestic users of heating fuel. The former group was required to conform to limits set on the emission of smoke, fumes, ash, soot, and cinders; toxic, radioactive, or malodorous substances; and others deemed to be air pollutants. Specific limits were placed on the emission of solid matter from steel furnaces and similar stacks. Home-owners were required to install efficient stokers if they wished to burn volatile bituminous coal; otherwise, they were restricted to oil, gas, anthracite coal, or less-volatile fuels.

A Bureau of Air Pollution Control was established to inspect furnaces, installations, and repairs (except domestic) and to monitor industrial emissions. Some concept of the magnitude of this task is gained by considering that there are over 1000 stacks in the area that must be controlled. The results of these labors and expenditures, however, have been a signal success. In a recent review, Pittsburgh, the former "Smoky City," was rated as having only half the sootfall of New York City in the same period, 30 tons/mile^2 compared with New York's 60.[5]

Visible coal smoke and fog, generally termed "London smog," have unfortunately proved to be far from the only hazardous air pollutant, and we have seen that valleys are not the only locations conducive to the accumulation of dangerous concentrations of these substances.

In the area that now includes the city of Los Angeles and its environs, atmospheric conditions unfavorable to ventilation have long prevailed. In 1542, Cabrillo, the discoverer of Long Beach and San Pedro harbors, must have noted the smoke lingering from Indian campfires when he named them Bahia de los Humos, "Bay of Smokes." In modern times, the open, suburban construction of Los Angeles and its lack of heavy industry utilizing coal, together with the use of natural gas for its modest home heating needs, obviated the buildup of any London-type smoke hazard, even when the city's population approached its current proportions.

The hazards of petroleum combustion, however, which in the form of automobile exhaust was recognized as an air pollutant as early as 1915,

1.1. Historical Perspective

proved to be Los Angeles's nemesis. The sprawling distances within the city, which kept down the concentration of heating combustion products, also stimulated the use of automobile transportation on what was then an unprecedented scale. Also, during World War II, many new industries sprang up in the area, and petroleum processing was greatly stimulated. Now, all the conditions were present for the genesis of "Los Angeles smog," once a comedian's byword and now a menace spreading with the growth of automobile transport to cities all over the globe.

The ingredients of Los Angeles smog, as we shall see, are petroleum vapors, in the form of hydrocarbons and exhaust fumes, exposed to sunlight and contained under an atmospheric "lid," which concentrates the mixture while so-called secondary reactions take place in the atmosphere. The results of these reactions cause irritation to the eyes and impairment of lung function in sensitive persons, damage to rubber and fabrics, and injury to many valuable crops and other plants, as well as lowered visibility. Rubber damage alone is so serious that tires sold in California must be specially formulated to combat disintegration.

The mechanism by which these damaging and irritating substances, termed *oxidants*, are formed was not unraveled for another decade. By the 1950s, it had been demonstrated by A. J. Haagen-Smit of the California Institute of Technology that secondary reactions were occurring between the hydrocarbon vapors and nitrogen oxides (from combustion processes), under the influence of ultraviolet light—a photochemical reaction triggered by sunlight. The resulting products include ozone and a number of highly reactive organic oxidants, including peroxyacetyl nitrate (PAN) and acrolein, which are irritating in concentrations of parts per billion.

The dominant cause of photochemical smog is the automobile, and as all the contributing factors—high-density traffic, sunny days, and atmospheric stability—are present in abundance in Los Angeles, this city still suffers most seriously from this problem. But other U.S. cities are following suit. It has been shown that most of California's major cities, and St. Louis, Denver, Philadelphia, Cincinnati, Chicago, and Washington, D.C., exceed eye-irritating concentrations of oxidants roughly half the days of the year.[6] As the automobile proliferates, in Rome, Tokyo, and even Honolulu, the Los Angeles smog travels with it around the world.

Currently, in the United States, major efforts are being made to combat various types and sources of air pollution. Local enforcement of regulations against stationary-source emissions is not effective in the face of atmospheric contamination by mobile and interstate sources and are being supplanted by federal and state laws—chiefly the Clean Air Act of 1963 and its amended form of 1970. The provisions and implications of these laws will be detailed further in the text. Air pollution is now a national, and even an international, problem; it is no longer confined to a single city or state nor even to a single country. Whether these new measures will be able to more than temporarily stem the rising tide of atmospheric pollution or whether entirely new economic and technical approaches will be required remains to be seen.

1.2. AIR POLLUTION AS A SOCIETAL SYSTEMS PROBLEM

The difficulty that arises in an attempt to study the problem of air pollution rationally is that the subject reveals very different aspects to different people, much as the proverbial elephant and the blind men. The point of view may vary from that of the professional meteorologist or research chemist to the nature cultist's. Each viewpoint is entitled to respectful consideration on its own merits. The "worship of nature," faction, which may be considered an extreme cult of the ecology movement, has a legitimate historical and philosophic basis in the writings of the eighteenth-century philosopher Jean-Jacques Rousseau and the poet Thomas Grey.[7] Those influenced by Rousseau's concepts included some of the men who drafted the Constitution of the United States.

In approaching a problem from the systems point of view, it is necessary to account for all of its aspects, even if some cannot be logically justified or quantified. They may be considered as part of its background or "environment." It is characteristic of urban and other social problems that economic, political, moral, ethnic, and even selfish or seemingly irrational elements must be considered and accounted for; otherwise, no viable or "salable" solution can be designed. In this respect, the systems approach to air pollution, which is clearly a problem of this class, differs widely from that of the aerospace and military. In the latter systems, the external and operating environments are rigidly defined and highly structured and the motivation is relatively simplistic.

On the other hand, air-pollution control is different from other important societal problems: poverty, ghetto housing, mass education, and street crime. The last are almost entirely functions of the social and economic interactions between people. Technical events, except those in the production base, which affect employment, enter only peripherally into their frequency and means of control. For example, a technical ability to provide more housing at low cost is, in itself, inadequate to solve the problem of deteriorating inner-city ghettos. This has been clearly shown in recent studies by Jay W. Forrester of the Massachusetts Institute of Technology, who utilized mathematical models and a specially designed computer language.[8]

1.2. Air Pollution as a Societal Systems Problem

The solution to air-pollution problems, on the contrary, lies largely in the technical domain. The actions of the average householder have little influence on the degree of pollution of the airspace surrounding his home; even the emissions of his heating furnace are determined by the technical (and economic) factors determining the availability of various types of fuel. The exhaust emissions of automobiles will be largely controlled by the specifications set by federal-government engineers and by the stringency of inspection and maintenance imposed by the states. These will, in turn, be almost entirely decided by the technical state of the art in the model year in which automotive regulations become effective. Likewise, industrial-plant emissions are a function of the interplay between the technical means of surveillance and measurement of pollutants available to regulatory enforcement officials and those technical controls available to plant operators. Of course, the goals for these technicians are set by the interplay of people in the economic and political arenas, but even in goal-setting, the technical tools that are available must be considered as prime factors. In any event, this interpersonal relationship is of an entirely different order than that between a street criminal and his victim.

Air pollution has become more than a local urban problem: it is regional, national, and even global in scope. Its control in any one country is a national problem, generally with international ramifications. Any effective means to control air pollution, consequently, must be a large-scale system, and its synthesis and design should follow the scientific and technical disciplines of systems engineering. That, at least, is the thesis of this book. In order to accept it, we must define "system" in the engineering sense and then examine the problem to see if the definition matches.

Following Machol[9] in this regard, we can state that there are seven criteria of an engineering system:

1. *A system is man-made* and involves equipment, or "hardware." Certainly the greatest part of urban air pollution is man-made, as are those elements by which we attempt to control it. Furthermore, both the equipment producing air pollution (such as the automobile) and that needed to abate it (such as catalytic mufflers) are hardware.
2. *A system has integrity*: "All components are contributing to a common purpose, the production of a set of optimum outputs from the given inputs."[9] The historical means of controlling air pollution within cities, we have seen, have been piecemeal solutions and lacked this integrity. The mere prohibition of a certain kind of fuel or smoke color has not eliminated the problem, nor have the results been optimal in the sense of obtaining the best societal results at the

least cost to society. As Machol states, the definitions of *optimum* and *inputs* (and here, even *outputs*) may be unknown at the start of systems design. It is a major purpose of the present work to elucidate some of the inputs and desirable outputs and, in some sense, to indicate the nature of the "optimum" even though it may not yet be capable of complete definition.

3. *A system is large*—in parts, functions performed, and cost. Certainly no one doubts that the totality of air pollution control elements will constitute a large, multifunctioned system; even in their current state, lacking coordination, they do.
4. *A system is complex*: a change in one variable will affect many other variables—rarely in linear fashion. This is a primary characteristic of air-pollution control. For example, the variables that affect formation of photochemical smog—the pattern of urban automotive traffic (itself a system of great complexity), the weather variables, the topographic environment, and the presence or absence of other hydrocarbon and nitrogen oxide sources—combine in the most complex nonlinear fashion. The description of these variables as a mathematical model is a work that can well engage the talents of some of our best analysts. One can then remember that what we are describing is merely one physicochemical phenomenon. The effects of the reaction products on human health, the ecology, and property, and the costs of control, including the comparison of alternate means, must be modeled on successively higher levels until we have arrived at a complete description of the process. Even at this point, we have considered only one class of air pollution.
5. *A system is semiautomatic*, involving the actions and decisions of both men and computers. Early means of local air-pollution control, of course, preceded computers, and perhaps this is one reason for their inadequacy. At the other extreme, an air-pollution control system cannot be completely automatic, although the control of pollution from individual sources, such as nuclear reactors, may be. The national air-pollution control system, then, meets this criterion also.
6. *A system is stochastic*. By this we mean that its inputs are random and the results are likewise random processes that, judging only from the present state, cannot be precisely predicted for any future time. The Donora and Meuse Valley air-pollution episodes are strictly random phenomena for which even the statistical model is not known. On a lesser scale, the meterology of regions and micrometeorology of cities can only be predicted on a probability basis. The randomness of air-pollution levels is the crux of the entire

problem and is the factor that obviates any simple solutions other than literal "zero emissions."

7. *A system is competitive.* This is true of air-pollution control in several ways. In one sense, it is a "competition with nature"—combinations of natural and man-made elements may combine stochastically to create harmful conditions, and it is the control system's function (in a game-theoretic sense) to devise the best strategy to minimize the harm and cost. In another sense, an air-pollution enforcement system competes with the pollutor, who may cheat or be merely noncooperative. In still another sense, there are competitive wants of society, which compete in the establishment of system goals and for the resources needed for their implementation. There is the economic desire to expand the production and consumption base versus the problem of disposing of the waste that inevitably accompanies such expansion.

It will not be possible in this work to derive, even in outline, a complete solution to air pollution. Having established the rationality of the systems approach, we will be satisfied if we can develop a description of the problem; approach the philosophy, logic, and functional components of possible solutions; and examine the present state of the art of some of the mathematical and scientific tools that may be needed to implement solutions. But to attempt to define the chronological phases of the design, which depend as much on political and economic factors as any other, and the administrative or managerial subsystems needed to implement the design, is beyond the scope of this book.

We will restrict our efforts further to the description of the more advanced technical tools that must be called into play. Much of the air-pollution abatement equipment used for many years in industry is identical to other process equipment for collecting dust and vapors, cleaning and cooling gases, and similar chores. There is little point in repeating detailed description of this equipment, which has in any case been well done by others.[10] At the other extreme, the use of mathematical models and computers to predict urban air-pollution levels is a rapidly emerging science; hence, we will attempt to describe the state of this art, among others.

Finally, we will attempt to show that the most socially beneficial air-pollution control will be that which meet the four principles of systems design, also set forth by Machol:[9]

1. *The system will maximize the expected value.* Once the necessary goals of the system with respect to health, ecological, esthetic, and long-range effects have been set and the probable occurrence of pollution levels above those that produce harm to persons and

property deterioration have been determined, the costs of various control and abatement strategies and their relative effectiveness should be assessed and traded off until the system that offers the most for the least (i.e., is the most cost-effective) is defined. In a high-order multivariable system such as we are considering, this is an exercise of the utmost difficulty and one that will require continual study and readjustment.

2. *The system design will not be compromised in favor of events of low probability.* This means that the most economical system is that which is designed to operate on a day-to-day basis within the true statistical range of the input events, not on the rarest possibility. If, for example, prevailing meteorological conditions allow adequate ventilation of a region so that fuel emissions of a given level from a power plant can be safely dissipated 99% of the time without causing harm, it is most wasteful of resources to demand "zero emission" all the time. On the other hand, the system should provide for fuel switch to natural gas, or shutdown and purchase of power from other utility regions in the rare event that local meterological conditions so demand, even though these actions are costly. In effect, this principle calls for two levels of operation or two separate subsystems. One is the least-cost solution to the most likely events. The other system, although much more costly to operate, and even disruptive of normal life, insures against disaster, but the very rarity of its operational need means that its average operating cost is low.

3. *Information and authority in the system shall be centralized.* In a situation where high air-pollution events can arise rapidly and unexpectedly and where danger of serious harm exists, it is insufficient to rely on local and fragmented authority to control the situation. Information must be processed and transmitted quickly utilizing computers as required. Authority must be located in the hands of those who can observe the situation for the entire affected region and make decisions that are timely and optimum for all facilities and populations.

4. *Suboptimum solutions shall not be implemented.* This principle states that the narrow solution to a local problem is more likely to harm rather than help the overall situation we are trying to combat. An increase in operating efficiency of a boiler in order to reduce emission of smoky particles may, at the same time, increase the combustion temperature and the production of nitrogen oxides. The ultimate result may be a reduction in London-type smog, at the expense of a more harmful increase in photochemical oxidants. An

optimum solution must take into account all elements of the problem. On other other hand, there are increasingly higher levels of systems that may be optimized. As Machol warns, a preoccupation with optimization on successively higher levels may lead to inaction and paralysis. In this case, it is well to remember that air pollution is not the only harmful product of our urban and industrial activities. It is true that dumping of garbage in the ocean is a cheap and effective way of eliminating smoke from city incinerators. It may be far from optimum in an ecological sense. There are no quick and easy answers to these dilemmas. If we are to remain a viable civilization, we must unquestionably strive continuously to extend our knowledge of the interactions between our technological society and our environment so as to enlarge the scope of our pollution-control systems.

1.3. SOCIETAL VIEWPOINTS

Having established that air pollution is a social problem, hopefully amenable to a systems approach, it behooves us to consider in somewhat greater detail what social effects have been noted and what conflicts arise so that we will be in a better position to design the structured part of our system in such a way as to constitute a solution acceptable to the majority of people and therefore one that is viable.

Although much has already been said on the subject, a few general observations are useful in establishing the scale of the air-pollution problem and its effect on the population.

Each person inhales 30–35 lb of air per day. Air quality has relatively more emotional and psychological impact than water or food, since we are continuously at the mercy of the former, wherever we are, but may exercise some choice in the selection and timing of food or drink.

The total emission of all contaminating gases and particulate matter into our atmosphere, including natural sources such as plant life and volcanoes, is estimated to be 10^{12} tons/yr. Only 500 million tons, or 0.05%, result from man's activities, but since these tend to remain concentrated near their urban and industrial sources, they are held responsible for the ills we are describing.

We are concerned with the constituents of the air on two levels: that of the global atmosphere and that near cities or industry—the "urban air."

Even small changes in the turbidity of the global atmosphere may cause severe world climatic alterations.[11, 12] Some evidence of such changes has been noted over the past 50 years and attributed to increased man-made particle emission. Extrapolations of present trends predict a substantial global temperature drop by 1990 and threaten world glaciation by 2000. Furthermore, evidence of a 30-fold increase in lead aerosol over that of the preindustrial atmosphere has been detected.[13] Average lead levels in the human body are already near those of lead intoxication cases.[14]

Although global considerations may be somewhat speculative or remote, the concentration of urban air pollutants is not. Increased precipitation (10%) and loss of the sun's radiation (30%) are consequences of urban smoke and dust.[15] Temporary upward fluctuations of gaseous and particulate pollution accompanied by rare, disastrous weather conditions (episodes) have been responsible for the well-documented "excess" deaths of up to 4000 persons already cited (London, 1952).[16] Definite correlation between mean sulfur dioxide concentration and the New York death rate has been recently established.[17]

In 1967 the direct cost of all human mortality and morbidity from diseases associated with air pollution, including emphysema, bronchitis, asthma, and lung cancer, was set by the economist Ridker at $2 billion annually in the United States.[18] In 1971 the Environmental Protection Agency (EPA) increased this estimate to $6 billion and in 1972 raised it again to $9.3 billion, accounting only for work loss and medical care and neglecting such effects as the anxiety caused by illness[19, 20] (these estimates exclude also those costs attributable to the major pollutants carbon monoxide, nitrogen dioxide, oxidants, and hydrocarbons, since appropriate data are lacking). Other costs of air pollution include losses in property value, estimated at $8 billion annually, agricultural losses (over $19 million during 1970 in the Los Angeles basin alone),[21] and deterioration of material (33% of power lines and steel rails;[22] severe deterioration of rubber, leather, paper, and building materials; personal cleaning bills adding up to $100 per year for each Washington, D.C., resident and to $180 for New Yorkers),[23] summing up to $7.6 billion yearly. All these costs add up to $24.9 billions or $125 per year for each inhabitant of the United States.

Even these crude approximations are thought to be a minimum cost, neglecting aesthetic values, discomfort, and effects of pollutants in combination or at low levels.

This is sufficient to show that air pollution is a societal problem approaching disastrous proportions—a statement that could not have been made one or two decades ago. The rapid deterioration of the environment can be correlated with the corresponding accelerated growth of productivity and the labor force, since production, including such items as electric power, means consumption and ultimately a waste-disposal problem. Edwin Dale, writing in the *New York Times*,[24] shows that U.S. consumption has risen 60% in the past 13 years and will inevitably rise another 50%

by 1980 (compounding at 4%). The labor force for the next 20 years is already born, so that the unpalatable alternate is massive unemployment and scarcity.

Management specialists have ascribed the equation of production and pollution in part to scientific ignorance of the side effects of technical innovation (as in pesticides) but more generally to the unbalance of the cost-benefit picture, since the costs of pollution are paid by the public, not the polluter.[25] Kenneth Boulding, University of Colorado economist, states that the "bads," including pollution costs, should be subtracted from the "goods" of production and services in order to obtain a net "gross national cost," which would be an index with more significance to the national welfare than the conventional gross national product (GNP).

Traditionally, the marketplace economy has set no value on air or water purity, since these are common property. This has been called by Garrett Hardin, a California biologist, the "tragedy of the commons," referring to the nineteenth-century English "commons" pasture land.[26] With grazing land held in common by a community, self-seeking economic logic leads each rational herdsman to add one more cow to the limited area, until the entire pasture fails through overgrazing. Recognition of this attitude of self-interest on the part of many industries has lead to well-publicized Earth Day demonstrations and consumer-protective activities on the part of many organizations.[27] Since in the main, marketplace forces and economic "payout" will not stop air pollution, it is realized by political and business leaders that government, and particularly federal government, must play an increasing role in setting standards, monitoring, and enforcement. Enlightened business managers, further, are well aware of the intangible side benefits to their corporate images that can result from antipollution activity. The trend toward strengthened activity is seen by the constantly growing projections of expenditures on behalf of U.S. government, industrial, and private sectors to enforce and implement the 1970 Clean Air Act. Recently these were estimated well in excess of $23 billions for the 1970-1975 period[28] and $105.6 billions for 1972-1981.[29]

On the negative side is the historical record in U.S. government of a 50% gap between authorization and the appropriation of antipollution monies.[30] Furthermore, there is a significant pollution backlash, including sociologists and antipoverty interests[31] who would prefer to see government priorities placed elsewhere; industry groups and industries which, in some cases, have shut down plants or threatened to do so rather than invest in costly pollution-control equipment; and representatives of economically backward areas who profess to admire smog as a mark of industrial progress.[32] Even more recently, of course, the "energy crisis" has been cited as a reason for slowing up or even reversing progress in air-pollution abatement.

These are some of the intangible social factors that must be considered as a background to system design.

To summarize, because the national cost of air-pollution damage and the price of abatement have not been fully reckoned and the rare killing episodes cannot be forecast or their probability calculated, there is no common agreement on the best strategy of control. Ignorance, an emotionally charged public, and pressures for performance and profits placed on power utilities and industry have engendered widely divergent philosophies. Some, fearing known and unknown dangers, reject any part-measures, advocate the strictest possible limitations of air-polluting emissions at the source, and call for a NASA-type crash program to develop even more effective controls.[33] At the other extreme, many U.S. industries have opposed any government regulation of air pollution and, to quote one prominent business leader, "have repeatedly put off action because all the facts were not in."[34]

Cost-effective measures, variable with conditions, have been rejected by both extremes, and not only because costs and benefits are (and in some respects, may always be) unquantifiable. Some argue for fixed, stringent abatement on medically conservative grounds. Sociological objections to variable emission standards that do not fall equally on all polluters have been voiced.[35] Authorities claim that cost-benefit evaluations are too complex to enforce and too sophisticated for our legal system.[36] One group at Argonne National Laboratories analyzed optimum abatement strategies for Chicago and adjudged industry activity curtailment to be economically infeasible.[37] Yet, it has also been shown that optimum controlled abatement schedules can provide desirable air quality in other cities at one-sixth the cost of fixed abatement.[38, 39]

On consideration, it should be clear that a mixed strategy is needed. For reasons of health, aesthetics, and economics the day-to-day levels of ambient pollution must be lowered, and these benefits can only be achieved by source abatement. But the costs of emission control can be enormous in some industries. Over $10 million spent for air cleaning in a single power plant is not unknown,[40] and the growth of population and energy use will compound these costs many times. Sheer economic necessity and the spectre of power "brownouts" or transportation breakdowns will force a cost–benefit optimization at some point short of "zero emissions." Our most valuable resources for achieving optimum control are the wind and the weather, the chief scavengers of air pollution.[41] To make the best use of the atmosphere, we must keep close tabs on the current levels of air pollution and the meteorological dilution capabilities in and around our cities. To prevent episodes of high concentration, we must not be content with alerts when they are imminent but must be capable of forecasting pollution levels well in advance—a capability beyond current practice, perhaps even current theory. We have much to accomplish before reaching a satisfactory compromise with our overburdened air environment. In the long term, this will only be achieved by a mixed strategy of optimal abatement, adequate ambient monitoring, and preventive episode control based on accurate forecasting.

1.4. DEFINITIONS AND SCIENTIFIC FUNDAMENTALS

The term *air pollution* has been defined in many ways, none of them completely satisfactory. In the beginning of this chapter we referred the pollution of the atmosphere to its effects on the biosphere, a global view that will be taken up in more detail in the next chapter. The problems of more immediate interest are those associated with the places where man and his technological works are concentrated. We might label these problems *urban air pollution*, because they are more or less localized effects that do not necessarily alter global balances. The effects may be only transient, the results of insufficient mixing, although nonetheless serious. Finally, there is the problem of pollution of air in enclosed spaces, such as mines, factories, laboratories, kitchens, or submarines. This class of problem has been treated for many years, primary by industrial health specialists, and might be termed *enclosed space hazards*. They are generally characterized by a limited time of exposure and defined by "toxic limits" or *threshold limit values* (TLV) specified by industry or government groups.

It can be seen that air pollution is best defined operationally. Within a given milieu, then, a pollutant is any substance other than the usual constituents of air (see Table I and Chapter 2) producing a measurable adverse effect on man, other life, or materials. Thus, we can only describe pollution by enumerating the pollutants and describing their harmful effects or by comparing their concentration against some standard or formula based on these effects.[42] This is the approach taken in the "Air Quality Criteria," which have been published as United States Government Reports.[12, 22] The criteria for individual pollutants, as defined in the United States by the National Air Pollution Control Administration (NAPCA) and its successor, the EPA, are descriptive of the effects and lead to standards of air quality that are prescriptive upon the states and control enforcement bodies. The operational definition of a pollutant is discussed at greater length in Chapter 2, and the legal definition in Chapter 4.

A *contaminant*, as distinguished from a pollutant, is any addition causing a deviation from the mean global atmospheric composition. Consequently, it is the more general term. For example, CO_2 is considered a contaminant but not a pollutant, because the harmful effects of the current global or urban concentrations have not yet been established.

Air pollutants may be in the form of gases, or *aerosols,* the latter consisting of microscopic solid or liquid particles, less than 100 microns (μm) in diameter. (1 μm = 10^{-6} meters.) Aerosols seldom occur as a single chemical species and may be physically complex as well. For example, water droplets usually contain dissolved materials, while solid aerosols contain absorbed gases. From the point of view of their origin we may put pollutants into two classes: *primary pollutants* are those from identifiable emission sources; *secondary pollutants* result from the interaction of two or more primary pollutants or from their reactions with constituents of the normal atmosphere, activated in some cases by solar radiant energy.

1.4.1. Primary Emissions

The primary pollutants have been classified in accordance with the following scheme.[43]

Fine solids (less than 100 μm in diameter) include metals, carbon, tar, resins, oxides, nitrates, salts, silicates, and their combinations, as well as particles of living matter such as pollen and bacteria. Fine solids have important light-scattering properties, leading to smog and haze. They also may provide catalytic surfaces for reactions of absorbed substances and charged nuclei for consolidation or coalescence of other particles and gases. They may also be corrosive, toxic, or merely a source of dust and soiling. A noteworthy fine particulate is aerosol lead from the combustion of gasoline containing antiknock additive, primarily in the form of lead bromochloride. The northern hemisphere contains 1000 times more lead than it would without this source. The median aerosol diameter is 0.25 μm but most of the material is in the respirable range (0.3–1.0 μm). The heavier particles (over 10 μm) do not enter the lungs, but may invade the digestive tract via the nasal passages.[44] Abnormally high lead concentrations have been noted in the blood of persons living in areas of heavy urban auto traffic, confirming these hypotheses. Another possible health threat from lead aerosol exists through concentration in food chains or in rainfall ending in water supplies. In major cities where large quantities of leaded gas are sold, rainfall contains over twice the U.S. Public Health Service standard for drinking water.[45] It has also been hypothesized that urban climate modifications result from lead aerosols that are converted to cloud-seeding particles.

Coarse particles (over 100 μm diameter) are a less serious problem because of their rapid removal by gravitational settling, the greater defense of the lungs and respiratory system against them, and their smaller surface–weight ratio, which reduces the area available to promote secondary reactions. But they are a more obvious source of soiling on settling, and so, perhaps attract undue public attention.

Sulfur compounds: Sulfur dioxide (SO_2) and trioxide (SO_3), originating from the combustion of sulfur-bearing fossil fuels, are among the most ubiquitous of the gaseous pollutants and are primary suspects in the fatal London smog episodes. They are known to cause breathing resistance (bronchostriction). The trioxide is converted in the presence of moisture to sulfuric acid, which readily attacks materials such as nylon and metals, or can be converted to particulate solids in the form of sulfate salts. Another sulfur compound, hydrogen sulfide (H_2S), is a highly toxic substance known to have caused death and illness in the vicinity of a Mexican oil refinery from which it was released. Other sulfur pollutants include the organic homologous of H_2S, the mercaptans (RSH, where R represents an organic radical). The mercaptans have a strong, nauseating odor.

Organic compounds other than mercaptans include many different types of aliphatic (chain) and aromatic (ring structure) hydrocarbons, both saturated and unsaturated, together with their oxides and halogen derivatives. They may appear in the form of vapor, liquid droplets, or solid aerosols. Some have objectionable odors; others, such as the polynuclear aromatics, are suspected carcinogens. However, except for formaldehyde (HCHO), formic acid (HCOOH), acrolein (CH_2CHCHO), and some of the phosphorus and fluorine compounds, most are harmless in their primary state but may enter into secondary reactions, which result in deleterious end-products. These reactions are discussed further in the following section.

Nitrogen compounds: The commonest of these pollutants include nitric oxide (NO), nitrogen dioxide (NO_2), and ammonia (NH_3). The first two result from high-temperature combustion in air. NO_2 is in itself an irritant, but the nitrogen oxides are primarily of interest as participants in the photochemical secondary reactions. Another nitrogen oxide, N_2O (nitrous oxide), is present in the air, chiefly as a natural constituent, but not exclusively so, as indicated in the literature.[46]

Inorganic carbon compounds include carbon dioxide (CO_2) and the monoxide (CO). The former (see Section 1.1) is produced globally in enormous quantities as an end-product of the combustion of all types of fossil fuels. Carbon monoxide (CO), on the other hand, results from incomplete fuel combustion, notably that of gasoline in automobiles. CO_2 is potentially a cause of world-wide temperature rise because of its ability to trap infrared radiation reflected from the earth ("greenhouse effect"), but it is not known to be harmful in other ways, below a few percent total concentration. On the contrary, CO inhibits oxygen uptake of hemoglobin in the blood and is dangerous from this viewpoint. CO is almost exclusively man-made; about 80% of the atmosphere's total is emitted from auto exhausts.

Halogen compounds, specifically HF (hydrofluoric acid) and HCl (hydrochloric acid), result from some metal smelting, fertilizer, ceramic, and a few other industrial processes. These pollutants can be severely irritating, corrosive, and destructive to plant and animal life.

Radioactive materials form another class of atmospheric pollutant. At present these are not considered a general hazard but should be considered in any potential trade-off between nuclear and fossil-fueled power plants. Because of the special nature of radioactive materials, with respect to both their effects and their means of detection, we exclude them from detailed consideration here.

1.4.2. Secondary Pollutants

Any taxonomic division of pollutants, such as primary and secondary, is bound to result in an incomplete description because of the highly dynamic character of the atmosphere–pollutant system. Photochemical and thermal reactions are constantly occurring at different rates, some being catalyzed by certain of the constituents, while still other reactions proceed on the surface of aerosols. The system is physically and chemically in a state of flux, tending to approach a state of equilibrium and minimum free energy through various routes whose intermediate steps and reaction rates are influenced by concentration, radiation, and moisture content. Thus, the mere collection and analysis of stable chemical species cannot provide a true picture of the state of the polluted atmosphere.

Ordinary *thermal* reactions, not involving photon energy, are among the simplest of secondary pollution reaction events.[80] For example, SO_2 may react in solution with dissolved O_2 in airborne water droplets to produce sulfuric acid mists. This reaction is catalyzed by oxides of Mn and Fe, among others (see Figure 1a). Acid mists may, in turn, react thermally with metallic oxides to form halide salts or sulfates in this example.

The absorption of very dilute gases on aerosol surfaces can concentrate them, accelerating reactions at that site and enhancing irritating or toxic effects on the lungs, for example. This is an example of *synergy*, where the combined action of two or more agents is greater than the sum of their separate actions.

Photochemical reactions have been mentioned in Section 1.1 in connection with the genesis of the Los Angeles smog. Irritating and harmful oxidants are formed in the atmosphere through the agency of nitrogen dioxide and various hydrocarbons, largely unburned gasoline. As noted in Figure 1b, it is possible for ozone to be formed directly from oxygen, but the intensity of ultraviolet energy required is such that this reaction only

1.4. Definitions and Scientific Fundamentals

(1) $$SO_2 + \tfrac{1}{2}O_2 \xrightarrow{\text{Catalyst}} SO_3$$

(Catalysts: Charcoal; graphite; nitrogen; ferric, manganese, or chromic oxides; and others)

(2) $$SO_3 + H_2O \rightarrow H_2SO_4$$

(3) $H_2SO_4 + B$ (Metals or base) $\rightarrow B\,SO_4$ (Sulfate salt aerosol)

(a)

$$O_2 + \tfrac{1}{2}O_2 \xrightarrow[\text{UV}]{\text{Sunlight}} O_3 \quad \left\{ \begin{array}{l} \text{Only in the} \\ \text{stratosphere} \end{array} \right.$$

(1) $$NO_2 \xrightarrow[\text{UV}]{\text{Sunlight}} NO + O^*$$

(2) $O^* + O_2 + M \rightarrow O_3 + M$

(3) $O_3 + NO \rightarrow NO_2 + O_2$

(b)

Figure 1 Secondary reactions in the atmosphere: (a) thermal reactions of sulfur oxides; (b) photochemical reaction: formation of ozone and peroxyacetyl nitrate (PAN) (a highly simplified version), O* is an energized atom. If M is a reactive hydrocarbon, (2) yields PAN and other oxidants.

occurs in the stratosphere. The rate of diffusion to the lower atmosphere is far less than that needed to account for the observed concentration of O_3 during smog episodes. The primary photochemical event that does occur at the required rate is the dissociation of NO_2 by radiation at wavelengths less than 450 nm, producing NO and energetic 0 atoms, labeled 0* in Eq. (1) of Figure 1b. The excess energy of the activated 0* can be removed [Eq. (2)] by collision with another atmospheric constituent, M, most likely N_2. In this event, ozone is formed and NO_2 regenerated in cyclic fashion, as in Eq. (3). However, if M should be a "reactive" hydrocarbon molecule, a complex set of events is initiated that may end in the production of a much more irritating oxidant, such as PAN (peroxyacetyl nitrate, $CH_3COOONO_2$). The exact nature of these reactions is still not completely understood, but the difficulty involved in unraveling their details can be appreciated by noting that 81 steps are required to describe the photo-oxidation kinetics of only one system: propylene–NO_x–air.[47] Another simplified reaction scheme is shown in Figure 2.[45] The full explanation of organic photoxidants formation may require as many as 10^6 steps.[48] (See also Chapter 8, section 8.4.)

The most harmful secondary pollutants produced by photochemical reactions include ozone, formaldehyde, organic hydroperoxides, PAN and its reactive homologues (such as peroxyproprionyl and peroxybutyryl nitrates), as well as short-lived free radicals existing in the presence of UV energy

NO₂	+ Light	→ NO + O
Nitrogen dioxide		Nitric oxide, Atomic oxygen
O	+ O₂	→ O₃
	Molecular oxygen	Ozone
O₃	+ NO	→ NO₂ + O₂
O	+ HC	→ HCO·
	Hydrocarbon	Radical
HCO·	+ O₂	→ HCO₃·
		Radical
HCO₃·	+ HC	→ Aldehydes, ketones, etc.
HCO₃·	+ NO	→ HCO₂· + NO₂
		Radical
HCO₃·	+ O₂	→ O₃ + HCO₂·
HCO$_x$·	+ NO₂	→ Peroxyacyl nitrates
Radical		

Figure 2 Simplified reaction scheme for photochemical smog.

and chemical reactants.[49] These are responsible for eye irritation, damage to vegetation, changes in lung function, and possible longer-range medical effects.

1.4.3. Concentration of Gases in the Atmosphere

Gas concentrations in the atmosphere are normally stated in parts per million by volume (ppm/v or simply ppm) or, where appropriate, parts per hundred million (pphm) or parts per billion (ppb). However, toxicological data are generally stated on a gravimetric basis, for example, micrograms per cubic meter or milligrams per liter (mg/liter); 1 mg/liter equals 10^6 μg/m³. Conversion from volumetric to gravimetric concentration can be accomplished with sufficient accuracy by applying the gas laws. The general equation for this conversion is

$$\mu g/m^3 = (\text{ppm}) \times \frac{pM}{RT} \times 10^3,$$

where p = total pressure (atm)
 M = molecular weight of gas of interest
 R = gas constant = 0.0821 l-atm/(mole)(°K)
 T = absolute temperature, °K.

For standard conditions of temperature and pressure, say, 0°C and a atm (760 mmHg), this formula reduces to

$$\mu g/m^3 = \text{ppm} \times M \times 44.64.$$

Leithe[81] gives a table of conversion factors for gases of interest with molecular weights from 1 to 200 and for two sets of conditions: 0°C, 760 mmHg and 20°C, 760 mmHg. Factors for the conversion from micrograms per cubic meter to parts per million are also given. The general equations for this conversion are

$$\text{ppm} = \left(\frac{\mu g}{m^3}\right) \times \frac{RT}{pM} \times 10^{-3}$$

$$= \frac{\mu g}{m^3} \times \frac{1}{M} \times 0.02240 \qquad (0°\text{C}, 760 \text{ mmHg}).$$

The concentration of pollutants is not static, owing to diffusion, source variability, and turbulence; hence, the averaging time of the instrument or sample will affect the readings. Measurements for various averaging times suggest the relationship $C_{max} \doteq Kt^{-a}$ for the maximum concentration indicated by an instrument having an averaging time t; K and a are constants. Thus, the shorter the averaging time, the higher the maximum reading to be expected. A log–log plot of concentration versus averaging time will approximate a straight line with negative slope. For example, maximum nitrogen oxides concentrations in Washington, D.C. (1962–1963),[50] approximated the line $C_{max} = 0.9\,(t)^{-0.29}$.

The frequency distribution of concentration readings for a given averaging time tends to approach the log-normal; for example, approximate straight lines are obtained by plotting the log of SO_2 concentration against the cumulative probability of exceeding that concentration (percent time concentration is exceeded) on normal probability paper.[51] The log-normal distribution appears commonly in pollution measurements and will be discussed more fully in a succeeding section.

Normal Air

Table I, compiled from several recent sources,[52-54, 77-79] lists the concentrations of the principal constitutents of normal dry air. The ranges of CO_2, methane, N_2O, and NO_2 are within the normal fluctuations expected with variations of time and location.

Water vapor from 1% to 3% is normally present in the atmosphere and is a most important ingredient. It takes part in many secondary reactions, such as the conversion of SO_3 gas to acid mist (Figure 1a) which is responsible for reduced visibility and material corrosion.

The Extent of Pollutants in the Atmosphere

Table II lists those major categories of contaminants of the airspace over the United States that are officially termed "pollutants" according to

Table I
Concentration of Gases Comprising Clean Dry Air from Sea level to 25 km

Molecular species	Concentration, ppm/V	Molecules/cc (approximate)
N_2	780,840.	2×10^{19}
O_2	209,480.	5×10^{18}
Ar	9,340.	
CO_2	314.–318.	
Ne	18.2	
He	5.24	
CH_4	1.0–2.0	3.8×10^{13}
Kr	1.1	
H_2	0.5	
N_2O	0.25–0.5	6×10^{14}
CO	0.1	3×10^{12}
Xe	0.087	
NH_3	0.01	1.5×10^{11}
O_3	0.01–0.07[a]	1.2×10^{12}
NO_2	0.001–0.02	3×10^{10} (NO + NO_2)[c]
NO	0.0002–0.002	
SO_2	0.0002–1.0[b]	
H_2S	0.0002	
H_2O (50% RH)		5×10^{17}
HNO_3		1×10^{10}
HNO_2		2×10^{8}
NO_3		1×10^{7}
N_2O_5		2×10^{6}
H_2CO (formaldehyde)		1.5×10^{10}
HO_2		5×10^{8}
OH		5×10^{6}
CH_3O_2		1×10^{4}

[a] Variable, increasing with height.
[b] Variable, decreasing with height.
[c] Rural. Urban A.M. $\approx 3 \times 10^{12}$.

current definitions. As our knowledge of their effects is enlarged, others may be added to this list. The total weight of emission does not, of course, indicate directly the extent or magnitude of the air pollution problem in any given place. Air pollution depends equally on the geographic location of the emitters and the rates of dilution and elimination of the emissions. It is nevertheless noteworthy that over the past 20 years, the data has shown an upward trend for most emissions; hence, it is to be expected that the concentrations in urban air will likewise show increases.

For many reasons it is not easy to make definite statements about pollutant concentrations, nor is it simple to express the extent of air pollution once they are known.[42, 61] This is in part because of the extreme variability of the pollutant concentrations themselves, which vary constantly with the local air turbulence and the strength of emissions, and in part because of the inadequacy of measuring instruments and the scarcity of sampling sites.

The Continuous Air Monitoring Project (CAMP) of the National Air Surveillance Network (now in EPA) has operated continuous monitoring stations in several major U.S. cities on a routine basis during most of the past decade. Summary reports of their measurements have been published for each city.[62] Table III reproduces some of the recent results of this program as well as the "National Primary Standards" issued by the EPA in 1971, which are considered the maximum tolerable ambient concentrations compatible with maintenance of health (see, also, Chapter 4). From this tabulation it is clear that all these cities suffer a severe problem from most of these pollutants, judged by the primary criteria, and also that the nature of the problem varies with local conditions. In general, the East Coast and the more industrialized cities exhibit the highest concentration of sulfur oxides and cities on the West Coast, the highest degree of photochemical oxidants, hydrocarbons, and carbon monoxide. This result is to be expected from the relative utilization of sulfur-bearing industrial and heating fuels, on the one hand, and reliance on the automobile for mass transport, on the other.

A few quantitative statements can be made about the concentration of the pollutants classified above and other air contaminants in the urban environment.

Table II

Estimated Emissions of Air Pollutants in the United States (Millions of Tons Per Year)[55]

Pollutant	Total emissions		
	1970	1969	1950
CO	147	154	103
SO_x (sulfur oxides)	34	34	24
HC (hydrocarbons)	35	35	26
NO_x (nitrogen oxides)	23	22	10
Particulates	25	27	26
Totals	264	272	189

Table III

Selected Maximum Concentrations of Air Pollutants Measured in United States Cities[56-60]

Pollutant	Averaging time	Primary standard	Chicago	Denver	Los Angeles	New York	Philadelphia	Washington, D.C.	Year
Photochemical oxidants	1 h	160 μg/m³ (0.08 ppm)	314	411	1274	359	333	509	1967
Hydrocarbons (nonmethane)	3 h (6-9 A.M.)	160 μg/m³ (0.24 ppm as methane equiv.)	—	1500	1000-1700 3500[a]	—	330-1150	270-1200	1966-1968 1967
Nitrogen dioxide	Annual (arith. mean)	100 μg/m³ (0.05 ppm)	301	59	292	187	241	129	1969
Carbon monoxide	8 h	10 mg/m³ (9 ppm)	90	68	—	—	73	88	1968
	1 h	40 mg/m³ (35 ppm)	36.8	34.5	43.7	24.5	26.5	26.5	1967
Sulfur dioxide	Annual	80 μg/m³ (0.03 ppm)	307	34	30[b]	—	212	97	1968
	24 h	365 μg/m³ (0.14 ppm)							
Particulate matter	Annual (geom. mean)	75 μg/m³	177	147	145	135	170	104	1969-1970
	24 h	260 μg/m³			(Average of 60 cities, 100)				

[a] Maximum, 6-9 A.M.[58]
[b] 1964.

Carbon Dioxide

CO_2 concentration will vary as a result of urban activity and with the presence of vegetation and its respiration cycle, thus showing large seasonal and diurnal fluctuations: Diurnal variations of 100 ppm in cities and a range of 300–500 ppm at various nonurban locations have been noted.[62] The global background has apparently increased an average 0.3 ppm, or 0.11%, per year since 1890, and over twice this amount, 0.27% yearly, between 1958–1963. The average concentration can be expected to grow from the latest value (320 ppm) so long as fossil fuel is used.

Carbon Monoxide

NAPCA data for a three years' average (1964–1966) of off-street sites in five major cities indicated an annual mean concentration of 7.3 ppm. Correlation with the degree of auto traffic was demonstrated by 1967 street measurements in New York City, where levels exceeded 15 ppm between 9 A.M. and 7 P.M. but dropped to 1–2 ppm in the early morning hours. Maximum instantaneous values exceeding 100 ppm (and up to 120 in Los Angeles) have been recorded, but the mean of peak measurements in U.S. cities is between 5 and 10 ppm, exhibiting an approximate log-normal frequency distribution.[63] Levels of 100 ppm have also been reported in enclosed spaces such as the heavily trafficked tunnels entering Manhattan,[64] although the air in automobile tunnels is generally held by forced ventilation to lower levels.

Polluted Los Angeles air averages 10–12 ppm, but during severe inversions, monitoring stations there and in other parts of the state have commonly recorded CO exceeding the California state air-quality standard, 30 ppm for 8 h. Table IV shows the published range of concentration measured in a number of U.S. cities.[65] It is seen that the current 8-h standards listed in Table III were exceeded in nearly every case.

Hydrocarbons

This general class or organic vapors (abbreviated "HC") enters the atmosphere from a wide variety of urban sources, including evaporation of fuels and solvents, petroleum-refining processsses, and incomplete combustion of organic fuels. Hydrocarbons are of particular interest as a factor in the formation of ozone with nitrogen oxides and sunlight; hence, they have been termed *oxidant precursors* in earlier literature. Their heterogeneous origin causes the concentrations reported to be a function of the compounds or mixtures selected for monitoring or of the analysis techniques. Freezing methods at liquid O_2 temperature ($-185°C$) yield from normal urban air about 0.5–1.0 ppm total organics, which includes over 80

Table IV
CAMP Measurements; Gaseous Air Pollutant Levels, Selected Cities: 1968 (Concentration, ppm)[65]

City	Maximum day	Minimum month	Maximum month	Yearly average	Maximum day	Minimum month	Maximum month	Yearly average
	Sulfur dioxide				Nitric oxide			
Chicago	0.51	0.03	0.27	0.12	0.23	0.04	0.11	0.07
Cincinnati	0.08	0.01	0.03	0.02	—[a]	—[a]	—[a]	—[a]
Philadelphia	0.36	0.04	0.16	0.08	0.37	0.02	0.10	0.05
Denver	0.05	0.00	0.03	0.01	0.21	0.01	0.07	0.04
St. Louis	0.16	0.01	0.06	0.03	0.13	0.02	0.05	0.03
Washington, D.C.	0.18	0.00	0.10	0.04	0.31	0.02	0.08	0.04
	Carbon monoxide				Total oxidants			
Chicago	16	4.9	7.1	6.2	0.11	0.01	0.04	0.02
Cincinnati	32	3.5	7.7	5.6	—[a]	—[a]	—[a]	—[a]
Philadelphia	—[a]	—[a]	—[a]	—[a]	0.08	0.02	0.03	0.02
Denver	21	4.3	7.3	5.4	0.08	0.02	0.04	0.03
St. Louis	9	3.8	5.6	4.6	0.05	0.02	0.03	0.02
Washington, D.C.	14	1.9	6.6	3.4	0.10	0.01	0.04	0.03
	Nitrogen dioxide				Total hydrocarbons			
Chicago	0.10	0.04	0.06	0.05	5.4	2.4	3.4	2.9
Cincinnati	0.10	0.02	0.05	0.03	5.6	2.2	3.2	2.6
Philadelphia	0.09	0.02	0.05	0.04	4.8	1.8	2.5	2.2
Denver	0.12	0.03	0.05	0.04	6.4	2.1	4.2	2.9
St. Louis	0.05	0.01	0.04	0.02	9.8	2.1	4.8	3.4
Washington, D.C.	0.08	0.04	0.05	0.05	5.7	1.9	3.0	2.2

Source. Dept. of Health, Education and Welfare, Public Health Service; Annual Data Tabulations, Continuous Air Monitoring Projects, 1969.

[a] Insufficient data.

distinct molecular species such as cresol, phenol, benzene, styrene, ethanol, and chlorinated hydrocarbons. Methane is not detected, since it does not condense at this temperature. However, it is typically more than half of the total organics, perhaps 70%. Methods to determine the total carbon of hydrocarbons, which include, but deemphasize, methane, have yielded a maximum hourly value of 10 ppm in Los Angeles and an instantaneous maximum of 40 ppm. Ranges of C_1–C_5 hydrocarbons measured by gas chromatography are listed in Table V.

Photochemical oxidation of unsaturated hydrocarbons containing a double bond (olefins) may result in the formation of formaldehyde, acrolein, acetaldehyde, propionaldehyde, isobutylaldehyde, and other aldehydes. Few quantitative data are available because of poor monitoring methods that do not differentiate between them. Formaldehyde is most commonly found in polluted air. Data for 1946–1951 in several U.S. cities range from 0 to 0.27 ppm, averaging 0.04–0.18 ppm. Data from Baltimore for 1964 gave a mean of 0.04 and a range of 0.01–0.20 ppm. Extensive data[66] taken in Los Angeles between 1951 and 1957 showed a range of 0.20–1.20 ppm maximum 1-h averages for total aldehydes and a formal-

Table V

Ranges of Some Hydrocarbon Concentrations Expected in Urban Air[83]

Component	Range, ppm in air	
	Minimum	Maximum
Methane	1.2	15
Ethane	0.005	0.5
Propane	0.003	0.3
Isobutane	0.001	0.1
n-Butane	0.004	0.4
Isopentane	0.002	0.2
n-Pentane	0.002	0.2
Ethylene	0.004	0.3
Propene	0.001	0.1
Butene-1	0.000	0.02
Isobutylene	0.000	0.02
Trans-2-butene	0.000	0.01
Cis-2-butene	0.000	0.01
1,3-Butadiene	0.000	0.01
Acetylene	0.000	0.2

dehyde range of 0.05–0.12 (for 1951). Later studies (1960) showed similar concentrations for these components but a maximum of 0.01 ppm (usually half this value) for acrolein. A maximum recorded for total aldehydes in Los Angeles is 1.87 ppm.[67]

Total hydrocarbon concentrations recently measured during the CAMP program are also shown in Table IV.

Sulfur Oxides

SO_2 is the most common primary sulfur oxide pollutant; SO_3 is formed in a secondary reaction and becomes sulfuric acid aerosol. The dioxide is of great significance because of the enormous quantities emitted into the atmosphere from heating and power plants using high-sulfur coal and other sulfur-containing fossil fuels (33 million tons in the United States alone during 1969) and because of its association with the most serious urban pollution disasters. The two-day average during the disastrous London smog of December 1952 was 1.34 ppm. U.S. locations reported an arithmetic average of 0.024 ppm for 36 urban stations during 1964–1965. SO_3 has been detected during high SO_2 periods but is estimated to be only one-hundredth the concentration of the latter. The oxidation SO_2 to SO_3 occurs in air, but slowly.

Recent U.S. urban concentrations are listed in Table IV.

Hydrogen Sulfide; Mercaptans

H_2S is emitted from both man-made sources and natural ones, *e.g.*, decaying organic matter and volcanoes. It is rarely observed in concentrations greater than 1 ppm and has a distinctive odor. Paper mills are sources of the organic sulfur compounds, mercaptans, which are readily detected by their powerful odor (organoleptic methods) as well as by quantitative methods.

Nitrogen Oxides

About 10% of the total world emissions of NO and NO_2 (collectively, NO_x) are man-made, chiefly the product of fuel combustion in furnaces and engines. In the United States, 53% (10×10^6 tons in 1969) was from stationary sources: electric utility boilers and industrial furnaces. Without controls, these sources will double their emissions by the year 2000. Nitric oxide, NO, is formed by a high-temperature chain reaction in furnace atmospheres, which can be represented by the overall equation

$$N_2 + O_2 = 2NO.$$

At flame temperatures, the equilibrium concentration of NO is about 3000 ppm, while the level of NO_2 is negligible.[68] The dioxide, NO_2, is formed by reaction with O_2 in the flue gases and, to a much greater extent, by a rapid reaction with ozone (O_3) stimulated by photochemical events in the atmosphere (Figure 1b).

As a consequence of its origin, the distribution of NO_x closely follows population concentrations; in the United States, over 60% of the emissions occur in urban areas.[69] As the residence time of NO_x in the atmosphere is 3–4 days, pollution from this source is regional rather than a global problem (see Chapter 2).

Other nitrogen oxides (N_2O_3, N_2O_4, N_2O_5, NO_3) are present in low concentrations and are not at present known to be harmful. Nitrous oxide (N_2O), almost entirely of natural origin, also represents a substantial portion of the nitrogen oxides emitted during some chemical processing with nitric acid. It exists in the atmosphere at about 0.25 ppm and is considered physiologically innocuous. However, other plumes emitted from nitric acid plants, which may be colored (high NO_2 to NO ratio) or purposely "decolorized" (converted to NO), represent high concentrations of NO_x (typically 3000 ppm) and are likely to be significant local pollution sources.[70] Recent values of the concentration of NO and NO_2 in some U.S. cities are given in Table IV. Equivalent NO_x levels range from 0.5 to 0.12 ppm on an annual basis. NO_2 concentrations appear in these data not to exceed the U.S. primary (health) standard of 0.05. It is possible, however, that the health data on which these standards were largely based[71] were influenced by peak concentrations and the long-term averages.[72]

Peaks of 0.68–1.35 ppm NO are reported in CAMP data for 1964. A combined reading of 3.93 ppm in Los Angeles (January 1961) is the record.[67]

Ammonia

The sources of ammonia air pollution are few and NH_3 does not currently represent a global or urban hazard. Ultimately, it may come under control of the EPA as an odorant, since organoleptic detection is the chief source of complaint. Background concentrations of NH_3 are on the order of 0.01 ppm (Table I). Tebbens[73] states that maximum background concentrations of 3 ppm that have been reported before 1956 are below the threshold of any known effect, including detection by odor.

Man-made NH_3 pollution largely stems from accidental industrial releases or from sewage waste treatment. (The latter is a good example of the coupling between the various forms of pollution—air and water in this case. The incineration of sewage waste sludge is another.)

Organic Nitrates (PAN and PBZN)

The process of sunlight-activated photochemical reactions in polluted urban air generates this class, once known as "compound X." Since 1960, they have been identified as peroxyacyl nitrate and related compounds (PAN), resulting from the union of an oxygenated hydrocarbon free radical and NO_2 portrayed in Figure 2. More recently the role of benzene-related compounds has been recognized in the formation of peroxybenzyl nitrate (PBZN), an end-product with similar eye-irritating effects. The urban concentration of these substances, even in severe pollution episodes, is probably very low.

Chromatographic separation of PAN from aldehydes and other atmospheric reaction products has yielded a maximum of 35 ppb. It is significant even in these low concentrations because it is highly toxic to plants (phytotoxic) as well as being a primary smog eye-irritant.

Ozone (Oxidants)

O_3 measured by CAMP (using potassium iodide methods) is the major component of "total oxidants" reported, which also includes NO_2 and alkylperoxides. Yearly urban averages (1964–1965) ranged from 0.019 to 0.033 ppm. Data prior to 1964 are inaccurate because of uncorrected interferences of SO_2. A maximum reading for O_3 alone in Los Angeles (where more precise methods were used) was 0.99 ppm during a 1956 smog, but total oxidant concentrations of 1.00 ppm have been reported in other places. Smog effects are noticeable at 0.15 ppm of oxidant and are irritating above 0.25 ppm. Increase in urban oxidants concentrations is demonstrated by the following maxima measured in Cincinnati: 1956, 0.06 ppm; 1956–1959, 0.29 ppm; 1964, 0.32 ppm.

In Philadelphia, for 1967 the yearly average was 0.026 and the 5-min maximum reading was 0.19 ppm. The 1968 averages for several cities are listed in Table IV.

Halides

Halogens (fluorine, chlorine, and bromine) and their acids can be released by accident leakage or from a number of inorganic manufacturing processes and from burning coal. Chloride salts, of course, are a constituent of ocean sprays and mists. Hydrofluoric acid has been the subject of most investigations, since it is highly phytotoxic. Sources include not only the relatively few hydrofluoric acid plants servicing the aluminum, petroleum, and plastic industries but also the far more common combustion of coal containing traces of fluoride minerals. Measurements of ambient concentration of HF conducted in recent decades have shown values of 0.8 to 18 ppb. Even in rural and residential areas, concentrations exceeding 25 ppb have been recorded. The highest value (prior to 1956) was 80 ppb, recorded in an industrial locality.

1.4.4. Particulate Matter in the Atmosphere

Fine particles, ranging from molecular size to that of dustfall, can cause a variety of harmful effects in the atmosphere, including lowering of visibility; effects on health; soiling; and synergistic enhancement of the effects of other pollutants. The physical, chemical, and biological effects of fine particles are very complex, which is apparent in the very profusion of nomenclature associated with particle science.

Aerosols are formed by *comminution* (pulverizing, grinding, and atomization), by *condensation* from supersaturated vapors, and by gaseous chemical reactions. Broadly speaking, particles less than 1 μm in diameter arise from condensation, and larger ones result from comminution. High surface energy prohibits dry grinding from producing particles smaller than a few microns. Combustion processes may form several types and sizes of particles: vaporized particles 0.1 to 1 μm; reaction products below 0.1 μm; and entrained fuel, ash, and soot particles, which may be either larger or smaller than 1 μm. Although the general expression *urban aerosol* is preferred in air-pollution work, specialized terms can be defined as follows:

Mist: liquid-particle aerosol of any size or origin.
Dust: solid-particle dispersion aerosol of any particle size.
Smoke: condensation aerosol with solid or solid and liquid particles.
Smog: smoke and fog.
Smaze: smoke and haze.
Cloud: any free aerosol system having well-defined boundaries.

Particle Size Range

The diameters of airborne particles range from about 6×10^{-4} μm, slightly above the size of small molecules, to about 10^3 μm. Particles larger than 10 μm are termed dustfall, since they rapidly settle out of the air by gravity (sedimentation). Above 1 μm, spherical particles will obey Stokes's law; the terminal or settling velocity in a gas or liquid medium is

$$V = \frac{2ga^2(d_1 - d_2)}{9\eta} \text{ cm/sec}$$

where a is the particle radius (cm), d_1 and d_2 are, respectively, the sphere and medium densities (g/cm³); η is the medium viscosity in poises (dyne-sec/cm²); and g is the acceleration of gravity. For a particle density of 1, at 10 μm diameter, the settling rate is significant (0.3 cm/sec), and at 1000 μm it is substantial (390 cm/sec), relative to the normal turbulence of air masses.

The Stokes's law velocity may be used as a measure of particle diameter. If a particle is not spherical, the relationship leads to a fictitious measure called the *aerodynamic* or *Stokes* diameter. If the density is not known, an arbitrary value of 1 (g/cm^3) is employed yielding the *reduced sedimentation diameter*.

Particles less than about 1 μm in diameter fail to obey Stokes's law, since the air molecules no longer act as a fluid but as individual particles. Particles with a diameter near the mean free path of the air molecules (0.06 μm at normal temperature and pressure) are suspended by Brownian movement. Between 0.1 and 1.0 μm, air motions are large with respect to their settling velocity; hence, particles tend to remain airborne. In addition, aerosols tend to coagulate or stick together when they come into contact. This mechanism is predominant in particles less than 0.1 μm, because of Brownian movement. At 0.05 μm, 50% of the particles are eliminated per day by coagulating into larger entities. Thus, larger particles, more than 5 or 10 μm, are removed through settling by gravity, and those less than 0.1 μm, by coagulation, leaving the vast majority of air-suspended particles that remain in a mature aerosol in the 0.1 to 1 μm range. This is termed a *self-preserving* aerosol.

Coincidentally, the average wavelength of daylight (0.524 μm) is in the middle of this range, greatly enhancing the optical effects of suspended particles. Reduction in visibility is the means by which the presence of particles is most readily perceived, and it is a measure of their concentration in the atmosphere. Light is scattered by particles in this size range according to the complex phenomenon known as *Mie scattering*. The loss in visibility according to the Mie theory is a function of the square of the particle radius. The more familiar *Rayleigh scattering,* which results largely from particles less than 0.1 μm in diameter, is of little importance in air-pollution phenomena. (The blue sky color results from the Raleigh scattering effect of air molecules.)

Particulate pollutants must be described statistically in terms of the frequency distribution of particle size. To be meaningful, the moment or statistical parameter specified should be that which is of the most significance relative to the effect studied. If we are concerned with light scattering, the distribution with respect to the second moment or square of the radius is important; in toxicology, we use weight or the third moment (r^3). The mathematical model best describing the size statistics of particulate matter is the log-normal distribution. This is similar to the familiar normal (Gaussian) distribution, except that the mean radius and standard deviation of the latter are replaced by the geometric mean, M_g, and the geometric standard deviation, σ_g. A log-normal distribution plotted on semilog paper will yield the bell-shaped curve; on probability-versus-log paper, the cu-

1.4. Definitions and Scientific Fundamentals

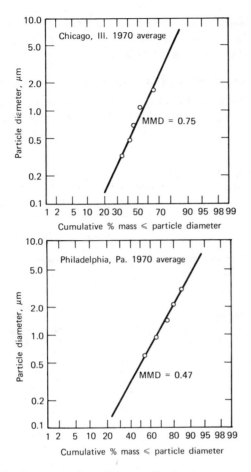

Figure 3 Typical urban aerosol particle-size distribution curves (*Research/Development*, courtesy of Dr. Robert E. Lee, Jr.).

mulative distribution will be a straight line. M_g is the same as the *median* particle radius, and σ_g is numerically equal to the ratio of the values of r at 84.2% and 50% on the cumulative distribution curve. Given these parameters, the area distribution can be computed from

$$\ln M_a = \ln M_g + 2\sigma_g^2$$

and the mass distribution

$$\ln \mu_m = \ln M_g + 3\sigma_g^2,$$

where M_a and μ_m are the area and mass means, respectively. The geometric standard deviation remains the same for all moments.

Some recent plots of the cumulative particle-size distribution in U.S. cities are illustrated in Figure 3.[82] It is seen from the above that the concept of particle size is by no means a simple one.

Concentration and Chemical Composition of Atmospheric Particulates

Some idea of range and average composition of suspended particulate matter in the atmosphere is given by Table VI. These data were taken

Table VI

Composition of Particulate Matter[63]

	Number of stations	Concentration in $\mu g/m^3$	
		Arith. avg. unless noted	Maximum
Urban stations			
Particles	291	105 (90)[a]	1254
Fractions:			
Benzene-soluble organics	218	6.8	—
Nitrates	96	2.6	39.7
Sulfates	96	10.6	101.2
Ammonium	56	1.3	75.5
Antimony	35	0.001	0.160
Arsenic	133	0.02	—
Beryllium	100	0.0005	0.010
Bismuth	35	0.0005	0.064
Cadmium	35	0.002	0.420
Chromium	103	0.015	0.330
Cobalt	35	0.0005	0.060
Copper	103	0.09	10.00
Iron	104	1.58	22.00
Lead	104	0.79	8.60
Manganese	103	0.10	9.98
Molybdenum	35	0.005	0.78
Nickel	103	0.034	0.460
Tin	85	0.02	0.50
Titanium	104	0.04	1.10
Vanadium	99	0.050	2.200
Zinc	99	0.67	58.00
Nonurban stations			
Particles	32	37(28)[a]	312
Fractions:			
Benzene-soluble organics	28	1.2	—
Arsenic	24	0.005	0.02

[a] Geometric mean for all stations.

biweekly from a large number of stations in the United States during 1964–1965. The average difference between urban and nonurban stations is about 3:1 in terms of total weight concentration. Recent (1969) nonurban data showed an average of 30 µg/m³ for 6400 samples in 1969, ranging from 13 µg/m³ in desert regions to 50 µg/m³ on the Atlantic and Pacific coasts, while at urban sites values clustered in the 60–140 µg/m³ range.[74] For 57 U.S. cities during the same period, the lowest annual arithemetic average was 32 µg/m³ (Cheyenne, Wyoming) and the highest were St. Louis, Missouri and Charleston, West Virginia, with 213 µg/m³ each. The extreme range of biweekly sampling for all cities sampled included a low of 7 µg/m³ and a high of 530 µg/m³.[75]

Generally, it is noted that the concentration of suspended particulates in urban atmospheres fluctuates to the degree expected of a moderately short residence time. The average 24-h maximum is about 300% higher than the annual mean. As much as a 700% increase is noted in 2% of the stations. Sundays and holidays fall 15–20% below weekdays. During pollution episodes, two or three times the normal amount of suspended particles may be noted.[76]

The organic compounds tabulated as benzene-soluble organics are of particular concern because many compounds detected, such as the polynuclear aromatic hydrocarbons, are carcinogenic. These compounds are derived from the combustion or pyrolysis of carbonaceous fuels. Measurements of this particulate fraction for the 57 cities monitored in 1969 range from a high of 12.5 µg/m³ in Los Angeles to 1.9 µg/m³ for Cheyenne, Wyoming, or from 12% to less than 5% of the total suspended particulate.[75] In 1966 the national average was 8%. A specific carcinogenic compound, benzo (a) pyrene, has been measured in concentrations ranging from 8 to 61 µg/1000 m³ in U.S. cities. As much as 146 µg/1000 m³ has been reported for a total of eight other polycyclics in urban samples (1958–1959). However, fewer than 5% of the compounds in these studies were identified, and the total concentration may have been much greater.

Dustfall, primarily particles 10–500 µm in diameter, is measured by collecting fallout in jars with specified mouth area; the observed weight is usually extrapolated to larger areas and time periods. City values average 0.35 to 3.5 mg/(cm²) (month), or 10–100 tons/(miles²) (month), although 20 times this amount may be recorded near a polluting source.

A commonly employed measure of suspended particulate matter is the staining, or soiling, index, determined by drawing a known quantity of air through a filter paper and measuring the discoloration. Results are expressed in terms of the transmittance optical density (OD) or calibrated to relate the reflectance of the stain to weight. The transmittance measurements are converted to COH's per 1000 linear feet of air passing through the filter. (COH is an acronym for "coefficient of haze.") One COH produces an OD of 0.01 measured by light transmission, where

$$OD = \log_{10} 100/(\% \text{ light transmission compared with clean paper})$$

The reflectance is measured similarly by reading a reflectometer set at 100 for clean paper. The units are RUDS ("reflectances of dirt shade") and are likewise based on the passage through the filter paper of 1000 ft of air. In European countries, calibration curves relating reflectance to weight concentration are used, but as these are based on arbitrary factors, they are not necessarily applicable to U.S. conditions.

COH units are used for air-quality criteria in U.S. communities; for example Allegheny County, Pennsylvania, soiling index classifications are

0–1.0 COH/1000 linear feet: slight pollution
1.0–2.0 COH/1000 linear feet: moderate pollution
2.0–3.0 COH/1000 linear feet: heavy pollution
3.0–4.0 COH/1000 linear feet: very heavy pollution.

Other Contaminants

Pollens (plant particles 10–50 μm or larger) are often collected in fallout, as are microorganisms (algae, rusts, fungi, molds, yeasts, and spores), bacteria (usually associated with dust), parts of insects (scales, hairs), and other material from living organisms.

Radioactive substances are often associated with fine particulates. They emit from natural sources—rocks, coal, and oil—as well as from nuclear fission. A 100-MW coal power plant for example, will produce about 400 mCi/yr of ^{226}Ra and ^{228}Ra and a similar oil-burning plant about 0.5 mCi. Reactor accidents have occurred, such as the Windscale incident in England, releasing sufficient local radioactivity to endanger milk supplies, but these have been fortunately rare. A voluminous literature on radioactivity exists but is considered outside the scope of this volume.

REFERENCES

1. A. I. Oparin. *The Origin and Initial Development of Life.* Medetsina: Moscow, 1966 (Tech. Trans. NASA TT F-488, NASA: Washington, D.C., 1968).
2. J. S. Shklovskii and C. Sagan. *Intelligent Life in the Universe.* Holden Day: San Francisco, 1966.
3. D. J. Garrod. *Symposium on Measurement and Control of Industrial Pollution.* Institute of Measurement and Control: Hollingbourne, England, 1971.

4. R. A. Papetti and F. R. Gelmore. "Air Pollution." *Endeavor*, **30**, 107–114 (1971).
5. G. F. Friedlander. "Airborne Asphasia—An International Problem." *IEEE Spectrum*, **2**, 56–59 (1965).
6. J. B. Heywood. *MIT Technol. Rev.*, **73**, 21–29 (1971).
7. K. Clark. *Civilization*, Harper and Row: New York, 1969, pp. 269–291.
8. J. W. Forrester. *Urban Dynamics*, MIT Press: Cambridge, Mass., 1969.
9. R. E. Machol. In R. E. Machol, Ed., *Systems Engineering Handbook*. McGraw-Hill: New York, 1965, Chapter 1.
10. W. Strauss, Ed. *Air Pollution Control (Part 1)* Wiley-Interscience: New York 1971.
11. E. Robinson. In Stern, A. C., Ed., *Air Pollution*, Vol. I, 2nd ed. Academic: New York, 1968, p. 397.
12. U. S. Dept. of Health, Education and Welfare (HEW). *Air Quality Criteria for Particulate Matter*. Washington, D.C., 1969, p. 43.
13. Ref. 11, p. 396.
14. Lead Industries Assn. *Lead and the Atmosphere*, New York, p. 4.
15. Ref. 12, pp. 40–41.
16. Ref. 12, p. 150.
17. *New York Times*, June 7, 1970.
18. R. Ridker. *Economic Costs of Air Pollution*. Praeger: New York, 1967, p. 54 (Table 3).
19. Council on Environmental Qualtiy (CEQ). *Environmental Quality, 2nd Annual Report*. Washington, D.C., 1971, p. 106.
20. Environmental Protection Agency (EPA). *The Economics of Clean Air, Annual Report to Congress*. Washington, D.C., 1972.
21. California State Dept. of Agriculture. *A Survey and Assessment of Air Pollution Damage to California Vegetation in 1970*, Report APTD-0694. Sacramento, Calif., 1971.
22. HEW. *Air Quality Criteria for Sulfur Oxides*. Washington, D.C., 1970, p. 56.
23. J. Marshall. *The Air We Live In*. Coward-McCann: New York, 1969, p. 32.
24. E. Dale. "The Economics of Pollution." *New York Times Magazine*, April 19, 1970, pp. 27–47.
25. J. C. Davies. *AMA Management Rev.*, May 1970.
26. G. Hardin. In S. Cotton, Ed., *Earth Day—The Beginning*. Bantam: New York, 1970, pp. 39–41.
27. D. Hayes, *Environ. Sci. Tech.*, **4**, 461 (1970).
28. Ref. 19, p. 111.
29. CEQ. *Environmental Quality, 1973*. Washington, D.C., 1973, p. 93.
30. Ref. 26, p. 62.

31. A. Etzioni. *Science,* **168,** 3434 (1970); see also *IEEE Spectrum,* **7,** 12–13 (1970).
32. *Philadelphia Evening Bulletin,* May 29, 1970, p. 33.
33. R. S. Berry. *Bull. Atomic Sci.,* **26,** 40 (1970).
34. J. K. Jamieson. *J. Air Poll. Control Assoc.,* **21,** 63 (1970).
35. D. H. Bowen. *Environ. Sci. Tech.* **4** (1970), editorial.
36. A. Atkisson and R. S. Gaines, Ed. *Development of Air Quality Standards.* Merrill: Columbus, Ohio, 1970, p. 208.
37. D. O. Parson and E. J. Croke. "An Economic Evaluation of Sulfur Oxide Air Pollution Incident Control," APCA Paper 69–20. *Air Pollution Control Association (APCA) Meeting,* New York, June 22–26, 1969.
38. R. E. Kohn. In Ref. 36, pp. 108–109.
39. J. R. Norsworthy and A. Teller. "The Evaluation of the Cost of Alternate Strategies for Air Pollution Control," APCA Paper 69–172. *APCA Meeting,* New York, June 22–26, 1969.
40. G. D. Friedlander. *IEEE Spectrum,* **7,** 71 (1970).
41. G. R. Hilst. "Utilizing Our Present Atmospheric Environment." *MIT Industrial Liaison Program Symposium on Air Pollution,* May 5–6, 1970.
42. W. A. Thomas; L. R. Babcock, Jr.; and W. D. Shultz. *Oak Ridge Air Quality Index* (Report ORNL-NSF-EP-8), Oak Ridge National Lab, September 1971.
43. L. A. Chambers. In Ref. 11, p. 13.
44. Office of Science and Technology (OST), *Cumulative Regulatory Effects on the Cost of Automobile Transportation (RECAT),* The White House: Washington, D.C., 1972, pp. 1–A14.
45. American Chemical Society (ACS). *Cleaning Our Environment—The Chemical Basis for Action.* Washington, D.C., 1969, pp. 27–44.
46. Ref. 45.
47. T. A. Hecht and J. H. Seunfield. *Environ. Sci. Tech.,* **6,** 47–56 (1972).
48. Ref. 45, pp. 154–157.
49. J. R. Goldsmith. *Sci. J.,* March 1969, pp. 44–49.
50. B. D. Tebbens, In Ref. 11, p. 24.
51. Ref. 22, p. 37.
52. L. A. Chambers. In Ref. 11, p. 13.
53. Ref. 45.
54. R. J. List. In *American Institute of Physics Handbook.* McGraw-Hill: New York, 1957, pp. 2–125.
55. CEQ. *Third Annual Report.* Washington, D.C., 1972, p. 6.
56. Ref. 19, p. 216.
57. Ref. 19, p. 214.
58. Ref. 44, p. 1-G4.

59. *Federal Register,* **36,** April 30, 1971, p. 8187.
60. Ref. 12, pp. 13–14.
61. L. R. Babcock, Jr., and N. L. Nagda. "Indices of Air Quality." 138th Meeting American Ass. for Advancement of Science, Philadelphia, 1971.
62. B. D. Tebbens. In Ref. 11, p. 29.
63. HEW (PHS). *Air Quality Data from National Air Sampling Networks and Contributing State and Local Networks 1964–65.* Cincinnati, Ohio 1966.
64. Ref. 17, July 9, 1969.
65. U. S. Bureau of the Census. *Statistical Abstracts of the United States: 1971,* 92nd ed. Washington, D.C., 1971, p. 170.
66. HEW. *Air Quality Criteria for Hydrocarbons.* Washington, D.C., 1970, pp. 3-9 to 3-12.
67. Los Angeles County Air Pollution Control District (LACAPCD). *Air Quality of Los Angeles County, Data Through 1959,* Vol. XI, 1961.
68. National Academy of Engineering/National Research Council (NAE/NRC). *Abatement of Nitrogen Oxides Emissions from Stationary Sources* (COPAC-4). Washington, D.C., 1972.
69. Ref. 68, p. 21.
70. Ref. 68, pp. 47–48.
71. C. M. Shy et al. "The Chattanooga School Children Study." *J. Air Poll. Control Assc.,* **20,** 539–588 (1970).
72. Ref. 68, p. 24.
73. B. D. Tebbens. In Ref. 11, p. 42.
74. Ref. 65, p. 171.
75. Ref. 65, p. 172.
76. M. Corn. In Ref. 11, pp. 47–90.
77. R. L. Grossman. In V. E. Derr, Ed., *Remote Sensing of the Troposphere* (NOAA 72091642). U.S. Dept of Commerce, National Oceanic and Atmospheric Administration: Boulder, Colo., 1972.
78. E. E. Ferguson. In Ref. 77, Chapter 27.
79. H. Levy. *Science,* **173,** 141 (1971).
80. S. Glasstone. *Textbook of Physical Chemistry,* 2nd ed. Van Nostrand: New York 1946, p. 1154.
81. W. Leithe. *The Analysis of Air Pollutants,* Ann Arbor: Ann Arbor, Mich., 1970, p. 265.
82. R. L. Lee. *Res. Devel.,* **23,** 21 (June 1972).
83. *Recommended Methods in Air Pollution Measurements,* California State Dept. of Public Health: Berkeley, Calif., 1967, Method 9-A, p. 8.

2

Air Pollutants:
Sources, Sinks, Residence Times

2.1. WHAT IS A POLLUTANT? BACKGROUND CONCENTRATIONS IN THE ATMOSPHERE

The discussion on pollutants in the atmosphere can be put into reasonable perspective by first considering the composition of "natural" background atmosphere—that is, the atmosphere not influenced by the industrial or agricultural activities of mankind. With this information as a base, we can then consider at what concentration level constituents usually present begin to have harmful effects. Finally, we can discuss those gases generated by human activity that are not usually present and the level at which they become pollutants. Implicit in this approach is the assumption that below a certain concentration, a material, whether present in the atmosphere naturally or artificially, is not a pollutant, while above a certain concentration, materials normally present in low concentrations may become pollutants (see Section 1.4).

The background composition of the lower atmosphere (up to 100 km) has been given by a number of authorities. The figures differ in unessential details, probably because of minor local differences in composition, statistical treatment of the data, or variations in analytical techniques. A representative set of data for the lower portion of the troposphere is given in Table I of Chapter 1. These data are fairly representative of conditions up to 90 km or so. Above 100 km, the composition of the atmosphere changes a great deal, but this region is of little interest in the present context. It is seen that in addition to oxygen, nitrogen, carbon dioxide, and the rare gases, a number of other gases are invariably present that are not ordinarily considered constituents of the atmosphere. Among these may be mentioned nitrous oxide (N_2O), methane (CH_4), and hydrogen (H_2). These enter the atmosphere through natural processes discussed in more detail below. The materials normally considered pollutants (SO_2, O_3, CO, NO_2, and NO) are naturally present in very small concentrations. If they exist in the atmosphere because of natural processes, can they be considered pollutants? What about excessive methane near marshes or excessive sulfur compounds near volcanoes and tectonically active regions? It is clear that we need to be careful in our definition of a pollutant. One typical statement is the following: "Air pollution means the presence in the outdoor atmosphere of one or more contaminants such as dust, fumes, gas, mist, odor, smoke, or vapor in quantities, of characteristics, and of duration, such as to be injurious to human, plant or animal life or to property or which unreasonably interfere with the comfortable enjoyment of life and property."[1] As an operational, scientific definition, this statement is not of much value, and we are left with the notion that an air pollutant is whatever people say is an air pollutant. Thus, the materials considered air pollutants are subject to change as the field grows and changes with more knowledge and better analytical techniques.

As a working definition of pollutant, we propose the following: A material present in the atmosphere, either from natural or man-made sources, may be considered to be a pollutant when its concentration reaches or exceeds a level such that (a) some adverse effect on human health or welfare is observed; (b) some deleterious effect on animal or plant life is observed; or (c) damage to materials of economic value to society is observed.

This level has not been well defined for many materials loosely called pollutants. However, its determination is the first requirement for a

Table I

Mass and Volume of Atmospheric Gases, Total and in Atmosphere[2]

Compound	Total supply of earth, g	Total supply per unit area, g/cm²	Total in atmosphere per unit area, g/cm²	Total in atmosphere, g	Total in atmosphere,[a] moles
H_2O	1500×10^{21}	300×10^3	2	0.01×10^{21}	0.0555×10^{19}
CO_2	250×10^{21}	50×10^3	0.45	0.00275×10^{21}	0.00625×10^{19}
N_2	5×10^{21}	1×10^3	800	4×10^{21}	14.3×10^{19}
O_2	15×10^{21}	3×10^3	200	1×10^{21}	3.12×10^{19}
Total	1700×10^{21}			5.01×10^{21}	17.5×10^{19}
Total[b]				5.3×10^{21}	
Total, tons				5.8×10^{15}	

[a] Calculated by present writer from Johnson's data.[2]
[b] Estimate of Robinson and Robbins.[3] Total volume of atmosphere calculated as ideal gas at 20°C, 1 atm: 4.2×10^{24} cc = 4.2×10^{21} liters = 4.2×10^{18} m³.

cost–benefit analysis. Such an analysis would consider the trade-off between the costs of keeping a material at or below its pollutant level (as defined above) and the benefits of permitting it to exceed that level by an specified amount. It is clear that one of the difficulties in any such analysis is to assign numerical costs to parameters on both sides of the cost–benefit equation, as we shall show in Chapter 3.

Table I of Chapter 1 gave the composition of the atmosphere both with respect to major constituents and trace compounds. Some other data about the atmosphere on a global scale are provided in Table I, since these data are inherently interesting and also provide the basis for calculations that will be useful later on. It would appear that the estimates of the weight of oxygen and of nitrogen are somewhat in error, since their molar equivalents do not give a ratio of 80% of nitrogen by volume (moles) but 82%. However, the total mass given by Johnson[2] agrees quite closely with the estimate of Robinson and Robbins.[3]

2.2. THE SOURCES OF MATERIALS PRESENTLY CONSIDERED POLLUTANTS

The materials which we consider in this section include sulfur gases (SO_2, H_2S), nitrogen gases (N_2O, NO, NO_2, NH_3), and carbon gases (CO, CO_2, HC). The sources for these may be divided into two primary categories: natural and man-made. Some authors refer to any man-made emission as pollution.[3] We will avoid this usage, because it does not follow that a man-made emission is harmful; according to the definition above, for example, steam and CO_2 are not. What we will attempt to do is provide data on the total global emission of operationally defined pollutants from all sources, in order to show what portion of the global total is caused by the activities of men. It will then become reasonably clear that for a number of pollutants, man-made sources constitute a negligible fraction of the total emission. Thus, on a global scale at least, man-made emissions may not appear invariably as a problem.

If all the gases emitted due to human activities were rapidly and uniformly distributed into the troposphere, there might be no problem, since their levels of concentration would probably remain below the limits where harmful effects are detectable. Whether or not this would be true is a function of the strength of man-made sources, added to the "background" concentration provided by natural sources, less the activity of natural "sinks," including precipitations, adsorption, and the many other physical, chemical, and biological mechanisms that tend to remove the added substances from the atmosphere. Of course, it is clear that the lower atmosphere is *not* homogeneous and that meteorological factors do not invariably provide instantaneous and reliable dispersion of the emissions concentrated around industrial and urban sources. This concentration has been increasingly exacerbated in this century owing to three factors—the growth of city populations, the dependence on automobiles as a means of mass urban transportation, and the concentration of power generation and industrial production facilities within a small number of limited areas, resulting in a vast increase of local emissions. Natural meteorological mixing mechanisms are usually able to disperse the low pollutant load in the countryside and frequently are adequate for the high pollutant load generated in large metropolitan areas. When the natural mechanisms of dispersal are inadequate in metropolitan areas, persistence of gases such as NO, NO_2, SO_2, and HC in the atmosphere above the area from day to day provides the opportunity for photochemical reactions leading to smog and eye-irritation[4] from which few large cities in the world are free. Thus, we conclude that air pollution is a "local mixing" problem for cities or heavily concentrated industrial and chemical manufacturing areas.

Nevertheless, it is necessary for us to examine the global picture to determine the total sweep of the air-pollution problem. The quantity and rate of natural and man-made emissions together, opposed by removal mechanisms establishing their ultimate fate, determine both their background levels and "residence time" during which high concentrations may persist in the atmosphere. Consequently, they bear directly on the problem of local pollution levels as much as does the geographic location of the man-made sources. We proceed now to a consideration of global sources.

2.2.1. Sulfur Compounds

The total annual emission of sulfur compounds into the atmosphere around 1965 amounted to 215×10^6 tons of sulfur. Of this, about one-third

Table II

Man-made SO_2 Emission by Hemisphere, 10^6 Tons SO_2/yr (1965)[8]

Source of sulfur	Total SO_2	Northern hemisphere	Southern hemisphere	Northern % of total
Coal	102	98	4	96
Petroleum	28.5	27.1	1.4	95
Smelting (Cu, Pb, Zn)	15.7	11.0	4.7	70
Total	146	136	10	93

Sources. United Nations Statistical Papers, 1963–1966, and U.S. Bureau of Mines, 1967.

2.2. The Sources of Materials Presently Considered Pollutants

Table III

Global Emissions of Sulfur by Hemisphere, 10^6 Tons S/yr.[9]

Source of sulfur	Total, as S	Northern hemisphere	Southern hemisphere	Northern % of total
Man-made SO$_2$	73	68	5	93
Biological H$_2$S				
Land	68	49	19	72
Marine	30	13	17	43
Sea spray	44	19	25	43
Total	215	149	66	69

was man-made and emitted as sulfur dioxide (SO$_2$), sulfuric acid (H$_2$SO$_4$), hydrogen sulfide (H$_2$S), and particulate sulfate. Hydrogen sulfide from natural sources accounted for 100×10^6 tons or 45% of the total.[5] The rest of the natural-source sulfur is emitted into the atmosphere as particulate sulfate from ocean spray with very little penetration into the atmosphere over land areas. More recent estimates are contained in the MIT report of the Study of Critical Environmental Problems (SCEP).[6] A total of 93×10^6 tons/yr of SO$_2$ emissions is reported, based on statistics for 1967–1968. Kellogg et al.[7] revise this figure to 100×10^6 tons/yr based on a difference of SO$_2$ emission factors between U.S. and non-U.S. SO$_2$ sources. Since the figures are based on estimates of fossil fuel consumption on a worldwide basis and assume some sulfur content for these fuels, we will choose the Robinson and Robbins figures, since they provide more detailed breakdowns of the total sulfur emissions. In any event, our main argument will not be much affected by these differing estimates.

Sulfur emission is the first example of our thesis that man is not the only pollution source on a global scale. However, he is a substantial contributor; 33% cannot be ignored.[7] Man's contribution is mainly in the form of SO$_2$ from the burning of fossil fuels. Let us look at the SO$_2$ emissions more closely. Table II shows SO$_2$ emission on a global scale, by hemisphere, due to activity of men. It is seen that the preponderant SO$_2$ emission (93% of the total) due to man occurs in the northern hemisphere. This exactly reflects the preponderance of urban population, industrial activity, and automobile use in the northern hemisphere: United States, Canada, Western Europe, the Soviet Union, and Japan. The localized aspect of sulfur dioxide emission is dramatically illustrated in this table. The situation is not much mitigated even when we consider the total of sulfur emissions including those from natural sources. These data are given in Table III. Nature helps balance the emissions somewhat with sulfur from the ocean but even so, over two-thirds of the sulfur comes from the northern hemisphere.

Sulfur Dioxide

We first examine the global emission figures for each of the sulfur compounds separately, starting with sulfur dioxide. Sulfur dioxide emissions result overwhelmingly from human activities—combustion of fuels and smelting of metals. Although natural sources of sulfur dioxide exist (e.g., volcanoes and fumaroles), the total amount of sulfur dioxide emitted is negligible compared to man-made sources.[3] Table IV gives estimates of sulfur dioxide emissions as a result of various human activities. It is seen that the major source of SO_2 (70%) is the burning of coal. The petroleum industry accounts for close to 20%, while smelting operations involving sulfide ores account for a little over 10%. It will be fairly clear from these figures that those operations involving the production of power (coal burning) and automobile transportation (petroleum refining and combustion) when brought together in large metropolitan areas will result in large local concentrations of sulfur dioxide. We have discussed these urban SO_2 levels in the previous chapter. Whereas the background SO_2 level is 0.2 ppb, urban levels are in the 0.1 ppm range, an increase of several hundred times over the background concentrations.

In view of the fact that SO_2 in urban atmospheres is preponderantly of man-made origin it may be useful to trace historically the emissions of this gas due to industrial activities and, if possible, to estimate future trends from the historical data. This exercise has great bearing on the general thesis that SO_2 is a pollutant due to man's industrial activity in congested urban areas. Robinson and Robbins[8] have examined the available data and have estimated SO_2 emissions from various sources going back to 1860. Furthermore, they have estimated global emissions of SO_2 to the year 2000 based on extrapolations of past and present industrial activities. They have assumed that no controls on SO_2 emissions will be introduced and that the sulfur content of fuels and ores will remain constant. These figures have been gathered in Table V, and the total SO_2 emission plotted in Figures 1 and 2. It is seen that SO_2 emissions have doubled in the period 1940–1965 (25 years), will triple in 1965–1985 (20 years), and quadruple in 1985–2000 (15 years) if SO_2 emissions are not controlled. It seems reasonable that the major part of SO_2 emissions due to man will have to be controlled in view of the demonstrated inability of natural dispersion mechanisms to handle present atmospheric loadings adequately.

Table IV
Worldwide Annual Emissions of Sulfur Dioxide (ca. 1965)[3]

Source of sulfur	Consumption or production/yr	SO_2 factor	SO_2 emission, 10^6 tons/yr	Subtotals	% of total
Coal	3074×10^6 tons	3.3 tons/100 tons		102	70
Petroleum combustion:					
Gasoline	379×10^6 tons	9×10^{-4} tons/ton	0.3		
Kerosene	100×10^6 tons	24×10^{-4} tons/ton	0.2		
Distillate	287×10^6 tons	70×10^{-4} tons/ton	2.0		
Residual	507×10^6 tons	400×10^{-4} tons/ton	20.3		
Petroleum refining	$11,317 \times 10^6$ barrels	50 tons/10^5 barrel		22.8	15
Smelting				5.7	4
Copper	6.45×10^6 tons	2.0 tons/ton	12.9		
Lead	3.0×10^6 tons	0.5 tons/ton	1.5		
Zinc	4.4×10^6 tons	0.3 tons/ton	1.3		
Total				15.7	11
				146.2×10^6 tons	

Table V

Sulfur Dioxide Emissions in the Years 1860–2000 (10^6 tons SO_2/yr)[10]

Year	Coal	Petroleum	Smelting (Cu, Pb, Zn)	Total	1940 = 100	1965 = 100
1860	5.0	0.0	0.22[a]	5.22	7	
1870	7.8	0.01	0.24[a]	8.05	11	
1880	12.2	0.07	0.26[a]	12.53	17	
1890	18.7	0.17	0.68[a]	19.55	27	
1900	28.1	0.33	2.22	30.65	42	
1910	42.1	0.70	3.71	46.51	64	
1920	49.5	1.79	4.02	55.31	76	
1930	51.3	3.12	4.91	59.33	82	
1940	61.0	4.62	6.96	72.58	100	
1950	66.0	8.30	7.38	81.68	113	
1960	95.7	19.9	12.28	127.88	176	
1965	102.0	28.5	15.7	146.20	202	100
Increase Projected:	0.2%/yr	6.2%/yr	4.6%/yr			
1970	103	38	19.7	161	222	110
1980	105	62	29.0	196	270	133
1990	107	100	42.7	250	344	170
2000	109	162	62.7	333	459	227

[a] Copper smelting only.

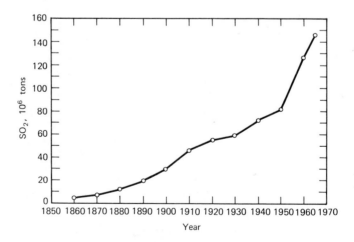

Figure 1 Estimated historical SO_2 emissions.[10]

2.2. The Sources of Materials Presently Considered Pollutants

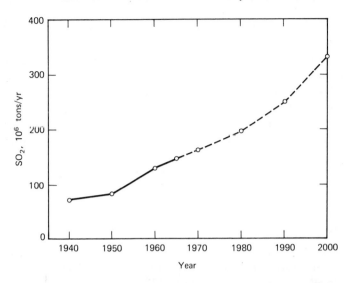

Figure 2 Projected global SO_2 emissions based on 1965 emission factors.[10]

Hydrogen Sulfide

Discussions on the sources of H_2S are speculative because well-confirmed analytical data on the ambient H_2S concentration are not available. Estimates of natural sources are made on the basis of balancing a sulfur cycle not completely proven. Eriksson[11] has given the figures shown in Table VI. These estimates were confirmed by those of Junge.[12] However, Robinson and Robbins[3] estimate ocean emissions of 30×10^6 tons/yr and land emissions of 70×10^6 tons/year. Thus, an emission of ca. 70×10^6

Table VI
Hydrogen Sulfide Emissions

	Eriksson,[10] Junge[11]		Robbins and Robinson[3]	
Source	Tons/yr as H_2S	Tons/yr as S	Tons/yr as H_2S	Tons/yr as S
Oceans	202×10^6	190×10^6	30×10^6	28×10^6
Land; decaying vegetation	82×10^6	77×10^6	70×10^6	66×10^6
Total	284×10^6	267×10^6	100×10^6	94×10^6

tons/year over land was agreed upon, although Eriksson has recently revised the figure upward to 112×10^6 tons/yr.[13]

Various industrial operations, such as kraft paper manufacture and oil refining, emit H_2S. Specific H_2S emission data, however, are seldom obtained; it is frequently counted with SO_2. Since there is a serious odor problem with H_2S, even from minor amounts, industrial emissions are generally well-monitored and the H_2S removed by scrubbing. Rough estimates of H_2S emission from kraft mills on a worldwide basis give a figure of 0.06×10^6 tons/yr. Industrial H_2S may amount to 2% of SO_2 emissions; if this is applied to the global SO_2 figure of 146×10^6 tons/yr, a total of 3×10^6 tons/yr of H_2S is obtained. These amounts for industrial H_2S emission are considered insignificant in the total sulfur cycle.

Sulfates

Sulfates (sulfuric acid and its salts) do not exist in the atmosphere in the gaseous state but rather as particulates in the form of aerosols. Eriksson[11] has estimated an annual production of 130×10^6 tons from sea spray but of this only 10% (13×10^6 tons) penetrates land areas. The remainder is precipitated back into the oceans in a closed cycle. Junge[12] also considers that sulfate from the oceans circulates essentially in a closed marine cycle.

The other sources of sulfate aerosols are atmospheric oxidation of SO_2 and H_2S to form H_2SO_4 or sulfate salts. These reactions will be discussed in more detail below. From a study of sulfur isotope ratios it was concluded that in industrial areas, sulfate aerosols are derived from oxidation of SO_2, while H_2S serves as the source of sulfates in nonindustrial areas.[3]

2.2.2. Nitrogen Compounds

Compared to sulfur, many more compounds of nitrogen are involved in the nitrogen cycle through the atmosphere. However, of the seven oxides of nitrogen, only three are of importance in the atmosphere: N_2O (nitrous oxide), NO (nitric oxide), and NO_2 (nitrogen dioxide). These compounds are gases at normal conditions. The most important reduced atmospheric nitrogen compound is NH_3 (ammonia), a gas under ordinary circumstances. Particulate forms of nitrogen compounds also exist in the atmosphere in the form of aerosols of nitric acid (HNO_3), and its salts (mainly ammonium nitrate, NH_4NO_3), nitrous acid (HNO_2) and its salts (MNO_2, where M stands for some univalent cation), and finally ammonium salts, mainly ammonium nitrate and ammonium sulfate, $(NH_4)_2SO_4$.

The major sources for the gaseous compounds are biological action and organic decomposition in the soil and perhaps in the ocean. Man-made emissions of gaseous nitrogen compounds are confined to NO and NO_2, totaling about 50×10^5 tons per year on a global basis. This compares to

2.2. The Sources of Materials Presently Considered Pollutants

natural annual emissions of 500×10^6 tons of NO_2, 5900×10^6 tons of NH_3, and 1000×10^6 tons of N_2O globally.[3] Thus, on a total mass basis, the man-made emissions account for only a small fraction of nitrogen emitted into the atmosphere. However, it is also true, as for SO_2, that the importance of the man-made emission is caused by local buildup of concentrations to a level many times the background when meteorological conditions are not conducive to their dispersal.

Aerosols containing ammonium ion (NH_4^+), nitrate ion (NO_3^-), and nitrite ion (NO_2^-) are formed by reactions of the gaseous nitrogen compounds in the atmosphere. There are no known direct sources of these materials.

Nitrous Oxide (N_2O)

The most abundant nitrogen compound in the atmosphere is N_2O, with a mean concentration of about 0.25 ppm. This relatively inert gas is apparently produced mainly by bacterial action on nitrogenous compounds in the soil.[3] This view is supported by laboratory experiments in which bacterial action on NH_4^+ and NO_3^- produced N_2O and N_2. Another possible source is the ocean; it is more likely, however, that the ocean is a sink, since the dissolved N_2O concentration is lower than the equilibrium value indicated by the atmospheric concentration. The average annual global production of N_2O appears to be about 590×10^6 tons per year.

Nitric Oxide (NO) and Nitrogen Dioxide (NO_2)

There is strong indication that some NO may be formed by reduction of nitrates by bacterial action under anaerobic conditions. The NO is then rapidly oxidized in air to NO_2. Nitrogen dioxide concentrations of several hundred parts per million have been noted in closed silos, apparently arising by the mechanism just described.

Fixation of gaseous nitrogen by action of lightning has been discussed for a long time. Close examination of this question by Junge[12] indicates that this is a negligible source for N compounds.

Man-made sources of NO and NO_2 are considered together because the data seldom distinguish between the two. The main sources of NO and NO_2 are combustion processes in which temperatures are high enough to oxidize atmospheric nitrogen and the hot gases are rapidly quenched, minimizing decomposition of the nitrogen oxides. Estimates of nitrogen oxide (NO + NO_2 as NO_2) production by man are given in Table VII. The preponderant production of NO_x (NO_2 + NO) is seen to be caused by combustion and/or production of coal and petroleum. The total of 53×10^6 tons/yr NO_x is a minimum figure, since it does not include estimates of natural NO_x. However, the man-made production of NO_x has the same

Table VII
Worldwide Annual Emissions of Nitrogen Oxides (NO and NO_2 as NO_2) (ca. 1965)[14]

Fuel	Use	Consumption or production per year	NO_2 factor	NO_2 Emissions $\times 10^6$ ton	Sub-totals	% of total
Coal	Power	1219×10^6 tons	20 lb/ton	12.2		
	Industrial	1369×10^6 tons	20 lb/ton	13.7		
	Dom./comm.	404×10^6 tons	5 lb/ton	1.0	26.9	51
Petroleum	Refinery production	$11,317 \times 10^6$ barrels	6 tons/10^5 barrels	0.7		
	Gasoline	379×10^6 tons	0.113 lb/gal	7.5		
	Kerosene	100×10^6 tons	0.072 lb/gal	1.3		
	Fuel oil	287×10^6 tons	0.072 lb/gal	3.6		
	Residual oil	507×10^6 tons	0.104 lb/gal	9.2	22.3	42
Natural Gas	Power	2.98×10^{12} ft^3	390 lb/10^6 ft^3	0.6		
	Industrial	10.72×10^{12} ft^3	214 lb/10^6 ft^3	1.1		
	Dom./comm.	6.86×10^{12} ft^3	116 lb/10^6 ft^3	0.4	2.1	4
Others	Incineration	500×10^6 tons	2 lb/tons	0.5		
	Wood	466×10^6 tons	1.5 lb/tons	0.3		
	Forest fires	324×10^6 tons	5 lb/tons	0.8	1.6	3
Total					52.9	100

characteristic as man-made SO_2: it is a result of human activity in high population density, high technology urban areas.

Ammonia (NH₃)

Ammonia enters the atmosphere mainly from natural sources rather than human activity. The predominant natural source is the biosphere: bacterial breakdown of organic nitrogenous matter releasing nitrogen as ammonia. In soils, the release of NH_3 depends partly on pH, a higher pH favoring release of gaseous NH_3. Studies of deposition of nitrogen compounds on the land and over the oceans and of the total nitrogen cycle indicate that the total natural NH_3 emission is 1160×10^6 tons/yr.

Man-made NH_3 emissions are only a small fraction of the estimated natural NH_3 emissions, some 4×10^6 tons/yr as indicated in Table VIII. Ammonia emissions arise primarily from combustion sources, but some arise from specific industrial processes. Like H_2S, NH_3 has a strong odor and is scrubbed from effluents where emission into populated areas would give rise to an odor problem.

Nitrite (NO_2^-), Nitrate (NO_3^-), and Ammonium (NH_4^+) Compounds

Since all of these materials are charged, they do not exist as ions in the atmosphere but form ionic salts, which are solids. In order to persist in the atmosphere, their particle size must be small, probably less than 1 μm, so that Stokes's law settling is very slow. Hence, these materials appear as aerosols in the atmosphere.

The sources for these ions are chemical reactions of the gaseous nitrogen compounds with substances already in the atmosphere such as oxygen, ozone, and water:

$$2NO + O_2 \rightarrow 2NO_2$$

$$3NO_2 + H_2O \rightarrow 2HNO_3 + NO$$

$$HNO_3 + 2NO + H_2O \rightarrow 3HNO_2$$

$$NH_3 + H_2O \rightarrow NH_4OH$$

$$NH_4OH + HNO_3 \rightarrow NH_4NO_3 + H_2O$$

$$NH_4OH + HNO_2 \rightarrow NH_4NO_2 + H_2O$$

$$NH_4OH + SO_2 \rightarrow NH_4HSO_3$$

$$2NH_4OH + 2SO_2 + O_2 \rightarrow 2NH_4HSO_4.$$

The lower oxides of nitrogen tend to be converted to nitric acid and nitrates as long as they persist in the atmosphere. Even ammonia may be

Table VIII
Worldwide Annual Emissions of Ammonia (ca. 1965)[15]

Fuel	Use	Consumption or production per year	NH$_3$ factor	NH$_3$ emissions $\times 10^6$ tons	Subtotals $\times 10^6$ tons	% of total
Coal	Power	1219 × 10^6 tons	2 lb/ton	1.2		
	Industrial and other	1773 × 10^6 tons	2 lb/ton	1.8	3.0	63.2
Petroleum	Catalytic cracker	2410 × 10^6 barrels	0.7 tons/10^5 barrels	0.02		
	Fuel oil	794 × 10^6 tons	2 lb/ton	0.8	0.82	17.2
Natural gas	All uses	20.56 × 10^{12} ft^3	0.01 tons/10^6 ft^3	0.2	0.2	4.2
Others	Incineration	500 × 10^6 tons	0.015 tons/100 tons	0.08		
	Wood	466 × 10^6 tons	0.12 tons/100 tons	0.6		
	Forest fires	324 × 10^6 tons	0.015 tons/100 tons	0.05	0.73	15.4
Total					4.75	100

converted to oxides by oxidation processes, but it seems more likely that most of the ammonia in the atmosphere is precipitated eventually as ammonium sulfate and ammonium nitrate.

2.2.3. Carbon Compounds

In this section we shall be discussing carbon dioxide (CO_2), carbon monoxide (CO), and hydrocarbons (for convenience we frequently abbreviate hydrocarbons as HC). Carbon dioxide is not considered a pollutant in most discussions. It exists in the atmosphere at a concentration of 320 ppm. It has no untoward effects physiologically until its concentration reaches about 1% at which point it induces hyperventilation by its action on the respiratory center. At very high concentrations (about 75%) it can cause death in a few hours.[16] However, as well shall see, CO_2 may very well be the material that requires the most serious attention because of its possible long-term effects on the climate, including temperature, and glaciation. The primary sources of CO_2 are respiration of plants and animals, decay of organic materials, and finally, but by no means least, combustion of carbonaceous fuels. The sinks for CO_2 are photosynthetic activity of plants and formation of calcium carbonate ($CaCO_3$).

The background concentration of carbon monoxide is 0.1 ppm. It is a colorless odorless gas that at concentrations of several hundred parts per million can cause loss of mental alertness, dizziness, and ultimately death. The primary man-made source of CO is incomplete combustion of fuel in automobiles. Several natural sources exist, such as forest fires, marine siphonophores, and reactions of terpenes emitted by conifers in forests. CO is unusual in that man-made sources predominate. The exact removal mechanisms for CO are not understood. It appears that the main sinks are oxidation in the atmosphere, absorption by plants and soil bacteria, and removal in the ocean.

Hydrocarbons and other organic materials form part of the natural background in the atmosphere. For example, methane (CH_4) at 1.5 ppm exists throughout the atmosphere. A wide variety of higher alkanes, alcohols, aldehydes, ketones, and olefins are found in the fractional parts per billion range even in the most remote locations. Methane is produced naturally by biological means in swamps and marshes; it also results from man's activity such as leaks in natural-gas lines and incomplete combustion. Overall, man-made hydrocarbons constitute only 15% of the total emitted into the atmosphere but again concentrated in centers of population, industry, and automobiles. Reactive hydrocarbons, those participating in photochemical reaction, constitute a special problem in urban atmospheres. Hydrocarbons are removed from the atmosphere by chemical

Table IX
Worldwide Annual Emissions of CO_2 Based on 1965 Fuel Usage[19]

Fuel	Use	CO_2 Factor	CO_2 Emissions, tons
Coal, lignite, etc.	3074×10^6 tons	4,700 lb/ton	7.22×10^9
Petroleum			
Gasoline	379×10^6 tons		
Kerosene	100×10^6 tons		
Fuel oil	287×10^6 tons		
Residual oil	507×10^6 tons		
Total	1273×10^6 tons	6,320 lb/ton	4.03×10^9
Natural gas	20.56×10^{12} ft^3	0.116 lb/ft^3	1.19×10^9
Waste	500×10^6 tons	1,830 lb/ton	0.46×10^9
Wood fuel	466×10^6 tons	2,930 lb/ton	0.68×10^9
Forest fires	324×10^6 tons	2,380 lb/ton	0.39×10^9
Total			14.00×10^9

reactions ending in aerosols, which ultimately are removed by rain or surface deposition.

Carbon Dioxide (CO_2)

The amount of CO_2 entering the atmosphere because of human activities in 1965 was 1.4×10^{10} tons annually on a global scale.[8] Sisler[17] gives a figure of 1.7×10^{10} tons for 1965; figures given in the MIT study[18] are consistent with these estimates. The breakdown is given in Table IX. The annual production of CO_2 due to respiration of plants and animals, decay of organic material, and release from the oceans is estimated at 1×10^{12} tons. Thus, man-made CO_2 is only about 1% of the total annual CO_2

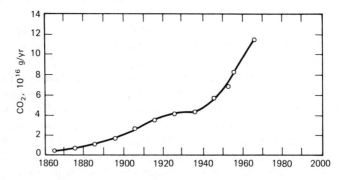

Figure 3 Average CO_2 emissions to the atmosphere from fossil-fuel combustion (decade average data 1860–1960).[20]

production; yet source and sink are so finely balanced that a 1% contribution by men appears to be having profound effects.

First, let us look at the trend in CO_2 emissions. Figure 3 shows CO_2 emissions for the period 1860–1960. A moderate increase from 1860–1920 is followed by a plateau in the production lasting until World War II, when there occurred an enormous acceleration in CO_2 production, which does not appear to be abating. The total production of CO_2 during the period shown has amounted to 3.58×10^{11} tons. This is equal to 12.5% of the total CO_2 in the atmosphere estimated at 2.48×10^{12} tons (ca. 1970).[2, 18] What was happening to the CO_2 concentration in the atmosphere during this period? An analysis of the historical data on atmospheric CO_2 concentration was made by Callendar,[21] whose results for the period 1870–1956 are summarized in Figure 4. Despite wide fluctuations, it can be seen that CO_2 production by man is accompanied by increase in CO_2 concentration in the atmosphere. The generally accepted value of 290 ppm CO_2 in the background atmosphere before large-scale burning of fossil fuels is supported by this data.

It is of interest to examine more recent experimental data, where the most advanced analytical techniques were employed, precautions were taken to screen local biasing, and locations were chosen to ensure good exposure to the free atmosphere. In Figure 5 the monthly average concentration for the period 1958–1963 is plotted for atmospheric CO_2 at the Mauna Loa observatory in Hawaii. The annual cycle is clearly seen, with maxima in the spring and minima in the fall. The long-term significance of this data is brought out when the identical data is plotted in the form of the 12-month running average. This is shown in Figure 6. The trend barely

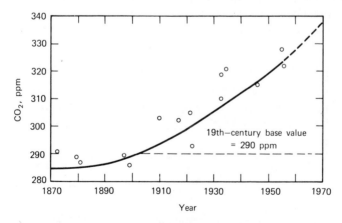

Figure 4 Average CO_2 concentration in North Atlantic region (1870–1956).[21]

observable in Figure 5 becomes obvious in Figure 6. The straight line indicates an annual rate of increase of 0.68 ppm. This rate of 0.7 ppm/yr appears to have been confirmed by CO_2 data in Scandinavia as well as Antarctica.[22]

Robinson and Robbins[23] point out that in the early 1960s, CO_2 concentrations were increasing by about 0.7 ppm/yr, as indicated above. This amounts to 0.22% of a 315 ppm average concentration. Emissions of CO_2 due to burning of fossil fuels amounted at this time to about 1.1×10^{10} tons/yr. This amounts to about 0.44% of the total in the atmosphere. Thus, it appears that about half of the man-made CO_2 emitted into the atmosphere remains there. This conclusion is supported by the MIT study.[24]

The increase in CO_2 content of the atmosphere is seen by a number of observers to have profound climatic effects that overshadow the limited range (both in time and space) of ill effects due to other pollutants such as SO_2 and NO_2. The idea that climatic change could result from changes in atmospheric CO_2 content was suggested by Chamberlain in 1899 and Arrhenius in 1903.[6] Carbon dioxide, along with ozone, water vapor, and clouds, exerts the so-called greenhouse effect—that is, solar radiation rich in UV reaches the surface but earth radiation, rich in IR, is absorbed. Thus, the overall effect is to raise the mean temperature of the troposphere. This effect is counterbalanced by the increase in the amount of particulates in the atmosphere which prevent solar radiation from reaching the surface. A recent study by Rasool and Schneider[25] indicates that increasing CO_2 does increase the surface temperature, but the rate of increase diminishes as CO_2 content increases. For particulates (aerosols), on the other hand, the net effect of an increase is to reduce the surface temperature. Because of the exponential dependence of the backscattering of solar radiation, the rate of temperature decrease is augmented with increasing aerosol content. These writers conclude that an increase in global aerosol content by only a factor of 4 may be enough to reduce the mean surface temperature by

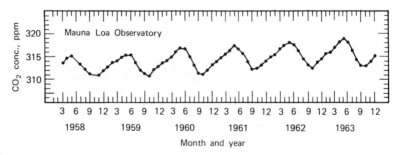

Figure 5 Monthly average concentration of atmospheric CO_2 at Mauna Loa Observatory versus time.[47]

2.2. The Sources of Materials Presently Considered Pollutants

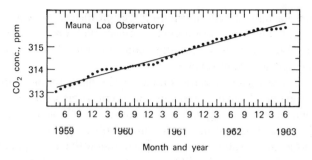

Figure 6 Twelve-month running mean of the concentration of atmospheric CO_2 at Mauna Loa Observatory. Means are plotted versus the sixth month of the appropriate 12-month interval. The straight line indicates a rate of increase of 0.68 ppm/yr.[47]

3.5°K and, if sustained over several years, may trigger an ice age. Further discussion of these points is deferred until Chapter 3, where we consider the global effects of air pollution in greater detail (Section 3.1).

Carbon Monoxide (CO)

Carbon monoxide (CO) is a product of almost all combustion processes and is therefore as widespread a pollutant as the processes that generate it. It is a colorless, odorless gas and therefore cannot be detected by unaided human senses. The internal combustion engine is the largest single source of CO. As a consequence, the highest concentrations are found in urban areas having the highest population as well as automobile density, thus exposing the greatest number of people to hazard.

The presence of CO in urban atmospheres in the parts per million range has been known for a long time. However, the presence of CO in the background atmosphere was recognized only in the late 1940s. The identification was made by detection of CO absorption lines in the IR region (4.7 μm) of the solar spectrum.[26] More extensive work with more sensitive instrumentation showed CO to be a definite, but variable, constituent of the atmosphere. Studies on the Greenland ice cap in 1967 with a field recording instrument showed the average CO concentration over a 2-wk period to be 0.11 ppm with a range of 0.09–0.5 ppm. This result was confirmed by work at Port Barrow, Alaska, as well as by analyses done during the course of a number of cruises in the Pacific Ocean.[26] All of the data indicate an average global concentration of 0.10 ppm. With this figure, a total of 6.2 × 10^8 tons of CO in the atmosphere can be calculated.

Carbon monoxide has been recognized as a man-made pollutant for many years. The man-made sources are detailed in Table X. The data confirms that automobile use is the major source of CO.; that is, the CO derived from gasoline is 68% of the total CO produced. As with SO_2, there is a strong imbalance in CO emission between the northern and southern hemispheres. On the basis of gasoline consumption, 95% of CO emission occurs in the northern hemisphere.

Aside from forest fires, natural sources of CO have not been investigated to any great extent. Volcanic gases do not seem to be a significant source. Some evidence exists for vegetation as a source of CO, while some bacteria appear to produce it as a result of their metabolism. Evidence of the oceans as a CO source has resulted from measurements of the atmosphere over the Pacific. It was noted that the CO concentrations showed marked diurnal variations, suggesting that the ocean was a source, a sink, or both. Analyses of CO in the air and ocean in the Caribbean area indicated that the ocean was supersaturated with respect to atmospheric CO, at least by a factor of ten.[28] Although the source of the CO has not been identified, it appears logical to suppose that it arises from biological action, with the siphonophores, perhaps, playing a major role. The siphonophores are an order of open-sea marine animals of the group hydrozoa (or hydromedusae)—in common language, polyps and jellyfish. The latter in turn belong to a larger group called coelenterates. The siphonophores all form colonies that float or swim with the aid of individuals in the colony that act as swimming-bells and others who form gas-containing floats.[29] Pickwell found that siphonophores expel bubbles containing nearly 80% CO.[30] He calculated a possible normal and maximum CO contribution to the atmosphere varying from 0.02 to 2 cc CO/(mile^2) (day). This would lead to 3.62×10^6 to 362×10^6 tons/yr, depending on the density of siphonophore

Table X

Estimated Annual Global CO Emissions, Man-made (ca. 1965)[27]

Fuel	Use	CO Factor	CO emission
Gasoline	379×10^6 tons	2.91 lb/gal	193×10^6
Coal			
Power	1219×10^6 tons	0.5 lb/ton	
Industry	781×10^6 tons	3.0 lb/ton	
Residential	404×10^6 tons	50.0 lb/ton	
Coke and gas plants	615×10^6 tons	0.11 lb/ton	
Total	3074×10^6 tons		12×10^6
Wood and non-commercial fuel	1260×10^6 tons	70 lb/ton	44×10^6
Incineration	500×10^6 tons	100 lb/ton	25×10^6
Forest fires	18×10^6 acres		11×10^6
Total			285×10^6 tons

Table XI

Annual Global Emissions of CO, Man-made and Natural[32]

Man-made		Natural		
Source	Emission, tons/yr	Source	Emission, tons/yr	Total
Gasoline	193×10^6	Forest fires	11×10^6	
Coal	12×10^6	Siphonophores	360×10^6 max.	
			4×10^6 normal	
Noncommercial fuel	44×10^6	Terpene reactions	60×10^6	
Incineration	25×10^6			
Total	274×10^6		$75 \times 10^{6\,a}$	$349 \times 10^{6\,a}$
			$431 \times 10^{6\,b}$	$705 \times 10^{6\,b}$
% of total	79^a		21^a	
	39^b		61^b	

[a] Normal production of CO by siphonophores.
[b] Maximum production of CO by siphonophores.

swarms. The lower figure would appear to be more normal, based on typical numbers of animals seen or captured. The diurnal CO cycle above the ocean atmosphere noted by Robinson and Robbins thus appears to be related to a similar cycle in the siphonophores.

A potential source of CO over land areas proposed by Went involves CO produced by photochemical reactions of terpenes emitted by plants.[30] Estimates of 10^9 tons/yr of volatile organics of plant origin have been made. Photochemical oxidation by ozone of HC could produce 61×10^6 tons/yr.

Another marine organism containing CO is the familiar kelp, or seaweed, with large floats or bladders. The gas in the latter has been found to contain up to 800 ppm CO.

Current estimates of CO emission, man-made and natural, are summarized in Table XI. Using the normal figure for siphonophore CO production, man-made CO accounts for 79% of the total annual emission; using the larger figures, man-made CO accounts for 39% of the total. Thus, the man-made CO is undoubtedly a substantial fraction of CO emission on a global scale.

Hydrocarbons and Other Organic Compounds

It is seen from Chapter 1 (Table I) that methane is a normal constituent of the atmosphere at approximately 1.5 ppm, while the nonmethane hydrocarbons constitute about 1 ppb in the background. An extensive series of analyses was carried out in 1967 under the auspices of the Stanford Research Institute at Point Barrow, Alaska. The purpose of this work was to obtain data on the background concentrations of the materials considered in this section. The technique used was gas chromatography, and a dozen analyses were done over a period of 24 h. The averaged results are given in Table XII. The data for methane and CO confirm estimates previously given. The wide variety of other types of organic materials in the fractional part per billion range is somewhat surprising. It is difficult to decide whether each of these is truly a natural background material or is industrial in origin and has diffused into the global background atmosphere, although of man-made origin. If we consider the atmosphere at Point Barrow to be reasonably free of short-term human influences and adhere to our working definition of background, then each of the materials listed would appear to arise from some natural source. The present writers are skeptical, although Robinson and Robbins[31] appear to favor the biosphere as the source of n-butanol. It would appear that one way to help answer this question is to follow the diffusion paths of a tracer compound (isotopic n-butanol, for example, or sulfur hexafluoride) released in, say, Chicago and analyzed at Point Barrow.

Table XII

Background Concentration, Organic Materials (Point Barrow, Alaska, September 1967)[a]

Compound	Concentration
Ethane, CH_3—CH_3	0.45 ppb
Ethylene, CH_2=CH_2	0.38 ppb
Butane, C_4H_{10}	~0.5 ppb
Pentane, C_5H_{12}	0.98 ppb
Hexane, C_6H_{14}	~0.8 ppb
Acetaldehyde, CH_3—CHO	0.48 ppb
Acetone, CH_3—CO—CH_3	0.75 ppb
Methanol + ethanol, CH_3OH + C_2H_5OH	0.15 ppb
Benzene, C_6H_6	~0.1 ppb
n-Butanol, C_4H_9OH	9.22 ppb
Methane, CH_4	1.50 ppm
Carbon monoxide, CO	0.104 ppm

[a] Adapted from Ref. 3, p. 95.

Table XIII

Emissions of Terpene-type Hydrocarbons from Plants[34]

Type of plant	Estimated emission, tons/yr
Coniferous forest	50×10^6
Hardwood forest ⎫ Cultivated land ⎬ Steppes ⎭	50×10^6
Carotene decomp. in organic material	70×10^6
Total	170×10^6

Methane (CH_4)

Methane is produced in nature by bacterial decomposition reactions in swamps, marshes, and other waters. The principal component of "marsh gas" is methane. Natural gas also contains methane as a major component; the methane therein may enter the atmosphere through seepage and leaks from natural-gas reservoirs. Isotopic analysis of atmospheric CH_4 indicates that it is predominantly "young" carbon, leading to the conclusion that the biosphere rather than the lithosphere is the main source. Robinson and Robbins,[35] considering natural emissions on a global scale, conclude that 1600×10^6 tons/yr of CH_4 are produced. As to man-made methane sources, emissions from automobile exhaust and from other carbonaceous combustions are important.

Other Organic Materials

The biosphere is a major contributor of terpenes to the atmosphere. Table XIII gives some estimates of terpene emissions from various types of plants. Activities of men also contribute organics to the atmosphere. Table XIV gives estimates of hydrocarbon emissions on a global scale from a variety of materials handled industrially, commercially, and domestically. The data of Table XIV are of interest in that, again, the contribution of the automobile to the urban pollution is emphasized: well over one-half of man-made organic emissions are caused by the petroleum industry, the primary source of automobile fuel. When reactive hydrocarbons are considered (i.e., those that participate in atmospheric smog-forming reactions), the contribution of the automobile is even more dramatic—close to two-thirds. Incineration is also seen to be a major man-made contributor to organic emission—17% of the total and 21% of the reactive component.

Table XIV
Global Hydrocarbon Emissions[35]

Source	Quantity ($\times 10^6$)	Emission factor	Reactive HC, %	Total emission, tons $\times 10^6$	Reactive emission, tons $\times 10^4$	Subtotals, tons Total HC $\times 10^6$	Subtotals, tons Reactive HC $\times 10^4$	Fractions % total	Fractions % reactive
Coal									
Power	1219 tons	0.2 lb/ton	15	0.1	0.02				
Industrial	1369 tons	1.0 lb/ton	15	0.7	0.10				
Dom. and comm.	404 tons	10 lb/ton	15	2.0	0.30	2.8	0.4	1.9	1.1
Petroleum									
Refineries	11,317 barrels	56 tons/10^4 barrels	14	63.4	8.88				
Gasoline	379 tons	180 lb/ton	44	34.0	15.00				
Kerosene	100 tons	0.6 lb/ton	18	0.1	0.01				
Fuel oil	287 tons	1.0 lb/ton	18	0.1	0.02				
Residual oil	507 tons	0.9 lb/ton	18	0.2	0.04				
Evap. and transfer loss	379 tons	41 lb/ton	20	7.8	1.56	105.6	25.5	72.7	72.5
Other									
Solvent use	3 tons	30 lb/(yr) (person)	15	10	1.50				
Incinerators	500 tons	100 lb/ton	30	25	7.50				
Wood fuel	466 tons	3 lb/ton	15	0.7	0.01				
Forest fires	324 tons	7 lb/ton	21	1.2	0.25	36.9	9.3	25.4	26.4
Total				145.3	35.2	145.3	35.2	100.0	100.0

Table XV
Total Annual Organic Emissions

	Tons/yr × 10^6	Subtotal	%
Natural			
Methane	1600		
Terpenes	170		
Forest fires	1.2	1771.2	92.4
Man-made			
Coal	2.8		
Petroleum	105.6		
Other	36.9	145.3	7.6
Total		1916.5	100.0

The reason for this is the poor combustion efficiency and a high emission factor of material burnt in incinerators. If incineration were as efficient as burning of wood fuel, these figures would be reduced to less than 1%.

It is instructive to put man-made emissions in the perspective of total organic emissions. This is done in Table XV. It is seen that man-made organics constitute about 8% of the total.

2.3. REACTIONS IN THE ATMOSPHERE AND SINKS FOR POLLUTANTS

We are concerned in this section with mechanisms that operate to remove pollutants from the atmosphere. The removal mechanisms are of three kinds:

1. Chemical reactions which transform the pollutant into another material and so lower the concentration of the pollutant. In the usual case, the transformation will result in a species that is not considered a pollutant. In the case where a new pollutant is created, we discuss the removal mechanism of that pollutant separately.
2. Physical mechanisms such as rain, absorption on solid surfaces and in the soil, or solubility in oceans and lakes. These mechanisms also act to reduce pollutant concentration in the atmosphere.
3. Biological mechanisms such as respiration of plants or action of soil bacteria.

Our viewpoint is that any mechanism that lowers the atmospheric concentration of a pollutant constitutes a sink for that pollutant, the concept of a sink being a very general one. We will see that some sinks close the cycle of movement of an element, while others keep the cycle open. For example, CO_2 ending up as seashells essentially removes it from circulation (closing the cycle), while CO_2 absorbed in plants during photosynthesis keeps the circulation going.

2.3.1. Sulphur Compounds

Sulfur Dioxide

Sulfur dioxide is very soluble in water (11.28 g SO_2/100 ml water at 20°C). In the presence of water, SO_2 is very reactive in the atmosphere, since it is both acidic and a reducing agent. The end product is nearly always sulfuric acid (H_2SO_4) or a sulfate salt.

The exact oxidation mechanism in the atmosphere is not entirely understood. In the absence of catalysts, the rate of SO_2 oxidation in small H_2SO_4 droplets is negligible as determined in laboratory experiments.[36] Other laboratory work showed appreciable rates of H_2SO_4 formation in the presence of radiation at 3130 A with SO_2 concentrations at less than 1 ppm and relative humidity less than 50%. Sulfur dioxide dissolved in fog or cloud droplets becomes sulfurous acid (H_2SO_3), which is rapidly oxidized by dissolved oxygen to form sulfuric acid (H_2SO_4).[7] The interest in SO_2 oxidation relates to the Donora and London episodes in which high SO_2 concentrations were found. The effect of catalysts such as ferric ion, manganese salts, and magnesium ion were studied. It seems unlikely, however, that these materials are present in the atmosphere in the form or in the amount necessary to explain the observed rates of SO_2 oxidation. A better candidate is ammonia (NH_3), which is emitted along with SO_2 when coal is burned. By increasing the pH of water, dissolved ammonia enhances the rate of solution of SO_2. The oxidation rate of ammonium sulfite, $(NH_4)_2SO_3$ appears to be greater than that of sulfurous acid.

Analyses of power plant plumes tracked by helicopters indicated very rapid oxidation rates, 0.1–2%/min. This rate was dependent on the moisture content of the plume as well as the ambient relative humidity, increasing with the moisture content from either source. It appears reasonable that in power plant plumes, both ammonia and particulate matter accelerate oxidation of SO_2.

Direct oxidation of SO_2 by molecular oxygen has been shown to proceed at a negligible rate. However, photochemical oxidation of SO_2 in the presence of NO_2 and hydrocarbons to form aerosols is probably a significant mechanism for SO_2 removal in polluted atmospheres. A great many studies of this system (SO_2, NO_2, HC, and air) have been done. The work

Doyle and coworkers at Stanford Research Institute seems to be most pertinent.[37] They showed that mixtures of olefins, SO_2, and NO_2 in air form particulate matter when irradiated by sunlight and that H_2SO_4 was the major constituent, besides water, of the aerosol particles. As little as 0.10 ppm SO_2 was sufficient to promote H_2SO_4 aerosol formation under these conditions. It appears that olefins are more effective in the reactions leading to H_2SO_4 aerosols in the presence of NO_2 than saturated hydrocarbons.

An integrated system to explain SO_2 reactions in the atmosphere does not yet exist. Several possible reactions appear to compete with each other and oxidation of SO_2 to H_2SO_4 can proceed very rapidly under favorable circumstances. In the daytime, at low humidity, photochemical reactions involving SO_2, NO_2, and unsaturated hydrocarbons are of primary importance in converting SO_2 to H_2SO_4 aerosol. At night under high humidity conditions, in fog, or during actual rain, the process involving solution of SO_2 into alkaline water droplets and oxidation within the droplet by dissolved oxygen appears to be the primary SO_2 removal process.

Up to now we have been discussing removal or, more precisely, reactions in which SO_2 is changed within the atmosphere. Other removal mechanisms are also known. For example, absorption of SO_2 by vegetation and resultant plant damage is well known. The rate of SO_2 absorption by plants is a complex function of plant physiology and weather. Experiments by Katz and Ledingham showed conclusively that SO_2 was rapidly absorbed by plants.[38] They covered plants with fumigation chambers and passed air having SO_2 in the range 0.8–1.0 ppm through the chambers. Analyses of the effluent air from the chamber along with a material balance proved that the plants absorbed SO_2. The rate of SO_2 absorption dropped markedly at night or in the shade. Apparently the plants oxidize the absorbed SO_2 to sulfate, thus removing sulfur from the atmosphere entirely.

Absorption of SO_2 into the soil and low-lying vegetation such as grass is another means of removal. The exact mechanism of this form of SO_2 removal is not yet understood.

Hydrogen Sulfide

It appears most likely that the main removal mechanism for H_2S is oxidation in the atmosphere by atomic oxygen (O), oxygen (O_2), or ozone (O_3).[7] The heterogeneous reaction of H_2S and O_2 on surfaces is very rapid. Cadle and Ledford[39] confirmed the simple reaction chemistry: $H_2S + O_3 \rightarrow H_2O + SO_2$. They also noticed the effect of surfaces and concluded that the reaction was at least partially heterogeneous. The reaction rate was found to be a function of the square root of the specific surface (area per volume), zero order in H_2S and $\frac{3}{2}$ order in O_3. Since there is a great deal of particulate matter in the atmosphere providing large amounts of surface area, rapid destruction of H_2S in the atmosphere is possible. Junge[12] reported particle concentrations over land as $15,000/cm^3$ and background counts of $200/cm^3$. It is possible to estimate specific surfaces from these figures and thus calculate the lifetime of H_2S concentrations at various O_3 concentrations. These calculations lead to an H_2S lifetime of 2 h over land and 28 h for the background case. Thus, H_2S is rapidly destroyed.

Sulfate

The sources of sulfate particles are reactions of SO_2 and H_2S that lead to H_2SO_4 aerosols; evaporation of fog and cloud droplets in which SO_2 was absorbed and oxidized; and evaporation of sea spray. In any case, sulfate can only exist in the atmosphere in particulate form. However, the lifetime of particulates in the atmosphere depends on their size. If they are heavy, gravitational settling according to Stokes's law will remove them relatively quickly. If they are light, they will persist in the atmosphere for longer periods. Junge[12] has shown that the tropospheric sulfate aerosol has 50% of its mass in particles smaller than 0.15 μm radius. If particles are very small, they will tend to coagulate. Thus, the processes leading to a fairly narrow size range (0.1–1.0 μm) in the atmosphere are coagulation, sedimentation (gravitational settling), and condensation and evaporation cycles of clouds, fog, and rain drops (see Section 1.4). Sedimentation will remove sulfate from the atmosphere. Rain and snow will also remove sulfate by "scrubbing" the atmosphere of particles. Rain washout efficiencies decrease rapidly with decreasing aerosol particular size, and washout is almost entirely caused by particles greater than 0.5 μm.

2.3.2. Nitrogen Compounds

Nitrous Oxide (N_2O)

Nitrous oxide is essentially inert at normal temperatures and pressures, taking no part in tropospheric reactions. In the stratosphere, photodissociation occurs:

$$N_2O + h\nu \rightarrow N_2 + 0$$

$$N_2O + h\nu \rightarrow NO + N (\lambda < 2500 \text{ A})$$

Nitric Oxide (NO) and Nitrogen Dioxide (NO_2)

The reactions of NO with ozone in the atmosphere is rapid. Since ozone is constantly introduced into the troposphere from the stratosphere it is always available, and NO is assumed to have a very short lifetime:

$$NO + O_3 \rightarrow NO_2 + O_2.$$

The reaction
$$2NO_2 + O_3 \rightarrow N_2O_5 + O_2$$
is 500 times slower than the NO–O$_3$ reaction. In spite of this, the half-life of 1 ppb NO$_2$ in the presence of 5 ppb O$_3$ is about 2 wk.

As we shall see below, the residence time of NO$_2$ in the atmosphere is about 3 days. This extremely short time is explained by the scavenging reaction:
$$3NO_2 + H_2O \rightleftharpoons 2HNO_3 + NO$$
The equilibrium constant for this reaction is only 0.004 atm^{-1} at 25°C[40] but with the great excess of water in the atmosphere, 10% of the NO$_2$ is converted. The nitric acid is rapidly removed by reaction with NH$_3$ and absorption into water droplets so that the reaction proceeds to the right in accordance with LeChatlier's principle. All of the HNO$_3$ eventually becomes nitrate salt aerosol.

We have mentioned the reactions of NO and NO$_2$ with olefins and organic oxides and peroxides in connection with SO$_2$ removal. Haagen-Smit and Wayne[4] suggest free-radical reactions involving all of these materials:

$$RCH = CHR + O_3 \rightarrow RCHO + RO\cdot + HCO\cdot$$

$$R\cdot + O_2 \rightarrow ROO\cdot$$

$$ROO\cdot + NO \rightarrow ROONO \xrightarrow{h\nu} RO\cdot + NO_2$$

$$ROO\cdot + NO_2 \rightarrow ROONO_2 \text{ (peroxynitrate)}$$

$$ROO\cdot + RCH = CH_2 \rightarrow ROOCHR\text{---}CH_2\cdot$$

$$ROO\cdot + SO_2 \rightarrow ROOSO_2\cdot$$

$$ROO\cdot + O_3 \rightarrow RO\cdot + 2O_2$$

Kellogg et al.[7] consider that oxidation of SO$_2$ to sulfate is most likely accomplished through a peroxy free radical such as written above.

Ammonia

Ammonia is removed from the atmosphere by three routes:

1. About 75% of the ammonia is converted to ammonium (NH$_4^+$) aerosol in water droplets:

$$2NH_3 + H_2SO_4 \rightarrow (NH_4)_2SO_4$$
$$NH_3 + HNO_3 \rightarrow NH_4NO_3$$
⎫ Direct reaction with acid aerosols

$$NH_3 + H_2O \rightarrow NH_4^+ + OH^-$$ Hydrolysis in water droplets followed later by neutralization.

2. About 25% is oxidized in the atmosphere, but the mechanism is not understood. The reaction with ozone is extremely slow, while photo-oxidation can only occur in the upper atmosphere. The direct gas-phase reaction between NH_3 and NO_2 is essentially a neutralization of the NH_3 and later oxidation of nitrite to nitrate by atmospheric oxygen.
3. A small fraction of NH_3 is reabsorbed from the atmosphere by soil and plants.

Ammonia converted to ammonium aerosols is finally removed from the atmosphere by coagulation, rain, and sedimentation, as previously described for SO_2.

In summary, most of the nitrogen in the atmosphere is converted to particulate material, after which it is removed by processes already described.

2.3.3. Carbon Compounds

Carbon Dioxide

An important sink for removal of CO_2 from the atmosphere is the photosynthetic activity of green plants. The total annual uptake of CO_2 by vegetation is 1.55×10^{11} tons/yr, of which about 59% occurs in the northern hemisphere. If this were the only removal mechanism on a global scale, the residence time (which is discussed in more detail below) of CO_2 would be 17 yr. Considering total CO_2 production on global scale and the amount of CO_2 in the atmosphere, a residence time closer to 3 yr is calculated. Thus, at least one other big sink must be operating.

Another important factor in the CO_2 cycle is the ocean. Various estimates of the residence time for CO_2 in relation to uptake by the ocean have been made. Some of these have been made by Young and Fairhill[41] using data on ^{14}C produced by atomic bombs. If only the troposphere is considered, the oceanic CO_2 rate is sufficient to give a residence time of 2.5 yr. If the troposphere contains 84% of the total atmospheric CO_2, this leads to an uptake of 8.7×10^{11} tons/year. In the ocean, CO_2 is consumed in plant and animal growth, just as on land. Most of the carbon goes to form protoplasm, but some of it ends up as shells and other calcareous materials. Decay on land results in restoration of carbon to the atmosphere. A great deal of decay in the ocean, however, removes carbon from the atmospheric cycle, because dead organisms settle into deep water, which is isolated from the atmosphere.

It is useful to discuss briefly atmospheric and oceanic CO_2 in relation to the secular increase in atmospheric CO_2 mentioned above. Since CO_2 is quite soluble in water, it would appear that rising CO_2 levels in the atmosphere would be adequately buffered by the ocean. However, it has been found that only the relatively shallow mixed layer near the surface is involved in CO_2 exchange with the atmosphere. This layer provides very little storage capacity for increase in atmospheric CO_2. This low storage factor has been examined[42] and found to be related to the ionic equilibria of CO_2 and water:

$$CO_2(g) + H_2O \leftrightarrows H_2CO_3(l) \leftrightarrows H^+ + HCO_3^-$$
$$\updownarrow$$
$$H^+ + CO_3^{2-}$$

Calculations indicate that a 10% increase in atmospheric CO_2 (such as from 290 to 320 ppm) results only in a 1% increase in oceanic CO_2 content. A two-layer model of the ocean was used, and calculations based on this model agreed with observed CO_2 changes in the atmosphere due to fossil-fuel combustion. It seems clear from the examination of the above equilibria that the limited capacity of the ocean to absorb CO_2 from the atmosphere is caused by the relatively low pH of the ocean, which is probably controlled by other equilibria in addition to the carbon dioxide equilibria. If the pH were raised, large amounts of CO_2 would be absorbed from the atmosphere. Similarly, if the pH were lowered, large amounts would be released.

Other factors have been considered in relation to the CO_2 storage capacity of the ocean. For example, changes resulting from temperature variation were studied. A one-layer ocean model was used, and it was concluded that an increase in ocean temperatures of 1°C would result in an atmospheric CO_2 increase of 6%.[42] If the two-layer model is used, the increase is only 0.4%. Thus, any reasonable temperature changes would only result in local fluctuations in atmospheric CO_2 level.

Carbon Monoxide (CO)

At present emission rates, CO concentration in the atmosphere would double in three years. Since no such concentration increase is being observed, it follows that some fairly efficient removal mechanisms must be operating. Paradoxically, these mechanisms have not been proven. We will, therefore, discuss some postulated mechanisms:

1. *Gaseous reactions*:

   $2CO + O_2 \rightarrow 2CO_2$ (gas-phase reaction with molecular oxygen)

   $CO + O_3 \rightarrow CO_2 + O_2$ (gas-phase with reaction with ozone)

   $CO + NO_2 \rightarrow CO_2 + NO$ (gas-phase reaction with nitrogen dioxide)

These reactions have been found to be extremely slow. The activation energy for the ozone reaction is 20 kcal/mole and for the nitrogen dioxide reaction 28 kcal/mole, both very high values. Reactions with the hydroxyl radical, generated by photochemical smog reactions, have been postulated. They have not been demonstrated.

2. *Absorption processes:* It is postulated that CO is absorbed on surfaces exposed to the atmosphere. Unless some reaction removing or changing absorbed CO is also postulated, surface absorption cannot be seriously considered. No such reaction has been found. The oceans, being a sink for CO_2, may also be a sink for CO. Although CO solubility in seawater is 20 ml/liter at 20°C,[43] one must also propose a mechanism for removal of CO from solution. In contrast to CO_2, no such mechanisms are known. The solubility based on Henry's law and a background of 0.10 ppm CO in the atmosphere would result in a total of 3.75×10^6 tons dissolved in the ocean, assuming uniforming mixing. This is only 1% of the yearly global CO production (see Table XI). Thus, the ocean cannot be a sink unless we find removal reactions for CO in the sea.

3. *Biological processes:* Two types of biological reactions can be postulated: the first involves utilization of CO by specific species of plants or animals in their metabolic cycles; the other involves CO binding by widely distributed organic compounds in a manner similar to binding by hemoglobin. Some species of bacteria metabolize CO. Estimates of the distribution and amount of bacteria have not been made so that their efficiency as CO sinks cannot be evaluated. Binding with porphyrin-type compounds universally distributed in living materials may be considered. Heme compounds, containing iron, could bind CO, just as hemoglobin in blood does. Permanent removal would depend on reaction of the bound CO to form CO_2. Plants could provide a site for such reactions. Plant respiration seems to provide potential for CO pickup, since plants are effective scavengers for CO_2, SO_2, and the like, as noted previously. Perhaps, binding to plant porphyrins could initiate removal of CO by plants.

The fact that the background concentration of CO is not increasing, combined with the known man-made emissions of CO (220×10^6 tons/yr globally), indicates a natural sink (or sinks) for the removal of CO from the atmosphere. We have discussed some possible sinks above: oxidation to CO_2 in the atmosphere, formation of CO–hemoglobin, absorption by vegetation, and action of soil bacteria. None of these sinks has been examined in sufficient detail to prove definitely that it is indeed a sink for CO. A recent paper by Inman et al.[44] describes a direct investigation of soil in removing CO from the atmosphere. Various soils were exposed to test atmospheres containing known amounts of CO in closed chambers. The chamber atmosphere was monitored for CO as a function of time, using gas-chromatographic techniques. Soils tested included greenhouse potting mixture and soils from California, Florida, and Hawaii. Most soils tested showed great ability to remove CO from test atmospheres. On sterilization by autoclaving, ability to remove CO was lost. This indicated that bacteria were responsible or some absorbing material in the soil whose activity was destroyed by autoclaving. A number of experiments were run that definitely showed that microorganisms were involved in the CO-removal property of nonsterile soils. Generally speaking, the rate of CO removal increased with percent of organic matter in the soil. On the basis of uptake rates measured in the laboratory, Inman et al. calculated the uptake capacity of the United States alone to be 626×10^6 tons/yr, 6.5 times the U.S. emission of CO and three times that of global CO emission. These authors conclude that the soil must be considered a major natural sink for CO.

2.4 RESIDENCE TIMES

We begin the discussion on residence time by examining in detail what this concept means. The atmosphere may be considered a vast container into which various pollutants are poured and from which they are withdrawn by various mechanisms. We have already considered sources in Section 2.2 and removal mechanisms in Section 2.3. If pollutant X is poured into the atmosphere at a rate of w_e tons/yr and its concentration in the atmosphere does not change, it is clear that w_e tons/yr must leave the atmosphere. From this it follows that a molecule of X resides for some average period of time before it is removed to be replaced by a different molecule of X. The effect of this emission-removal mechanism is to keep the concentration constant. If the concentration does remain constant, then, the greater the emission of pollutant X into the atmosphere, the shorter the average dwell time of a molecule must be. Thus, if τ is the residence or dwell time, τ is inversely proportional to w_e; that is,

$$\tau = \frac{k}{w_e}, \tag{1}$$

where k is some proportionality constant. Rewriting Eq. (1),

$$k = w_e \tau. \tag{2}$$

where τ is residence or dwell time in years and w_e is emission of X in tons per year. Therefore, k has the dimensions of tons (tons per year $\times$ year). It is easy to see that k embodies the condition of constant concentration, expressed not as a concentration but as the total amount of X in the atmosphere; that is, the condition of constant concentration implies for an atmosphere of fixed volume a constant total amount of X. This amount we have called k. From Eq. (1), we can calculate τ if we know k and w_e. We can also calculate τ if we know k and w_r, the weight removed per year, since for constant amount or constant concentration, $w_r = w_e$.

The concept of residence time is very useful in carrying out the material balances required in going through the cycle for an element such as S or N. In the usual case, we are better acquainted with sources than with sinks. Thus, having calculated τ from the sum of the sources, we are in a better position to look for sinks of the correct order of magnitude and thus account for the complete cycle of an element or pollutant. This is a rather complex problem with many unknowns, as the above discussions have indicated. We will not discuss the complete cycles for this reason. For those who wish an introduction to this subject, the work of Robinson and Robbins is recommended.[2,3,8]

At this point we will carry through a few calculations of τ for the pollutants we have been discussing. Some authors give figures on the total amount of a pollutant in the atmosphere. If the total annual emission is known, simple substitution into Eq. (1) will give τ. However, the total amount in the atmosphere is not immediately accessible; it can be calculated from a knowledge of the average concentration and the total volume or total moles of the atmosphere. The former is obtained by repeated careful analyses of the background atmosphere, as described previously. The latter is obtained by estimates of the total contents of the atmosphere such as given in Ref. 2 and summarized in Table I (page numbers refer to Ref. 2 and data in the examples below are from the reference.)

Let us take an example CO. The emission given is 2.57×10^8 tons/yr (p. 42). The average background is 0.13 ppm (p. 62). The total amount in the atmosphere is

$$k(CO) = 0.13 \times 10^{-6} \times 1.75 \times 10^{20} \text{ moles}$$
$$= 2.27 \times 10^{13} \text{ moles}$$
$$= 6.36 \times 10^{14} \text{ g}$$
$$= 7.00 \times 10^8 \text{ tons}$$

$$\tau(CO) = \frac{7.00 \times 10^8 \text{ tons}}{2.57 \times 10^8 \text{ tons/yr}} = 2.7 \text{ yr.}$$

The data for SO_2 are as follows: Emissions given are 1.47×10^8 tons/yr.; the average background is 0.2 ppb (p. 53).

$$k(SO_2) = 2 \times 10^{-10} \times 1.75 \times 10^{20} \text{ moles} = 3.50 \times 10^{10} \text{ moles}$$
$$= 2.24 \times 10^{12} \text{ g} = 2.44 \times 10^6 \text{ tons}$$

$$\tau(SO_2) = \frac{2.44 \times 10^6 \text{ tons}}{1.47 \times 10^8 \text{ tons/yr}} = 0.0166 \text{ yr} = 6.0 \text{ days.}$$

Table XVI
Residence Times

Pollutant	Background concentration[a]	k, total in atmosphere, tons[a]	w_e, total global emissions, tons/yr[a]	τ, residence time[c]	τ^a	τ^d
SO_2	0.2 ppb	2.46×10^6	146×10^6	6.2 days	4 days	4 days (55)
H_2S	0.2 ppb	1.31×10^6	103×10^6	4.6 days	2 days	
N_2O	0.25 ppm	2.12×10^9	590×10^6	3.6 yr	4 yr	
NO	0.2–2 ppb	1.16–11.6×10^6	$457 \times 10^{6\,b}$	0.9–9.3 days		
NO_2	0.5–4 ppb	4.43–35.4×10^6	$684 \times 10^{6\,b}$	2.4–19 days	>5 days	3 days (57)
NH_3	6–20 ppb	19.6–65.5×10^6	1164×10^6	6.1–20.5 days	7 days	
CO_2	320 ppm	2.71×10^{12}	1.01×10^{12}	2.7 yr	2–4 yr	5 yr (15)
CO	0.1 ppm	5.40×10^8	3.49×10^8	1.5 yr	<3 yr	2 yr (62)
CH_4	1.5 ppm	46.3×10^8	16.0×10^8	2.9 yr	16 yr	

[a] Data from Ref. 8, p. 5.
[b] Total man-made emission of 53×10^6 tons/yr NO–NO_2 divided equally between NO and NO_2.
[c] Calculated from figures in table.
[d] Estimates of Robinson and Robbins. In Singer, Ref. 2, pp. 50–63 (page no. in parenthesis).

All of the residence-time data based on estimates discussed in previous sections are given in Table XVI. It is seen that τ for the various pollutants ranges from a few days to several years. Furthermore, the pollutants divide themselves into two classes based on residence time:

1. Long dwell time (years): N_2O, CO_2, CO, CH_4
2. Short dwell time (days): SO_2, H_2S, NO, NO_2, NH_3.

The carbon compounds mentioned are essentially chemically inert in the atmosphere and relatively long-lived. Of the nitrogen compounds only N_2O is chemically inert and has a long lifetime in the atmosphere. NO is rapidly oxidized and converted to NO_2, which is a strong oxidizing agent and acidic at the same time. Under natural conditions, it would not be reasonable to expect NO_2 to survive long. NH_3 is strongly basic and soluble in water and could be expected to react rapidly with any acidic materials in the atmosphere, such as CO_2 and NO_2. Finally, we come to the sulfur compounds. SO_2 is both a strong reducing agent and an acid, while H_2S is very unstable in the presence of oxygen and is a weak acid. Thus, the dwell times of all of these materials can be understood on the basis of their chemical reactivities in the presence of commonly available materials in the atmosphere and their acid–base properties.

2.5. SUMMARY

We summarize all the information in this chapter in condensed form in Table XVII. This table is modified from one of Robinson and Robbins,[45,46] to bring out the points that we have been emphasizing. Of the pollutants that are a major concern (SO_2, NO_x, CO, CO_2, and HC) only SO_2 and CO are caused preponderantly by activities of man. Even with these, if they were dispersed, they would constitute no problem. Thus, the problem of air pollution is primarily one where, on a local scale, the meteorological mechanisms, along with the natural scavenging mechanisms and sinks, are too slow to deal with man-made emissions. It would appear then, that the problem of air pollution could be solved by one or more of the following measures:

1. Disperse emission sources so that the rate of natural dispersal and scavenging is equal to the rate of emission and the pollutants remain close to background or below some defined safe level.
2. Decrease local emissions to meet the condition in (1) by scrubbing techniques and the like.
3. Devise mechanisms to assist and speed up natural dispersion and scavenging.
4. Decrease use of energy sources contributing pollutants.

Table XVII

Summary: Sources, Sinks, Residence Times

Pollutant	Major man-made source	Natural source	Total global emission, tons/yr	% caused by man-made	Background concentration	Residence time	Sinks
SO_2	Combustion of fossil fuels	Volcanoes	146×10^6	100	0.2 ppb	4 days	Oxidation to SO_4^{2-} pption in rain
H_2S	Chemical industry sewage treatment	Volcanoes, biological action in swamp	103×10^6	3	0.2 ppb	2 days	Oxidation to SO_2
N_2O	None	Biological decay	590×10^6	0	0.25 ppm	4 yr	Photodissociation in stratosphere, biological in soil
$NO-NO_2$	High-temperature combustion	Bacterial action in soil	1141×10^6	0.5	NO 0.2–2 ppb NO_2 0.5–4 ppb	5 days	Oxidation to NO_3^- after sorption by aerosols aided by light and HC
NH_3	Waste treatment	Biological decay	1164×10^6	0.3	6–20 ppb	7 days	Reaction with SO_2 to form $(NH_4)_2SO_4$
CO_2	Combustion	Respiration, decay, release from oceans	1.014×10^{12}	1.4	320 ppm	2–4 yr	Photosynthesis, oceans
CO	Combustion	Forest fires, terpene reaction, oceans	349×10^6	79	0.1 ppm	<3 yr	None proven
HC	Combustion chemical industry	Biological processes	1916×10^6	7.6	CH_4 1.5 ppm non-CH_4 < 1 ppb	2.9 yr (CH_4)	Non-CH_4: photochemical with $NO-H_2O$

5. Remove pollutants from energy sources or alter manner of use so as to decrease pollutants.
6. Use nonpolluting energy sources.

REFERENCES

1. W. L. Faith. *Air Pollution Control.* Wiley: New York, 1959, p. 6.
2. S. F. Singer, Ed. *Global Effects of Environmental Pollution,* American Association for the Advancement of Science Symposium, Dallas, Texas, Dec. 1968. Springer-Verlag: New York, 1970; F. S. Johnson. *The Oxygen and Carbon Dioxide Balance in the Earth's Atmosphere,"* p. 4.
3. E. Robinson and R. C. Robbins. *Sources, Abundance, and Fate of Gaseous Atmospheric Pollutants,* Project PR-6755, Final Report. Stanford Research Institute: Menlo Park, Calif., February 1968. See also Ref. 46.
4. A. J. Haagen-Smit and L. G. Wayne. "Atmospheric Reactions and Scavenging Processes." In A. C. Stern, Ed. *Air Pollution* 2nd ed. Vol. I. Academic: New York, 1968, pp. 149–179.
5. Ref. 3, p. 3.
6. Massachusetts Institute of Technology. *Man's Impact on the Global Environment, Report of the Study of Critical Environmental Problems.* MIT Press: Cambridge, Mass., 1970, Tables 1.2, 5.6, 7.2, 7.3, and 7.5.
7. W. W. Kellog, R. D. Cadle, E. R. Allen, A. L. Lazarus, and E. A. Martell. "The Sulfur Cycle." *Science,* **175,** 587–596 (1972).
8. Ref. 3, Supplemental Report, June 1969, p. 51.
9. Ref. 8, p. 52.
10. Ref. 8, pp. 54–58.
11. E. Eriksson. *Tellus,* **11,** 375–403 (1959); **12,** 63–109 (1960).
12. C. E. Junge. *Air Chemistry and Radioactivity.* Academic: New York, 1963.
13. E. Eriksson. *J. Geophys. Res.,* **68,** 4001 (1963).
14. Ref. 3, p. 69.
15. Ref. 3 p. 71.
16. N. I. Sax. *Handbook of Dangerous Materials.* Reinhold: New York, 1951, p. 80.
17. F. D. Sisler. "Impact of Land Sea Pollution on the Chemical Stability of the Atmosphere." In Ref. 2, p. 12ff.
18. Ref. 6, p. 49.
19. Ref. 8, p. 9.
20. Ref. 8, p. 10.
21. G. G. Callendar. *Tellus,* **10,** 243 (1958).
22. Ref. 6, p. 47.
23. Ref. 8, p. 14.

24. Ref. 6, p. 11, 49.
25. S. I. Rasool and S. H. Schneider. "Atmospheric Carbon Dioxide and Aerosols: Effects of Large Increases on Global Climate." *Science,* **173,** 138–141 (1971).
26. Ref. 8, p. 27–35.
27. Ref. 8, p. 38.
28. Ref. 8, p. 40.
29. *Encyclopedia Britannica,* Vol. XII. Encyclopedia Brittanica: Chicago, 1961, p. 5 ff.
30. Ref. 8, p. 42.
31. Ref. 3, p. 94.
32. Ref. 8, p. 44.
33. Ref. 3, p. 95.
34. Ref. 3, p. 97; also F. W. Went. *Proc. Nat. Acad. Sci.,* **46,** 212 (1960).
35. Ref. 3, p. 99–101.
36. E. R. Gerhard and H. F. Johnstone. *Ind. Eng. Chem.,* **47,** 972–976 (1955).
37. Ref. 3, p. 19.
38. Ref. 3, p. 24.
39. R. D. Cadle and M. Ledford. *Air and Water Poll. Int. J.,* **10,** 25–30 (1966).
40. Ref. 3, p. 75.
41. J. A. Young and A. W. Fairhill. *J. Geophys. Res.* **73,** 1185 (1968).
42. Ref. 8, p. 18.
43. E. Douglas. *J. Phys. Chem.,* **71,** 1931 (1967).
44. R. E. Inman, R. B. Ingersoll, and E. A. Levy. "Soil: A Natural Sink for Carbon Monoxide." *Science,* **172,** 1229–1231 (1971).
45. Ref. 8, p. 5.
46. E. Robinson and R. C. Robbins. "Emissions, Concentrations, and Fate of Gaseous Atmospheric Pollutants." In W. Strauss, Ed., *Air Pollution Control.* Wiley: New York, 1972, pp. 1–93.
47. J. C. Pales and C. D. Keeling. *J. Geophys. Res.,* **70,** 6066 (1965).

3
Societal and Economic Costs of Air Pollution

3.1. EVALUATION OF AIR ENVIRONMENTAL INSULTS

The system engineering approach toward air pollution introduced in Chapter 1 requires that we attempt to "optimize" in some way the nature and degree of controls placed on man-made pollution sources, with the objective of achieving maximum societal benefit at the least expenditure. We have already seen that there are many obstacles along this path. First, there are the compromises that must be made to achieve the necessary cooperation of diverse groups, such as automobile drivers, homeowners, manufacturers, and voters. Second, there are the natural sources (and sinks) of pollutant emissions, discussed in Chapter 2, which are presently beyond our control. And third, there is the fact that our culture allows us no means to quantitatively compare benefits and expenditures except by monetary evaluation; that is, there are no commonly accepted units by which to measure social utility.

It is clear that many of the possible adverse effects of air pollution cannot be quantified in terms of money: changes in the global air resource and climate, the human and emotional cost of the increased death rate during episodes, and the destruction of scenic views or art treasures. Just as clearly, there are many effects which could be quantified but for which the needed data are sparse or lacking altogether.

The existence of unknowable and unknown factors in the evaluation of abatement benefits or the costs of harm done by pollution, as contrasted with the (monetary) costs of controls, which are often all too apparent, has been put forward as a reason to forego any attempt to rationalize air-pollution control. This argument has often suited those who, on the one hand, wish to continue unhampered the free use of the common air resource as a means of disposing of wastes and those, on the other hand, who wish to inhibit production or growth on grounds other than pollution.

At the same time, it has bolstered the simplistic "zero emissions" view put forward by others who are merely appalled by the sheer complexity of the task of evaluating the effects of emissions.

For purposes of this work, we can consider the adverse effects of air pollution as composed of three classes. For those events known or suspected to be irreversible in effect, such as changes in global climate, air composition, or significant modifications of the biosphere, or those potential catastrophic events such as the occurrence of high death rates accompanying urban episodes, no costs can be assigned and the avoidance of these events must constitute absolute constraints on the polluting systems. There are other effects, conversely, which can be quantified more or less precisely and to which costs can be assigned or at least estimated. These include such items as the deterioration of materials, costs of cleaning clothing and buildings, and the direct cost of health care and the productive labor lost by air pollution-induced illnesses. A third category, aesthetic degradation, is semiquantifiable, perhaps representing widely different degrees of utility to different individuals but certainly capable of being traded off for more fundamental human needs in less than affluent societies.

In this chapter we shall attempt to summarize information on effects of air pollution to which costs can ultimately be assigned and that represent areas where pollution-abatement savings can be directly traded against control expenditures. In addition, we shall discuss the possibility of certain irreversible effects that must be prevented, regardless of cost, and because of which monitors of the environment and pollution levels must be established.

3.2. GLOBAL EFFECTS

We consider here changes in the atmosphere caused by the injection of man-made pollutants that have possible climatic effects. It has already been suggested in the preceding chapter that the rise of CO_2 concentration levels in the atmosphere may cause a corresponding rise in the average worldwide temperature. There is no question that CO_2 has increased in recent decades. The evidence examined by G. Robinson[1,2], Watt[3], and that in the last chapter indicates that the level has risen approximately 40 ppm since 1880 and 0.2%/yr since 1958. This increase is caused by man's fuel-burning activities, which have added CO_2 to the atmosphere at an accelerating rate (see Chapter 2).

More than 7.5×10^9 tons/yr, or one-third of these man-made additions, remain in the atmosphere, the remainder being removed by the mechanisms discussed in the previous chapter. To put this increase in perspective, Watt calculates that if all the current economic fuel deposits, amounting to 5.5×10^{12} tons of carbon, were burned in one year and absorption remained constant, the CO_2 level would increase to 1000–2000 ppm, less than half the threshold limit value (TLV) established for safe industrial atmospheres.[4] (See Table IV.) Since CO_2 is beneficial to plant life up to 30,000 ppm, he concludes that climatic alteration is the only deleterious possibility of CO_2 increase. This argument neglects possible long-term physiological effects of levels below TLV of which there is no evidence at present.

It is also true that an increase in global temperature of about 0.4°C, apparently correlating well with CO_2 rise, has been documented for the 30–40 years prior to 1950.[9] But it is not too easy to establish a cause-and-effect relationship. Climate, defined by Robinson[1] as the mean and variance of weather observations over 30 yr, has undergone changes during the 250 yr of weather recording at England's Kew observatory. Furthermore, climatic changes have obviously preceded man's global pollution activities, since, as long ago as A.D. 1000, an agricultural colony existed in southern Greenland that could not be supported by the present climate.

The heat-capturing capability of CO_2 has been cited as the reason for expecting an influence on ground temperature, the so-called greenhouse effect. This depends on the ability of the atmosphere to pass solar radiant energy, which is peaked at a wavelength below 1 μm, and to trap that which is reflected back from the earth. The reflected energy is at uniform intensity from about 5 μm to above 30 μm. In fact, CO_2 absorbs primarily in two fundamental bands, corresponding to molecular vibration modes centered at 4.35 μm (2349 cm^{-1}) and 15 μm (667 cm^{-1}). It is the latter band, spreading from 13 μm to 17 μm because of pressure broadening, that is the primary absorber of reradiation from the earth's surface. The limitations of this band restrict the heat blanketing effect. As shown by Watt,[3] the amount of energy absorbed by water vapor in two broad absorption bands, one extending from 5 to nearly 8 μm and one from 17 μm upward, greatly exceeds that of CO_2. Furthermore, he concluded that the steepness and depth of the band at 15 μm make the absorption rather insensitive to even large increases in CO_2 concentration. This belief is not entirely shared by Robinson,[1] who states that doubling the CO_2 would increase the earth surface temperature 3–4°C by the greenhouse effect or, if the relative humidity remained constant, by as much as 10°C. (Otherwise, cloud cover could reduce the increase by 60%.) The uncertainty of the role of water vapor and cloud formation and the inadequacy of the model make more-refined calculations useless. On balance, it is easy to agree with Watt's conclusion (quoting Möller)[5] that "the theory that climatic variations are affected by CO_2 content becomes very questionable."

Another reason for failure to confirm the influence of CO_2 is that after 1960 the global temperature fell by a few tenths of a degree centigrade despite the continued CO_2 rise. Concurrently, the amount of particulate matter in the atmosphere rose sharply, largely as a result of the volcanic activity at Agung (Bali) in 1963. Volcanic dust is known to be associated with temperature drops as great as 2–3°C. Watt computes the global decrease to be as much as $0.027°C/10^6$ tons of high altitude dust in the 1-μm range. Krakatoa emitted (ca. 1882) an estimated 55×10^6 tons and Agung, more than 16×10^6.

Watt's postulated mechanism involves a reduction of the earth's *albedo* (reflecting power), which directly affects the balance of heat interchange with the sun and space, and hence the mean temperature. The albedo of the earth averaged over its surface is estimated to be 0.33–0.36; corresponding to an atmospheric temperature of 250–253°K.[6, 9]

The ratio of the mean surface temperature, 286°K, to the radiating temperature of the atmosphere, 253°K, is known as the *greenhouse coefficient*, which is thus 1.13 for the earth. A reduction of only 2% in the absorbed radiation (either by change in albedo or reduction of the solar constant) could lower the global mean temperature by 2°C, possibly inducing worldwide glaciation.[9] The albedo of clouds varies widely, from a few hundreds to 0.7–0.75 for thick clouds, while that of the earth's surface is about 0.15–0.2.[7] This implies that the extent of aerosols in cloud form in the atmosphere can greatly affect the albedo, and hence the global temperature. In general, the presence of particulate matter increases scattering, and it has already been noted (Chapter 1) that Mie scattering of visible light (solar radiation) is most effectively implemented by the 0.1–1.0-μm size range of the "self-preserving aerosols" generated by air pollution.

Calculation of the corresponding temperature changes is most complex and beyond the scope of this discussion. For example, particulate matter may also absorb energy, tending toward an increase in temperature (by greenhouse effect), offsetting the scattering. Not only particle size, but the altitude of the scattering layer above the earth is important. It must suffice to examine the trend of global airborne particulate concentration, knowing that large variations will have an equal or greater influence on world climate than will CO_2.

Atmospheric particulates from natural sources include, not only volcanic emissions, but also wind-raised dust and water-soluble nuclei arising from combustion, seaspray, and chemical reactions in the unpolluted atmosphere. Man-made sources include combustion products and local industrial and construction operations. The background concentrations vary from 0.6 to 1.7 $\mu g/m^3$ in isolated localities (Hawaii; Boulder, Colorado) to

5–20 $\mu g/m^3$ over the East Coast of the United States and the industrialized cities of Western Europe. At an altitude of 16–18 km, there is a layer of particles consisting mostly of sulfates (SO_4^{-2}). In 1968–1969 this layer was measured on flights over Central America and the United States and was found to range between 0.1 and 0.4 $\mu g/m^3$. This concentration was 10 times that measured (by Junge) in 1960–1961. It is this increase of sulfate particles which was caused by the 1963 Agung-Bali volcanic activity and to which the post-1960 global temperature fall is ascribed. More recent (1970–1971) stratospheric flights have shown a return toward Junge's originally measured levels.[8]

The sulfate particles predominant in the particulate stratospheric layer, possibly influencing climate, arise from a number of sources in addition to volcanic emissions, as discussed by Kellogg.[8] Airborne SO_2 is one of the most likely. In the presence of moisture, SO_2 is rapidly converted to sulfurous acid:

$$SO_2 + H_2O \rightarrow H_2SO_3.$$

In cloud or fog droplets, the sulfurous acid is rapidly oxidized by dissolved O_2 to form sulfuric acid (H_2SO_4). This oxidation reaction can be catalyzed by metallic airborne dusts, which then react with the acid to form sulfate salts. The catalyzed gas-phase reaction illustrated in Chapter 1,

$$SO_2 + \tfrac{1}{2}O_2 \xrightarrow{\text{catalyst}} SO_3,$$

is probably less important in the formation of sulfuric acid and sulfates.

A three-body reaction with atomic oxygen (a product of the photolysis of O_3 and NO_2 in polluted air) can also produce SO_3 by

$$SO_2 + O + M \rightarrow SO_3 + M$$

(where M is an O_2 or N_2 molecule), and then sulfuric acid by

$$SO_3 + H_2O \rightarrow H_2SO_3$$

to form droplets. This is a very likely source of the 16–18-km sulfate layer. In the presence of any atmospheric ammonia, these will form ammonium sulfate

$$2NH_3 + H_2SO_4 \rightarrow (NH_4)_2SO_4$$

or bisulfate, NH_4HSO_4. The acid droplets will also react directly with sodium chloride particles with the release of HCl gas:

$$H_2SO_4 + 2\,NaCl \rightarrow HCl + Na_2SO_4.$$

Finally, in polluted (smog) atmospheres, the SO_2 may be oxidized to SO_3 directly by organic peroxy free radicals, as generated during the production of photochemical PAN (Chapter 1).

Man-made SO_2, it was concluded in Chapter 2 (Table XVII), amounts to 146×10^6 tons/yr injected into the global atmosphere—some 66×10^6 tons/yr due to electrical power generation and 27×10^6 tons/yr from other major industrial sources. Kellogg et al.,[8] after examining the evidence, estimate that volcanic activity on the average, over the last few hundred years, is two orders of magnitude less than this, only 1.5×10^6 tons/yr. A much greater natural source of sulfates is the ocean. The annual amount of sulfate sent into the atmosphere by ocean spray is estimated at 130×10^6 tons.

All of the estimated SO_2 is believed to be converted to H_2SO_4 aerosol before leaving the atmosphere. In the future, the already significant contribution of man-made SO_2 to atmospheric sulfate particulates will probably be increased. It estimated by Kellogg et al. from data produced by the MIT Study of Critical Environmental Problems (SCEP)[9] that SO_2 production from man-made sources will be 275×10^6 tons/yr by 2000. Table I summarizes these findings. It is seen that SO_2, converted to sulfate particulate, will nearly double by the year 2000 if human industrial activity is maintained at the growth rates estimated by SCEP (4% to 1980 and 3.5% thereafter).

When it is further noted that the very small volcanic contribution has exhibited a noticeable climatic effect as a result of its concentration in time and geographic area, then the further statistic from SCEP, that 93.5% of the man-made SO_2 is concentrated in the northern hemisphere, becomes of more than passing interest.

Before concluding consideration of global particulate effects, the possible role of supersonic aircraft fleets in contaminating the atmosphere must be considered, even though the cancellation of the U.S. program of SST development somewhat reduces this threat. The effect of 500 SST's flying 7 h every day in the stratosphere, as envisioned for the 1985–1990 time period, comprises a "massive environmental change," according to the SCEP

Table I

Amount and Sources of Atmospheric Sulfate Aerosols (Calculated as SO_4^{-2} Equal to $1.5 \times SO_2$, in 10^6 Tons/yr)[8]

Year	1965	2000
Man-made	150	410
Volcanic (average)	2	2
Oceanic	200	200
Totals	352	612

group. Based on U.S. data, it found that the stratospheric water content would increase from 3.0 to 3.2 ppm with possible temperature effects. The particulates arising from SO_2, hydrocarbons, and soot, could double the pre-Agung eruption averages and peak at 10 times this figure in regions of dense traffic. The effect is not certainly known, but it could be as large as that of the Agung explosion. The latter caused a 6–7°C rise in the equatorial stratospheric temperature, which remained 2–3°C higher for several years. No corresponding temperature decrease in the earth or lower atmosphere was noted, however (see Ref. 9, p. 16).

One reason for the uncertainty in predicting climatic effects is the complexity of the global atmosphere and its climatic statistics, Robinson has pointed out that these statistics are not stationary, that is, the distinction between "signal" and "noise" cannot be resolved by integration over a period of time. Furthermore, mathematical climate models are composed of nonlinear, partial differential equations, which may produce radically different solutions from minute variations in input and boundary conditions. The nature of the resulting statistics is termed *almost intransitive* (i.e., they are not a unique solution of the equations).

Current atmosphere models must be solved by numerical methods on digital computers in order to untangle the numerous variable interrelationships and loop feedbacks. This involves the specification at each point of at least seven quantities: pressure, temperature, density, water vapor, and the three velocity vector components. These are related by four "primitive equations" stemming from the basic physics: horizontal and vertical momentum equations, continuity equations for total and for water mass, energy balance, and the gas law relationship. The boundary conditions and constraints include solar radiation incidence, infrared emission, and turbulent transport of heat, water vapor, and momentum. These "primitive" models suffer from many deficiencies, not the least being their inability to distinguish between the change of state of water vapor to precipitation, or as cloud formation, and consequent changes in albedo.

To determine the effect of the rather small injections of man-made pollutants on the climate, runs of at least 100 yr must be made (we have already noted that temperature records extend back 250 yr). In integrating a numerical model, it is necessary to match the computed time steps to the geographical grid spacing considered; otherwise, propagating computational errors will invalidate the results. If the horizontal grid spacing is 300 km, the time step must not exceed 5 min. At this rate an advanced model requires 2 or 3 h of computation per model day, using available computers. In order to obtain reliable predictions on the climatic effects of pollutants, it is necessary to run these models long enough to overcome the natural "noise level" of the atmosphere. It may prove necessary to utilize large parallel processors such as the Illiac IV to obtain these results. This machine is able to process a weather model such as that of the National Weather Service (18,000 points) at 100–200 times the current speed of 1 h/model day.[10]

Other than CO_2 and aerosol particulates, air pollutants may have geophysical effects, but they are of much less concern. The potential reduction of oxygen reserves has been substantially disproven by Broecker[11] and others. SCEP estimated that combustion of all recoverable fossil fuel on earth would merely reduce the oxygen content from 20.946 to 20.800%. Robinson[1] considered CO to be an unlikely environmental threat because of its low volume concentration and low production rate by combustion (2 $\times$ 10^8 tons/yr, less than 2% of CO_2). Nevertheless, it cannot be ignored because even though its concentration has remained constant, its removal mechanism is not certain. Similarly, SO_2, NO, NO_2, and O_3 are not expected to cause geophysical changes, although their local and urban effects are pronounced. SO_2 has a negligible radiative effect at its low atmospheric concentration of 0.0002 ppm (Table I, Chapter 1) and short residence time (1–3 h in polluted atmospheres and 2.5 days at background concentrations).

NO_2 is rapidly formed from NO by oxidation with ozone even at the reduced concentration of O_3 observed in the background troposphere, 0.01 ppm. NO_2 absorbs most wavelengths in the visible range between 0.3 and 0.6 μm, accounting for the reddish brown color of NO_2-rich plumes and of certain urban smogs and haze. Although it may cause significant reduction of local solar radiation, the global effects are not likely to be large.[12] Likewise, O_3, being a secondary reaction product, may have serious local effects in both the short and long terms but is unlikely to result in global geophysical changes.

So far we have only discussed climatic changes. At this stage of our knowledge, we can say very little about ecological changes other than those implied by the climate. Environment can cause ecological changes: certain species of the moth in England have undergone a darkening of their wing color in order to gain better protective coloration against the soot-colored background that has become their environment. It is certain that ecological mechanisms are just as complex as geophysical ones, so that we can anticipate the occurrence of such effects but cannot predict their direction or nature. At the present time, the biosphere is probably more vulnerable to such pollutants as insecticides (e.g., DDT, effectively banned in the United States as of this writing) and contaminants of the water environment, including nitrates.

What are we to conclude from these considerations? It is certain that our industrial society would suffer disruption if we were compelled to reduce or cease production of CO_2 in the near future. Currently, we have no alternate to burning fossil fuels for the generation of domestic and industrial power and heat other than nuclear energy, which in itself implies an entire universe of by-product environmental problems. Much lead time is required if we are to be adequately warned of such an event and to adjust without social dislocation. The lead time is also conditioned by the residence time of the pollutant involved. In the case of CO_2, 2–4 yr represents the time lag before we could make significant changes in the concentration trend. For particulates, the residence time is only a matter of 6 days to 2 wk in the troposphere, but in the stratosphere it is 1–3 yr for micron particles.[9] Consequently, urban and industrial contamination control could be effected much more rapidly than that from any high-altitude source.

The lead time we need can only be achieved through the mechanisms of an adequate predictive model and a reliable monitoring network able to detect change signals in stochastic noise. The number of monitoring stations required for a given accuracy depends on statistical considerations and will vary for different contaminants. CO_2 is well mixed in the atmosphere and has a long residence time; consequently, it can be adequately monitored with a smaller number of stations than the shorter-lived lower-atmosphere particles.

The conclusions of the SCEP group relative to global air pollution are pertinent and may be summarized as follows:

1. Improve computer models of climate with more realistic simulation of clouds, air–sea interaction, and particles; study effects of pollutants and climatic change.
2. Improve estimates of global sources of CO_2 from combustion and natural sources; improve monitoring of CO_2 concentration and its distribution by the establishment of 12 research stations.
3. Study fine particles in the atmosphere and their optical properties and effects on radiative transfer; monitor particles and trace gases by means of about 110 stations; and monitor the whole earth's albedo by means of satellites.

It is gratifying to note that the United Nations Conference on the Human Environment (Stockholm, June 5–16, 1972) approved the proposal to establish a network of 110 stations, including 10 "baseline" locations to establish background. It also recommended the evaluation of climatic effects of pollutants by the responsible nations and consultation with other nations when contemplating actions that involve such emissions.[13]

3.3. URBAN EFFECTS AND PUBLIC HEALTH

Although the degree, if any, to which current levels of man-made air pollution can alter world climate and the biosphere remains moot, there is no such debate as to their impact on urban life and health.

As we have seen, human activities inject at least 150 million tons of particulates and over a half-billion tons of pollutant gases (excluding CO_2) into the atmosphere annually. But the entire mass of the atmosphere is so large that even this huge quantity would not constitute an acute threat if it were uniformily mixed. The total atmospheric mass is approximately 6×10^{15} tons,[14] so that 10^9 tons of pollutant is equivalent to an addition of 0.16 ppm/yr. Since half of this is CO, with a residence time of 3 yr (see Chapter 2), the equilibrium concentration of CO should be about 0.24 ppm (0.08 ppm $\times$ 3 yr).

This is substantially below the concentration (9 ppm) considered as acceptable quality in the United States but is nevertheless far above the background level of "normal dry air." As stated earlier, the acute hazard of man-made pollutants lies in the fact that they are not uniformly mixed but tend to concentrate near their place of origin—man's industrial and urban living spaces.

Since the pollution problem is one of mixing and of the ventilation above factories and cities, air-pollution levels fluctuate with atmospheric changes. The ventilation is at a low level when there are little or no winds, and the stagnant air is confined to a restricted volume because of a "temperature inversion" aloft. This is a temperature distribution that inhibits the upward passage of polluted air and acts as a "cap" above the area. In subsequent chapters we shall discuss meteorological stability and inversions quantitatively. At present, it is sufficient to note that the pollution concentration will tend to rise so long as the weather encourages stagnation and the emissions from polluting sources continue.

3.3.1. Acute Episodes of Air Pollution

When stagnant air conditions persist, the pollutant concentrations rise to high levels, and, in densely populated areas, acute illness and mortality from respiratory causes increase sharply. These incidents are known as *episodes of high air pollution,* or simply *episodes.* Such occurrences previously mentioned include the London episode in 1952, responsible for 4000 "excess" deaths; that in 1948 at Donora, Pennsylvania, blamed for 20; and the Meuse Valley incident of December 1–5, 1930, killing 63 persons. The term *excess* is somewhat ambiguous but refers to the excess over the statistical mean expected during the period. In most cases, the number is not an exact figure but merely a statistic computed long after the event. Nevertheless, the correlation between such disasters as the 4000 deaths in London in 1952 with the SO_2 and particulate levels at that time is beyond reasonable doubt. The fact that deaths were of "susceptible" people, the elderly and those with pulmonary or cardiac disease, does not alter the relationship.

More recently, on November 18–20, 1971, an episode during stagnant conditions (lack of ventilating air movement conducive to pollutant buildup) in Birmingham, Alabama, was of such magnitude that the EPA was required to invoke emergency powers to curtail the emissions of 23 industries. This was the first time that this federal power under the Clean Air Act of 1970 was invoked.[15] At the same time, because of the high barometric pressure prevailing over the entire East Coast, stagnation occurred in other areas. New York City, for example, was placed on standby alert to switch electric power generators from coal to natural gas. The inversion cap dropped to 500–1000 m from its normal 1500; the wind at 5 knots was below the 8 knots needed to insure ventilation.[16] Similar episodes occurred in the New York–New Jersey area[17] on May 7, 1970, and in Northern New Jersey[18] during the period October 1971–February 1972. In one typical incident in Sayersville, New Jersey, on September 16, 1970 (as reported in *Life* magazine), over 125 young high school athletes suffered respiratory difficulties ascribed to air pollution, some being hospitalized. The episode covered an area of 40–50 miles,[2] within which high oxidant levels (O_3 and NO_x) resulting from automobile traffic were recorded; the wind was low, at 2 knots; and relative humidity was at 78%. However, nearby New York City, although also experiencing inversion at the same time, was spared high oxidant concentrations as a result of brisk local winds.

The nature and concentration of pollutants during past episodes is in general not known, since pollutant monitoring instruments are of fairly recent origin. The constituents of coal smoke held responsible for the London episode were considered to be SO_2 and SO_3. A peak concentration of 1.34 ppm of sulfur oxides was recorded during the third and fourth days.[20] We now know that particulate matter will enhance the ill effects of sulfur oxides. In Donora, where a zinc plant and a sulfuric acid plant were among the industries emitting fumes during the 1948 episode, an industrial air filter showed a very high concentration of zinc and zinc ammonium sulfates, $ZnSO_4$ and $Zn(NH_4SO_4)_2$, amounting in combined total to over 15% of the airborne dust sample. The combined effect of inhaling these particles (0.3 μm in diameter), together with concentrations of 2.5 ppm SO_2 estimated to be present at Donora, was found severely irritating to guinea pigs and is believed to have been a primary cause of the human health effects.[19] No precise measurements were made at the Meuse Valley, but it is estimated that SO_2–SO_3 reached 9 ppm.[20]

In the Birmingham incident, the particulate level was over 750 $\mu g/m^3$ for two days (771 and 758 peak), compared to a 275-$\mu g/m^3$-24-h health standard considered acceptable by the EPA.

A strong casual relationship between mortality and high sulfur oxide plus particulate concentration has been established,[21] but this is not the case with photo-oxidants. The primary short-term effects noted up to 1.0 ppm are severe eye irritation and some pulmonary function impairment.

No permanent morbidity was known to result from the New Jersey oxidant episode described in *Life*. Levels of nearly 0.1 ppm were noted during the afternoon in the vicinity, but in New York, where no problem was experienced, the oxidant level remained at 0.02 ppm during the same period.

Some idea of the oxidant rise during heavy smog episodes in Los Angeles can be obtained from Table II.

The possibility of urban episodes caused by high carbon monoxide levels also exists. These will be associated with heavy automobile traffic in almost every case, since nearly 74% of carbon monoxide on a national basis (151.4 $\times$ 10^6 tons in 1969) is emitted from internal combustion engines.[23] In Boston, Chicago, Denver, and New York, transportation is responsible for 90–95% of the CO in the atmosphere and in Los Angeles and Washington, D.C., over 98%. Since CO concentration follows the level of automobile traffic, it will fluctuate during the day, peaking at morning and evening rush hours, and will be greatest near the ground and highly traveled roads. The presence of meteorological stagnation conditions and local effects such as turbulence caused by airflow around automobiles and buildings will also be factors in determining the CO level at any point, as well as that of other pollutants.

Standards for permissible levels of CO differ from those of other pollutants, such as sulfur oxides and particulates, because so far as is known, the effect of moderate exposure to CO is reversible. CO is absorbed into the bloodstream through the lungs, forming a stable compound with hemoglobin in preference to oxygen. The blood concentration of this compound, carboxyhemoglobin (COHb), builds up slowly to some equilibrium value, depending on the amount of CO in the air and on the physical

Table II

Oxidant Levels in Los Angeles (ppm)[22]—Data Through 1959

	Smog condition		Peak recorded
	Episode	Normal	
NO_x (NO + NO_2)	0.25–2.00	0.05–1.30	3.93
O_3	0.20–0.65	0.05–0.30	0.90
Total oxidant	0.20–0.65	0.10–0.35	0.75

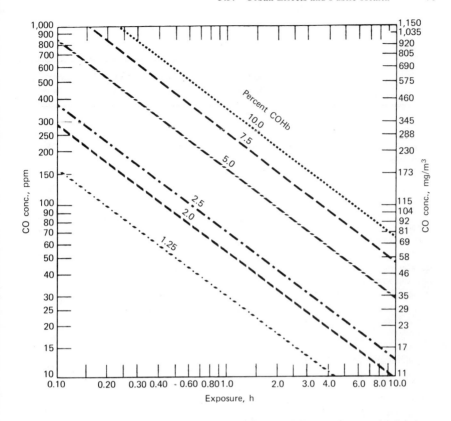

Figure 1 Concentration and duration of continuous CO exposure required to produce blood COHb concentrations of 1.25, 2.0, 2.5, 5.0, 7.5, and 10% in healthy male subjects engaging in sedentary activity.[25]

activity being performed. Performance of heavy work can increase the CO respiratory uptake by as much as 2–3 times. The formation of COHb is reversed by breathing oxygen or fresh air. In cases of serious overexposure, the reversal can be hastened by breathing a mixture of oxygen and 7–10% CO_2.[24]

A concentration of 30 ppm of CO will result in blood levels of 5% COHb after about an 8-h exposure, and 10–15 ppm will produce 2.0–2.5%, following the formula[25] %COHb = 0.16 × CO (ppm) + 0.5. The normal or background level for nonsmokers is represented by the factor 0.5%. The effect of shorter exposures is seen in Figure 1. At 5% COHb, it has been determined that adverse health effects result, evidenced by impaired performance on psychomotor tests, increased reaction time, and reduced visual

Table III

Significant Levels for Episode Control

	Significant harm level	Alert level	Warning level	Emergency level
SO_2: 24-h average				
$\mu g/m^3$	2620	800	1600	2100
ppm	1.0	0.3	0.6	0.8
Particulate: 24-h average				
$\mu g/m^3$	1000	375	625	875
COH	8.0	3.0	5.0	7.0
SO_2 particulate (product)				
$\mu g/m^3$ ($\times 10^3$)	490	65	261	393
ppm $\times$ COH	1.5	0.2	0.8	1.2
CO: 8-h average				
mg/m^3	57.5	17	34	46
ppm	50	15	30	40
4-h average				
mg/m^3	87.3			
ppm	75			
1-h average				
mg/m^3	144			
ppm	125			
Photochemical oxidants $(O_3)^a$				
4-h average				
$\mu g/m^3$	800			
ppm	0.4			
2-h average				
$\mu g/m^3$	1200			
ppm	0.6			
1-h average				
$\mu g/m^3$	1400	200	800	1200
ppm	0.7	0.1	0.4	0.6
Nitrogen dioxide				
1-h average				
$\mu g/m^3$	3750	1130	2260	3000
ppm	2.0	0.6	1.2	1.6
24-h average				
$\mu g/m^3$	938	282	565	750
ppm	0.5	0.15	0.3	0.4
Meteorological condition		Stagnation for 12 h		

[a] Proposed change to single "significant harm" level of 1200 $\mu g/m^3$ (0.60 ppm), 1-h average, and "Emergency" level of 1000 $\mu g/m^3$ (0.50 ppm). (*Federal Register* **39,** March 13, 1974, pp. 9672–9675.)

discrimination, which may cause an increase in automobile accidents.[26, 27] Above 5% there is evidence of physiological stress in heart disease patients. At 3% COHb, some changes in visual acuity and visual brightness threshold have been noted.[27] At even higher levels (20% and above), intoxication, unconsciousness, and eventually death take place.

It has been concluded by the EPA, as a result of these studies, that substantial risk to health can occur at 7% COHb and that episode controls should be designed to prevent this level from ever being reached by the nonsmoking population. Furthermore, sufficient functional impairment of motorists exists at 5% COHb to control CO exposure so that this value is not exceeded. (It should be noted that COHb levels of one-pack-a-day smokers frequently reaches these levels.) An episode-control strategy can be devised based on these limits and Figure 1 that will ensure that the equilibrium value of 7% COHb cannot be reached and that the time–concentration relationship will keep exposure below levels leading to a 5% COHb.[28]

Because the federal Clean Air Act of 1970 acknowledges that the statistical fluctuations of emissions and meteorological conditions can lead to severe episodes of concentrated air pollution with consequent acute illness and mortality, it requires each state to provide an Episode Control Plan for regions subject to significant pollution, to prevent ambient concentrations from reaching levels causing "significant harm to health of persons." These levels are defined in Table III. Episode control plans are discussed in more detail in the next chapter.

3.3.2. Air-Pollution Epidemiology

The chronic effects of low concentration of air pollution are more difficult to establish quantitatively than those of an acute nature, which are of primary consideration in emergency episodes. Although health effects are the most intensely studied pollution research, restrictions on direct experimentation and human variability make precise data rare. Generally, both duration of exposure and pollutant concentration must be considered, as in the case of CO, when establishing limits of toxic effects. A time–concentration plot will exhibit these limits as broad bands or areas, rather than narrow lines of demarcation.

It is difficult to generalize on health effects owing to their broad range among individuals, the large number of pollutants, the unknown nature of many atmospheric constituents, and the interaction or enhancement (synergy) that may occur with the presence of two or more pollutants. A review by Cassell[29] of the correlation between symptoms such as "colds," sore throat, and eye irritation, and urban environmental factors, encountered in the Cornell Family Illness Study (which represents a 1747 person cross section of New York City inhabitants) concluded that CO and particulate matter were positively correlated with respiratory symptoms, coughs and colds (coefficients of 0.2+), and SO_2, less so (0.10–0.11 correlation coefficient). In general, he concluded:

"(1) Air pollution has definite, but varying adverse effects upon health;
(2) no single pollutant appears solely responsible for the observed effects;
(3) the effects seem to come from the totality of the atmospheric environment with weather and season playing their part;
(4) the effects are influenced by variations in individual habits and disease as well as by social determinants."

In other studies, a causative relationship between average SO_2 concentration and death rate has been established (in New York City). Based on the 1960–1964 data, it was shown that a mean reading of 0.4 ppm SO_2 caused 10–20 more deaths per day than reading below 0.2 ppm.[30] Previously, such correlation has been denied. Thus, it is not only the abnormally high episodes but the day-to-day concentrations that must be controlled.

Chronic disease is highly probable in many cases but even harder to pin down statistically. It has been shown that 43% more U.S. urban males contract lung cancer than nonurban males, but no direct correlation with SO_2 and smoke levels has been demonstrated. There is some proof of a causal relationship between pollution levels and bronchitis or emphysema. Asthma in urban dwellers is apparently caused less by general air pollution than by pollen. CO and lead are still under study, and no simple cause–effect relationships have yet been established at normal urban levels.[31, 32]

A distinction should be made between air-pollution levels and industrial limits, the so-called threshold limit value (TLV) and maximum allowable concentration (MAC). These represent experience with specific chemicals, on generally healthy male individuals without special idiosyncrasy for the given exposure, and for a 40-h work-week. Even so, measured community air-pollution levels have on occasion approached or even exceeded the TLV. Examples include carbon monoxide, 360 ppm (London) compared to a TLV of 50; SO_2, 3 ppm (Chicago) compared with 5; lead up to 0.034 mg/m^3 in New York[32] and over 0.054 in Los Angeles,[33] compared with a TLV of 0.15 mg/m^3. Table IV is a list of TLV's for representative materials present in polluted urban atmospheres.

Table V, excerpted from Ref. 20, illustrates the gradation of health effects and synergistic and concentration–duration relationships as well as can be expressed with current data for SO_2 and particulates. It is interesting to compare Table V with some of the tabulated statistics for American cities. Criterion H, associated with increased death rates among

Table IV
Selected Threshold Limit Values (TLV)[4]

	ppm V (25°C/760 mm)	mg/m³
Acrolein	0.1	0.25
Ammonia	25.	18.
Beryllium (p)[a]	—	0.002
Carbon dioxide	5000.	9000.
Carbon monoxide (t)	50.	55.
Chlorine (t)	1.	3.
Fluoride (as F) (p)	—	2.5
Fluorine	1.	2.
Formaldehyde (n)	2.	3.
Hexane (n-hexane)	500.	1800.
Hydrogen chloride (n)	5.	7.
Hydrogen sulfide	10.	15.
Lead (p) (tent. 1972)	—	0.15
Nitrogen dioxide	5.	9.
Ozone	0.1	0.2
Sulfur dioxide	5.	13.
Sulfuric acid (p)	—	1.
Dusts	*Million particles/ft³*	
Silica (amorphous)	20	
Asbestos	5[b]	
Mica	20	
Portland cement	30	
Nuisance particulates (<1% quartz[c])	30 or 10 mg/m³	

[a] p = particulate; t = tentative value (1966); n = not to be exceeded.
[b] Fibers/cc > 5μ in length (intended change).
[c] Nontoxic particles, such as starch, limestone, and glass.

persons over 50, was exceeded by Chicago, Cincinnati, Denver, St. Louis, Philadelphia, and Washington, D.C., in 1968. Criterion G, which may increase deaths from bronchitis and lung cancer, was exceeded by a factor of 2 in Philadelphia and by nearly three times (311 μg/m³ of SO_2) in Chicago.

In Denver there were 26 days in 1968 when the particulate level exceeded criterion E, for worsened acute bronchitis; there were 9 days in St. Louis, and 7 in Philadelphia.

Criterion D, a level of SO_2 responsible for increased hospitalization for respiratory diseases and for absenteeism, was reached in Chicago on 131

Table V

Health Effects of Sulfur Dioxide and Particulates[20]

	Concentration				
	Sulfur dioxide		Particulates		
Criterion	μg/m³	ppm	μg/m³	Duration	Effect
A	1500	0.52	(6 COH units)	24 h	Increased mortality
B	715	0.25	750	24 h	Excess deaths, increased illness
C	500	0.18	low	3–4 days	Increased mortality
D	300–500	9.18	low	3–4 days	Increased respiratory hospitalization and absenteeism
E	630	0.22	300	24 h	Worsened acute bronchitis
F	250	0.09	200	6 months	Increased absences
G	115	0.04	160	Annual mean	Increased deaths, bronchitis, lung cancer
H	Low		100	Annual mean	Increased death rate, persons over 50 yr

days, on 75 days in Philadelphia, on 17 in Washington, D.C., and on 7 days in St. Louis, during the same year.[34]

Table VI summarizes the time–concentration relationship of those pollutants directly affecting health for which ambient standards have been determined by the EPA for the United States. The standards themselves are discussed in the following chapter. These are not static, since the 1970 Clean Air Act on which they are founded sponsors continued research by the EPA leading to improved standards. For example, the previously mentioned paper entitled "Health Hazards of Lead"[33] was issued by the EPA on February 22, 1972, based on a comprehensive report of the NRC-NAS (National Research Council and the National Academy of Sciences, acting jointly).[35] This paper contains all the available information necessary to establish an ambient standard for lead in the atmosphere. Such a paper may serve as an advance warning that a federal standard is about to be issued.

The natural background of airborne lead, according to this study, is about 0.0005 $\mu g/m^3$. But in 1968, over 184,300 tons were emitted from man-made sources—over 99% resulting from the use of gasoline with lead antiknock additive. In consequence, the airborne lead concentration near heavy traffic routes has been found to range from 7 to 54 $\mu g/m^3$ with mean values of 14–24 during rush hours.

Since lead is not known to serve any useful biological function, any metabolic effects of lead are considered to be adverse and any increase in the body burden of lead a health hazard. Signs of lead poisoning do not appear below 50–80 $\mu g/100$ g of blood in otherwise healthy adults, but above 40 $\mu g/100g$, the function of certain metabolizing enzymes is inhibited.

The body absorbs about 30% of inhaled lead, some three times more than that which is ingested. However, human blood levels are not known to rise with exposure below 2–3 $\mu g/m^3$. Above this point, the dose–response curve assumes that 30% of the lead in the volume respired (about 23 m^3/day) is absorbed. The expected blood levels as a function of air lead concentration are computed in Table VII, taken from the report.

Table VI

Minimum Concentration–Duration Conditions Defining Pollution Effects in Current Federal Criteria

Pollutant	Concentration, ppm	Averaging time	Duration	Effects
Sulfur dioxide	0.11	24 h	3–4 days	Health
	0.04	Annual mean		Health
	0.03	Annual mean		Vegetation[c]
Particulates	300[a]	24 h	1 day	Health[b]
	80[a]	Annual mean		Health[b]
	60[a]	Annual mean		Materials
Photochemical oxidants	0.03–0.3	1 h	1 h	Health
(Ozone)	0.1		Peak	Health (eye irritation)
	0.5	—	4 h	Vegetation
Carbon monoxide	10–15	—	8 h	Health
	8–14	Week	Week	Health

Source. Adapted from pertinent NAPCA "Air Quality Criteria," publications AP 49, 50, 62–64.

[a] Micrograms per cubic meter.
[b] With approximately 0.22 ppm SO_2.
[c] The effects of long-term exposure of vegetation to SO_2 have been subsequently re-evaluated and ascribed instead to short-term exposure at higher concentrations.

Table VII

Relationship of Adult Blood Lead Levels to Airborne Lead Exposure[33]

Air lead exposure, μg/m³	Daily lead absorption, μg/day			Expected blood lead [c] μg/100 g	Relative excess in blood lead, %
	Air[a]	Diet[b]	Total		
2.0[d]	13.8	30	43.8	20.7	0%
2.5	17.3	30	47.3	22.5	9%
3.0	20.7	30	50.7	24.2	17%
3.5	24.2	30	54.2	25.8	25%
4.0	27.6	30	57.6	27.2	31%
4.5	31.1	30	61.1	28.6	38%
5.0	34.5	30	64.5	29.9	44%
10.0	69.0	30	99.0	40.1	94%
20.0	138.0	30	168.0	52.7	155%
50.0	345.0	30	375.0	71.7	246%
100.0	690.0	30	720.0	87.3	322%

[a] Assumes inhalation of 23 m³/day and 30% lung retention.
[b] Assumes 10% gastrointestinal absorption of the average adult daily total dietary intake (300 μg) from food and water.
[c] Computed from regression formula: blood lead = $-69.2052 + 54.7605 \times \log$ μg Pb absorbed daily, as given in Chapter 3 of *Airborne Lead in Perspective* by the National Research Council, National Academy of Sciences.
[d] According to the National Academy of Sciences, "It is not possible, on the basis of available epidemiological evidence, to attribute any increase in blood lead concentration to exposure to ambient air below a mean lead concentration of about 2 or 3 μg/m³" (p. 64). It follows that a relative excess in blood lead may be associated with ambient air exposure above 2.0 μg Pb/m³.

The report concluded that "airborne lead levels exceeding 2 micrograms per cubic meter averaged over a period of three months ... constitute endangerment of public health." Current levels range from 2 to over 5 μg/m³ in many urban areas (e.g., maximum quarterly levels of 3.6 in Philadelphia, 2.8 in New York, 5.7 in Los Angeles). Since airborne lead comes almost entirely from gasoline additives, controls are indicated. Subsequent regulations require the availability of one grade of "lead-free" gasoline by July 1, 1974 (91 Research Octane Number or more), containing not more than 0.05 g lead or 0.005 g phosphorus per gallon.[36] The purpose of this action was stated to be the attainment by 1977 of an airborne lead level of less than 2 μg/m³, consistent with the research findings.[37]

Similar study documents were planned by NAS to define the health effects of airborne cadmium, nickel, vanadium, manganese, arsenic, and polychlorinated biphenyls (PCB). Other materials will be reviewed at the rate of about five per year by the same organization. These are listed in order of priority in Table VIII.

The object of each of these studies is to discover the critical level of ambient concentration of that pollutant, if any exists, below which no unfavorable health or other effects can be found. With some other substances, either no lower concentration limit defining ill-health effects has been found or the material is so hazardous that only the strictest emission controls at the source will protect human health. Federal standards may be established for such materials, designated as "hazardous pollutants." The Clean Air Act provides for federal emission control of substances that the EPA finds

Table VIII
Suspected Airborne Health Hazards in Order of Priority for Studies Planned by National Academy of Sciences[38]

Cadmium
Nickel
Vanadium
Manganese
Arsenic
Polychlorinated biphenyl (PCB)
Sulfur oxides
Fine particulates
Copper
Zinc
Chlorine
Hydrocarbons
Nitrogen oxides
Carbon monoxide
Photochemical oxidants
Selenium
Iron
Infectious aerosols
Boron
Barium
Aero-allergens
Phosphorous oxides
Pesticides
Tin
Antimony
Titanium
Tellurium

Table IX

Federal Emission Standards for Hazardous Pollutants[a]

Asbestos:	No ambient limits, since there is no suitable sampling and analysis method.
Beryllium:	Ambient limit, 0.01 μg/m^3
Mercury:	Ambient limit, 1 μg/m^3
	Applies to any new or modified plant.
	License required from EPA.
	Emission tests required.

CONTROLS

Asbestos:	1. Fabric filter or cyclone equivalent required.
	2. Visible emissions prohibited.
	3. Spraying emissions prohibited.
	4. Surfacing roadways with asbestos prohibited.
Beryllium:	1. Total emissions shall not exceed 10 grams per 24 hour day; or
	2. No more than 0.01 μg/m^3, 30 day average, measured by a sampling network.
	Emission tests and periodic tests may be required.
	Alternately, continuous monitoring with filters may be substituted; 0.03 μg/m^3 or 0.01 μg/m^3 in 30 days must be reported to EPA.
Mercury:	1. Emission limit is 5 lb/24 h.
	2. Stack sampling is required.

[a] See *Federal Register*, **36**, 23239–23256 (December 7, 1971).

are contributing causes of death or illness. Currently, those so designated are asbestos, beryllium, and mercury.

The evidence assembled by the EPA[39] shows that occupational exposure to asbestos is responsible for diseases, including pulmonary fibrosis (asbestosis) and bronchogenic carcinoma. Since asbestos fibers have been found in the lungs of nonoccupationally exposed persons also and a safe exposure level has not been established, it has been implicated as a "serious air-pollution threat." Beryllium, on the other hand, is known to be safe for 30 days exposure at levels at or below 0.01 μg/m^3 and as high as 25 μg/m^3 for 30 min. This has been found as a result of experience with rocket firings by the military. Finally, elemental mercury (vapor or aerosol) is known to be absorbed by man more or less completely when inhaled at concentrations below 350 μg/m^3 and to affect the central nervous system when absorbed into the blood. Symptoms include tremor, psychological disturbances, loss of weight, and others. The danger of chronic disease with long-term exposure places a limit on the average intake to 1 $\mu g/m^3$ of air over 30 days to avoid ill effects.

"Hazardous pollutants" differ from those of Table VI, for which ambient standards are established, since there is no safe level that can be permitted in urban atmospheres. They represent hazards that are too great to be tolerated, even if confined to a restricted area such as a single industrial plant. The means chosen to control them is to set limits on the amount of hazardous material that can be emitted by a single facility—hence the phrase "emission standards." The ambient limits refer to the airborne concentration in the vicinity of the polluting industry, which the operator can elect to monitor as an alternate to keeping track of the material emitted from the plant stacks. Table IX summarizes these requirements.

3.3.3. Threshold of Biological Effects

The difficulty of establishing relationships between ambient air-pollution levels and health effects is increased by the hypothesis of "no threshold." This concept holds that there is no dosage so small that it does not have some biological effect deleterious to health and that our inability to detect such effects at very small concentrations are merely caused by the limitations of observing technique. This theory has been followed in the case of the CO–COHb relationship and its physiological effects. The linear relationship between dose and response (decrement in temporal perception) has been extrapolated down to zero (see Ref. 25, pp. 8–21, for example).

This line of reasoning imposes a philosophical and moral stumbling-block on the whole concept of cost-effective tradeoffs for air-pollution controls, for if there is always a deleterious health effect concurrent with any level of pollution, even down to the entrance of one molecule into a cell, responsiveness to the ideal that no single individual shall suffer harm will continue pressures for pollution reduction to the point of "zero emissions." The containment of *all* by-product emissions is, in the case of real processes and plants, a goal that we would not expect to achieve short of infinite expenditures of effort or zero production.

This difficulty has been examined by Dinman[40] with encouraging results. He has examined the probability of a foreign atom entering a cell and the joint probability of it having a biological effect, given that circumstance. There are a great many ways for an atom of lead or mercury, as examples, to become bound in a protein molecule such as a cell membrane or enzyme. But the function of active atomic sites on these molecules is very specific, as brought out by the molecular biologist Jacques Monod.[41] Many of the binding positions for the foreign atom will have no biological significance whatever. Therefore, the consequence of a harmful effect is not a certainty.

Furthermore, there are a very large number of places where foreign atoms may be diverted around the cell (including the substrate) before they can enter the cell. Finally, if a cell molecule is damaged, it is replaced in the normal activity of the cell, just as the death and replacement of the cell itself is a normal event.

Although an exact stochastic model cannot be constructed, consideration of the total number of atoms in a typical cell (over 10^{14}) and the probable activity and concentration of those atoms of which it is normally made up leads to the conclusion that a threshold for biological activity exists at about 10^4 atoms/cell. Stochastic factors, according to this concept give the biological system an inherent protection by limiting the range of its response to environmental challenge, such as the entry of deleterious atoms.

3.4. ECONOMIC COSTS OF AIR POLLUTION

From the system viewpoint, as defined in Chapter 1, the most important criterion for the design of air-pollution controls is that of maximizing their expected social and economic value. In the economic sense, this means that the ratio of monetary benefits to the costs of control is to be maximized. The most recent estimate of the cost of implementing current and anticipated air-pollution controls in the United States is $106.5 billion for the 10 yr following 1972.[42] What benefits will accrue to the nation that will justify costs exceeding $10 billion/yr?

The benefits most readily computed will be in the reduction in tangible costs that would be obtained if air pollution was reduced. Among these are the dollar costs of health care and mortality ascribable to air pollution, the conversion of material and loss of crops, and the reduction in the value of property due to soiling, odors, and environmental degradation.

For the purpose of optimizing a pollution control system, such data would be most useful in the form of a plot of damage cost as a function of the degree of pollution. This damage curve would be a composite of a large number of functions of the type discussed by the economist Ronald Ridker:[43]

$$CP = \sum_i C_i Q_i F_i(S),$$

where CP is the direct cost or "damage function" of a single pollutant (i), C_i is the cost per unit of damage, Q_i is the number of receptors affected, and $F_i(S)$ is the damage per unit of concentration of the pollutant i and would presumably be a nonlinear relationship. The benefit curve would be the inverse of the damage plot, since it represents the reduction in cost over that with no controls.

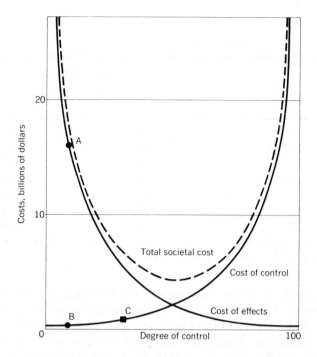

Figure 2 Cost–benefit relationship (courtesy of *IEEE Spectrum*), **8**, 26 (1971).

A similar curve representing the cost of control versus its degree or effectiveness could then be combined with the first plot to generate a net cost, as in Figure 2. The minimum net cost then defines the most desirable degree of control. The cost–benefit relationship will be discussed at greater length in Chapter 5.

Unfortunately, there is not yet sufficient data to arrive at these plots, and furthermore, they are not static with time. The cost of control will change as the technology is improved or more widely utilized, and the cost of damage will increase with the number of individual receptors (e.g., the population).[44]

The direct costs are, of course, minimum estimates, since they will include (in the case of death or illness) only the earnings loss of the worker, given in Table X, plus treatment and associated costs, not the "psychic cost" of his loss to his family and community. Furthermore, they will exclude "aesthetic costs," such as the loss of visibility and the mortality of noncommercial plants and animals. Likewise, the cost of synergisms, or the nonadditive effects of two or more pollutants present simultaneously, are missing.

Table X

Economic Cost of Premature Mortality

	Present value of future earnings[a]	
Age	Male	Female
10	42,800	12,100
20	66,300	17,600
30	72,100	18,800
40	59,200	16,400
50	39,410	9,900
60	16,300	3,500
70	1,700	350

[a] After Ridker,[43] 5% discount, 1960 dollars.

The paucity of factual information on air-pollution damage costs is pointed up by the small number of studies that have been conducted—about 36, of which only a third have been published, according to a recent survey.[45] The costs of control, specifically the cost of implementing the U.S. Clean Air Amendments of 1970, are better known, since they are required by law to be studied comprehensively and reestimated annually. The EPA publishes them annually as a report to Congress, *The Economics of Clean Air*.[46]

One of the earliest reports on U.S. damages was published in 1913 by the Mellon Institute.[47] The estimate for Pittsburgh alone at that time was $10 million, $20 per capita, and included such items as the following (in millions):

Cost (to the polluter) of imperfect combustion	$1.5
Individual laundry and cleaning bills	$2.3
Household cost, painting, corrosion, wallpaper, cleaning, and lighting.	$2.3
Stores, damaged goods, maintenance, lighting	$3.7
Cost to buildings	$0.2

The 1913 Mellon Institute study was extrapolated on the basis of commodity price index and population, resulting in the figure of $11 billion for 1959; or $60 per capita,[48] which has been widely cited in popular literature as a national cost. Ridker has objected to this statement as being unsupported by facts,[43] and in an unpublished report to the U.S. Public Health Service (1966) predicted a total cost between $7.3 billion and $8.9 billion for 1970.[49] A similar figure was obtained by Gerhard also in unpublished report for NAPCA (1969) cited by Barrett and Waddell.[50] Gerhard's estimate for 1968 was $8.1 billion and a range of $6.8–15.2 billion. Each of these later studies utilized a procedure that might be inferred from Ridker's "damage function" equation:

1. Identification of air-pollution damage categories
2. Estimate of total value of the category regardless of air-pollution effects
3. Application of a damage factor to the total value of the category
4. Summation of the estimates over all damage categories.

Barrett and Waddell followed a similar methodology in their critical review of past work[45] and arrived at a total estimate of $16.1 billion for 1968 U.S. air-pollution damages. The breakdowns by type of pollutant and by type of source for these costs are given in Tables XI and XII.

These approximations are considered by their authors as "realistically conservative and defensible." Some of the detailed assumptions will be discussed below. Many effects were omitted because of lack of data. Regardless, these estimates cannot be equated with benefits of pollution control. For example, all sources of emission are not subject to control by existing air-quality management programs. Structural and forest fires are stated to account for half of the "miscellaneous" category of Table XII and about 8% of the total pollution assumed (Table XIII). Deducting this uncontrollable percentage from the potential benefit of total control leaves $14.8 billion.

The above methodology is based on an assumption of a linear function relating the pollution cost and the weight of pollution emissions. The ac-

Table XI

National Total Annual Costs of Pollution for Types of Pollutants and Effects in 1968 (Millions of Dollars)[45]

Effects	SO_x	Particulates	Oxidant	NO_x	Total
Residential property	2808	2392	—[a]	—	5,200
Materials	2202	691	1127	732	4,752
Health	3272	2788	—	—	6,060
Vegetation	13	7	60	40	120
Total	8295	5878	1187	772	16,132

[a] No known value.

Table XII
National Total Annual Costs of Pollution for Types of Sources and Effects in 1968 (Millions of Dollars)[45]

	Stationary-source fuel combustion	Transportation	Industrial processes	Solid waste	Miscellaneous	Total
Residential property	2,802	156	1,248	104	884	5,200
Materials	1,853	1,093	808	143	855	4,752
Health	3,281	197	1,458	119	1,005	6,060
Vegetation	47	28	20	4	21	120
Total	7,983	1,474	3,534	370	2,765	16,132

Table XIII
Estimates of Nationwide Emissions, 1968 (10^6 tons/yr)[45]

Emission source	CO	Particulates	SO_x	HC	NO_x
Transportation	63.8	1.2	0.8	16.6	8.1
Fuel combustion in stationary sources	1.9	8.9	24.4	0.7	10.0
Industrial processes	9.7	7.5	7.3	4.6	0.2
Solid waste disposal	7.8	1.1	0.1	1.6	0.6
Miscellaneous	16.9	9.6	0.6	8.5	1.7
Total (1968)	100.1	28.3	33.2	32.0	20.6

Source. Division of Air Quality and Emissions Data, BCS, NAPCA, after Ref. 45.

curacy of this assumption has not been tested; for example, a ton of SO_x is considered to cause the same damage cost as a ton of hydrocarbons (HC). The influence of locality, also, has not been studied. A forest fire, to use the previous case, will not harm vegetation that escapes the fire because of CO emission, since large concentrations of CO are benign to plant life, furthermore, it is unlikely to cause health damage to a large population. An urban fire, on the contrary, may cause considerable health damage with CO, particulates, and carcinogenic hydrocarbons. Nevertheless, the assumptions of linearity and regional uniformity are the best available and are currently being utilized to compute cost–benefit relationships.

Table XIV
Projected National Annual Damage Costs[a] by Pollutant in 1977 (1970 Dollars in Millions)[46]

	Pollutant					
	Particulates	SO_x	Oxidant[b]	NO_x	CO	Total
Health	3,880	5,440	—[c]	—[c]	—[c]	9,320
Residential property	3,330	4,660	—[c]	—[c]	—[c]	7,990
Materials and vegetation	970	3,680	1,700	1,250	—[d]	7,600
Total	8,180	13,780	1,700	1,250	—[c]	24,910

[a] Based on Ref. 45.
[b] Assumed proportional to HC emissions.
[c] Not available because of lack of data.
[d] Assumed to be negligible.

Table XV

Projected National Annual Damage Costs[a] by Source Class in Fiscal 1977 (1970 Dollars in Millions)[46]

	Damage class				
Source class	Health	Residential property	Materials and vegetation	Total	%
Mobile	220[b]	190	1,740	2,150	8.6
Solid waste	230	190	200	620	2.5
Stationary fuel combustion	4,950[c]	4,240	3,650	12,840	51.5
Industrial processes studied	1,780	1,530	920	4,230	17.0
Industries not studied	1,260	1,080	460	2,800	11.2
Miscellaneous	880	760	630	2,270	9.1
Total	9,320	7,990	7,600	24,910	100.0

[a] Based on Ref. 45.
[b] Health damage costs due to CO, NO_w, and O_x not included due to lack of data. Entry is health-damage cost ascribed only to minor amounts of vehicle-related particulate and SO_x and, therefore, considerably understates probably health damage costs due to mobile-source emissions.
[c] Health damage costs due to NO_x from stationary fuel combustion not included because of lack of data.

The 1968 estimates have been extrapolated one step further to 1977.[46] The result is that without application of controls, the national air-pollution cost would rise to nearly $25 billion. Tables XIV and XV break down this projection in terms of pollutants and source class. Again, this total is considered conservative, since (1) lack of data eliminated any health costs due to CO, HC, NO_x, and oxidants (O_x); (2) no estimated costs of effects on industrial, commercial, and real property were included; and (3) the costs of aesthetics and visibility losses, odor, and soiling were omitted. In these tables, the allocation of damage to source class assumes that each unit weight of pollution is uniformily damaging, regardless of source. It is considered that this method understates mobile-source health damage, since auto exhausts are closer to the level of people's breathing than are the emissions from stationary-source stacks.

3.4.1. Health Costs

Since the annual costs mentioned above are so staggering, it serves to examine Barrett and Waddell's sources and assumptions in greater detail.

3.4. Economic Costs of Air Pollution

As stated earlier, there have been only a few studies in this field. Barrett and Waddell cite three for their health cost data, of which only Ridker's work[43] and that of Lave and Seskin[51] are published and relevant. In all these studies the methodology explained above was chosen: (1) the total health losses are estimated, (2) a coefficient is determined for the share of this value attributable to air pollution, and (3) the value of health loss due to air pollution is the total loss multiplied by the coefficient.

Ridker estimated disease costs by applying the cost of premature death based on future earnings lost (Table X) premature burial costs, treatment, and absenteeism. His disease costs for the year 1958 totaled $2 billion (Table XVI). He applied coefficients of 18–20% of the total to obtain health damages due to air pollution of $360–400 million. The coefficients were derived from studies relating respiratory and lung-cancer mortality rates in rural and urban areas.

Lave and Seskin[51] extended Ridker's study to include other forms of internal cancer and heart disease. Their total costs were derived from a study by Rice;[52] breaking down aggregate costs (such as "respiratory system diseases") by the proportion of patient hospital days utilized. They arrived at the coefficients for allocating responsibility to air pollution by regression analysis. For example, in one study the index of bronchitis mortality showed a probable reduction of 50% caused by a 50% reduction in air pollutant concentration. (See Table.)

Their results can be summarized by Table XVII. The total benefits obtained by a 50% improvement in air quality amount to $2.08 billion annually—approximately 20% of the cost of the named diseases or 4.5% of the national cost of all morbidity and mortality. Only the cost of wage-earner's illness are included.

Lave and Seskin's conclusions are the subject of their own criticism for omitting of such factors as smoking and urban life styles, unrelated to general air pollution, and of Barrett and Waddell's on the strength of excessive faith in their regression coefficients. Nevertheless, all agree that the percentage estimate is a "robust" figure, not sensitive to the exact numbers used to calculate it, and that the figure of $2 billion is justified and even conservative. On this basis, the latter authors compute their 1968 estimate. The total cost of 1963 air-pollution-related illness is $4.16 billion, assuming the linear relationship of dose–response, which is 0.7% of the GNP for that year, $590.5 billion. They assume the same percentage of the 1968 GNP of $865.7 billion (thus accounting for inflation, population increase, and real growth) to arrive at a 1968 health cost of air pollution equal to $6.06 billion.

Table XVI
Resource Costs[a] of Disease Associated with Air Pollution (1958)[43] (Millions of Dollars)[45]

Type of cost	Cancer of the respiratory system	Chronic bronchitis	Acute bronchitis	Common cold	Pneumonia	Emphysema	Asthma	Total
Premature death	518	18	6	na	329	62	59	992
Premature burial	15	0.7	0.2	na	13	2	2	32.9
Treatment	35	89	na	200	73	na	138	535
Absenteeism	112	52	na	131	75	na	60	430
Total	680	159.7	6.2	331	490	64	259	1,989.9

[a] Using a discount rate of 5%.

Table XVII

Annual Health Cost of Air Pollution (1963)[a]

	Total annual cost, $ × 10^6 [b]	Coefficient[c]	Benefit, $ × 10^6
Bronchitis	1,000	0.50	500
Other respiratory diseases	2,888	0.25	722
Lung cancer	132	0.25	33
Other cancer	2,380	0.15	357
Cardiovascular diseases	4,680	0.10[d]	468
Total	10,180		2,080

[a] After Ref. 51.
[b] Cost of premature death (earnings loss at 6%), treatment, and absenteeism.
[c] For 50% reduction in air-pollution levels in 114 U.S. urban areas (from regression analyses).
[d] Uncertain, may be 0.20.

3.4.2. Materials

Considerable data is available on deterioration of metals, fabrics, rubber, and other materials as a result of air contaminants, and costs have been more readily reckoned because of this cause. One accepted factor is a one-third reduction in the life of power-line metal hardware due to SO_2 in polluted areas. Particulates accelerate steel corrosion caused by the presence of moisture and SO_2. An increase in particulate concentration from 60–65 to 176 $\mu g/m^3$ results in three times greater corrosion of outdoor steel samples. When the SO_2 and particulate concentrations increase further, the corrosion rate again doubles. In tests in six cities, a straight-line relationship between SO_2 and corrosion rate was shown. A list of nonmetallic materials damaged by air pollutants would include building materials, paint, leather, paper, textiles, dyes, rubber, and ceramics. Some corrosion and building damage is an irretrievable cultural loss: the marble carvings on the Parthenon in Athens (a dangerously polluted area) suffered little damage in 2240 yr but in the last 136 yr of urban air exposure have become unrecognizable.

Paint drying is inhibited by 1–2 ppm of SO_2; paint is also blackened by H_2S. Pollution incidents often damage automobile paints. Leather, initially free from sulfuric acid, absorbs as much as 7% of the SO_2 present in the air. The resultant deterioration of book bindings is well known. Likewise,

SO$_2$ causes embrittlement of paper at 2 ppm. Textile loss arises primarily from the additional cleaning necessitated by airborne particles.

Rubber and electrical insulation are attacked by airborne ozone, requiring protection of stored tires, for example. Other damage includes degradation of electrical and electronic parts, microfilm, and many other items.

Barrett and Waddell[45] surveyed the literature of these costs and adopted the 1970 report of the Midwest Research Institute as the most comprehensive. This work was accomplished in the following manner: Data on materials, corrosion rates, and corrosion damage was gathered by literature surveys, interviews, and mailings. First, the economic value of the fraction of each material exposed to air pollution was estimated. Then the value of "interaction per year," expressed as a fraction of the total worth of exposed material lost annually, was calculated for each material from the difference in deterioration in "polluted" and "unpolluted" atmospheres. The product of these two numbers yields the annual economic loss. Summing these products results in a measure of the annual cost of materials lost by air pollution.

Table XVIII summarizes the MRI results, modified as explained below. The third column of the table was generated by computing the annual production of the material, multiplying by its economic life in years and by an "exposure index," representing the estimated fraction exposed to the atmosphere, and multiplying this product by a "labor content" factor to account for the value of the fabricated products. For example, the annual production of copper was listed as $2.041 billion, the economic life 22 yr exposure index 0.5, and the labor content factor 2.4. Multiplying these factors results in a value of the in-place material exposed to the air of $54.88 billion.

The values obtained by MRI must be considered as relative rather than absolute values. For only a few materials and pollutants—zinc and SO$_2$, rubber and O$_3$, and cotton and SO$_2$—were there dose–response calculations. For many materials such as concrete, paints, fibers, and plastics, little or no quantitative information was available. In several other studies surveyed by Barrett and Waddell there was a more detailed correlation for specific materials. One of these cited was the unpublished work of Haynie[53] on the corrosion of galvanized steel. He computed a wide range of possible values of loss, ranging from $1.4 billion to $13 billion annually. The survey authors chose to accept Haynie's minimum value in substitution for MRI's results for zinc and for carbon and alloy steels.

Likewise, the MRI estimate for synthetic and natural rubber losses, totaling $194 million, was raised to $380 million on the basis of a comprehensive study of the subject conducted by Battelle Memorial In-

Table XVIII

Annual Material Deterioration Costs of Air Pollution

Material	Value of interaction per yr	In-place value of materials exposed, (millions of dollars)	Value economic loss, (millions of dollars)
Galvanized steel[a]	—	—	1,400.1
Paint	0.50×10^{-1}	23,900	1,195.0
Elastomers (rubbers)[b]	—	—	380.0
Cement and concrete	0.10×10^{-2}	316,210	316.0
Nickel	0.25×10^{-1}	10,400	260.0
Cotton (fiber)	0.40×10^{-1}	3,800	152.0
Dyes (fabric fading)[c]	—	—	206.0
Tin	0.26×10^{-1}	5,530	144.0
Aluminum	0.21×10^{-2}	54,080	114.0
Copper	0.20×10^{-2}	54,880	110.0
Wool (fiber)	0.40×10^{-1}	2,480	99.2
Nylon (fiber)	0.40×10^{-1}	950	38.0
Cellulose ester (fiber)	0.40×10^{-1}	820	32.8
Building brick	0.10×10^{-2}	24,150	24.2
Urea and melamine (plastic)	0.10×10^{-1}	2,270	22.7
Paper	0.30×10^{-2}	7,530	22.6
Leather	0.40×10^{-2}	5,150	20.6
Phenolics (plastic)	0.10×10^{-1}	1,980	19.8
Wood	0.10×10^{-2}	17,610	17.6
Building stone	0.23×10^{-2}	7,650	17.6
PVC (plastic)	0.10×10^{-1}	1,540	15.4
Brass and bronze	0.42×10^{-3}	33,120	13.9
Polyesters (plastic)	0.10×10^{-1}	1,370	13.7
Rayon (fiber)	0.40×10^{-1}	330	13.2
Magnesium	0.20×10^{-2}	6,500	13.0
Polyethylene (plastic)	0.10×10^{-1}	1,170	11.7
Acrylics (plastic)	0.10×10^{-1}	1,000	10.0
Polystyrene (plastic)	0.10×10^{-1}	850	8.5
Acrylics (fiber)	0.40×10^{-1}	190	7.6
Acetate (fiber)	0.40×10^{-1}	190	7.6
Polyesters (fiber)	0.40×10^{-1}	160	6.4
Polypropylene (plastic)	0.10×10^{-1}	640	6.4
ABS (plastic)	0.10×10^{-1}	610	6.1
Epoxies (plastic)	0.10×10^{-1}	470	4.7
Cellulosics (plastic)	0.10×10^{-3}	400	4.0
Bituminous materials	0.10×10^{-3}	22,450	2.2
Gray iron	0.50×10^{-3}	3,860	1.9
Nylon (plastic)	0.10×10^{-1}	170	1.7
Polyolefins (fiber)	0.40×10^{-1}	40	1.6
Stainless steel	0.85×10^{-4}	18,900	1.6

Table XVIII—Continued

Material	Value of interaction per yr	In-place value of materials exposed, (millions of dollars)	Value economic loss, (millions of dollars)
Clay pipe	0.10×10^{-2}	1,440	1.4
Acetate (plastic)	0.10×10^{-1}	120	1.2
Malleable iron	0.16×10^{-2}	580	0.9
Chromium	0.75×10^{-3}	1,080	0.8
Silver	0.12×10^{-1}	570	0.7
Gold	0.10×10^{-3}	5,800	0.6
Flat glass	0.10×10^{-4}	28,590	0.3
Lead	0.11×10^{-3}	2,180	0.2
Molybdenum	0.25×10^{-3}	510	0.1
Refractory ceramics	0.10×10^{-4}	1,930	0.02
Carbon and graphite	0.10×10^{-5}	300	0.00
Total			4,752.02

Source. Midwest Research Institute. Systems Analysis of the Effects of Air Pollution on Materials. EPA Contract No. CPA-22-69-113. Final Report January 1970.

[a] Ref. 53.
[b] Ref. 54, cited in Ref. 55.
[c] Ref. 56.

stitute.[54] Battelle surveyed the rubber manufacturers and totaled their added costs for pollution-resistant polymers, antiozonants, waxes, and other product-protection measures. The aggregate was $56 million for the entire industry. Assuming a retail price approximately three times manufacturing costs, the retail price of pollution-damage prevention amounts to $150–170 million. Another $217 million is estimated to be the annual replacement cost of rubber products at the consumer level resulting from air pollution. This estimate includes $100.5 million for medical goods, $34.5 million for hoses, and $32.2 million for tires. Thus, $380 million is the total cost borne by the consumer for both initial and replacement costs.

MRI did not consider the cost of dye-fading on textiles, but this was estimated in a study by Salvin.[56] He has established a total cost of $122 million due to NO_x, causing the fading of dyes on acetates, rayons, and cotton and the yellowing of other fabrics. Similarly, O_3 costs nearly $84 million from fading of these materials and of carpets and garments. The total of $206 million appears in Table XVIII.

3.4.3. Vegetation

The deleterious effects of ozone, organic oxidants, and other components of photochemical smog and of SO_x on plant life and crops have been documented for many years. An authoritative report in the form of a pictorial atlas was published by the Air Pollution Control Association.[57] Earlier dose–response studies have shown that the growth of rye grass exposed to sulfur dioxide is reduced at levels as low as 0.01 ppm, but acute symptoms of injury do not generally appear below 0.3 ppm (for 8 h), equivalent to a yearly average of 0.03–0.05 ppm. Short-term (4 h) damage at lower concentration is noted when synergistic reactions occur with O_3. SO_2 affects corn, squash, and alfalfa, also.

Other pollutants injuring plants include ethylene (affecting orchids); PAN, ozone, and other photochemical oxidants (lettuce and tobacco); fluorides (gladiola); and H_2S, HCl, and NH_3.

The 1968 estimate of vegetation losses used by Barrett and Waddell are based on two studies—one, a statewide survey in Pennsylvania,[58] using state university personnel, foresters, and nursery inspectors who were trained in a special course, and the other, a study by Benedict of Stanford Research Institute[59] in which a theoretical model of plant damage was developed based on the fuel consumption in large metropolitan areas (SMSA).

Sixty field investigations out of the 92 conducted in the Pennsylvania study revealed air-pollution damage. Direct losses exceeding $3.5 million and another $8 million of indirect damages were counted.

The Stanford study, on the other hand, proceeded by identifying 90% of the damage as being caused by PAN, ethylene, NO_x, SO_x, and fluorides. Then, in the 500 most important agricultural counties, a count of the crop production and the emissions of these substances from adjacent SMSA's resulted in an estimate of $64 million lost from hydrocarbons damage, $4 million from SO_2, and $2 million from fluorides. If this is taken as 90% of the country's damage, the national total would be $80 million. If, however, the total is revised by adjusting for the ratio of Pennsylvania's survey to the SRI theoretical figure for the same state, and so modifying the entire estimate, it becomes $120 million for the national total in 1968. This is the sum shown in Tables X and XI.

The Pennsylvania data have been updated by newer studies conducted in other states. One such study, conducted by the California Department of Agriculture in cooperation with county agricultural commissioners, assessed the 1970 air-pollution damage to agricultural plant crops. The efforts were concentrated in areas with known histories of air-pollution problems. These agricultural specialists, on the basis of their observations, estimated a loss approaching $25,690,680 for the 1970 crop.[60] These losses, however, were confined to only 15 of the state's 58 counties. Citrus plantings in the Los Angeles basin alone, accounted for $19,533,400. Plant injury, on the other hand, was observed in 22 of the counties. Further, these losses do not include reductions in crop yield resulting from invisible damage, except in the case of citrus and grapes, nor the losses in forests and other native vegetation or to landscape plantings. The survey found that most of the losses were the result of photochemical smog and allocated the following percentages: ozone, 50%; peroxyacetyl nitrates (PAN), 18%; fluorides, 15%; ethylene, 14%; sulfur dioxide, 2%; and particulates, 1%.

3.4.4. Residential Property

Losses in the value of real property due to air pollution were calculated by Barrett and Waddell, basing their work on original studies by Ridker.[43] The latter used multiple regression analysis to determine the variation in mean property value (MPV) in the St. Louis SMSA, as a function of the SO_x concentration of the air (SUL) expressed in sulfation units. Sulfation is measured by exposing a "candle" or "plate" of known area, prepared from lead peroxide paste, to the atmosphere for periods up to a month and analyzing the product for sulfate. The result, a measure of the long-term average of sulfuric acid, H_2S, and other sulfate-forming air impurities, has ranged from a few hundredths to 8 mg $SO_3/(100$ cm$^2)$ (day).[61]

Ridker included 11 other variables in his regression analysis, reflecting the housing density, age, accessibility, population density, and composition and was able to explain 90% of the MPV variation. The best estimate of the MPV–SUL relationship was $100 per 0.1 mg/100 cm^2-day change in sulfation. For all single-family dwelling units in the St. Louis SMSA, this amounted to a potential benefit of $251 million if the pollution level were to be reduced to the 0.49 mg/100 cm^2-day level.[62]

Barrett and Waddell reported other studies similar to Ridker's in St. Louis and four other localities—Toronto; Kansas City; Washington, D.C.; and the Delaware Valley area of Philadelphia. These correlated variables similar to Ridker's, adding family income and suspended particulates. All studies agreed on the inverse relationship of sulfation to MPV, and results clustered about a probable value between $100 and $300 for marginal decrease of 0.1 mg $SO_3/100$ cm^2-day. The review authors chose a middle value ($200), and on the basis of desirable rural and suburban areas such as Pensacola, Florida, selected a background level of 0.1 mg $SO_3/100$ cm^2-day as an objective. For all SMSA's, current mean sulfation rates were estimated or averaged from the measurements of monitoring stations, corrected for bias due to the station's location. The capital cost of the national property loss was then computed by applying the marginal value of $200 to a uniform reduction of sulfation pollution value to 0.1 mg in each SMSA.

The resulting relationship becomes

$$\text{National damage} = \sum_{ALL\ SMSA} (\text{SUL}) \left[\frac{\text{SMSA sulfation} - 0.1}{0.1}\right] (NH),$$

where SUL = \$200, NH is the number of housing units (households) in the SMSA as of January 1969, the and expression in the center brackets is taken as the next larger integer. Applying an average return rate of 0.10 to this capital sum, representative of all real property, they arrived at a national annual cost of \$5.2 billion.

3.4.5. Other Costs of Air Pollution Not Included

Although there are other obvious sources of damage, no others were included in the Barrett–Waddell review. For example, livestock are also affected by air pollutants. Ingested fluoride, dustfall from industrial plants, is harmful if concentrated in feed over 30 ppm for cows and 150 ppm for chickens. Arsenic pollution from smelters has killed hundreds of sheep grazing as far as 15 miles away from the plant. Cattle and other domestic animals have been affected by the outputs of lead smelters and molybdenum refineries. Some of the reported damages are diseases, such as asthma, emphysema, and similar respiratory troubles; others include malnutrition and lowered fertility rate of the animals.

Wild animals and those in zoos have been affected by air pollution; lung cancer, for example, has been reported by the Philadelphia Zoo in a number of cases. It is probable, however, that the damage to wildlife is greater from insecticides containing chlorinated hydrocarbons and similar herbicides. Even more serious may be the ecological consequences of the introduction of heavy metals into the food chain of wildlife. But lack of any hard data on these effects force us to consider them in the same category as aesthetic damage.

Soiling of household and commercial goods and properties is another cost not included in the national 1968 estimates. Some damage from this cause was estimated in the 1913 Mellon Institute survey, which has already been cited. A similar study prepared in England, the Beaver Report,[63] estimated direct cost (1954) at about £150 million annually, plus another £100 million in efficiency loss. This breaks down as follows:

Laundry	£ 25 million
Painting, etc.	30
Cleaning and depreciation of buildings	20
Metal corrosion	25
Textile and goods damage	52.5
Total	£152.5 million

Other studies cited by the Barrett–Waddell review include those of Michelson,[64] Ridker,[43] and other studies in Philadelphia[65] and in Japan. None of these are considered to be definitive or consistent enough to procedure quantitative answers, although it is acknowledged that many annual costs (laundry, hairwashing, building cleaning, streetlamp washing) may amount to millions. Residential cleaning costs, on the other hand, are included in the residential property values computed by the reviewers.

3.5. AESTHETIC DEGRADATION FROM AIR POLLUTION

The aesthetic degradation resulting from air pollution includes such items as damage to art works and historic sites; reduction in visibility of scenic views; modification of urban weather to cause cloudiness and haze; the loss of animal pets, wildlife, and wilderness foliage, previously mentioned; as well as the general discomforts caused by odors, eye irritation, and other nondisabling malaise. Some of these items are quantifiable, or partly so, although it would be a difficult exercise in utility theory to determine the value of damages to the Parthenon and Venetian art work, to say nothing of that of a Mount Hood view to the inhabitants of Portland, Oregon.

3.5.1. Willingness to Pay

One quantitative approach to the problem is to survey individuals to determine their "willingness to pay" for the reduction of air pollution in their home, neighborhood, or city. Another is to measure public awareness as a dose–response function of complaints. Another dose–response measure that has the benefit of objectiveness is visibility. Although visibility unquestionably has economic implications, such as the cost to airlines and possible increase in auto accidents, the aesthetic cost, as in the Mount Hood example, is high but unquantified.

Other studies relate the precipitation and cloudiness over cities to air pollution, so that poor weather is an aesthetic cost. Ridker[66] cites surveys of the aesthetic value of pollution control. In a high-pollution area the residents were willing to pay $0.55–0.62/month to solve the problem. In a low-pollution area, this willingness dropped to $0.43–0.46. The difference, $0.12–0.46/month, is statistically significant and represents the nuisance cost of air pollution to these residents. The data were gathered in Philadelphia during 1965 by Daniel Yankelovich, Inc. A more recent (1970–1971) study by Trendex, Inc., determined that half the families queried in Pennsylvania would pay up to $90 in added living costs to meet the state's clean-air goals. This willingness was independent of savings to the individual. The study also revealed that 68.4% of families in a statistically balanced sample believed there was a need to improve their local air purity. This was roughly equivalent to the ratio of urban to rural families in the sample.

Ridker queried victims of a severe soot fallout in Syracuse, New York, and found them willing to pay a "psychic cost" up to $11 to avoid such an incident, an average of 27% over the actual cost of cleanup. Another willingness-to-pay study conducted in 1966 by Lawyer,[67] in Morgantown, Pennsylvania (unpublished), revealed that the average inhabitant would pay more than $16 annually to eliminate perceptible air pollution.

3.5.2. Public Awareness

The costs of odor, dirt, and mental concern are indirectly measured by public awareness of the nuisance. A study conducted in the St. Louis area showed a linear mathematical relationship between the particular concentration and the percentage of population complaining. The expression was

$$Y = 0.3X - 14,$$

where X is the percent of population concerned about air pollution and Y is the annual geometric mean particulate concentration in micrograms per cubic meter.[68] In accordance with this formula, the community would be completely satisfied with a particulate concentration of 43 $\mu g/m^3$ or less.* Other studies have shown public awareness and concern to be correlated with high daily local pollution levels rather than longer-term averages and to increase with socioeconomic status.

3.5.3. Visibility

Public protest against air pollution rises sharply with the reduction in visibility that accompanies particulates and sulfur compounds. This problem has not been confined to urban areas, as evidence by the difficulties being encountered by the installation of power plants totaling 2,208,500 kW, the largest coal-burning facility in the country, at Four Corners near the Colorado, New Mexico, Arizona, and Utah common boundaries. A recent article[69] describes the drastic reduction of visibility in this area, once famous for 100-mile vistas of the Continental Divide and desert views. A 700-foot monolith, Shiprock Peak, is now veiled by haze at 20 miles. Approximately 200–250 tons of emitted particulates per day cause smoke palls over thousands of square miles. An additional 236 tons/day of SO_2 and 240 tons/day of NO_x may form acid mists and photochemical smog.

Public protest has caused the establishment of new regulations requiring reduction of particulates to 30 tons/day, which will cost the power com-

---

* cf., National Ambient Air Quality Secondary Standard of 60 $\mu g/m^3$ (Table I, Chapter 4).

panies involved $14.5 million for wet scrubbers and $24 million for baghouse filters. The eventual increase in capacity to 13,895,000 kW will effectively multiply these expenditures manyfold. Since the population in this area is sparse, the health costs are minimal, and these costs can be attributed largely to the aesthetic damage to visibility (see Chapter 12 on "clean area" degradation).

Visibility can be quantitatively related to particulate concentration under certain conditions, in a dose–response relationship.

Visibility is reduced strongly by aerosols (particulates or mists) in the range 0.1–1.0 μm in radius, which scatter light into and out of the line of sight; this reduces contrast through loss of light from the object and its background and by illumination of the intervening air. Urban particulate matter, including most of the sulfuric acid and sulfates, are in this size range and thus contribute substantially to reduction in visibility. The correlation between particulate concentration and visibility is particularly good for relative humidities below 70%; fresh smoke plumes and perhaps photochemical reaction products are excluded.

Because of the complex effects of Mie scattering, the light measured at any given angle from arbitrary size particles might vary widely. Regardless of this fact, it has been shown by Middleton[70] that the "visibility" (i.e., the visible range in hazy atmospheres) is a simple function of particle scattering,

$$L_V = 3.9/b_{\text{scat}}, \tag{1}$$

where L_V is the visual range in meters and b_{scat} is the scattering coefficient of the particles (the light absorption by gases and aerosols and the molecular scattering are assumed to be subordinate).

Airborne particles are classified by their diameter in a number of ways (see also Chapter 1). For example, *Aitken particles* are those less than 0.1 μm and bigger than large ions (0.01 μm); *large particles,* ranging from 0.1 to 1 μm, are of most interest as causes of haze and reduced visibility; and *giant particles,* are over 1 μm and may range up to 100.

Because larger particles settle out rapidly per Stokes's law and smaller one agglomerate into large units, the main aerosol mass ultimately consists of suspended particles ranging from 0.1 to 10 μm in diameter. This range includes those most affecting visibility, since they bracket and interact most strongly with the wavelengths of visible radiation, 0.3–0.7 μm. It is also identical to the range of condensation nuclei (but not of the "giant" precipitation or freezing nuclei), which explains the formation of mists in polluted atmosphere. For these reasons (see also Chapter 1, Section 1.4) the mass and size distribution of atmospheric aerosols other than fresh smoke and fumes tends to be concentrated in the range 0.1 to 10 μm. Mie-

3.5. Aesthetic Degradation from Air Pollution

scattering calculations show that the overall value of b_{scat} is controlled by particles from 0.1 to 1 μm and that the variations caused by refractive index differences of particles found in the atmosphere have little overall effect. For a several-decade range of atmospheric particle-size distributions and volume concentrations, measured in such diverse localities as the Austrian Alps and Seattle, it has been calculated that a linear relationship exists between concentration and b_{scat} for any single wavelength.[71] The relationship confirmed by measurements using an integrating nephelometer,[72] a total scattering instrument operating at a wavelength of 500 nm is

$$(\text{Mass}) \text{ Conc. } (\mu g/m^3) \cong 3 \times 10^5 \times b_{scat}, \tag{2}$$

which can be combined with Eq. (1) to write

$$L_V \times \text{Conc.} \cong 1.2 \ (g/m^2). \tag{3}$$

Because of density and other variations, the accuracy of Eqs. (2) and (3) are within a factor of 2; this is confirmed by visibility–mass measurements in New York and West Coast cities; that is, the coefficient in Eq. (2) will be between 6×10^5 and 1.5×10^5 whereas that of Eq. (3) is between 2.4 and 0.6. The mass–concentration relationship must be verified for each city. Further caveats in using these relations must be noted:

1. Relative humidity must be below 70% (due to volume change in hygroscopic aerosols).
2. Fresh plumes or fumes in the line of sight of the sample should be avoided.
3. Mass measurements should be made at the same point as b_{scat} measurements.

Where Eq. (3) holds, it can be applied directly to air quality criteria, since the reduction of visibility to less than 5 miles results in severe restrictions on the operations of aircraft by Federal Aviation Administration rules. On the average, a particulate concentration of 150 μg/m³ will result in 5-mile visibility, but this may also occur as high as 300 or as low as 75 μg/m³, applying the upper and lower limits of Eq. (3). In terms of SO_2 concentration, this will occur at 0.10 ppm and 50% relative humidity, as the result of acid mist formation.

3.5.4. Weather and Climate

The weather directly over cities is substantially influenced by suspended particulate pollution. The NAPCA particulate criteria[71] states that a typical U.S. urban area, having a geometric mean annual particulate concentration of 100 μg/m³, receives 5% less sunlight for every doubling of this concentration. Ultraviolet is most heavily cut. Other sources estimate losses in polluted cities as one-half of visible radiation and two-thirds of UV. Suspended particles can also act as condensation nuclei to cause rain, but evidence is that cloudiness is more often the result. Smoke control has resulted in a decrease of fog in central London.

The average weather over cities was investigated by Landsberg[73] in 1962 and reviewed by Peterson in 1969.[74] The two studies substantially agree in their conclusions cited below.

The significant climatic changes due to cities found by Landsberg are summarized by Peterson in Table XIX. These effects are largely caused by the emission of Aitken nuclei and other particulates from urban sources and by thermal emission. Condensation and ice nuclei, for example, are formed when lead particles in auto exhaust combine with iodine. Although the city is recognized as a good source of nuclei, their effects on

Table XIX

Climatic Changes Produced by Cities[73,74]

Element	Comparison with rural environs
Temperature	
Annual mean	1.0–1.5°F higher
Winter minima	2.0–3.0°F higher
Relative Humidity	
Annual mean	6% lower
Winter	2% lower
Summer	8% lower
Dust Particles	10 times more
Cloudiness	
Clouds	5–10% more
Fog, winter	100% more
Fog, summer	30% more
Radiation	
Total on horizontal surface	15–20% less
Ultraviolet, winter	30% less
Ultraviolet, summer	5% less
Wind Speed	
Annual mean	20–30% lower
Extreme gusts	10–20% lower
Calms	5–20% more
Precipitation	
Amounts	5–10% more
Days with <0.2 in.	10% more

precipitation are not so certain. Rainfall may increase or decrease. Recent work on the relationship of industrial pollution to rain clouds (in the Niagara Falls–Buffalo area) has been conducted by Cornell Aero Labs (now Calspan).[75] The large number of Aitken nuclei produced by sources in this area are matched by a similar increase in cloud nuclei. The former tend to produce tiny cloud droplets that do not coalesce into rain, while the latter are more conducive to precipitation. The probability is being investigated that stable clouds, rather than rain, are being formed, explaining the 5–10% increase in cloudiness reported by Landsberg.

3.5.5. Odor

Odor is one of several ways in which air pollution affects the senses directly, with the effect of aesthetic degradation rather than disability, so long as the pollution does not reach the point of causing nausea. By definition, odor is an individual physiological response, and the organoleptic method (using the human nose) is the only currently reliable method of measurement. Consequently, there is no objective means of measuring odor other than the use of specially selected and trained individuals, or "odor panels." These persons are exposed to predetermined dilutions of standard substances and to the unknown substance under carefully controlled test conditions, in order to establish the threshold of odor detection and hence the strength of the original sample.[76]

Objective chemical instruments for the indirect measurement of odor have not proved to be of much value to date. Much of this failure can be ascribed to the lack of complete understanding of the psychophysiological mechanism of odor perception. At present this problem and the larger question of damage to the physiological mechanism by air pollutants are being investigated in a number of research laboratories. A complete explanation of olfaction, the sense of smell, would contribute directly to the development objective measurement means.

Heist and Mulvaney, two scientists employed by the Honeywell Corporate Research Center, have conducted research in this basic problem and have reported some of their conclusions in a recent paper.[77] Their investigations indicated, for example, that odor molecules enter directly into the bipolar sensing cell of olfactory tissue, raising the question of permanent damage from air pollutants. A direct quotation from the cited paper indicates the state of the art in this respect:

"The immediate problem concerning us at this time is the relation between air pollution and odor perception. Fortunately our odor sensing cells do not appear in serious danger of being injured by the major air pollutants; however, this is not a certainty at present. Much has been written and

stated about effects of pollutants to our overall health. Certain compounds cause eye irritations, some cause various degrees of coughing, others antagonize emphysematous conditions, etc. Little is actually known, though, about the pollutant effects at the cellular level such as the chemical reaction of polluting compounds with the cells of the bronchial tubes and lungs. Even less is known about the effects on olfactory cells, since little specific research has been conducted on them. . . . *Following destruction of the olfactory tissue by a 1% zinc sulphate solution, almost complete regeneration of the epithelium occurred within one month. This is encouraging in the light of possible cell injury by air pollutants. We doubt very much that injury by pollution would ever be as severe as that caused experimentally by zinc sulphate. Public reaction to the odor–air pollution problems will undoubtedly be more concerned with the nuisance and psychological aspects than with actual effects on the odor sensing cells at this time.*"

3.6. GROWTH OF AIR POLLUTION

The foregoing data seem to demonstrate amply that air pollution in the United States is already a significant problem and requires much more control than has been applied in the past. However, it can be demonstrated that the sources of air pollution will continue to increase and will require more controls for each of the existing sources as well as new ones, in order to maintain even the present levels of air quality.

It can be readily seen that the requirement for air-pollution controls—that is, pollutant emissions without controls or other changes—is directly proportional to the national production. About 50% of the air pollutants in the nation are emitted from transportation, largely automobiles; over 27% from power generation and general industrial plants; and the remainder from space heating, refuse disposal, and miscellaneous sources (1969 data). All of these items increase with the national output simply because they are its components. Barring a large change in the ratio of services to production or in the proportion of control, the amount of emissions will increase with productivity and number of workers. It will be seen that even strict national control of automobile pollution will not greatly change the picture (except in the case of carbon monoxide, which amounts to more than 75% of automotive emissions).

A survey by Edwin Dale for the *New York Times Magazine*[78] shows that U.S. consumption of goods has risen 60% in the past 13 years. He forecasts a 50% increase in GNP over 1969 in the next 10 years, or $500 billion rise (in constant dollars) by 1982. This is an inevitable consequence of two factors; a century-old trend toward increase of productivity, currently forecast at 3%/yr, and a 1% annual increase in the labor force, adding up to a 4% compounded rate of growth. Even a drastic reduction in national workweek standards with its accompanying costs, according to Dale, could not reduce this estimate below 2% annual growth in the next decade.

It is then clear that the degree of control required in the United States will have to be increased by 25-50% over the next 10 years if the air-pollution situation is not to deteriorate badly. It follows that the controls and instruments for prevention and policing of air pollution must grow by that amount over what is now required to keep pollution from growing (not what controls now exist).

Detailed estimates published by NAPCA[79] support this general view. The estimated growth of SO_2 emission is given in Table XX. In the decade 1970-1980 the increase is 105% for power plants and 40% overall. A more recent estimate sees SO_2 emissions tripled in the next 30 years, assuming similar controls, largely from a 300% increase in electric power consumption.

The rate of increase of air pollution is faster, not slower, in other countries that are becoming more industrialized or striving to approach the per capita income of the United States. Japan, for example, already has a severe problem in the Tokyo-Yokohama and Osaka areas. Singapore has had to take stringent air-pollution control measures against automobiles, which are increasing on the island at the rate of 14% annually.[80] The auto increase in Spain is 8% annually, compared to only 3.5% for this country. The Netherlands already has been forced to put a highly developed air-quality control system in operation. A serious air-pollution situation exists in Athens because of automotive and industrial wastes, which will in 10 years require a total of $182 million/yr to control.[81]

Quantitative expressions of the growth of chemical contaminants have been derived by the authors from figures published on the estimated production of olefins and aromatics in 1975 and 1980.[82] The 10-years worldwide increase in olefins production will be 178% (51.4 million metric

Table XX

Potential SO_2 Emission Growth (Millions of Tons Annually)[79]

	Power plants	All sources
1960	8	21
1970	18	37
1980	37	52
1990	43	60

Table XXI

Growth of Organic Chemical Production

	Increase (10^6 metric tons), 1970–1980	% Increase
United States and Canada	16.2	128
Western Europe	19.6	188
Japan	8.1	184
Others	7.5	545

tons), broken down as in Table XXI. Note that the increase in Europe is greater than in the United States, both absolutely and in percent, while the increase in the smaller producers ("others") is four times as rapid.

It can be concluded that the overseas need for air-pollution controls will grow at a rate equal to, or greater than, that of the United States.

3.6.1. Photochemical Smog

Analysis of these growth figures can be carried a step further to show that, although today automobile hydrocarbon and nitrogen oxide emissions are primarily blamed for photochemical oxidant smog, even the effect of full compliance with the latest federal automobile emission standards (as of 1972) will not significantly reduce the total of these pollutants in the atmosphere by 1980 and that both will continue to grow thereafter through 1990. The nationwide emission of hydrocarbons should be temporarily down 4% by 1980, but as nitrogen oxides will nearly double in same period, the net result may be to increase, rather than decrease, the total oxidants. The reason so little benefit is anticipated from automobile controls is the simultaneous growth of organic chemical production in the United States as well as overseas. However, the shift of onus from the auto to the chemical industry implies a movement of oxidant pollution sites toward Eastern and Southwestern communities.

Superficially, it may appear that photochemical production of air-polluting oxidants will cease to be a problem when projected federal automobile exhaust control standards are implemented. Photo-oxidant pollution, it may be recalled, is caused primarily by reactions between nitrogen oxides (NO_x) and reactive (olefinic) hydrocarbons (HC) in the atmosphere and sunlight. Hydrocarbons are largely emitted by gasoline vehicles (47.5%), with smaller percentages from fires and other miscellaneous sources and from industry (see Table XXII). A substantial amount of NO_x is emitted with auto exhaust, but over half of this pollutant comes from stationary combustion sources, accidental fires, and fuel handling.[83]

Automobile-exhaust control programs in the past have been primarily aimed at reducing CO and hydrocarbon emissions. It has been a feature of CO and HC control measures that the increased combustion efficiency they promote tends to favor an increase of NO_x. This is also true of smoke control measures for stationary sources and is suspected to occur as a result of low-sulfur fuel use in power plants. If no specific control is exerted on autmotive NO_x emissions, the changes in total emissions of NO_x and HC through 1990 would be as shown in Figures 3a and 3b.

Recognizing this fact, the EPA, after much debate as to the availability of technology, mandated in August 1972 a 90% reduction in new automobile NO_x emissions for the 1976 model year, for a maximum of 0.4 g/mile. The estimated result of this action (allowing for the increase in numbers of automobiles and the number of old cars left on the road) will be to reduce the concentration of NO_x in urban areas from a factor of 1.16 in 1970 (on a 1967 base) to 0.55 of the 1967 emissions—a reduction of 52% from 1970 to 1980.[85]

Table XXIII shows the resultant estimates of total HC and NO_x emissions for the years 1970 and 1980. The 1980 estimates are explained as follows: Automotive HC emissions are taken from Figure 3b and NO_x is 48% of the 1970 value from Figure 3a.

The increase in industrial organic production for 1970–1980 was stated in Table XXI as 128% for the United States against 185% for Japan and

Table XXII
Nationwide Emissions (1968)[83]

	HC		NO_x	
	Tons × 10⁶	%	Tons × 10⁶	%
Gasoline vehicles	15.2	47.5	6.6	31.9
Stationary fuel combustion (power, heat; coal, oil, etc.)	0.7	2.2	10.0	48.3
Industry sources	4.6	14.4	0.2	1.0
Miscellaneous (fires, handling)	8.5	26.5	1.7	8.2
Other	3.0	9.4	2.2	10.6
Total	32.0	100.0	20.7	100.0

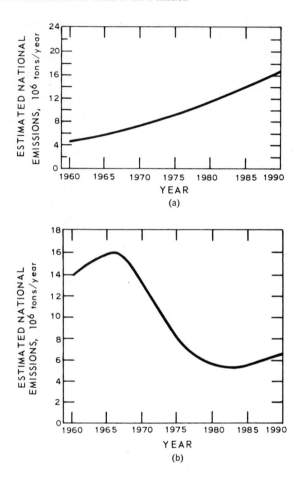

Figure 3 (a) Estimated NO_x emissions from gasoline-powered automobiles and trucks through 1990 if federal standards as of 1971 remain unchanged. Total urban and rural data used. (b) Estimated HC emissions from gasoline-powered automobiles and trucks through 1990 if federal standards as of 1971 remain unchanged. Total urban and rural data used.[84]

Western Europe. Since potential emissions are proportional to the throughput, this results in 11.3 million tons for industrial hydrocarbons in 1980. Other sources of HC, primarily accidental fires and evaporation losses, can be assumed to remain at least the same. For NO_x, automotive emissions are obtained as above. Power source and heating emissions are estimated to increase three-fold by the following reasoning: SO_2 emissions will triple, if uncontrolled, in this period[86] because of the increase in plant capacity. Control of SO_2 will not reduce NO_x emission factors and may even enhance them, thus resulting in at least a factor of 3 increase from this source of NO_X. Other sources (fires, waste) are presumed to remain unchanged.

In summary, total nationwide emission of hydrocarbons to the atmosphere will not decrease more than 4%, even with full application of federal automobile standards, as planned (even complete elimination of all automotive HC emission would leave 77% of the current amounts in the air). On the other hand, NO_x emissions will continue to grow, even with controls on auto exhausts and power plants, so that they will exceed 75% increase by 1980. This implies that photo-oxidant potential will not be significantly reduced on a national basis, and depending on the spatial distribution, the thermodynamics and the reaction rates, photo-oxidants may even increase as a result of the large additional amounts of NO_x.

An interesting corollary to the shift in proportion of auto-exhaust hydrocarbons to the nationwide whole is a concomitant shift in the site of photo-oxidant reactions, possibly from Western to Eastern and Southwestern cities. This would follow from the increased importance of industrial organic and fossil-fuel power sources.

It can then be predicted that the need to monitor photo-oxidant air pollution in the United States will exist at least as much in 1980 as in 1970. After 1980 the growing numbers of automobiles will cause a further

Table XXIII

Estimated Change in Nationwide Photo-oxidant Pollution Precursor Emissions (Millions of Tons)

	1970	1980
Hydrocarbons (HC):		
Automobile[a]	13.0	5.5
Industry	5.0	11.3
Other (power, heating, waste, fires)	12.0	(12.0)
Total	30.0	28.8 (−4%)
NO_x:		
Automobile[a]	7.0	3.3
Power and heating	10.0	(30.0)[b]
Other	4.0	(4.0)
Total	21.0	37.3 (+76%)

[a] Ref. 84.
[b] Ref. 86.

increase in oxidant pollution. The geographical base of oxidant pollution may well shift eastward and southward. In Europe and Asia, oxidant pollution will increase at much higher rates than in the United States because of a much more rapid growth of organic chemical production capacity as well as the automobile population.

3.6.2. Population and Air Pollution Growth

Air-pollution growth and environmental degradation in general have been attributed to many basic causes, including population growth; urbanization or population spatial distribution; economic growth; consumer preferences for such items as automobiles, disposable containers, television, and electrical air-conditioning; heedless use of technology, such as internal combustion engines; and pricing and legislation that provides insufficient economic incentive to recycle or curb emissions.

Ridker[87] has subjected these factors to an elementary analysis, considering the time scale between the immediate policy-planning range (5–10 yr) and the longer time frame (>50 year) in which population growth could overwhelm finite earth resources, subject to the postponement of the consequences of biological laws offered by technology.

The primary simplifications introduced in his basic analysis are a one-dimensional view of parameters and variables that are truly multielement vectors, such as "waste" or "pollution concentration," and an assumption of linear relationships between variables if not othewise designated, at least for a limited range.

With this assumption we can write, following Ridker,

$$\text{Waste:} \quad W = AQ \tag{4a}$$

$$= A(gN) \tag{4b}$$

where W represents the residuals of economic processes that are not recycled; Q is the total national output, or the GNP, to a sufficient approximation; g is the per capita output; N is the population; and A is a positive constant. Then the "pollution", P, in tons per year, is

$$P = BW \tag{5}$$

where B is the fraction of waste causing effluent pollution and is an inverse function of the capital invested in pollution control.

The average concentration of pollution, C, is an important measure of potential damage and depends on the volume, V, of the medium in which it is dispersed; the initial concentration, Co; and on r, the rate of pollution decay and absorption (by natural sinks):

$$C = Co - rCo + \frac{P}{V} \quad \text{(approximately).} \tag{6}$$

The per capita damage, d, is assumed to be a function of the concentration. As we have discussed earlier in this chapter, the nature of these functions is not well known in the majority of cases, but within a limited range, the proportionality constant can be considered an increasing function of the concentration (rather than simply proportional or exhibiting a threshold or synergistic relationship between pollutants); hence,

$$d = \delta C \tag{7a}$$

and

$$\delta = f(C). \tag{7b}$$

Total damage to the entire population of receptors (enclosed in the volume V) is

$$D = \delta C N. \tag{7c}$$

Substituting Eqs. (4), (5), and (6) into Eqs. (7a) and (7b), the per capita damage function is

$$d = \left(\delta \frac{AB}{V}\right) gN + (1 - r) Co \tag{8a}$$

and the total damage

$$D = \left(\delta \frac{AB}{V}\right) gN^2 + (1 - r) GN. \tag{8b}$$

The conclusion is that per capita output, g, and population, N, are equally significant factors in determining per capita damage, given that the parameters (A, B, V, δ) remain constant. Total damages, however, increase much more rapidly than the population.

Ridker's caveat for taking this abstract analysis too literally is that the parameters that were assumed constant are equally as likely to change as the major variable—population. A few examples will suffice. Urbanization has decreased the value of parameter A in this century because a concentrated population permits greater production without commensurate expansion of transportion (which produces much material waste). But the parameter V has been reduced by the reason of the same urbanization. In future years, will suburban growth have opposite effects? Similarly, the composition of consumption has altered and will change again in the future as consumer tastes, incomes, and household size and ages change. The consumption of electricity, he points out, generates 15 times more sulfur compounds and twice as many particulates as that of bread, on a dollar for dollar basis.

Other changes in the parameters will accompany changes in capitalization, including that for waste treatment; demographical changes due to decreased birthrate; improved technology in the use of materials; and the effect of government policy on prices and other incentives. Consequently, the entire subject is open to continued analysis.

REFERENCES

1. G. D. Robinson. *Long-Term Effects of Air Pollution—A Survey*. Center for the Environment and Man: Hartford, Conn., 1970.
2. "Global Environmental Monitoring." *MIT Technol. Rev.*, **73**, 19–25 (1971).
3. A. D. Watt. *IEEE Spectrum*, **8**, 59–72 (November 1971).
4. *Threshold Limit Values for Chemical Substances in Workroom Air*. American Conference of Government Industrial Hygienists: Cincinnati, Ohio (1973).
5. F. Möller. *J. Geophys. Res.* **68** (1963).
6. E. W. Hewson. In R. E. Machol, Ed., *System Engineering Handbook*, McGraw-Hill: New York, 1965, Table 5-1.
7. D. J. Portman et al. In Machol, Ref. 6. See also C. Sagan and G. Mullen. *Science*, **177**, 52–55 (1972), for an excellent review of the entire greenhouse effect on the earth.
8. W. W. Kellog et al. "The Sulfur Cycle." *Science*, **175**, 587–596 (1972).
9. Massachusetts Institute of Technology. *Man's Impact on the Global Environment* (*SCEP*). MIT Press: Cambridge, Mass., 1970.
10. D. L. Slotnick. *Sci. Amer.*, **224**, 76–87 (February 1971).
11. W. S. Broecker. *Science*, **168**, 1537–1538 (1970).
12. NAPCA. *Air Quality Criteria for Photochemical Oxidants*. Government Printing Office: Washington, D.C., 1970, pp. 2–11, 12.
13. *Philadelphia Evening Bulletin*, June 15, 1972.
14. The weight of the atmosphere is 14.7 lb/in^2 or 10,300 kg/m^2. The area of the earth is approximately 200×10^6 miles^2 or 520×10^{12} m^2; hence, the atmospheric mass is 5400×10^{15} kg or 6×10^{15} tons.
15. EPA. *Annual Report to Congress*. March 1972, p. 3.
16. *New York Times*, November 19, 1971.
17. Ref. 16, May 7, 1971.
18. N.J. State Department of Environmental Protection. *Environmental Times*, April 1972.
19. H. E. Stokinger and D. L. Coffin. In A. C. Stern, Ed., *Air Pollution*, Vol. I, 2nd ed. Academic: New York, 1968, p. 509.
20. NAPCA. *Air Quality Criteria for Sulfur Oxides*. Government Printing Office: Washington, D.C., 1970, p. 119–120.
21. Ref. 20, pp. 120–123.
22. *Technical Progress Report, Air Quality of Los Angeles County*, Vol 11. Los Angeles County Air Pollution Control District. (1961).

23. U.S. Bureau of the Census. *Statistical Abstracts of the United States: 1971 (92nd ed.)*. Washington, D.C., 1971.
24. F. A. Patty, Ed. *Industrial Hygiene and Toxicology*, Vol. II. Wiley-Interscience: New York, 1949 p. 621.
25. NAPCA. *Air Quality Criteria for Carbon Monoxide*. Government Printing Office: Washington, D.C., 1970, pp. 8–10.
26. Ref. 25, pp. 9–18.
27. Ref. 26, pp. 8–52.
28. A. D. Rossin and J. J. Roberts. *J. Air Poll. Control Assn.*, **22**, 254–259 (1972).
29. E. J. Cassell. *IEEE Trans. Geosci. Electron.*, **GE-7**, 220–226 (1969).
30. Ref. 16, June 7, 1970.
31. S. K. Hall. *Environ. Sci. Tech.*, **6**, 31–35 (1972).
32. R. L. Bazell. *Science*, **174**, 574–576 (1971).
33. EPA. *Health Hazards of Lead*, Rev. ed. Research Triangle Park, N.C., 1972.
34. HEW. *Danger in the Air*. Washington, D.C., 1970.
35. National Research Council–National Academy of Sciences. *Airborne Lead in Perspective*. Committee on Biological Effects of Atmospheric Pollutants of the Division on Medical Sciences: Washington, D.C. 1971.
36. *Code of Federal Regulations*, Chapter 1, Title 80; and *Federal Register*, **38**, 26449 (September 21, 1973).
37. *J. Air Poll. Control Assn.*, **22**, 303–305 (1972).
38. G. F. Hueter (Office of Criteria Development, National Environmental Research Center, EPA). American Management Association Meeting, Chicago, April 19, 1972.
39. EPA. *Background Information—Proposed National Emission Standards for Hazardous Air Pollutants: Asbestos, Beryllium, Mercury*. Research Triangle Park, N.C. 1971.
40. B. D. Dinman. *Science*, **175**, 495–497 (1972).
41. J. Monod. *Chance and Necessity*. Knopf: New York, 1971.
42. CEQ. *Third Annual Report*. Government Printing Office: Washington, D.C., 1972.
43. R. G. Ridker. *Economic Costs of Air Pollution*. Praeger: New York, 1967.
44. R. G. Ridker. *Science*, **176**, 1085–1090 (1972).
45. L. B. Barrett and T. E. Waddell. "The Cost of Air Pollution Damages: A Status Report" (April, 1971). In *Cumulative Regulatory Effects on the Cost of Automotive Transportation (RECAT)*. Office of Science and Technology: Washington, D.C., 1972.
46. EPA. *The Economics of Clean Air*. Government Printing Office: Washington, D.C., March 1972.

46. EPA. *The Economics of Clean Air.* Government Printing Office: Washington, D.C., March 1972.
47. J. J. O'Conner. *The Economic Cost of the Smoke Nuisance to Pittsburgh.* Smoke Investigation Bull. No. 4. Mellon Institute: Pittsburgh (1913).
48. I. Michelson and B. Tourin. *Public Health Reports,* **81,** 505–511 (1966).
49. Ref. 45, p. 1-J49.
50. Ref. 49.
51. L. B. Lave and E. P. Seskin. *Science,* **169,** 723–733 (1970).
52. D. B. Rice. *Estimating the Cost of Illness.* Public Health Service: Washington, D.C. Publication No. 947-6. 1966.
53. F. H. Haynie, unpublished, cited in Ref. 45, p. 1-J18.
54. Battelle Memorial Institute. *A Survey and Economic Assessment of the Effects of Air Pollution on Elastomers,* Final Report, EPA Contract CPA 22-69-146 (June 1970).
55. V. S. Salvin, cited in Ref. 45, p. 1-J19.
56. V. S. Salvin. *Survey and Economic Assessment of the Effects of Air Pollution on Textile Fibers and Dyes,* Final Report, EPA Contract PH-22-68-2 (June 1970).
57. J. S. Jacobson and A. C. Hill. *Recognition of Air Pollution Injury to Vegetation: A Pictorial Atlas,* Air Pollution Control Assn.: Pittsburgh, 1970.
58. N. L. Lacasse and T. C. Weidensaul. *J. Air Poll. Control Assn.,* **22,** 112–114 (1972).
59. Ref. 45, pp. 1-J 24–25.
60. A. A. Millecan. *A Survey and Assessment of Air Pollution Damage To California Vegetation in 1970.* Rept. APTD-0694. California State Department of Agriculture: Sacramento, 1971.
61. NAPCA. *Air Quality Criteria for Sulfur Oxides,* Publication No. AP-50. Government Printing Office: Washington, D.C., 1969, pp. 24–25.
62. Ridker, Ref. 43, p. 138.
63. H. Beaver. *Committee of Air Pollution Report.* Great Britain, 1954.
64. Ref. 48.
65. Booz-Allen and Hamilton, Inc. *Survey to Determine Residential Soiling Costs of Particulate Air Pollution,* Final Report, EPA, Contract CPA-22-69-103 (June 1970).
66. Ref. 43, p. 82.
67. Ref. 45, pp. 1-J 34–6.
68. J. Schusky. *J. Air Poll. Contr. Assn.,* **16,** 17–76 (1966).
69. A. Wolff. *Saturday Review,* June 3, 1972, pp. 29–34.
70. W. E. K. Middleton. *Vision Through the Atmosphere,* Univ. of Toronto Press: Toronto, 1952.

71. NAPCA. *Air Quality Criteria for Particulate Matter*, Publication AP-49. Government Printing Office: Washington, D.C., 1969, pp. 56–59.
72. R. J. Charlson et. al. *Atmospheric Environment,* **1,** 469–478 (1967).
73. H. E. Landsberg. "City Air—Better or Worse." In symposium *Air Over Cities,* USPH, Taft Sanitary Eng. Center, Cincinnati, Ohio. Tech. Report A62-5, 1962, pp. 1–22.
74. J. T. Peterson. *The Climate of Cities.* Public Health Service, NAPCA: Raleigh, N.C., 1969.
75. *Research Trends,* Summer, 1971 Cornell Aeronautical Laboratory, Inc., (Calspan) Buffalo, N.Y., pp. 34–35.
76. J. F. Bird and A. H. Phelps, Jr. "Odor and Its Measurement." In A. C. Stern, Ed., *Air Pollution,* Vol. II. Academic: New York, 1968, pp. 305–327.
77. H. E. Heist and B. D. Mulvaney. *Clean Air,* **4** (1970).
78. E. Dale. *New York Times Magazine,* April 19, 1970, pp. 27–47.
79. NAPCA. *Control Techniques for Sulfur Oxide Air Pollutants,* Publ. AP-52. Government Printing Office: Washington, D.C., 1969, p. 3.
80. Ref. 16, June 21, 1970.
81. Ref. 16, June 26, 1970.
82. *Chem. Eng. News,* June 8, 1970, pp. 24–25. See also: March 8, 1971, p. 15.
83. NAPCA. *Control Techniques for CO, NO_x, HC from Mobile Sources,* Publ. AP-66. Government Printing Office: Washington, D.C., 1970, pp. 2–15.
84. Ref. 83, pp. 2–16.
85. EPA. "Regulations for State Implementation Plans." *Federal Register,* August 14, 1971.
86. NAPCA. *Control Techniques for SO_2 Air Pollutants,* Publ. AP-52. Government Printing Office: Washington, D.C., 1970, p. 3 (Figure 3-2).
87. Ref. 44.

4

Legislative Basis of Air-Pollution Control Systems

Air-quality and antipollution laws are the most immediate factors influencing the design of control systems, since they represent (1) government funding for researching and enforcement; (2) penalties and coercion, counterbalancing the costs of control to industries and individuals; and (3) a codification of standards and goals. But it has been difficult to express the precise requirements of this legislation in the past because it has been going through a process of evolutionary change as a response to the growing public awareness of the deteriorating environment and because of the diversity of local regulations and authorities that have been allowed to proliferate in the United States.

During its evolution, legislation at the federal and local levels has been influenced by three major concepts: (1) local (state) versus central (federal) responsibility for control, (2) homogeneous air basins as air-control regions, and (3) the control of emissions at the source versus control based on ambient measurements.

Historically, air-pollution control has been considered a local problem, no doubt because local effects (such as sootfall and smoke-plume visibility) were the first to be perceived by the public. Only as the more widespread and subtle effects have been observed—the result of better data-gathering and knowledge of what to look for as well as the growing intensity of pollution—have the larger geographic and political entities become involved. This situation contrasts sharply with the evolution of water pollution laws, where federal entities, concerned with navigation in waterways and harbors, have been involved in pollution control since the nineteenth century.

The air basin concept has grown up in analogy to the water basin—an enclosed volume in which entering streams and pollution outfalls are mixed to a homogeneous quality. It would seem reasonable to extend local control action to a geographical and topological entity, such as a valley, where the effect of weather and pollution emissions should be reasonably uniform. When town or village ordinances provided inadequate, it seemed a logical level at which to establish state or federal control. But it is technically difficult to describe and delineate such basins in the light of the state of meteorology and diffusion theory. This, together with the time and cost to gather and analyze data on a nationwide basis, has forced federal and state legislation to retreat at least temporarily, from this rather elegant concept.

The debate between proponents of source control versus control through ambient measurements is one of long standing, with no clear decision as yet. Legislation based on source control has the apparent benefits of fairness and simplicity. It can be applied to all sources uniformly and so avoid the danger of "pollution havens" for industries seeking to avoid control. But control standards based on emissions do not touch the heart of the matter, which is the quality of the ambient air in which we live and breathe. By controlling emissions at an inappropriate place in the chain of events, emission controls expose us to the danger that they will be inadequate at times when air basins are not naturally ventilated and will at other times be too costly and will overcontrol, since they do not take full advantage of the natural cleansing processes—sinks and ventilation—which effectively remove most of the pollution most of the time.

In order of their significance to the air-pollution control-system designer, this section will discuss federal laws, state and local laws, and foreign and international control.

4.1. EVOLUTION OF FEDERAL LEGISLATION

We have seen that local air-pollution ordinances providing fines for smoke emission were initiated in the United States in Chicago during the year 1881, followed by similar laws in Pittsburgh, Cincinnati, Cleveland, and St. Paul. The validity of these ordinances was first tested and affirmed in the federal courts on a test appeal in 1915. Thereafter, smoke ordinances on a local level proliferated.

By 1947 the new menace of photochemical smog resulting from automobile traffic had its first legislative effect in the creation of the Los Angeles Air Pollution Control District (LAAPCD), with the authority to promulgate and enforce regulations to establish an increased level of control over the entire air basin. In response to the seriousness of the effects of smog pollution in this area, LAAPCD regulations have been precursors to federal laws and have anticipated them in stringency of control, especially in control of automotive emissions.

The first federal act was not passed until July 1955, when Congress

4.1. Evolution of Federal Legislation

enacted Public Law PL84-159, providing for research by HEW at the rate of $5 million annually for five years. The Air Pollution Prevention and Control Act (Title 42, Section 1857, U.S. Code) is the basic law now on the books, to which admendments have been added up to the present time. In its original form, however, the act did little but provide research funds to aid states and local governments in the actual task of pollution control.

In 1959 and again in 1962, funding was added in two acts of Congress (PL86-356 and PL87-761), keeping the law active until 1965. But by 1963 the first major piece of federal control legislation was passed, the Clean Air Act (PL88-206). Although it affirmed the previously stated priciple that air pollution was a state and local problem, it recognized the right of the federal government to enforce control measures in interstate situations. Congress noted the growth of air pollution emitted from industrial and automotive sources but, with the exception noted, did not infringe on the state and local responsibilities beyond offering help and encouragement. This was provided in the 1963 approach in the form of funds to aid operating state and local enforcement agencies. Federal intervention in interstate pollution disputes was limited to an HEW survey, hearings, and federal court action if the compliance deadlines so established were not met.

However, the principle of federal control was now established and continued to grow stronger. In 1965, PL89-272, for the first time, established federal authority to regulate emissions from new motor vehicles and engines. A National Air Pollution Control Adminstration (NAPCA) was established within HEW.

The national standards for 1968 automobile pollution emissions were the same as those previously set by the city of Los Angeles in 1963. In 1968, new federal standards were issued to apply to 1970 models.

In 1966, by executive order, the president established policy for the control of air pollution originating from federal installations. These agencies were required to plan for the prevention of pollution from existing and future installations, conforming to the more stringent local or HEW standards. Regulations for visible and particulate emissions and for SO_2 were issued.

In 1967 the Air Quality Act (PL90-148) was passed. It recognized the inadequacy of past efforts and established a timetable for the states to establish air-quality standards on a regional basis. The air quality control region concept contemplates an area with a homogeneous air resource and uniform pollution problems, such as the metropolitan area of New York–New Jersey or Philadelphia–Delaware Valley. It can be inter- or intrastate. PL90-148 also required states to control emissions for which NAPCA issued criteria.

The criteria are comprehensive reports that reflect "the latest scientific knowledge useful in indicating the kind and extent of all identifiable effects on health and welfare which may be expected from an air pollutant agent." In addition to naming the pollutants and publishing criteria, HEW also published recommended pollution-control techniques necessary to reach the levels set forth in the criteria. Comparative cost and current feasibility are also included. It is important to note that criteria are descriptive; that is, they describe the effects at various pollutant levels. They were not intended to be legally enforceable standards that would be prescriptive for a given geographical area and political jurisdiction.

It is of interest to review the provisions of the 1967 act, not only because it is basic to the current legislation (1970), which amends rather than supplants it, but to note the fragmented authority and lack of decisiveness, which inhibited the development of an effective nationwide control system. The 1967 federal law had the following effects:

1. HEW supported national research and development, training, and demonstration activities, giving grants and contracts to many kinds of individuals and organizations (including private industry). In particular, fuel evaporation and combustion processes were studied.
2. HEW was authorized to hold a conference on any *potential* air-pollution problem. These hearings were only advisory to the local agency but became an official record for future abatement proceedings.
3. HEW made grants-in-aid to assist state or local pollution-control agencies, up to two-thirds of the cost of planning, development, and improvement and half that of maintenance. Multimunicipal (regional) agencies could obtain up to three-fourths the cost of agency establishment and three-fifths of maintenance costs.
4. Federal enforcement authority could be invoked against any air pollution that "endangers" health or welfare. Proof of actual injury was not needed. Federal authority is intended to be operational in interstate cases and is subject to state and local action within any state's boundaries. Enforcement is through two approaches: by conference, or by application of regional air-quality standards.

4.1.1. Federal Enforcement by Conference

The conference-hearing approach could be initiated by the Secretary of HEW or a state agency, if interstate, but must be requested by state and affected municipality if intrastate. The steps involved hearings, reports, and recommendations to all public and private parties involved, and ultimate suit by the U.S. attorney general if necessary.

Of particular interest was the authorization of the conference technique to control air pollution emitting from this country and endangering foreign persons. This, however, must be based on reciprocal arrangements with the country involved. Thus, the concept of international air-pollution control was broached.

The conference method of enforcement, up to 1968, was applied to 9 interstate cases involving 13 states and the District of Columbia.

4.1.2. Regional Air-Quality Standards

This technique was the best publicized air-pollution activity of HEW and involved well-defined steps:

1. HEW designated air-quality control regions. These are based on the concept of a homogenous air shed (or air basin) centered about a major urban area.
2. HEW published the air-quality criteria and control-technique documents mentioned above, based on research and evaluation of all available knowledge and technology.
3. Following publication of these documents and based on them, the states were required to indicate within 90 days their intention and within another 180 days to adopt, after public hearings, standards for air quality within each designated region. Within still another 180 days, the states were required to adopt a detailed plan for enforcing and maintaining these standards. The intent was for the state to develop detailed procedures, after considering the particular meteorology of its region, its current pollutant level, number of sources, and other local factors in addition to the generalized criteria and technique recommendations published by HEW.
4. HEW then reviewed the state's plans, and if they were consistent with the Air Quality Act and its documentation, they were approved and became, in effect, the federal standard for that region as well as that of the state.
5. If the state did not take the prescribed action or if its standards and plans were found inconsistent within the act, the Secretary could promulgate his own standards in 6 months, unless the state has complied meanwhile or called for a hearing.
6. If the Secretary found that the air-quality standard established by either of these methods was not being enforced by the state, he could initiate further action via the federal courts.

These steps could not be implemented until HEW designated the air-quality control regions. By April 1970, 57 control regions were proposed, but only 28 had been designated. Considerable pressure was being exerted for early completion of the regional definition task, since without it, none of the rest of the control program could proceed, but it was not finished until the law was amended.

The other prerequisite for control authority, compilation of the criteria documents, was completed somewhat more expeditiously. In January 1969, two *Air Quality Criteria* were issed (AP-49, *Particulate Matter,* and AP-50, *Sulfur Oxides*) along with their companion *Control Techniques* reports. By March 1970, five of the six criteria required for the air pollutants currently the subject of ambient standards were issued. The additional documents were *Carbon Monoxide,* AP-62 (March 1970); *Photochemical Oxidants,* AP-63 (March 1970); and *Hydrocarbons,* AP-64 (March 1970). The remaining criteria document on *Nitrogen Oxides* (AP-84) was not completed until January 1971, under the amended law and successor agency (APCO). NAPCA had scheduled 14 more criteria through 1975, in addition to NO_x. None of these were formally issued by NAPCA or APCO as of this writing, but an equivalent document has been published for lead, and three others (asbestos, beryllium, and mercury) have come under the control of new legislation. NAPCA did publish four other *Control Technique* documents in March 1970, these are *Control Techniques for Carbon Monoxide Emissions from Stationary Sources* (AP-65); *Control Techniques for Carbon Monoxide, Nitrogen Dioxide, and Hydrocarbon Emissions from Mobile Sources* (AP-66); *Control Techniques for Nitrogen Oxides from Stationary Sources* (AP-67); and *Control Techniques for Hydrocarbons and Organic Solvent Emissions from Stationary Sources* (AP-68).

4.1.3. Motor-Vehicle Air-Pollution Control

Direct federal control of automotive-engine pollution standards is an exception to the above procedures. Under the 1965 amendment to the Clean Air Act, HEW issued standards for exhaust and evaporative emissions. They apply to all new motor vehicles and engines introduced into interstate commerce, including those imported. Prototype engines or vehicles are tested and a type certificate is issued for the same model of that year. Violators may be fined up to $1000 per vehicle. Control of old vehicles is left to the states or localities. They are preempted from regulations pertaining to new vehicle emissions, except for California, which had 1969 and 1970 standards more stringent than the federal standards. HEW was also authorized to register all fuel additives and to require information on their purpose, their chemical composition, and related data. A federal automobile laboratory was operated in Ypsilanti, Michigan, and a new building constructed in Ann Arbor to implement these regulations.

4.1.4. Inadequacies

As suggested above, the federal control system under the 1967 law was not able to respond adequately to the worsening air pollution and to the pressure of various interests—conservationists versus the power industry, for example. One reason was suggested by Arthur C. Stern[1] of North Carolina University, formerly of NAPCA, who stated that the air-quality control region concept is a "snare and a delusion," since regions, particularly interstate regions, are not viable administrative bodies. More important, the region designation proceeded too slowly; he pointed out that none had been designated in North Carolina in three years under the 1967 act. Stern believes that the states should be the controlling factors in administrating research, training, planning, subsidy, and tax relief. From a technical view, moreover, the region concept is based on the idea of planning for the meteorology of a particular geographical location, the air shed. But it is believed that dangerous air-pollution episodes do not occur during normal weather variations but result from rare or unpredictable conditions, such as a stable mass or fog lasting several days. These combinations may occur once or twice per decade, and their statistics are too poorly defined to serve as a basis for planning and to correlate with a topographical entity or air shed.

While Stern advocated state control, other environmental authorities declared that "the 50 states are even less capable of carrying out major policies than the federal government." But a need for greater centralized control than the 1967 Act was apparent.

At one point, a strong argument was pressed to centralize all environmental control in the Department of the Interior. Two presidential commissions opposed this view in favor of an entirely new agency that would take over authority from both Interior and HEW. The Council of Executive Organization recommended a single cabinet-level Environmental Protection Agency (EPA), which would oversee radiation-, water-, and air-pollution standards and aircraft, automotive, and interstate agricultural wastes and would acquire the existing programs from HEW and Interior. The Presidential Council on Environmental Quality (CEQ), reporting to the cabinet, concurred in the view.

4.2. THE CLEAN AIR AMENDMENTS OF 1970

In consequence of the rather general dissatisfaction with the Air Quality Act of 1967 on the part of both executive and legislative branches of the U.S. government, as well as the time lag experienced in its implementation, a new set of amendments, the Clean Air Act of 1970, were enacted and signed into law in December 1970.[2] This law proved to be a rather complete reversal of the principles of the prior legislation, although some of the structure of the latter remains.

In the same period, the air-pollution control authority was given to the new executive agency, EPA, taking over the personnel and tasks formerly assigned to NAPCA and the Public Health Service in HEW.

The chief provisions and effects of the new act are as follows:

1. Nationwide, uniform air-quality standards are set by the federal government rather than by each state (guided by the *Criteria*), as under the earlier law.
2. All air-quality control regions under consideration (approximately 100) were established immediately and any residual geographic areas in a state that had not already been so constituted by virtue of the air shed concept were also designated as control regions (the right to redesignate these regions after further study is reserved.)
3. It became immediately compulsory for each of the states to establish "implementation plans" to conform to the federal air-quality goals (ambient standards) by a fixed date. The states still maintained the right—in fact, were required—to establish their own methods of reaching the goals, so long as they could show how they would succeed. In practice, the time period allowed to submit their plan was so short that the federal guidelines issued as "Regulations for State Implementation Plans"[3] were closely followed. In cases where the state control plan was deemed inadequate, the EPA administrator had the right to modify it or to substitute his own plan, thus assuming a certain degree of uniformity (55 plans were submitted by states and territories; 14 were fully approved and the remained were supplemented by the EPA administrative).
4. Federal control of automobile emissions by annual certification of new engine types was continued by the new law.
5. The authority to control vehicle fuel additives (such as lead and phosphorus in gasoline) that may be harmful to health was also included.
6. The law provides for two classes of national emission standards. The first is for the emissions of new stationary sources categorized by the EPA as contributing significantly to air pollution. The second is for "hazardous air pollutants," defined as those for which no ambient air-quality standard is applicable and that may contribute to mortality or irreversible illness.

4.2. The Clean Air Amendments of 1970

In addition to the above, other sections of the act are significant and worth noting. These include grants or funds for research, training, planning, and control (Sect. 103-5); provision for federal enforcement if the state plan fails (Sect. 113); federal right of inspection, monitoring, and entry into stationary sources and the public availability of emissions records so obtained, except where they would reveal trade secrets (Sect. 114); federal emergency powers to immediately restrain the cause of any air-pollution emissions that may be hazardous to health (Sect. 303); right of citizen suits, including those against the government or its agencies, if standards or limitations under the act are believed to be violated (Sect. 304); and a requirement for continuing, comprehensive economic studies of the cost of controlling pollution to the government, industry, communities, and others (Sect. 312).

4.2.1. Air-Quality Standards

Those substances for which the Air Pollution Control Office (APCO) of the EPA or its predecessor, HEW's NAPCA, had already issued criteria were the air pollutants for which the initial federal ambient standards were set. These are sulfur oxides, suspended particulate matter, carbon monoxide, photochemical oxidants, hydrocarbons, and nitrogen oxides.

Ambient air-quality standards are designated as "primary" or "secondary," depending on whether they are health hazards or have any other adverse effects. These standards were promulgated by EPA[4] on April 30, 1971, and are listed (as revised) in Table I.

Nine months after this action, the states were required to submit a plan for approval by EPA that provides for the implementation, maintenance, and enforcement of the primary standards in each control region (an additional 18 months were allowed for submission of a plan for secondary standards). The plan was required to provide for attainment of primary standards within 3 yr after EPA approval (1975), with a 2-yr extension to meet primary standards if it is shown that necessary technology will not be available in time. Secondary standards must be met "within a reasonable time."

4.2.2. State Implementation Plans

The states are required to do the following:

1. Classify each control region by priority category, depending on the degree of air pollution.
2. Submit their implementation plan to the EPA for approval, showing how they will meet national standards.

Table I
National Ambient Air-Quality Standards[a]

	Primary standards		Secondary standards	
	$\mu g/m^3$	ppm[c]	$\mu g/m^3$	ppm[c]
Sulfur oxides (SO_2):				
Annual arithmetic mean	80	0.03	—[d]	—
Max. 24-h concentration[b]	365	0.14	—[d]	—
Max. 3-h concentration[b]	—	—	1300	0.5
Particulate matter:				
Annual geom. mean	75	—	60	—
Max. 24-h concentration[b]	260	—	150	—
Carbon monoxide:				
Max. 8-h concentration[b]	10,000	9.0	Same as primary	
Max. 1-h concentration[b]	40,000	35.0	Same as primary	
Photochemical oxidants (corrected for NO_x and SO_2):				
Max. 1-h concentration[b]	160	0.08	Same as primary	
Hydrocarbons:				
Max. 3-h concentration 6–9 A.M.[b]	160	0.24	Same as primary	
Nitrogen dioxide:				
Annual arithmetic mean	100	0.05	Same as primary	

[a] *Federal Register,* **38,** 11355–11356 (May 7, 1973).
[b] Not to be exceeded more than once annually.
[c] By volume, at 25°C, 760 mm.
[d] Standard eliminated. Ref. *Federal Register,* **38,** 25678–25881 (September 14, 1973).

3. Show in the plan, legal authority to adopt emission standards; enforce the laws and standards; enforce abatement in emergencies; regulate new sources or modifications of old sources; gather data on sources to determine compliance with laws, including right to inspect and test; require stationary sources to install emission monitors and submit reports; and show plans for testing of motor vehicles where required by the control strategy.
4. Develop a control strategy for each pollutant that will comply, as required, with federal examples of emission limits obtainable with "reasonably available technology," for particulate emissions, sulfur compounds, organic vapors, CO, and NO_x.
5. To prevent air-pollution emergency episodes, each Priority I region (see below) shall provide a plan for emission control to prevent ambient concentrations from reaching levels of "imminent danger to health," as specified by EPA.

6. Provide for ambient air-monitoring stations in at least the quantity shown by subsequent tables and governed by the classification of the control region.

Classification of Control Regions

States are required to classify each control region within their borders according to the quality of air with regard to each of several pollutants. For sulfur oxides and particulate matter the categories I, II, or III are established by the concentrations shown in Table II. Where there is a difference between maximum and average concentration, the more restrictive (lowest) classification is chosen.

Note that because of questions regarding the adequacy and accuracy of the methods used to measure nitrogen dioxide (the Jacobs–Hochheiser method), the classification of regions based on this pollutant was deferred until after July 1973.[5] Reclassification and measurements based on other methods—arsenite, chemiluminescence, and Saltzman colorimetric—are given in Ref. 6.

Table II
Air-Quality Control-Region Priority Classifications

Pollutant	Priority class		
	I[b]	II	III
Sulfur oxides:			
Annual arithmetic mean[a]	>100 (0.04)	60–100 (0.02–0.04)	<60 (0.02)
24-h maximum	>455 (0.17)	260–455 (0.10–0.17)	<260 (0.10)
3-h maximum		≥1300 (0.50)	<1300 (0.50)
Particulate matter:			
Annual geometric mean	>95	60–95	<60–95
24-h maximum	>325	150–325	<150
Carbon monoxide:			
1-h maximum	≥55000 (48)		<55000 (48)
8-h maximum	≥14000 (12)		<14000 (48)
Nitrogen dioxide:			
Annual arithmetic mean	≥110 (0.06)		<110 (0.06)
Photochemical oxidant			
1-h maximum	≥195 (0.10)		<195 (0.10)
Hydrocarbons	Same as photochemical oxidants		

[a] All concentrations in $\mu g/m^3$. Number in parentheses is equivalent ppm (by volume).
[b] Classified 1A if air quality predominantly from a single-point source.

Where pollution-concentration-measured data is not adequate, estimating procedures based on simple mathematical models for both point and area sources are given. Models will be discussed in Chapter 8. In the case of carbon monoxide (CO), nitrogen dioxide (NO_2), and photochemical oxidants, if measured data does not exist, the priority is based on population. Each urban region over 200,000 is automatically classified Priority I.

Table III

Minimum Monitoring Requirements

Region class	Pollutant	Measurement method	Sampling frequency	Minimum number of sites
I	Suspended particulate	Hi Vol or Tape sampler	—[a] —[b]	4–24[c] 1–8[d]
	SO_2	Pararosaniline or equivalent (Table IV)	—[a] Continuous	3–14[e] 1–9[e]
	CO	NDIR or equivalent	Continuous	1–9[e]
	Photo-oxidant	Chemiluminescent or equivalent	Continuous	1–9[e]
	NO_2	Jacobs–Hochheiser[h]	—[f]	1–10[g]
II	Suspended particulate	Hi Vol or Tape sampler	—[a] —[b]	3 1
	SO_2	Pararosaniline or equivalent	—[a] Continuous	3 1
III	Suspended particulate	Hi Vol	—[a]	1
	SO_2	Pararosaniline	—[a]	1

[a] One 24-h sample/6 days.
[b] One sample/2 h.
[c] Add 0.16/100,000 population above 5 million.
[d] Proportional to population up to 2 million.
[e] Add 0.05/100,000 population above 5 million.
[f] One 24-h sample every 14 days.
[g] Proportional to population to 1 million.
[h] Superseded; see text.

Table IV

Methods for Air Quality Monitoring

SO_2 Equivalent	Pararosaniline (reference)
	GC^a + FPD^b (using Teflon)
	FPD after removal of interfering sulfur compounds.
	Coulometric after removal of O_3, NO_2, H_2S.
	Automated pararosaniline
Suspended particulates	Hi Vol samples: 24-h sample/6 days
	Tape sampler: 1 sample/2 h
CO Equivalent	$NDIR^c$ (reference)
	GC + cat. conversion + FID
Photo-oxidants Equivalent	Gas chemiluminescence (reference)
	KI colorometric, corrected for SO_2 + NO_2
	UV photometric determination of ozone (corrected)
	Other chemiluminescent
NO_2	See note (h), Table III
	One 24-h sample/14 days

[a] GC = gas chromatograph.
[b] FPD = flame photometric detector (hydrogen flame).
[c] NDIR = Nondispersive infrared analyzer.

Ambient Air Monitoring

Ambient air-pollution measuring stations are necessary to monitor the quality of the control regions and determine their status with regard to the standards, as well as to activate the episode warning and emission control system. The number of stations required for regions of each priority class is shown in Table III.

All of these stations must be in operation within two years after approval of the plan. Those stations required to give episode warning alerts must be in operation within one year. At least one station must be in the area of greatest estimated pollution.

Table IV lists approved methods of air-quality monitoring, including acceptable equivalents.

Table V lists the specifications for monitoring instruments.

Source Emission Control

For new or modified sources, the state shall require submission of the nature and amounts emissions, locations, and other data so that it can be

Table V

Specifications for Monitoring Instruments

	SO_2	CO	Photo-chemical oxidant[a]	NO_2	HC[b]
Range (ppm)	0–1.0	0–50.	0–0.5	0–1.0	0–5.0
Sensitivity (ppm)	0.01	0.5	0.01	0.01	0.20
Rise time, 90% (min)	5	5	5	5	5
Drift/3 days (span, %)	±2	±2	±2	±2	±2
Precision (%)	±2	±4	±4	±4	±2
Linearity (% full scale)	2	2	2	2	2

[a] Corrected for NO_2 and SO_2.
[b] Less CH_4.

determined if the proposed source will interfere with attainment of quality goals.

Owners of existing sources are required to monitor, record, and report periodically the nature and amount of emissions from stationary sources; periodic testing of sources may be required.

Compliance with abatement schedules necessary to meet national standards was to be negotiated with operations of stationary sources and reported to EPA within 45 days after the first state semiannual report.

"Reasonably available control technology" is to be used as a guide to emission limitations and abatement for stationary sources, except that the states are to consider (1) the necessity of these limits with regard to attainment of national standards, (2) social and economic impact of these limitations, and (3) alternate means of achieving standards. Emission limits are listed in detail for various industries in the federal regulations.[7] A brief summary of those considered "reasonably attainable" is as follows:

Particulate: 20% opacity or no. 1 Ringelman
    Incinerators—0.20 lb/100 lb refuse
    Fuel burning—0.30 lb/10^6 lb Btu
    Processes—∼0.1 lb (0.5–0.005)/100 lb processed.

Sulfur oxides: Fuel—depends on local conditions
    Refining—10 grains H_2S/100 ft^3
    Acid—6.5 lb SO_2/ton acid
    SO_2 recovery—0.01 lb/lb S
    Smelters—90% reduction

Nitrogen oxides: Fuel—0.2–0.3 lb/10^6 Btu (175–230 ppm)
    Acid—400 ppm.

4.2. The Clean Air Amendments of 1970

State emission-control plans require an extensive inventory of the sources in each region (particularly data on the emissions, abatement equipment efficiency, and tests on stacks) for "point sources." These are defined as (1) any stationary source emitting more than 100 tons/yr in an area population of 1 million; (2) any stationary source emitting more than 25 tons/yr if the urban population is less than 1 million; or (3) any source, regardless of amount of emissions, on the list of Table VI (e.g., acid plants cement kilns, large incinerators and boilers, and the like).[8]

The effect of the "control strategy" on the reduction of emissions from point and area sources is to be shown by the state.

Episode Control

The Air Quality Act recognizes that statistical fluctuations of emissions, and meteorological conditions can lead to severe episodes of concentrated air pollution regardless of the ambient standards (Chapter 3, Section 3.3). States are required to submit an Episode Control Plan for each Priority I region, to prevent ambient concentrations from reaching levels causing "significant harm to health of persons" as defined by Table III of the last chapter.

Basically, the plans call for continuous surveillance of air-pollution monitors and the application of increasing degrees of severe abatement as the pollution concentration and forecasts rise. Public alerts and rapid communication with authorities responsible for source abatement are essential features.

Briefly the plan includes the following:

1. Two or more stages of episode criteria (pollution concentration), together with public announcements and emission control actions, should be specified.
2. Episode emission control (preplanned) for each stationary source exceeding 100 tons/yr should be specified. Controls actually imposed during episodes should be applied to known causes (sources) if they can be identified.
3. Atmospheric stagnation should be monitored by meteorogical means daily and updated at least every 12 h during episodes.
4. Procedures should be developed for timely communication with source abatement controllers and the public, during alerts.

Suggested abatement procedures for air-pollution emergencies (keyed to Table III of Chapter 3) are as follows:

1. *Alert:*
   Prohibit open burning
   Limit incinerator use to noon to 4 P.M.

Table VI
Major Pollutant Sources

Chemical Process Industries
Adipic acid
Ammonia
Ammonium nitrate
Carbon black[a]
Charcoal[a]
Chlorine
Detergent and saop[a]
Explosives (TNT and nitrocellulose)[a]
Hydrofluoric acid[a]
Nitric acid
Paint and varnish manufacturing[a]
Phosphoric acid[a]
Phthalic anhydride
Plastics manufacturing[a]
Printing ink manufacturing[a]
Sodium carbonate[a]
Sulfuric acid[a]
Synthetic fibers
Synthetic rubber
Terephthalic acid

Food and Agricultural Industries
Alfalfa dehydrating[a]
Ammonium nitrate
Coffee roasting[a]
Cotton ginning[a]
Feed and grain[a]
Fermentation processes
Fertilizers[a]
Fish mean processing
Meat smoke houses[a]
Starch manufacturing[a]
Sugar cane processing[a]

Metallurgical Industries

Primary metals industries:
   Aluminum-ore reduction[a]
   Copper smelters[a]
   Ferroalloy production[a]
   Iron and steel mills[a]
   Lead smelters[a]
   Metallurgical coke manufacturing[a]
   Zinc[a]

Secondary metals industries:
   Brass and bronze smelting[a]
   Alluminum operations[a]
   Ferroalloys[a]
   Gray iron foundries[a]
   Lead smelting[a]
   Magnesium smelting[a]
   Steel foundries[a]
   Zinc processes[a]

Mineral Products Industries
Asphalt roofing[a]
Asphaltic concrete batching[a]
Bricks and related clay refractories[a]
Calcium carbide[a]
Castable refractories[a]
Cement[a]
Ceramic and clay processes[a]
Clay and fly ash sintering[a]
Coal cleaning[a]
Concrete batching[a]
Fiberglass manufacturing[a]
Frit manufacturing[a]
Glass manufacturing[a]
Gypsum manufacturing[a]
Lime manufacturing[a]
Mineral wool manufacturing[a]
Paperboard manufacturing[a]
Perlite manufacturing[a]
Phosphate rock preparation[a]
Rock, gravel, and sand quarrying and processing[a]

Petroleum-Refining and Petrochemical Operations[a]

Wood Processing[a]
Petroleum storage (Storage tanks and bulk terminals)

Miscellaneous
Fossil-fuel steam electric powerplants[a]
Municipal or equivalent incinerators[a]
Open burning dumps[a]

[a] Major source of sulfur oxides and/or particulate matter.

Limit boiler stack cleaning
Curtail motor vehicles
Curtail sources:
   Fossil-fuel electric power: switch fuel or buy outside power.
   Process steam: switch fuel, reduce load.
Curtail operations:
   Defer trade waste deposit, reduce heat demand for processing of primary metals, petroleum, chemicals, minerals, paper, pulp, and grain.

2. *Warning:*
Prohibit incinerator use.
Enforce use of car pools.
Activate all plant emission reduction plans.

3. *Emergency:*
Shut down mining and quarrying; all nonemergency construction; all manufacturing not having an air emergency plan; all trades except medical and food; government offices, except emergency; movies; schools; and use of motor vehicles.

4.2.3. Automotive Vehicle Emissions

The direct federal control of automobile emissions, continued by the new law, has been and continues to be one of the most controversial aspects, even within government circles (see Ref. 9, for example). But while the EPA has upheld stringent standards, they continue to be applied on the federal level to prototype engines that are given type certification rather than to vehicles actually in use. The control of emissions from vehicles in operation continue to be the responsibility of the states as part of their overall control strategy (see Chapter 6).

The 1970 law required a 90% reduction in automotive hydrocarbon and carbon monoxide emissions by 1975 and in nitrogen oxides by 1976. Automobile manufacturers are able to meet the first two requirements without too much difficulty but claimed to incur severe technical problems in trying to eliminate nitrogen oxides. The basic dilemma lies in the trade-off of these three emissions while adjusting the combustion cycle. Devices that reduce CO and HC by increasing the engine efficiency also raise the NO_x output, because of the higher combustion temperature. In California, during 1970–1971, when these controls were first tested, the worst ozone pollution in the state's history resulted.[10] The rise in nitrogen oxides triggered the photochemical reactions described in Chapter 1, producing PAN and ozone in increased proportion. Considering the difficulties, an extension of one year has been allowed manufacturers to meet the 90% reduction criteria. One technical solution, which the EPA determined to be feasible, appears to be a catalytic converter, using precious metals

Table VII

Automotive Vehicle Exhaust Emissions Requirements

Emission	1968	1970–1971	1972	1973[a]	1974[b]	1975[c]	1976[d]	1977	Typical uncontrolled engine
			(g/vehicle mile except as noted)						
1. California (light gas engine)									
HC	275 (ppm)	2.2	3.2[e]	3.2	3.2	—		[a]	11–17
Evaporation (g/test)	—	6.0	2.0						
CO	1.5 (% vol.)	23.0	39[e]	39	28	9.0		[a]	70–125
NO$_x$	—	4.0	3.2[e]	3.0	2.0	2.0		[a]	6
2. Federal (light gas engine)									
HC (tailpipe)		4.1	3.4	3.4	3.5	1.5	0.41	—	
HC (crankcase)		—	0				0	—	
Evaporation (g/test)		—	2				2	—	
CO		34	39	39	39	15	3.4		
NO$_x$		—	—	3.0	3.0	3.1	2.0	0.4	

3. Federal (heavy gas engine)
   HC (tailpipe) 275 (ppm) ⟶ (180)[f] ⟶
   CO 1.5 (% vol.) ⟶ 1.0 (%) ⟶ 1.0
   NO$_x$ (1800)[f]
   NO$_x$ + HC (g/BHP-h) 16 ⟶ 16
4. Federal (Diesel)
   HC (tailpipe) (g/BHP-h) 3.0 ⟶ 3.0
   CO (tailpipe) (g/BHP-h) 7.5 ⟶ 7.5
   NO$_x$ (tailpipe) (g/BHP-h) 12.5 ⟶ 12.5
   Opacity-peak (%) 50 ⟶ 50
   Lugging (%) 15 ⟶ 15
   Accel. (%) 20 ⟶ 20

[a] Ref. 15.
[b] Ref. 16.
[c] Ref. 17.
[d] Ref. 18.
[e] Optional test procedure.
[f] Approximate.
[g] See Federal (light gas engine).

(platinum and palladium), installed in the muffler. Since these catalysts cost from $130–160/oz and must be imported from the Soviet Union and South Africa, the cost to the consumer has been estimated at over $150 per car.[11,12] Furthermore, the use of catalytic mufflers will dictate the use of lead-free gasoline in order that they not be "poisoned" by the heavy metal during the 50,000-mile life that EPA has mandated. The reduction of lead content (to 0.05 g/gal or less) will have other direct health advantages, as discussed in the previous chapter.

The most recent regulations are summarized in Table VII from those published in the *Federal Register*.[13]

Even when these standards are met, however, we will not be able to consider the system problem as satisfactorily answered. In the 1980s the estimated automobile population growth will again push up the smog production. Alternate strategies, such as public transit or alternate propulsion means, will be sought. In urban localities such as Los Angeles and Philadelphia, restraints on automobile traffic or gasoline rationing have been considered as part of the overall strategy to meet the Air Quality Standards. Regulations requiring the states to submit a Transportation Control Measure plan, including reduction of vehicle use, changing traffic flow patterns, and reduction of in-use vehicle emissions, were published in the *Federal Register* in 1973.[14]

4.2.4. Aircraft and Airport Emissions

The Clean Air Act of 1970 empowered the EPA administrator to establish standards for aircraft and aircraft engine emissions deemed to contribute to air pollution. These regulations are approved by the Office of Transportation to assure air safety.

On December 12, 1972, a study by the EPA was published[19] that determined that the national ambient air-quality standards were indeed endangered by aircraft in the vicinity of major airports and proposed regulations to control the emission of engines were issued. These regulations[20,21] were similar to those already in effect for automotive engines but applied to older (in-use) engines as well as new ones.

These standards promulgated were not derived quantitatively from known aircraft emissions, because sufficient data was not available to relate them to the ambient standards. Instead, they were based on the "best practical levels" achievable now and over the 5-yr period extending from 1974 to 1979.

The regulations for airport operations are for large airports (over 1 million passengers annually) in Priority I regions for HC and CO pollution (see Table II). A list of airports meeting these qualifications is given in Table VIII.

Table VIII
Airports[a] with Over 1 Million Enplaned Passengers in 1970

Chicago O'Hare	Cleveland
Los Angeles	Seattle–Tacoma
Atlanta	Houston
John F. Kennedy	Kansas City
San Francisco	New Orleans
LaGuardia	Las Vegas
Dallas	Baltimore
Washington National	Memphis
Miami	Phoenix
Boston	Tampa
Detroit	Cincinnati
Newark	Portland
Denver	Buffalo
Philadelphia	Indianapolis
St. Louis	Salt Lake City
Pittsburgh	Dulles
Minneapolis	

[a] With the exception of Miami, Detroit, Atlanta, and Tampa, the airports listed are in air-quality control regions classified as Priority I for hydrocarbons or carbon monoxide.

The regulations propose to cut airport contamination by limiting the aircraft to use of only half their engines at a higher and more efficient power setting during taxi and idle modes. It is shown that the CO and HC emissions of a 707 aircraft with 4 gas turbine engines of type JT3D can be reduced to 32% and 21% as follows:

JT3D Taxi—Idle Emissions

	4 engines (lower power)	2 engines (high power)
CO (lb/h)	436	138
HC (lb/h)	392	82

A substantial fuel saving is achieved as a beneficial by-product, estimated by the EPA as high as $10 million/yr for the entire industry.

Subsequently, final standards for engine emissions were issued. In general, these require control of fuel-venting emissions; crankcase emissions; hydrocarbons, CO, and NO_x in exhaust gases; and the visibility of exhaust smoke. Typically, turbine engines manufactured after January 1, 1979 are required to reduce HC emissions by 70–80%; CO by 60%; and NO_x by 20–50%, depending on engine thrust.

Piston engines built after 1979 are required to reduce HC and CO emissions by 30% and 50%, respectively. After January 1, 1981, the turbine engine restrictions become more stringent, and by 1983 all aircraft must meet the 1979 standards by retrofit. The EPA estimates the total cost of engine modifications to the aircraft industry at $141 million over a 10-yr period, partially offset by a $29 million saving in fuel for piston engines. The final regulations for engine emissions were published by the EPA in the *Federal Register* in 1973.[22]

4.2.5. Federal Emission Standards for New Stationary Sources

For certain types and sizes of stationary sources of pollution, direct federal regulations for new plant emissions were issued on December 23, 1971. The following summarizes the constraints on these plants:

Plants Controlled (*New and Modified*)	*Size*	*Number of New Plants Expected Annually*
Fossil-fueled steam	250×10^6 Btu/h 25,000 kW	75
Sulfuric acid		2
Nitric acid		5
Portland cement		3
Large incinerators	50 tons/day (except those burning sewage sludge)	20–25

Pollutants Controlled
Particulate matter
SO_2
NO_x
Sulfuric acid mist
Standards   See Table IX.

Enforcement
   Owners of new plants shall permit EPA to perform tests of emissions, providing sampling equipment, platforms, access, and utilities.
Steam Plants
   The owner shall provide for each plant, the following equipment:
   1. Photoelectric or similar smoke monitor and recorder
   2. SO_2 monitor–recorder
   3. Isokinetic stack sampling.
Incinerators
   Require isokinetic stack sampling.
Nitric Acid Plants
   Require NO_x monitor.
   Accuracy ± 20% at 95% confidence level.

Calibrated once every 24 h.
Isokinetic sample sampling required.
Sulfuric Acid Plant
   SO$_2$ monitor required.
   ±20% at 95% confidence level and checked every 24-h.

In the above, "isokinetic" sampling is specified in order to assure that the proportion of particles 3 μm and larger is truly represented by the

Table IX
New Federal Plant-Emission Standards[23] (December 23, 1971 and October 15, 1973).

Plant type	Emission limits	Emission without control	Emission using prior normal control practice
Steam plant:		(lb/10^6 Btu/h)	
Particulate	0.1	6–10	1–4
NO$_x$	0.2–0.7^a	0.3–2.0	—
SO$_2$ (coal-fired)	1.2	1–7	—
SO$_2$ (oil-fired)	0.8	—	—
Opacity	20%	—	—
Incinerators:		(g/ft^3 exhaust)	
	0.08	1 g	—
Cement:		(lb/ton feed)	
Kiln particulate	0.3	to 45	—
Clinker cooler part.	0.1	to 30	—
Opacity	10%	—	—
Nitric acid:		(lb/ton acid)	
NO$_x$	3.0	43	—
Opacity	10%	—	—
Sulfuric acid:		(lb/ton acid)	
SO	4.0	21.5–85	—
Opacity	10%	—	—

^{a} A state may impose more stringent limits.

Table X

New Stationary Sources Performance Standards (March 8, 1974).[24]

Plant type	Emission limitations	
Asphalt concrete plant	Particulate:	90 mg/Nm^3 [a]
	Opacity:	20%
Petroleum refineries:		
Fluid catalytic cracking	Particulates:	1.0 kg/1000 kg coke
Catalyst regenerator	Opacity:	30%
	CO:	0.050 vol. %
Process gas burner	H_2S	230 mg/Nm^3
	(equivalent to 15–20 ppm SO_2 in combustion product)	
Petroleum liquid storage vessels	Floating tank roof required	
Secondary lead smelters or refineries	Particulates:	50 mg/Nm^3
	Opacity:	20% (10% for pot furnaces)
Secondary brass and bronze	Particulates:	50 mg/Nm^3 (for reverberatory furnaces)
	Opacity:	10% (20% for reverberatory)
Iron and steel:		
Basic oxygen furnace	Particulates:	50 mg/Nm^3
	Opacity:	(to be determined)
Sewage treatment:		
Sludge incinerator	Particulates:	0.65 g/kg dry sludge
	Opacity:	20%

[a] Nm^3 = normal cubic meter.
[b] Ref. Federal Register, **39,** Part 2, 9308–9323 (March 8, 1974).

sample. The term *isokinetic* means that the flow velocity through the sampling nozzle is that same as that of the main stream in the stack. Thus, larger or smaller particles are not deflected in the streamlines of flow entering the sample orifice, and an undistorted particle-size spectrum is achieved.

Standards for other categories of new sources have been proposed and others will come under federal emission control in the future. Table X lists the emission standards for the seven plant types that (as of this writing) are the most recent candidates for performance standards.[24]

4.2.6. Federal Emission Standards for Hazardous Pollutants

Because of the direct hazard to health, certain types of industrial plants, regardless of size or age, were made subject to direct federal controls, published in the *Federal Register* on December 7, 1971.[25] The criteria on which these standards were based included health, meteorology factors, control capability, and economic impact. The following data summarizes the controls on plants producing these substances (see also Section 3.3.2):

Asbestos: No ambient limits were established, since there is no suitable sampling and analysis method.
Beryllium: Ambient limit 0.01 $\mu g/m^3$.
Mercury: Ambient limit 1 $\mu g/m^3$.

These standards apply to any new or modified plant, for which a license is required from EPA. Emission tests are required. Test facilities (platforms, etc.) must be furnished by the operator. The direct controls placed on these plants are as follows:

Asbestos:
1. Fabric filter or cyclone equivalent required.
2. Visible emissions prohibited.
3. Spraying emissions prohibited.
4. Surfacing roadways with asbestos prohibited.

Beryllium:
1. Total emissions shall not exceed 10 g/24-h day; or no more than 0.01 $\mu g/m^3$, 30-day average, measured by an ambient sampling network.
2. Emission tests may be required periodically. Alternatively, continuous monitoring using filters may be substituted.
3. Any emissions equal to or exceeding 0.03 $\mu g/m^3$ or 0.01 $\mu g/m^3$ in 30 days must be reported to EPA.

Mercury:
1. Emission limit is 5 lb/24 h.
2. Stack sampling is required.

4.3. STATE LEGISLATION AND REGULATIONS

By 1971 all 50 states had some kind of air-pollution legislation on their books. The earliest of these had been passed in 1949 and the most recent in 1971—the greatest number (32) dating from 1967 and subsequently, coincident with or following the federal Air Quality Act of 1967. These laws varied widely in their scope, enforcement regulations, and mode of administration. The titles of the cognizant state agencies exhibited equal variety: some were styled "commission," "board," "bureau," or "council"; others were sections or offices of older agencies. Seventeen were under the state's board of health and only 9 were associated with an independent air-pollution or environmental-control organization. Their enforcement powers included penalties ranging from fines of $100 (2 states) to those of $10,000/day or more (4 states), as well as imprisonment. As these penalties were seldom, if ever, enforced, a more pertinent gauge of commitment was their administrative powers. Although 33 of the state agencies at this time were empowered to set air-quality and emission standards, merely 4 were permitted to establish air quality regions; the authority to collect data, institute monitoring, or declare an air-pollution emergency was granted to only one state for each of these.

The 1970 federal law authorized the EPA administrator to end this chaos by specifying clearly the requirements and a time scale for state plans to implement the federal air-quality standards. These regulations (Part 51 of the *Code of Federal Regulations*) required the states to acquire the authority and to develop a strategic plan to meet the federal goals by specific dates.[8] These plans were duly submitted and by May 31, 1972, the administrator had approved 14 in toto, and, as required by the law, supplemented the inadequate state regulations by those of his own so as to bring them all up to the federal requirements.

As discussed in the previous section, the target date for meeting federal air-quality standards was set May 31, 1975, with a statutory extension of two years permitted in the event that technology to control emission sources was not available. A recent federal court decision[26] determined that extensions claiming inability to meet CO and oxidant standards, granted to 17 states, would not be allowed but that state plans as approved or modified by the EPA administrator must meet the original target, as well as show the states' ability to maintain standards after that time. In addition, the court ruled that plans submitted by April 15, 1973, must include controls on transportation, including such alternate strategies as traffic controls or mass transit, exhaust controls on in-use automobiles, vehicle inspection, or car pooling, which controls had also been postponed.

As a consequence of these new state plans, approved to uniform federal standards, the older regulations have been or are being changed and their provision should ultimately bear a strong resemblance to one another. The major differences in regulations are now (or should be) a function of the control strategy chosen to be most appropriate to the nature of the state's and region's air-pollution problems and their severity. Nevertheless, there are still substantial differences in form and content between the various state and regional regulations. In the following paragraphs, the provisions of some typical regulations are summarized, but within the scope of one chapter it is impossible to be comprehensive.[27]

4.3.1. Titles and Organization

A great many of the air-pollution control organizations below the federal level come under the jurisdiction of existing public health departments. The Bureau of Air Quality Control of Maryland, the Metropolitan Boston Air Pollution Control District, the Air Pollution Commission of Pennsylvania, and the Kansas Air Quality Conservation Commission, to cite a few, all come under their respective states' departments of health. Arkansas, however, has an independent Department of Pollution Control and Ecology and New Jersey, a Department of Environmental Protection, aided by a Clean Air Council including representatives of engineers, manufacturers, labor, and municipal organizations.

4.3.2. Air Pollution: Definition and Ambient Standards

Many state regulations depend on the public health hazard or nuisance aspects of air pollution to define their subject. New Jersey, for one, defines air pollution as "the presence in the outdoor atmosphere of substance in quantities . . . injurious to human, plant, or animal life or [interfering] with comfortable enjoyment of life and property." Industrial hazards falling under the employer–employee relationship are excluded from this definition. Boston, in similar words, defines air pollution of the ambient airspace as the concentration and duration of air contaminants that (1) cause a nuisance; (2) are injurious to human or animal life, vegetation, or property; or (3) interfere with comfortable enjoyment of life, property, and the conduct of business. Other states use similar language. Some also specify ambient quality standards, which are permitted by the 1970 federal act if they are not less stringent than the national requirements. Table XI lists those adopted by Pennsylvania, issued 18 months prior to the federal standards (shown for comparison). The latter are not only exceeded in many cases, but the state also provided for ambient control over contaminants not yet recognized by the federal authorities. The same can be said of the Kentucky ambient standards (Table XII), which exceed the federal standards with respect to CO, for example, and by the inclusion of an odor standard.

4.3.3. Plans, Permits, and Registration

Many states require that no air-contamination source or abatement device may be constructed or installed without submission of plans to the cognizant agency and approval of these plans, signified by the issuance of a permit. In many of these states, Maryland, for example, the plan for compliance with the regulation is legally binding after approval of the control actions and timetable by both parties. The operator of the source is free from prosecution so long as he proceeds with implementation of the plan, but if he falls behind on the compliance schedule, the plan can be revoked and the operator prosecuted.[28]

When the polluting equipment is part of a secret process, in the commercial sense, only such parts of the process of manufacturing method as relate to direct emission of contaminants to the open air need be revealed

Table XI

Ambient Air-Quality Standards

Substance	Federal standards Primary[a]	Federal standards Secondary	Pennsylvania standards (apply to single point)
Suspended particulates			
AGM,[b] $\mu g/m^3$	75	60	65
24-h max,[d] $\mu g/m^3$	260	150	
SO_2			
AAM,[c] ppm	0.03		
24-h max.,[d] ppm	0.14		0.10
3-h max.,[d] ppm		0.5	
AGM, ppm			0.02
1-h max.			0.25
CO			
8-h max.,[d] ppm	9	9	
1-h max.,[d] ppm	35	35	
24-h max., ppm			25
Photooxidants			
1-h max.[d]	0.08	0.08	0.05
Hydrocarbons (non-CH_4)			
3-h max.,[e] ppm	0.24	0.24	
NO_2			
AAM, ppm	0.05	0.05	—[f]
Settled particulates, AGM, mg/cm^2			0.8
30-day max., mg/cm^2			1.5
Lead 30-day max., $\mu g/m^3$			5
Beryllium 30-day max., $\mu g/m^3$			0.01
Sulfates, as H_2SO_4 30-day max., $\mu g/m^3$			10 (24-h 30 $\mu g/m$)
Sulfuric acid mist			—[f]
Fluorides, soluble, as HF, 24-h max, $\mu g/m^3$			5
H_2S, 24-h max., ppm			0.005
1-h max., ppm			0.1

[a] Primary standards define levels necessary to protect public health. Secondary standards define levels necessary to protect the public welfare from any known or anticipated adverse effects.
[b] Annual geometric mean.
[c] Annual arithmetic mean.
[d] Not to be exceeded more than once a year.
[e] 6–9 A.M.; not to be exceeded more than once a year.
[f] Standard being considered.

Table XII

Kentucky Regulation 9—Ambient Air-Quality Standards

Settable particulates (dustfall)
(measured as total water solubles and insolubles):

Three-month average not to exceed	15 tons/(mile^2)(month)
	(5.25 gm/(m^2)(month))

Soiling index:

Annual geometric mean not to be exceeded	0.4 COH/1000 linear ft
Three-month average not to exceed	0.5 COH/1000 linear ft

Sulfur dioxide:

Annual geometric mean not to exceed	0.02 ppm (52 µg/m^3)
Maximum 1-month average	0.05 ppm (131 µg/m^3)
Maximum 24-h average	0.1 ppm (262 µg/m^3)
Maximum 2-h average	0.3 ppm (786 µg/m^3)
Maximum 1-h average	0.42 ppm (1100 µg/m^3)

Hydrogen sulfide:

Maximum 1-h average	0.01 ppm (14 µg/m^3)

Carbon monoxide:

Maximum 8-h average	8 ppm (9 mg/m^3)
Maximum 1-h average	30 ppm (34 mg/m^3)

Total oxidants (measured as Ozone):

Maximum 24-h average	0.02 ppm (39 µg/m^3)
Maximum 1-h average	0.05 ppm (98 µg/m^3)

Hydrogen fluoride:

Maximum 1-month average	1 ppb (0.817 µg/m^3)
Maximum 1-wk average	.2 ppb (1.636 µg/m^3)
Maximum 24-h average	3.5 ppb (2.863 µg/m^3)
Maximum 12-h average	4.5 ppb (3.681 µg/m^3)

Total fluorides:

Dry weight basis (as F$^-$) in and on forage for consumption by grazing ruminants
The following concentrations are not to be exceeded:

Average concentration of monthly samples over growing season (not to exceed 6 consecutive months)	40 ppm (w/w)
Two consecutive months average	60 ppm (w/w)
Any 1-month average	80 ppm (w/w)

Odors:

At any time not to exceed	7 dilutions[a]

---

[a] *Dilution* is the number of volumes of odorless air that must be added to a like volume of odor-bearing air to reduce the odor of the resulting mixture to just below the threshold level.

(e.g., Arkansas regulation, Section 3C). Such information will include the location of the process, name of the equipment, control means, rate of discharge, and dimensions, including height and location, of stacks discharging the contaminant. This information, which is disclosed or discovered to the state authority, is required to be kept confidential. However, information relating to emissions may be published as part of statistical summaries or analyses that do not identify the source.

The criteria for regulation vary, but the Boston regulation may be taken as an example:

1. The registration must be received annually.
2. Fuel-burning facilities with a rated input exceeding 3 million Btu/h must register.
3. Industrial facilities must register if their emission capability is:
   (a) 5 lb/h of particulate matter
   (b) 4 lb/h of SO_x
   (c) 40 lb/day or organic material
   (d) 1 lb/h of NO_2.
   (Industries included are asphalt-batching plants, foundries, chemical products manufacturing plants, petroleum products and manufacturing plants, aggregate manufacturing plants, food and food products plants, dry-cleaning establishments, paint and varnish manufacturing plants, paper manufacturing plants, leather manufacturing plants, concrete manufacturing plants, and metal coating and treating plants, and such other facilities.)
4. Incinerators reducing more than 1000 lb/h of waste must register.

Engineering guides, for the preparation of control plans, are provided by many states, but these are not necessarily binding. Advance consultation is offered by some (Boston, for example).

4.3.4. Inspection and Sampling

Registration and permits are often contingent on approval of a sampling plan submitted by the applicant. The sampling plan will not require that the equipment be provided with:

1. Sampling ports of sufficient size, number, and location as to provide representative sampling of the emissions.
2. Safe access to each port, including staging and ladders to support personnel and sampling equipment.
3. Other facilities, including a suitable power source for operation of the sampling equipment.

Representatives of an air-control board will have the right to enter and inspect premises at any reasonable time for the purpose of investigating a source of air pollution or ascertaining the state of compliance. No person is allowed to obstruct an inspector and operators must provide reports on emissions if requested (Kansas, 65-3009).

In some areas such as Boston, operators of emission sources specified by the department are required to install, maintain, and use emission-monitoring equipment of approved design. Reports derived from these monitors may be used for both emission control and public information.

The following paragraphs relate to specific forms of pollution.

4.3.5. Open Burning

Nearly every state prohibits open fires with some exceptions. Maryland forbids burning of leaves or refuse where public collection is available or any burning within specified distances of buildings or roadways. In other states, exceptions are made for barbeques and other noncommercial recreational fires; burning of some agricultural wastes, such as stubble, or controlled forest management fires; waste gas and hydrocarbon waste flares; or other closely controlled activities (Arkansas). Agricultural and similar controlled fires are only permitted under conditions of good atmospheric ventilation, without causing a nuisance, or under permit (Boston).

4.3.6. Sulfur Oxides and Fuel Regulation

The rules controlling the emission of sulfur oxides are extremely diverse, owing to the length of time they have been in effect under community control. Ambient controls are the most difficult to apply, because of the dispersion effected by tall stacks and the consequent problems of source identification. Since most SO_x emissions result from fuel burning, emission standards or regulations applied to the sulfur content of fuels are more common. Duprey[29] cites about 200 individual regulations on coal and oil sulfur content in the various states and regions and his list is not complete.

Permissible weights of sulfur in coal and oil have ranged as high as 2% (as in parts of Florida) and are still 1% in areas such as Connecticut, Delaware, Maryland, and Chicago. At the other extreme, Colorado requires 0.15% sulfur coal and 0.25% oil for new sources now and for all sources in 1975. The equivalent limits based on pounds per million Btu range from 2.3 in Minneapolis (reduced to 2.0 during an air-pollution alert), down to 0.28 in Boston's center city (0.17 for no. 2 oil during adverse conditions). Permissible sulfur has been reduced in stages in some areas and will be reduced further. However, shortages of low-sulfur fuel may require relaxation of these stringent rules (as in Boston) under some conditions.

Table XIII
SO_2 Concentration at Ground Level (Montana)

Concentration, ppm (v/v)		Duration, min		Period, min	Frequency
2.0		Anytime		—	—
1.0		2.5		60	Twice per 8 h
	or	15 (total)	per	day	—
0.5		5		60	Twice per 8 h
	or	30 (total)	per	day	—
0.2		10		60	Twice per 8 h
	or	60 (total)	per	day	—
0.1		30		60	Twice per 8 h
	or	150 (total)	per	day	—
0.1		(average)	per	day	—

Several states have SO_x regulations based on ambient concentration beyond the premises. Arkansas's rule is not to exceed for more than 30 min a limit of 0.20 ppm for SO_2 or 30 µg/m³ for SO_2 and acid mists. Violations may be determined by ambient sampling or calculations based on stack concentration and dispersion equations. Montana's regulations limit SO_x ambient concentrations at ground level as shown in Table XIII.

Persons responsible for the emission of more than 2000 ppm SO_2 must provide four continuous recording SO_2 measuring instruments, located so as to monitor the ground concentration in an approved manner. The rule applies to primary nonferrous smelters and kraft pulp mills as well as others. This regulation is clearly based on the duration–concentration and dose–response properties of SO_2 as applied to health (see Chapter 3). A similar ambient regulation, but less complex, is in effect in the San Francisco Bay area air-pollution district.

Boston's regulations are typical of direct SO_2 stack emission limits. They include the following:

Contact Sulfuric Acid Plants:
New plant 4 lb SO_2/ton acid produced
Existing plants 27 lb SO_2/ton acid produced
Other Sources: 25 lb SO_2/hour
Absolute Limit on Concentration: 500 ppm.

The New Jersey law is unique in its complexity as to SO_x emissions. Only specified compounds of sulfur may be emitted. If SO_2, the concentration may not exceed 2000 ppm (2-years' grace was given for 3000-

ppm emissions). Exceptions are granted for stacks discharging less than 3000 ft^3/min if the hourly average is less than 50 lb SO$_2$ and the instantaneous rate below 100 lb/h. Monitoring and recording facilities (SO$_2$ concentration, total volume, temperature, and pressure) must be provided for stacks over 1000 ft^3 min. For emissions of SO$_3$ and H$_2$SO$_4$ the total, as H$_2$SO$_4$, may not exceed 10 mg/ft^3 at standard conditions. For other sulfur compounds, the limits are calculated from stack conditions (height, exit velocity, temperature) and a graph of adjusted stack height versus sulfur emission in pounds per hour.

4.3.7. Particulates

State particulate emission regulations are at least as complex as those for sulfur gases and reflect an even longer history. Some of the criteria applied to suspended particulates and fallout are the following:

1. Contribution to the total ambient concentration
2. Fallout from the source
3. Reduction in potential emission rate
4. Process weight
5. Stack concentration.

In addition, specific rules are applied to particular industries.

The Arkansas regulations, as amended in 1972, utilize all five of these criteria. For example, no equipment is allowed to contribute (1) an amount (above background) greater than 75 μg/m^3 of suspended particulate averaged for 24 h or 150 μg/m^3 for 30′ min (measured by ambient air sampling or calculated by dispersion formula); (2) a fallout of 15 tons/(mile^2)(month); or (3) more than 120 particles above 60 μm in diameter in 24 h.

(The last provision is remarkable in that it apparently favors the emission of small particles, which are relatively more harmful to health, if less so to laundry!)

In addition to the above, the stack emission rate of any new or altered equipment must be so controlled than it does not exceed a fraction of its potential emission. The percentage emission allowed decreases rapidly with increased equipment size, from 20% for a nominal 50-lb/h to 1% for 50,000-lb/h potential.

Alternately, the particulate emission controls may be specified on the basis of "process weight." This is the total weight of all materials entering into the process that may cause any emission of particulate matter. For example, solid fuels are included in the process weight but not gaseous or liquid fuels. For batch operations, process weight is based on the operating time, with no credit for the idle periods.

The Arkansas rules can be expressed by the empirical equations

$$E = 3.59\, p^{0.62}$$

for p less than 30 tons/h, and

$$E = 17.31\, p^{0.16}$$

for p more than 30 tons/h.

In Arkansas, stack concentration constraints are exemplified by the incinerator code, which limits particulate emissions to 0.2 gr per standard cubic foot (SCF) of dry flue gas, corrected to the volume of excess air that would result in 12% CO_2 at the flue. For small incinerators, under 200 lb/h, the stack concentration may rise to 0.3 gr/SCF.

The above rules may be compared with the the regulations in Boston (Table XIV), where it is seen that fossil-fuel facilities are controlled on the

Table XIV
Particulates Emissions Limitations (Boston APCD)

Facility size	New	Existing	Critical city or town
Fossil-fuel facilities		lb particulate/million Btu[a]	
3–250 × 10⁶ Btu/h	0.1	0.15	0.12
>250 × 10⁶ Btu/h	0.05[b]	0.15	0.12
Ferrous cupola foundaries:		lb particulate/1000 lb flue gas	
Production (centurion)	0.10	0.25	0.10
Jobbing (intermittant)	0.40	0.40	0.40
Nonferrous foundries	0.10	0.15	0.10
Asphalt batching		lb particulate/h	
Production weight, 100–400 tons/h (process weight curve)	4.5–18.1	9.0–36.2	4.5–18.1
Industrial sources			
Process weight 50–3000 lb/h	0.12–2.55	0.24–5.10	0.12–2.55
3100–60,000 lb/h	2.59–20.0	5.18–40.0	2.59–20.0
Over 60,000 lb/h: $E =$ [c, d]	$55p^{0.11}-40$	$\tfrac{1}{2}(55p^{0.11}-40)$	$\tfrac{1}{2}(55p^{0.11}-40)$
Incinerators		Grains/SCF @ 12% CO_2[a]	
Municipal	0.05	0.10	0.10
Other types	0.10	0.10	0.10

[a] Must be tested by isokinetic sampling as specified in federal regulations.[30]
[b] May be 0.10 if SO_2 is controlled at the same time.
[c] E = emission in pounds per hour.
[d] p = process weight in tons per hour.

basis of Btu production. Some industrial sources are controlled as a function of gas weight, others by emission rate, and still others on the basis of process weight formulas. This diversity is typical of many states and represents the differing practices of industry groups. For example, the particulate regulations specifying the concentration (mass per unit volume) of effluent gas are based on a "model smoke ordinance" developed by the ASME in 1948. Typically, the limits range from 0.2 to 0.3 gr/SCF (at 60°F and 1 atm), depending on the definition of particulate, the sampling method, and the gas composition, but as Table XIV shows, newer regulations are more stringent.

4.3.8. Visible Emissions

As we observed in the first chapter, the oldest "prescientific" air-pollution regulations were concerned with visible smoke for obvious reasons. We also noted in Chapter 1 that smoke-plume visibility is not directly related to particulate concentration, this being only one factor contributing to optical obscuration. (Particle size is an important factor affecting plume visibility, those particles having a diameter in the vicinity of visible radiation wavelengths, about 0.5 μm, being relatively more effective in light obscuration. Similarily, the index of refraction of the particles is important.)

Nevertheless, the classic standard that is applied to the control of black smoke, and the only regulation in many communities, is that of the Ringelmann chart[31] first introduced in 1897 (Figure 1). It is a subjective measure, performed by visually comparing black smoke plumes with four cross-hatched charts exhibiting various proportions of black line width to white spacing. The charts are stationed at such distance as to merge the black lines into a uniform gray. The basic concept was expanded, first in Los Angeles, to include the "equivalent opacity" of white and colored plumes, a comparison that is even more subjective. However, it has been possible to train inspectors to agree within a half Ringelmann number, and the method is still firmly entrenched in enforcement practice despite its obvious shortcomings.

Some examples: Pennsylvania deems any open burning or emission of smoke darker than Ringlemann no. 2 (or its equivalent for obscurring visibility) to be air pollution. For incinerators, no emission greater than Ringelmann no. 1 may be discharged from any source. Exceptions to this rule are the following:

1. Greater than Ringelmann no. 2 if not for more than 6 min out of any 60 min
2. If the emission is water vapor

Spacing of Lines on Ringelmann Chart

Ringelmann chart no.	Width of black lines (mm)	Width of white spaces (mm)	Percent black
0	All white		0
1	1	9	20
2	2.3	7.7	40
3	3.7	6.3	60
4	5.5	4.5	80
5	All black		100

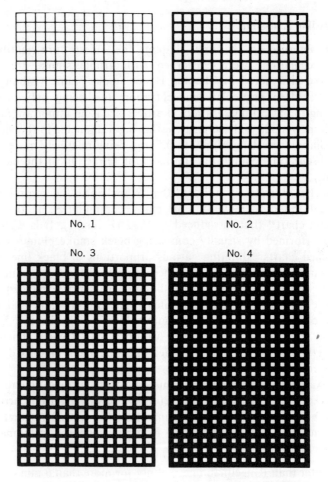

Figure 1 The Ringelmann chart (courtesy of Academic Press and *IEEE Spectrum*).

4.3. State Legislation and Regulations

3. Fire set by public officers for instruction or prevention
4. Motor-vehicle emissions
5. Equipment for controlling insects
6. Preparation of food for less than 25 people/day or 25 lbs/day
7. Discharge from battery of coke ovens if greater than Ringelman no. 2 for no more than 12 min in any 60 min
8. Emission from foundry cupolas greater than Ringlemann no. 2, provided it meets other standards on allowable emissions.

Chapter 4 of the New Jersey regulations, "Control and Prohibition of Air Pollution by Smoke," defines *visible smoke* as "smoke which obscures light to a degree readily discernible by visual observation" and *opacity* as "the property of a substance which renders it partially or wholly obstructive to the transmission of visible light, expressed as percentage to which the light is obstructed."

Section 2 prohibits visible smoke emission from combustion of fuel in a stationary indirect heat exchanger of less than 2×10^8 Btu gross input or through a stack of less than 60 in. internal dimension. For sources greater than the ones mentioned, Ringelmann may not be darker than no. 1, or opacity greater than 20%. These prohibitions do not apply for smoke visible less than 3 min in any 30-min period. Section 3 applies the Ringlemann no. 1 and opacity 20% limits to marine sources. Section 4 prohibits smoke darker than Ringelmann no. 2 or opacity over 40% from mobile sources for more than 10 sec. Section 5 applies Ringelmann no. 1 and 20% opacity to stationary internal combustion and turbine engines for 10 sec. Section 6 calls on the owner and operator of a source to provide facilities to determine density or opacity of visible smoke, using methods approved by the Department of Environmental Protection.

Boston's regulations, among the more recent (June 1972), limit visible emissions to Ringelmann no. 1 and, except for incinerators, allow a no. 2 shade for 6 min during any one hour. (Maryland reduces this to 4 min.) Furthermore, for any fossil-fuel facility that is classified as a "high-pressure system," a continuously operating smoke-density sensing device is demanded, complete with a recorder and an audible alarm that will be activated by emissions greater than no. 2.

4.3.9. Nitrogen Oxides

While states and localities have had many years of experience in the application of smoke, particulate, and sulfur compound regulatory controls, the problem of nitrogen oxide has only recently been recognized. NO_x is not only a health hazard in its own right but is a precursor of photochemical smog, as explained in Chapter 1. Stationary sources, of course, are not the only emitters of NO_x, but control of the automobile has been, to date, preempted by the federal government.

The state implementation-plan regulations issued by the EPA in conjunction with the National Ambient Air Quality Standards did not mandate particular controls on stationary sources but merely directed a reduction of the total emissions in an amount proportional to the existing above the standard, using the so-called "proportional" model:[32]

$$\frac{A - C}{A - B} \times 100 = \% \text{ reduction needed}$$

where A is the existing air quality at the location having the highest measured or estimated concentration in the region, B is the background concentration, and C is the national standard.

Based on the existing state of the art in boiler firing operations, the federal agency suggested that oil or gas-fired boilers of more than 250 million Btu/h capacity could be limited to 0.3 lb NO_x emission/(million Btu) (h) and similar sized coal units to 0.9 lb/(million Btu)(h). These numbers are equivalent to about 200 and 500 ppm by volume, respectively, at 3% excess oxygen combustion conditions.

The states' inexperience with these controls, however, coupled with the unknown factors associated with the control of automobile emissions, caused most of the populous states to omit NO_x controls from their proposals completely. The EPA administrator then was required to issue his own regulations, which were binding on the state. As a consequence, many of the state NO_x control regulations have substantially identical format. The following quotation is typical.

"*Control Strategy: Nitrogen Dioxide.*

"*Regulation for control of nitrogen oxides emissions. (1) No owner or operator of any stationary source shall discharge or cause the discharge of nitrogen oxides (expressed as nitrogen dioxide) into the atmosphere in excess of:*[33]

(i) 0.20 lb per million B.t.u. (0.36 g per million cal.) from gas fired fuel burning equipment of more than 250 million B.t.u. per hour heat input.

(ii) 0.30 lb per million B.t.u. (0.54 g per million cal.) from oil fired fuel burning equipment of more than 250 million B.t.u. per hour heat input.

(iii) Where gaseous and liquid fossil fuels are burned simultaneously in any combination in fuel burning equipment of more than 250 million B.t.u. per hour heat input, the applicable emission limitation shall be determined by proration. Compliance shall be determined using the following formula:

$$z = \frac{x(0.20) + y(0.30)}{x + y}$$

Where:

x is the percent of total heat input derived from gaseous fossil fuels.
y is the percent of total heat input drived from liquid fossil fuels.
z is the allowable emissions in pounds per million B.t.u.

"Where solid fossil fuels are burned simultaneously with gaseous and/or liquid fossil fuels in fuel burning equipment, the emission limitations of sub-paragraph (1) of this paragraph shall not apply.

"Compliance with this paragraph shall be in accordance with the provisions cited below.

"The stack sampling test method applicable to a source subject to this paragraph shall be Method 7 (phenoldisulfonic acid procedure).[34]

"Compliance Schedules.

"(1) Except as provided in subparagraph (2), the owner or operator of any stationary source shall comply with such regulations on or before December 31, 1973.

"(2) Any owner or operator of a stationary source, may, no later than December 31, 1972, submit to the Administrator for approval a proposed compliance schedule that demonstrates compliance as expeditiously as practicable but no later than July 31, 1975. The compliance schedule shall provide for periodic increments of progress towards compliance. The dates for achievement of such increments shall be specified. Increment of progress shall include, but not be limited to: Letting of necessary contracts for construction or process changes, if applicable; initiation of construction; completion and startup of control system; performance tests; and submittal of performance test analysis and results."

4.3.10. Hydrocarbons

Control of hydrocarbons is relatively new to states other than California, and for the same reasons given in the NO_x discussion—namely, they are required to be controlled only because of their recently recognized part in the photochemical reactions forming smog. Hydrocarbons per se have little or no toxic properties in the ambient ranges encountered.

The federal implementation plans suggested a number of specific methods for the states to adopt; consequently, many of these were taken nearly verbatim from the federal text.[35] A summary of these rules gives a profile of the states' requirements in this regard:

1. *Storage tanks* holding 40,000 gal of volatile hydrocarbons must be sealed and be capable of withstanding the highest working pressure without vapor loss, or (if the vapor pressure is less than 11 psi), they may be equipped with floating pontoon-type roofs. (Note that the "Performance Standards for New Sources" also requires pontoon-

type roofs.) Otherwise, a vapor recovery system preventing emissions to the atmosphere and gas-tight (except while gauging or sampling) is specified.

2. *Organic-compound–water separators* handling more than 200 gal/day of organics must follow similar rules.
3. *Volatile organic loading facilities* such as gasoline tank-truck loading racks are required to have vapor recovery systems and vapor-tight, quick-sealing release fittings. A vapor balance line connection between tank and tank truck is specified.
4. *Pumps and compressors* must be equipped with approved mechanical seals.
5. *Waste gases* from vapor blow-down systems must be burned by smokeless flares. In ethylene plants, the water gas stream must be burned at 1300°F for at least 0.3 sec. A restriction of 15 lb/day is placed in the evaporation of organic solvents for baking and polymerizing equipment, while other equipments are limited to 40 lb/day of "photochemically reactive" solvents. This limit applies also to machines processing moving sheets or webs (such as printing presses). The emission weight limitation is to be accomplished by incinerating 90% of the excess carbon content or by an equally efficient absorption process.
6. *Architectural coatings*, such as paints, may not be sold or used in containers larger than 1 qt if they contain photochemical reactive solvents, and these solvents may not be used as thinners. For the purpose of these rules, a photochemically reactive solvent is any solvent that has an aggregate of more than 20% of its total volume composed of the chemical compounds classified below or that exceeds any of the following individual percentage composition limitations, referred to the total volume of solvent:

   A. A combination of hydrocarbons, alcohols, aldehydes, esters, ethers, or ketones having an olefinic or cyclo-olefinic type of unsaturation: 5%.
   B. A combination of aromatic compounds with either six or more carbon atoms to the molecule except ethylbenzene: 8%.
   C. A combination of ethylbenzene, ketones having branched hydrocarbon structures, trichloroethylene, or toluene: 20%.

Whenever any organic solvent or any constituent of an organic solvent may be classified from its chemical structure into more than one of the above groups of organic compounds, it shall be considered as a member of the most reactive chemical group—that is, that group having the least allowable percent of the total volume of solvents.

4.3.11. Motor-Vehicle Regulations

Prior to mid-1973, all controls on motor-vehicle emissions were applied to new engines by type certificate and, with the exception of California, were promulgated by federal authority. The first state to require emissions testing on all light-duty automobiles was New Jersey, under a law passed in June 1971.[36] Although California applies only assembly-line emission tests for 1973 and subsequent vehicles, New Jersey requires emission tests on all automobiles coming under the state's annual inspection procedure.

New Jersey is particularly vulnerable to pollution from automotive sources because it has the highest density of automobiles in the United States—440/mile^2 (1969 figures) and as high as 7480/mile^2 in the metropolitan areas. The New Jersey experience with annual testing is of interest as a component of overall pollution-reduction strategy and will be considered by many other states.

The New Jersey standards are in three parts. The first prohibits visible emissions from the engines or crankcases of vehicles operating on the public highway. Fines ranging from $25 to $100 are provided.

The second part requires dealers to inspect new cars and certify that they meet state inspection standards, including those for smoke, CO, and unburned hydrocarbons (HC).

The third requirement is the inspection standards, which are applied annually at the 40 stations throughout the state. The procedure requires about 30 sec in addition to the normal annual inspection, and checks both CO and HC at idle, as well as visible emission under acceleration. The former are measured by a nondispersive infrared analyzer (NDIR) connected to a sampling probe inserted in the vehicle's exhaust pipe.

Table XV
Inspection Standards for New Jersey Motor Vehicles

	Enforcement date					
	July 1, 1973		July 1, 1974		July 1, 1975	
Model year	CO, %	HC, ppm	CO, %	HC, ppm	CO, %	HC, ppm
Up to 1967	10	1600	8.5	1400	7.5	1200
1968–1969	8	800	7.0	700	5.0	600
1970–1974	6	600	5.0	500	4.0	400
1975–later	—	—	—	—	—[a]	

[a] To be promulgated by amendment.

The inspection standards are listed in Table XV. They provide for three steps of increasing stringency with enforcement of reinspection or rejection requirements beginning in July 1973 and apply to older vehicles but to a progressively lower standard. It is expected that 12–15% of the state's 3.5 million cars will fail the 1973 tests, and 20% or more in subsequent years. The effectiveness of the procedure will depend on the ability of mechanics to "tune" the failed cars to acceptable emissions. The strategic implications of this type of road test are important as an element of overall transporation control plans (see also Chapter 12).

4.3.12. Episode-Control Regulations

The requirement of the 1970 Clean Air Act for states to have means of preventing high levels of air pollution that may harm health during periods of atmospheric stagnation was discussed earlier in this chapter. Some areas have had such controls for some time; for example, the New York–New Jersey Committee on Air Pollution put a regional warning system into effect in 1964.[37] Others have only recently developed such rules in response to the federal requirement and, like Maryland (in May 1972), now require major sources to file standby episode emission cutback plans to be executed on notice of an emergency. Criteria for declaring such emergencies are listed for three typical regions in Table XVI, together with the suggested federal values (Section 4.2).

Air-pollution emergency watches are initiated by such meteorological events as (1) a high-pressure air mass over the area, (2) low wind speed, or (3) a temperature inversion (warm air forming a "lid" above a cool air mass, trapping pollutants). Other atmospheric factors, such as humidity, may contribute to the emergency.

The stagnation advisories (HAPPA, or High Air Pollution Potential Advisory) are issued by the NOAA National Weather Service. In Illinois, the watch is initiated on receipt of a HAPPA for the next 24 h that includes Chicago, St. Louis, or other cities of 50,000 or more, and an average yellow alert value (see Table XVI), for any monitoring station during the preceding 2 h.[38] The New York–New Jersey–Connecticut criteria are a 36-h stagnation advisory with at least 12 h remaining, plus contamination at criterion level at four stations. (The monitoring stations may be those official stations located in accordance with the requirements of Table III.) However, the New York area does not demand the stagnation advisory if local conditions build up pollution at fewer than four stations with an expectation of spreading throughout the region. This allows accidents as well as natural episodes to be anticipated. In addition, they have added a modified weather advisory, comprised of the combination of stagnation and forecast sunshine, which is coupled with gradient levels to predict a photochemical smog alert.

Table XVI
Episode Criteria for Three Regions

	First alert (yellow)			Second alert (red)			Emergency			
	N.Y.-N.J.-Conn. (as of 7/71)	Los Angeles	Illinois (11/24/70)	N.Y.-N.J.-Conn.	Los Angeles	Illinois	N.Y.-N.J.-Conn.	Los Angeles	Illinois	Federal[a]
SO_2 (ppm)	0.3 (6-h avg)	0.5	0.3 (4 h)	0.5 (6 h)	1.0	0.35 (4 h)	0.6 (24 h)	1.5	0.4 (24 h)	0.8 (24 h)
Particulates (COH)	5.0 (6 h avg) or 3.0 (24-h avg)	—	—	8.0 (6 h) 6.0 (24 h)	—	—	8.0 (24 h)	—	—	7.0 (24 h)
$SO_2 \times$ particulates (ppm × COH)	0.5	—	1.0 (4 h)	1.6	—	2.0 (4 h)	2.0 (24 h)	—	2.4 (24 h)	1.2 (24 h)
CO (ppm)	15 (8-h avg)	100 (1 h) 200 ($\frac{1}{2}$ h) 300 (10 min)	—	30 (8 h)	100 (2 h) 200 (1 h) 300 (20 min)	—	50 (8 h)	200 (2 h) 200 (1 h)	—	40 (8 h)
Oxidants (ppm)	0.15 (4-h avg)	3 (NO_x) 0.5 (O_3)	—	0.25 (4 h)	5 (NO_x) 1.0 (O_3)	—	0.40 (4 h)	10 (NO_x) 1.5 (O_3)	—	1.6 (NO_2-1 h) 0.4 (NO_2-24 h) 0.6 (O_3-1 h)[b]

[a] Table III, Chapter 3.
[b] 0.5 ppm (O_3-1 h) proposed[41].

Table XVII
Worldwide Air-Pollution Regulations

	Date	Approval permit required	Survey monitoring provisions	Construction stack standards	Inspection (entry)	Emissions standards	Ambient standards	Emission controls on: Smoke	Industry	Vehicles	Fuel controls	Emergency controls
Australia (Victoria)	1970	X		X		X			X		X	
(N.S. Wales, Queensland, W. Australia)	1961–1963	X X										
Belgium	1964		X		X			X	X	X		
Brazil	1967					X	X		X			X
Bulgaria	1963	X		X				X	X	X		
Canada (Ontario/Saskatchewan)	1958–1965	X	X					X	X			
(Nova Scotia)	1960					X		X	X			
(Manitoba)	1960					X						
(Alberta)	1961	X						X	X			
(Toronto)	1957	X		X	X			X	X			
Chile	1961	X							X			
Czechoslovakia	1966–1967	X X				X	X		X	X		
Denmark	1858	X				X						
Eire	1962	X	X		X	X			X	X		
France	1961	X					X	X	X	X		

182

Country	Year							
Germany (Democratic Republic)	1968	X				X		
Germany (Federal Republic)	1959–	X			X	X	X	
Honduras	1966	X		X				
Italy	1966	X		X	X	X		
Jamaica	1961							X
Japan	1962	X	X	X	X	X		
Malta	1967	X	X	X	X			
Netherlands	1963	X						
New Zealand	1956–		X		X			
Philippines	1964	X		X	X			
Poland	1966–1967	X	X		X	X		
Portugal	1966	X	X		X			
South Africa	1965	X		X				
Sweden	1967–1969	X						
Switzerland	1960	X						
United Kingdom	1906–1966	X	X		X	X	X	
USA	1970	—[a]		X	X		X	
USSR	1949	X	X	X	X	X		
Yugoslavia	1965	X		X	X	X	X	

[a] See text.

It will be noted that the alert criteria for the three regions illustrated are quite different and only partially conform to the federal recommendations. To some extent this is because of local conditions; for example, Los Angeles is much more concerned with CO than Illinois and much less so with particulates emitted from domestic coal heating plants. But the differences are also related to control philosophy. In cities like Chicago, where in the past there has been a great deal of SO_2 and particulate pollution, the concentration of these contaminants has been taken as an indicator of general air-pollution conditions; in other words, SO_2 has been considered as a "tracer" gas, and its concentration a measure of overall ventilation. But the increased use of low-sulfur coal combined with the use in automobile traffic can invalidate this historical relationship. This has happened in the New York area and has resulted in three changes in alert criteria. Thus, the numbers given in Table XVI will be subject to further modification in the future.

4.4. WORLDWIDE AIR-POLLUTION REGULATIONS

Within the scope of this chapter it is impossible to give a detailed analysis of air pollution regulations in other countries, although this has been attempted elsewhere.[39, 40] Table XVII is a summary profile of the legislation in the countries listed, extracted from these and other sources. In some cases, the provisions listed are not yet firm, but have merely been authorized by enabling legislation, the details to be completed after further investigation. Except for the United Kingdom, it can be seen that most of these rules are of recent origin. A greater spread of air-purity regulations can be expected in developing countries as well as those reaching full industrialization, such as Italy, Japan, and the Soviet Union. It is to be hoped that economic and social pressures will force worldwide uniformity on constraints, so as to eliminate "pollution havens." With the spread of industry, power generators, jet aircraft, and the automobile, air pollution is becoming more of an international problem as well as a worldwide one. The 1972 UN conference held at Stockholm conference gives reason to believe that international control is a possibility in the future.[42,43]

REFERENCES

1. Cited in *Chem. Eng. News,* June 8 1970.
2. *Clean Air Act.* Environmental Protection Agency: Washington, D.C., 1970.
3. *Federal Register,* **36,** 22398–22417 (November 25, 1971).
4. Ref. 3, **36,** 8187 (April 30, 1971) and **38,** 11355–11356 (May 7, 1973).

5. Ref. 3, **37**, Pt. 3, 15094 (July 27, 1972).
6. Ref. 3, **38**, 15180 (June 8, 1973).
7. Ref. 3, **36**, App. B (August 14, 1971).
8. Ref. 3.
9. *Cumulative Regulatory Effects on the Cost of Automotive Transportation* (RECAT), OST, The White House: Washington, D.C., 1972.
10. T. Wicker. "Making Haste Too Fast" *New York Times*, March 5, 1972.
11. *New York Times*, "Deadline for Detroit—II" (editorial), January 20, 1973.
12. *Business Week*, April 21, 1973, p. 30.
13. *Federal Register*, **37**, 24250–24320 (November 15, 1972); **38**, 17441 (July 2, 1973). See also "Statement by Acting EPA Administrator R. W. Fri" (on suspension of NO_x standards). EPA: Washington, D.C. July 30, 1973.
14. Ref. 3, **38**, Pt. 3, 15194 (June 8, 1973).
15. Ref. 3, **37**, 24253 (November 15, 1972).
16. Ref. 3, **38**, 30494 (November 8, 1973).
17. Ref. 3, **38**, 17441 (July 2, 1973).
18. Ref. 3, **38**, 22474 (August 21, 1973).
19. *Aircraft Emissions: Impact on Air Quality and Feasibility of Control*. EPA: Washington, D.C., 1972.
20. Ref. 3, **37**, 26502–26503 (December 12, 1972).
21. Ref. 20, 26488–26500.
22. Ref. 3, **38**, 19088–19103 (July 17, 1973).
23. Ref. 3, **36**, 24876–24895 (December 23, 1971) and **38**, 28564 (October 15, 1973).
24. *Background Information for Proposed New Source Performance Standards*, Vol. I. EPA: Research Triangle Park, N.C., 1973; also *Federal Register*, **38**, Pt. 2, 15406–15414 (June 11, 1973).
25. Ref. 3, **36**, 5931 (December 7, 1971). See also: **38**, Pt. 2, 8820–8850 (April 6, 1973).
26. Ref. 11, February 2, 1973.
27. See, for example, the publications of the Bureau of National Affairs, Inc., Washington, D.C., for compilations of state regional environmental laws.
28. J. Schueneman. In *Compliance Timetables to Attain Air Quality Standards Emission Standards*. ASME: Cincinnati, Ohio, 1972, pp. 2–4.
29. R. L. Duprey. *Chem. Eng. Prog.* **68**, 70–76 (1972).
30. Ref. 3, **35**, 24888–24890 (December 23, 1971).
31. U.S. Bureau of Mines. *Circular 844*, revised in *Circular 7718*, Washington, D.C. (1955).
32. *Federal Register*, **36**, 22401 (November 25, 1971).
33. Ref. 3, **36**, 15486 (August 14, 1971).

34. ASTM D-1608-60 in *1968 Book of ASTM Standards*, Pt. 23. Philadelphia, 1968, p. 725–729.
35. National Ambient Air Quality Standards, Preparation, Adoption, and Submittal of Implementation Plans. *Federal Register,* **36**, (April 7, 1971).
36. New Jersey Department of Environmental Protection. *Air Pollution Control Code.* Trenton, N.J., 1971, Chapter 15.
37. A. I. Mytelka. "Criteria for the Air Pollution Warning System in the NJ-NY-Conn. AQCR." *64th Meeting of the APCA.* Atlantic City, N.J. June 27–July 2, 1971.
38. *Illinois Pollution Control Board Rules and Regulations,* Pt. IV: Episodes (Nov. 24, 1970) Chapter 2.
39. A. Lanteri. In W. Strauss, Ed., *Air Pollution Control,* Pt. 2. Wiley-Interscience, New York, 1972.
40. S. Edelman. In A. C. Stern, Ed., *Air Pollution,* 2nd ed., Vol. III. Academic: New York, 1968.
41. Ref. 3, **39**, 9272–9273 (March 13, 1974).
42. Ref. 11, "Text of the Environmental Principles" June 17, 1972.
43. Scientific Committee on Problems of the Environment (SCOPE), Global Environmental Monitoring (1971), Stockholm: Swed. Nat. Sci. Res. Council, (1972).

5

Control Strategies and Systems Design

5.1. OVERALL STRATEGIES

Thus far we have discussed the chief scientific and societal–economic aspects of the air-pollution problem and the nature of governmental responses to the danger. It is now appropriate to examine the adequacy of the control measures specified thus far in the light of the system design principles laid down in Chapter 1. From that viewpoint, it is at once clear that the problem scope is too broad to yield to a fragmented approach, even though the geographical, political, and meteorological factors involved seem to favor the establishment of control elements at the local or regional level.

Air-pollution effects on man, his works, and fellow inhabitants of this planet are so pervasive that consideration of a "grand strategy" is not unwarranted, for insofar as air-pollution control is one integral aspect of "ecological design," what is at stake is nothing less than our entire industrialized way of life—our ability to produce and distribute goods to support the earth's rising population. In the extreme view, what is threatened may not only be individual life and health but those of our entire species. The ultimate goal of the grand strategy is to maximize production and global distribution of goods at the least cost (in energy), without degrading the environment or stocks of unreplaceable resources.

The question of what grand strategy to use has been considered by many sources—unfortunately, often with considerable emotion, a particular bias, or a local slant. Others have taken a more rational and global view; among these, Koenig[1] and his associates at Michigan State University have identified several basic and independent strategies appropriate to air-pollution control that are worth examining. Three of these are (1) spatial distribution, and (2) material recycling, and (3) obsolescence rate.

The first strategy utilizes the local environmental capacity for dispersion and assimilation of pollutants by distributing sources of various strength and receptors in accordance with that capacity. If this distribution is geographic, as considered by the authors, this might be termed a *land use strategy*. One can also envision a distribution of source strength in time, given any fixed geographical distribution, in which the source strength is modulated in phase with the temporal variations in assimilation capacity. Since this capacity is in large part dependent on weather variables, this approach might be termed a *meteorological strategy*.

These two subsets taken together lead, as Koenig states, "to maximum utilization of the unaltered natural environment for processing biodegradable wastes—the only component of the total life machine that operates without man-made auxiliary sources of energy."

Material recycling is also a valid and necessary strategy. To the extent that materials can be reused, they need not be dispersed into the environment (as in the incineration of solid wastes or venting of refinery waste gases) and so tax its assimilation capacity. It is obvious also that, ultimately, recycling is the only way by which living beings can survive indefinitely in a closed system; the only question is whether man can manage this evolutionary process so as to "minimize the number of industrial and urban species ending up as fossils."

Recycling, although necessary in the equilibrium state, is a less desirable process than the first strategy, because it must utilize energy in an expensive and concentrated form. This is a consequence of the greater disorganization (entropy) of waste materials as contrasted with their initial ordered form as manufactured goods and of the thermodynamic laws that require energy input to reverse the situation. It can be argued, of course, that the dispersion of gaseous wastes by atmospheric processes also requires energy, but this energy is provided by the sun; it is free to mankind. This use of atmospheric circulation thus constitutes a net gain to our energy resources that would otherwise be wasted.

Reduction of the obsolescence rate of manufactured goods is the third strategy enumerated by Koenig. Clearly, the longer such commodities as clothing, wrapping materials, and books can be used, the less pollution they will cause in a stated interval of time and the less energy will be required to recycle or refabricate them. But consumables such as daily newspapers, food, electrical energy, and gasoline are not as amenable to this strategy and so constitute a large part of the pollution problem. Furthermore, although it is true that "planned obsolescence" as a means of stimulating the economy is nonproductive in the longer-range societal view, it is also true that a contrary policy is inhibiting in terms of flexibility and technological advances.

Much as he may wish or think it needful, the air-pollution control-system designer cannot adapt the entire societal structure to his specifications but must live in the real world. The social and economic factors that decree the

5.1. Overall Strategies

obsolescence rate of goods and the extent of recycling can be influenced by air-pollution considerations and should be, but it is unrealistic to believe they will be the sole determinant. It follows that the primary strategy available to the system's designer is the first: the modulation and distribution of sources of air pollution in accordance with local meteorological factors and atmospheric assimilation capacity. The tools to be used then are land use or zoning, permits, emission standards, long-term and short-term abatement, and monitoring and forecasting of the meteorological factors: dispersion and dilution. The legislation described in the preceding chapter is another and more immediate part of the system designer's constraints, but as we have seen, they conform in general to the land-use, zoning, and permit implementation of this strategy. Where our knowledge has been least, and the source of systems improvement greatest, is in the monitoring and forecasting aspects of control. These factors will be considered in greater detail in the remainder of this chapter and subsequent chapters.

In general, the relationship between abatement of source emissions by various means and the benefits to be obtained must be determined by a systems analysis. This will require a model that predicts a quantitative outcome as a function of input and that can be altered to reflect the various means and mixes of control suggested in the previous chapters; an objective function describing the benefits of the various model outcomes in an order that is based on some value-oriented criteria; and a means of gathering data with which to input the model.

Black,[2] considering these requirements in 1970, concluded that none of these problems had yet been completely solved for an energy–air-quality relationship. He did conclude, however, that air pollution must invariably be considered as a meteorological phenomenon, in which airborne wastes are transmitted from the emitters to the receptor with consequent loss in benefits to the latter. This process can be modified by moving the source (horizontally, as in zoning, or vertically, as in the use of tall smoke stacks) or the receptor. The common measure of disbenefit by which these alternates can be measured is the cost. Alternately, the emissions can be reduced by various control means or a lesser degree of source activity, also at some cost. To select a means of achieving a certain degree of air quality at the recipients' locale, it appears that cost can be the needed value-oriented objective function. (However, this does not answer the problem of how to select the degree of air quality as a goal; the cost of effects is discussed in Chapter 3).

Black, then, sees the necessary analytical model as consisting of three submodels. One represents the sources of pollution, another the receptors (or effects on the receptor), and the third is primarily meteorological. The first two submodels correspond roughly to the source inventory and the dose–response relationships (effects). The last sub-model includes as parameters the transport time from source to recipient (distance and wind velocity), wind direction and variability, the relative temperature of pollutants and ambient air, and the physical and chemical properties of the pollutant affecting its potency and lifetime.

A meteorological model such as this does not yet exist in final form, although, as we shall see, substantial advances have been made since Black's analysis. An unanswered and critical question is how many input-data-gathering points are required to give the model a meaningful output? If the model is merely required to produce a coarse average of pollution levels over long periods of time, a few monitoring stations may suffice. For monitoring fluctuations in urban concentration in real time, it is not apparent that sufficient data points are currently available. This point and the question of what kinds of instruments may be needed will be discussed further on.

However, once these problems of model and data are solved, there is a least-cost strategy that can be devised for any desired level of air quality, for any given locality. If, as Black points out, the climatological conditions of a city are such that the need for abatement is continuous, the least-cost strategy will be one that has a high capital cost and a low operating cost. If the need for abatement occurs only at rare intervals because of favorable meterological factors, then a high-operating-cost–low-capital-cost strategy will be optimum. But since each city has a unique location of pollution sources, population centers, process technologies, and climatological factors, it follows that no one strategic plan will serve for all. To achieve the least-cost strategy for each urban center is a task that can occupy a generation of system designers, after which the abatement and control technology may well improve to the point where another set of optima may be possible.

5.2. OVERVIEW OF EXISTING U.S. CONTROL SYSTEM

We have seen that the main sources of air pollution in the United States are those generating waste products from combustion processes. The burning of coal and oil in power plants, domiciles, and industrial furnaces results in sulfur and nitrogen oxides and the fine particles, or particulates, that are the ingredients of a London smog. The gasoline engine emits copious quantities of carbon monoxide, as well as unburned hydrocarbons and nitrogen oxides; converted by sunlight to the Los Angeles type of smog. Atmospheric processes and meteorological factors, including precipitation, wind, and air turbulence, are the agents that dilute, mix, and disperse air contaminants away from urban areas. In many areas, and for much of the time, these factors follow *normal* patterns of air movement, suggesting the existence of regional airsheds, or zones within which the buildup of dangerous concentrations can be prevented by control of contaminant emission at the source. Localities in deep valleys or those where atmospheric layering (inversions) promotes stagnation of the atmosphere can limit the number, strength, or geographic arrangement of pollution sources. The establishment of air-quality control regions by federal law and the policy regulation of emissions on the state, local, and regional level has frequently followed the airshed philosophy and has been useful in reducing chronic air pollution in many of these places.

The catch to this approach lies in the word *normal*. It is unfortunately true that most severe episodes of air pollution are associated with meteorological conditions that are unusual for the region. Most of these cases have occurred in prolonged conditions of large-scale atmospheric stagnation characterized by calm, or by light, variable winds; great stability; and sometimes fog. These conditions are not in themselves rare, but their persistence over several days is uncommon. For several decades prior to the December 1952 London episode (Chapter 1), only one such fog was recorded. The meteorological conditions conducive to the Meuse Valley disaster were found only five times in 30 yr.

It is possible to forecast large-scale atmospheric stagnation conditions that increase the short-term probability of episodes, over many areas of the United States, at least 36 h in advance, and such a service has, in fact, been operative since 1960[3]. It is not possible, however, to predict by current statistical techniques the probability or frequency over a long period of the atypical meteorological and pollution-emission events that can lead to a severe local episode. Zoning, air-quality regions, and regulation of sources to fixed emission standards cannot, therefore, be used as a basis of preventing these episodes with any predetermined amount of risk. This applies to any degree of emission control or "rollback," receding as far back as the equivalent population and power levels of 1873, when an episode in London similar to the 1952 event occurred.

Emission control is expensive if applied to existing sources, as evidenced by the cost of electrical precipitators for fly-ash removal or tall chimney stacks for smoke dispersal, which runs into five or six figures. Limiting the number of sources by regional zoning costs society in other ways, by limiting utilization of energy, transportation, and housing. Granting these points, it is clear that an overly effective source control cannot be an optimal means of controlling the everyday concentrations of air pollutants and, at the same time, provides no clear solution to the problem of dangerous episodes. No matter what reasonable (or unreasonable) degree of source emission control is imposed, there is a finite and unpredictable risk of a dangerous concentration buildup in some localities—unpredictable in the sense that its probability is unknown and, therefore, the risk cannot be calculated. It is a major premise of this chapter, however, that it will become possible to forecast the onset of air-pollution episodes in a given locality by acquiring sufficiently good information about the current state of source emissions and meteorological conditions. This will be an extension of the stagnation weather forecasts just mentioned, supplemented in great detail by mathematical models simulating the local meteorology and the dispersion of pollutants in and around each problem area. Forecasting the event will not of itself be sufficient to prevent it from happening. It will be necessary to utilize this information in reducing the current pollution emissions to a level that the atmosphere can disperse. This latter step itself is not new but is the basis of the air-pollution alert systems in Los Angeles, New York, Chicago, and the Rijnmond system in the Netherlands. In each of these, current levels of ambient air pollution are sampled and compared with meteorological forecasts. If unfavorable, orders are issued to industrial and other sources to curtail their emission. The novel element will be in the forecasting, in "feedforward control." By being able to compute air-pollution levels for each portion of a city sufficiently well in advance (several hours or a day), given the source emissions as input, it will be possible to select an emergency abatement schedule that will minimize both the health risk to all inhabitants and the disruption of other priorities of urban life.

The full capabilities to achieve feedforward control of air pollution do not yet exist, other than in an abstract way. The instrumentation, meteorological and chemical, that will be required may be much more than has been allowed for in current budget planning. The dynamic mathematical models of the polluted atmosphere that will permit such detailed forecasts have not yet been completely developed. It will be expensive, complex, and hard to sell to political leaders in these times of conflicting urban priorities. But we have already seen why we should strive to achieve this capability.

In the first place, it permits implementation of the most economical mix of abatement strategies, both for the short and long terms. Reduction of all emissions to an absolutely safe level will be impractically expensive, burdensome, or, in the most critical localities, impossible. The contrary policy of minimal emission control except during forecasted emergencies, however, would involve frequent "slamming on the brakes" as average pollution levels creep upward and random fluctuations push them more frequently over the critical margins of safety. Furthermore, it would be no solution for the multibillion-dollar problems of material and vegetation damage, visibility, chronic illness, and soiling. The optimum strategy is clearly a mix of these two extremes. Chronic air pollution should be kept down to something approaching a cost–benefit level, where the incremental costs of improved air quality equal those of air-quality damage less the societal cost of episodes. Admittedly, we do not yet know where this level is. At the same time, the feedforward system of emergency control will give us the highest possible assurance of safety for the most vulnerable recipients—children, old people, and the chronically ill.

The second benefit of the mixed strategy is growth potential. Air pollution, nationally and globally, must continue to grow because it is intermixed with increasing per capita energy utilization and population growth, and a quadrupling of the latter is forecast by 2030.[4] The only viable alternative is strict rationing of energy use and production. Automobile emissions of hydrocarbons will be reduced by enforcement of current federal satutes but will rise again after 1980 because of the growth of the vehicle population.[5] As seen in Section 3.6, production of hydrocarbons will increase to the exent that their increased emissions may largely offset the automotive reductions.[6] Automotive nitrogen oxide emissions, the other ingredient of Los Angeles smog, will continue to increase all during this period. The potential for sulfur oxide emissions from fossil fuel (coal and oil) used in power and industrial plants will continue to rise at a rate of about 50% of its 1960 level each 10 yr, until 1990, when it will (hopefully) fall off a result of nuclear power installations.[7] It will be partially offset to the extent that the less available natural gas, low-sulfur oils, or expensive coal desulfurization processes can be utilized.

Perhaps improved knowledge of the human health effects will, more than any other factor, increase our awareness of the problems of air pollution. As researchers delve more deeply into the effects of air contaminants, the levels considered to be dangerous to health are lowered and the number of suspect pollutants rises. Correspondingly, the extent of day-by-day control or the number of emergency control situations forecast will tend to increase.

A rigid strategy of fixed abatement levels, aside from being more costly, does not possess the requisite flexibility to cope with emergency situations. More and more, we will be forced to take an optimum path and implement sophisticated solutions to this deeply rooted problem.

Although the remaining chapters will be devoted to descriptions of the technical problems to be overcome and the current state of the art, all of the problems are not technical. First and foremost is that of national will—the recognition of the real nature of the problem and the willingness to implement cooperatively a difficult solution. The problem is nationwide, not because of any concept of a centrally located control center but because the effort can only be coordinated and directed at the national level. The legislation needed to implement control must be nationally applicable, since the problem itself is nationwide, requiring implementation of controls in all major cities.

The hardware and software that must be developed under this kind of program will be applicable nationwide, with appropriate tailoring to the locale, topography, climate, and source distribution. Instruments that are economical and appropriate for the task must be deployed in considerably larger quantities, but this constitutes only part of the problem. Returning to the technical gap in the state of the art of atmospheric models appropriate to cities and built-up areas, E. S. Savas[8] cites experiments in New York City supporting the concept of small pockets of high pollution concentration that may move about to random locations under the influence of winds and turbulence. The atmospheric model, then, should make it possible to track and predict the location of these high-concentration pockets rather than merely to monitor the average over the city or a few suspected "worst points." However, the necessity for a much finer scale of measurement and prediction is compensated, as Savas points out, by the possibility of finer control over the emission sources, with correspondingly less cost to the emitters and inconvenience to the population.

We believe then that the national air-pollution control establishment requires a total systems solution, much as does military air defense. The analogy is doubly apt, since without warning and counteraction, more episodes similar to the 1948 Donora tragedy can be expected whenever statistical fluctuations from rising urban pollution levels coincide with unfavorable weather variables. For defense against such disasters, crowded urban areas will require a real-time capability for surveillance, prediction, decision, and quick control reaction; in short, an effective control system capability.

As a first step in the effort to expose the problems of air-pollution control in the United States to the discipline of systems analysis, we will attempt to describe current "real world" activities, including the legislation outlined in the previous chapter, in such a manner as to define them as functional parts of a system. The organization, or "architecture," of this system and the constraints on its operation can then be identified. When this is done, the relationship between subsystems may be clarified, and we can proceed further to specify the detailed objectives and functions of each of the many parts, such as sensors, computing hardware or software, in more idealized form. Further analysis can then proceed to tell us where and in what manner each subsystem may be improved for more effective operation of the whole.

Over 14,000 air-quality monitoring devices are currently operated by various local and federal agencies. Approximately 10% are continuous automatic analyzers; some of the remainder are as simple as dust-collection jars. They may be considered as an embryonic national air-quality control system insofar as their operations have been supported and loosely coordinated by federal agencies.

The chief element of nationwide scope is the National Air Surveillance Network (NASN) under the EPA. In the ten years prior to 1969, NASN operated in a total of 370 U.S. sites, sampling for particulate concentration on an intermittent basis. Most of these data were taken from a single station per city.[9] The parameter of particulate-size distribution, important to health, was not measured until 1970, when NASN began collecting data from a 6-station network; 50 such stations were planned for the nation by 1974.[10] In 1971, the national data on SO_2 concentration was obtained by NASN and cooperating activities from a total of only 66 sites, each taking one 24-h sample approximately biweekly.[11] In contrast, the city of Cleveland in 1971 was operating 21 stations, taking 24-h samples each 6 days and measuring three parameters (SO_2, NO_2, and particulates).[12] California in the same period operated 70 continuous stations (35 for the state and the remainder in Los Angeles and the San Francisco Bay area); the city of Chicago and Cook County, 12 continuous stations; New Jersey, 18; and New York State, 11.[13] The total effort is thus highly fragmented and in an early formative stage, so that care must be taken in ascribing "system goals" or objectives to this collection.

Nevertheless, there are three functions that must be performed by any such system if it is to be meaningful. These are *identification* of air pollutants and their current concentrations and distribution; *prediction* of their future status in a timely manner; and *control* over pollution sources in the same time scale, to nullify undesirable trends.

In general, current efforts are unable to satisfy more than the first of these needs. Severe fiscal and technological barriers hinder the development of adequate systems; most can be removed within this decade, granted sufficient priority. Longer-term constraints on the extent of control over air pollution will be cost–benefit decisions, basically, the social and economic trade-offs between air purity and power or production efficiency. Given the present imbalance of air quality and control measures, it is safe to guess that these limits will not be reached for a long time.

Current EPA planning projects less than 800 air-quality monitoring stations monitoring 8–10 constituents and some 7000 industrial source installations monitoring 1 or 2. Statistical sampling considerations, introduced later, suggest that this number is one or two orders of magnitude less than that needed to perform the nationwide system functions suggested above. It can be assumed, then, that a second-generation control system must be planned and implemented and that this ultimate system will be far more comprehensive and less fragmented.

This is not to imply that a nationwide system should be a single, monolithic entity, controlling air-pollution sources from a central location. The unit of current technical and legislative efforts (see Chapter 4) is the self-contained airshed or air-quality control region. Its boundaries are theoretically defined so that there will be a minimum of air-pollution transfer in and out of the area (although in practice, compromises are made to suit political and demographic realities). Although the concept is admittedly relative in that pollution transport across some region boundaries does occur, and we have questioned as to the statistical validity of the airshed idea (considering the rarity of meteorological conditions associated with disasterous high pollution episodes), there is no question that local weather is a prime factor in air pollution. Local weather phenomena, of course, are strong functions of location, topography, and time. Local, or at least decentralized, air-quality monitoring, prediction, and control, therefore, are certainly requirements, with the added provision for exchange of data between local control centers or through a master data bank (cf. Sections 1.2 and 6.2.3).[14]

Organization and Goals

Since current efforts to control air pollution are so highly fragmented and responsibility is divided among state, local, and federal agencies, it is not yet correct to speak of a *national* air-pollution control system. Therefore, identification of system technical goals on this level is not strictly justified. The federal government (through the activities of the EPA), does, nevertheless, establish the limits and direction of the state and local control efforts through the regulations detailed in the previous chapter. The bulk of the day-to-day monitoring and control effort, as we have just noted, is left to the states and to the local control agencies. The physical abatement of pollutant emissions is necessarily accomplished at each source by the operators of the vehicles, municipal incinerators, industrial plants, and other local activities that cause them in the first place. These persons act under the spur of social pressures, enforcement agencies, and the penalties provided by law.

In this sense, a *nationwide* control system can be envisioned, at least in the ideal. Given a system, its goals, objectives, and the functional relationship of its parts can be examined. Without this viewpoint, the activities at the various levels appear chaotic and powerless, which, in fact, they are without some form of unifying directive.

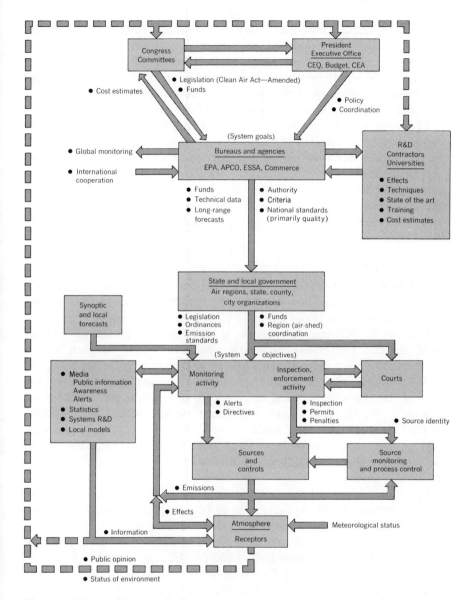

Figure 1 Nationwide air-pollution control (courtesy of *IEEE Spectrum*).

A generalized diagram of the system would appear something like Figure 1. This is clearly a loosely coupled miscellany of new and old political activities, technical functions, public and private organizations, all of which affect in some way the status of the air environment and of the animate and inanimate receptors. Nevertheless, feedback loops, which are the feature of any control system, can be discerned in the diagram.

The goals of the system, impelled somewhat loosely by public opinion and defined more precisely through the elective and legislative process, may be summarized as follows: to identify and oppose adverse trends of national and local air-pollution concentration affecting the receptors (humans, agriculture, and husbandry), climate or visibility, and materials. Less explicity, the system should strive to identify and monitor similar changes in the global atmosphere adversely affecting the air resource, world climate, or other environmental factors.

System Functions

We have stated that the operating functions of the system are identification, prediction, and control. We will attempt to define these in greater detail, with reference to our nationwide "system block diagram."

Identification implies casual or statistical knowledge of the dose–response and time–concentration behavior of all receptors in the nation to all significant pollutants as well as their synergetic combinations. This knowledge must be built into the criteria and standards that form the core of the operative technical functions. In turn, this is generated for the system by parallel research activities that form an auxiliary loop and that process data from the real world as well as the laboratory. In the U.S. system, this identification function is embodied in the EPA regulations defining the pollutants and their permissible concentration in ambient air of acceptable quality.

Identification further requires a monitoring network of such temporal, spatial, and analytical capabilities as to be capable of measuring the current concentrations and trends of all pollutants of interest.

Prediction implies an extrapolative capability or mathematical model of the atmosphere and pollution system and the necessary computing capacity to implement a timely prediction of future pollutant distributions in both space and time. In turn, these capabilities require a sufficient knowledge of the chemical and physical reactions of pollutants in the atmosphere; their sources, transport, diffusion, and decay; inputs of meteorological theory and data; and the means of acting on this information—instruments, computers, and software. The time scale of the prediction must be matched to the control function performed; only for emergency control of local episodes need it be in real time—that is, a calculation rapid enough to permit effective control action to be taken prior to acute damage to receptors.

Control activity results from information obtained through the predictive subsystem or through registration, inspection, or self-monitoring of the sources in response to an emission standards code. The latter is based on the air-quality standards, which in turn result from the identification subsystem criteria. Physical control requires techniques to abate pollutant emissions or authority to permit or regulate operation of a facility. When local concentrations cannot be kept below standards, control also requires an interface with the community, such as the communications media, so as to alert sensitive receptors to take appropriate action on their own behalf. A data- and information-retrieval capacity also serves to generate long-term trend statistics and to evaluate local system operation or improve predictive models. In some instances, the latter will also be used directly in the control loop, in order to evaluate the best means of abating the threat of severe pollution episodes—for example, shutdown or alternate sources. Ideally, this will be done on a cost–effectiveness basis, perhaps using linear programming techniques. It is implied that this kind of predictive simulation (during episodes, at least) must be computed a high speed so as to allow for timely decision and action.

In exercising the control function, the public monitoring activity or a cooperating enforcement branch acts to identify violators of the emission code by processing data from fixed sensors, by mobile or remote sensors, or by cooperative use of private stack sensors located at the sources.

5.3. CONSTRAINTS AND COST EFFECTIVENESS

Table I lists these system functions along with the constraints that, in our opinion, will limit the degree of control possible or desirable in both the near and long terms (by *near term* is meant the next 2–5 yr and by *long term*, 10 yr or more).

In the near term, many of the system constraints will be technical, but given sufficient priority, they should be relieved well before the end of the decade. Although research into long-range effects of trace pollutants on human health and on the balance of ecosystems will be a continuing task, other current technical problems lie within present capabilities. These include the development of reliable low-cost sensors, the refinement of transport and diffusion theory, adequate photochemical reaction models, and improvement of some control techniques. Other problems, involving legal or political organization and funding, show some signs of resolution in the foreseeable future. We do not minimize the decisive effects of societal

Table I
Control System Functions and Constraints

System function	Near-term constraints	Long-term constraints (>1980)
Identification:		
Identification of pollutants and synergistic combinations	Ignorance of effects on receptors and ecosystems, atmospheric reactions, and combinations.	Natural background. Cost–benefit limits to pollution control.
Definition of dose–response and time–concentration behavior	Experimental limitations, long-term effects, reliance on statistics and epidemiology.	Continuing problem, especially long-term effects on humans.
Existence of monitoring network	Funds to establish networks. Network design criteria ill defined. Poor fit to political divisions.	—
Spatial placement of sensors	Problems in defining air regions. Weak design criteria. Lack of survey data. Sensor cost per unit area.	Cost–benefit limits to control.
Chemical capability of sensors	Adequate sensors unavailable or costly.	—
Time response of sensors	Some sensors are very slow.	—

Prediction:		
Mathematical model		
Physical theory	Incomplete physicochemical theory, especially of photochemical dynamics.	—
Software	Cost of programming.	—
Computing hardware	May need large parallel processors.	—
Meteorological data	Data criteria incompletely defined. Some data (e.g., profile) inadequate.	Cost–benefit limits of small-scale meteorological monitoring.
Source survey and data	Detailed surveys costly.	Continuing problem. Ultimately, cost of telemetering network.
Control:		
Air-quality and emission standards	Conflict with technology growth. Self-interest of individuals and groups. Energy needs.	Limitation on public education, economic practices, and social motivation.
Legal authority to control	Authority mostly adequate.	Same as above.
In-stack, mobile or remote detection gear and organization	Inadequate or costly in-stack gear. Few remote sensors developed. Cooperative system incomplete.	Ultimately, cost of telemetering network.
Community communications	—	—
Optimizing control model or simulation	Model theory incomplete. Programming, possibly hardware costs.	—
Physical control techniques	Control techniques costly, inadequate in some cases. Possible shortage of low-sulfur fuels. Inadequate municipal incinerator funding. Inadequate alternates to automobile transport.	Cost–benefit limits. Population distribution and growth limits on mass transportation effectiveness.

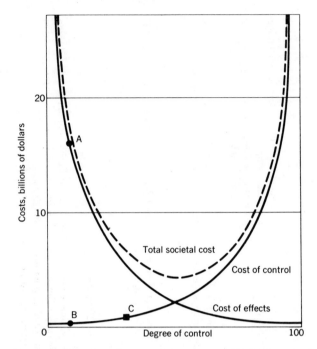

Figure 2 Cost–benefit control (courtesy of *IEEE Spectrum*).

and economic factors—the need for community and national will to overcome air pollution and to allocate an economic priority to this task among other pressing urban problems. Nevertheless, environmental conservation in general and the need to reduce air pollution in particular are now accepted national goals.)

In the longer term, there remain three basic types of constraints. One involves social motivation, which eventually is reflected in legislation. These limits involve fundamental choices on the amounts and sources of power to be produced, on mass transportation, on urban population distribution, and on the means for allocating pollution costs between the public and private sectors. Of these constraints, systems analysis on this level has little to offer. A second problem, that of assessing the medical and human effects of trace contaminants, may continue to be studied for a long time. The remaining constraints are, in theory, amenable to cost–effectiveness analysis.

The cost–benefit approach toward establishing the desirable degree of air-pollution control has been criticized, as noted in Chapter 3, on the grounds that we cannot yet plot curves of the costs of air-pollution effects

5.3. Constraints and Cost Effectiveness

or of the costs of controls as a function of the degree of control. The pioneer work in this field (see Section 3.4) was only published in 1967, after a three-year study by the economist R. Ridker.[15] We can predict the general shape of these curves, however, since the costs of effects will certainly decrease and the costs of control will rise as a greater degree of control is imposed. Following Ridker, we can draw the hypothetical graph, as in Figure 2 (repeated from Section 3.4). Here, the sum of effects and control costs represents the total societal cost, regardless of who is to pay it. Clearly, the optimum degree of control will correspond to the minimum total cost. This will be achieved when the marginal costs, the slopes of the effects and control cost curves, are equal.

It is difficult to quantify either curve for many reasons. The direct costs of the effects of a concentration level of one pollutant are a function of the damage to each receptor (person or object) attacked at that level, multiplied by the cost of damage and by the number or receptors. These costs must be computed for various levels, for different pollutants, and for their combinations. The data are not known for the most part and, as indicated in Section 3.4, will change with population. Furthermore, the "utility," or the dollar cost put on the loss of such aesthetic intangibles as scenic visibility or art works, will vary widely from person to person. The cost-of-control curve is equally elusive, since the costs will change with the growth of technology and may not be a smooth function of control. Chapter 3 described some of the work underway to amass the missing information; meanwhile, we may make some deductions from a few data.

It is useful to know, at least, on which side of the optimum control point we are now. If to the left of the optimum, how far should we go before it does not pay to clean up the atmosphere further? Barrett and Waddell, in 1970, cited $16 billion for the cost of effects in 1968, giving one clue[16] (see Table XII, Chapter 3). An estimate of $200 million for expenditures in air-pollution control during the same year, only one-eighteenth of the effects cost, is quoted by J. Yocum and R. McCaldin.[17] These are plotted as points A and B on Figure 2. A third clue, cited by the CEQ in their 1970 report to Congress, gives a cost of $4.5 billion through 1975, or $900 million/yr, to implement the 1967 Air Quality Act in 100 metropolitan centers, using assumed emission standards similar to those of New York and San Francisco.[18] This represents the cost of a remission of 70% from the 1967 levels of sulfur oxides, particulates, and hydrocarbons, but it by no means represents the total cost of "clean air" (cf. Section 3.4). Furthermore, it includes only the control expenditures, primarily the costs of gas-cleaning equipment and low-sulfur fuel, with no allowance for monitoring or prediction instrumentation. If, accepting these shortcomings, we plot this point

(C) on the hypothetical curves of Figure 2, the need for further progress toward optimum conditions is strongly suggested.* It is apparent that the nation could have spent at least five times the 1968 amount on air-pollution control without concern about overshooting the optimum. This sum, far from being excessive, represents less than 1% of the annual output of the industries concerned.

5.4. SYSTEM OBJECTIVES

Although many of the state and local components of the current national air-pollution control system have anticipated Federal law, the unifying force is now the Clean Air Act enacted by Congress in 1963 and amended in 1970. The provisions of this legislation were detailed in Chapter 4. In the 1970 amendments the trend toward direct federal control was confirmed, and authority was granted for federal air-quality standards (Table I, Chapter 4). Authority for federal emission standards for some stationary sources has also been provided. In general, the 1970 Clean Air Act Amendments speeded up the formation of a working, nationwide air-pollution control system by arbitrarily assigning as air-quality control regions any areas not already so designated, by establishing federal air-quality standards and by demanding that the states implement these standards without delay through legislation of appropriate emission standards and controls. In doing so, the problems of defining airsheds by diffusion modeling were momentarily sidestepped, and questions of cost–benefit in setting standards were held in abeyance but not, as has been inferred, put aside permanently.[20] As we have seen, there is little risk in setting the highest possible goals, since there is small risk of overshooting them. There can be no argument with the idea of mandating quality standards well below the level required to protect the public health—the so-called primary standards. But when we speak of air-quality goals, the "secondary standards," which are established to prevent "any known or anticipated effects" of a pollutant, we are immediately constrained to weigh the cost of *any* effects and led back to the area of cost–benefit analysis.

The system objectives, as a result of this division of responsibility, must be considered on the level of the federal government (the EPA), that of the states, regional, and local agencies; and of the sources which are to be controlled. In control-systems terminology, "supervisory control" and adaptive goal setting occurs at the federal level; "feedback control" is exerted at the

---

* That the control cost estimates are becoming more realistic is seen by comparing the 1970 CEQ estimate with that in 1972, which is 10-fold greater on an annualized basis.[19] Cash flow required for air pollution control during 1971–1980 is $106.5 billion in 1971 dollars which, however, includes $48.2 billion as capital investment.

Table II
Detailed Subsystem Objectives

I. Federal Subsystem Objectives[21]
   A. To require states to take responsible social, technological, and legislative action to protect the public from adverse air-pollution effects or to take federal action if the states fail to do so in a timely and effective manner
      1. To publish a list of those air pollutants that affect public health and welfare and result from "numerous and diverse sources."
      2. To publish criteria promulgating the latest scientific knowledge of effects of these air pollutants, including such factors as atmospheric conditions, interaction between pollutants, and known or anticipated effects on the environment and the economy.
      3. To designate air-quality control regions (AQCR) based on jurisdictional boundaries, urban–industrial concentrations, existing air-quality levels, and any other factors (such as meteorological or topographical factors) necessary to provide for effective implementation of air-quality standards. (States may further subdivide these AQCR into more suitable units.)
      4. To publish national ambient air-quality standards for the guidance of states. These include "primary" standards, deemed necessary to protect health and air-quality goals, or "secondary" standards, to protect the public welfare from any known or anticipated effect. Along with standards, to publish reference methods of analysis to determine the concentration of each pollutant.
      5. To require from states, subject to federal approval, plans to implement the air-quality standards by means of legislation, organization, and emission standards ("control strategy"). If state plans are not forthcoming or are inadequate, to implement federal plans.
   B. To publish direct emission standards for specified categories of new stationary sources, including new sources owned by the federal government. These standards are based on the "greatest degree of emission control achievable through the latest technology." To provide for certification of such sources and to enforce compliance with standards.
   C. To publish national emission standards for hazardous or other selected air-pollution agents for which there are no federal air-quality standards but which are adverse to health and welfare, and to provide for certification and enforcement of these standards or for prohibition of their emission. Initially, three hazardous agents were identified: asbestos, beryllium, and mercury.
   D. To establish emission standards for vessels, aircraft, and vehicles; certify new vehicles or engines; establish emission standards and certify controls for used vehicles; and control fuel additives.
   E. Exercise emergency powers to immediately enjoin emission of air pollution endangering health.
   F. To supply meteorological warnings of conditions over large areas conducive to high air-pollution potential within AQCR's.

Table II—(Continued)

II. State, Regional, and Local Subsystem Objectives

   A. To prepare and implement EPA-approved plans for the attainment of air-quality standards (primary) and goals (secondary) for each pollutant in accordance with the following detailed requirements:

   1. To classify each AQCR in accordance with its present air quality into one of three priorities (Table II, Chapter 4) and to maintain surveillance over the degree of pollution in these regions.
   2. To prepare an adequate "control strategy," consisting of emission requirements, timetables for conformance, and other measures needed to attain the primary air-quality standards in 3 yr and the secondary goals within a reasonable time, in each region, allowing for current levels of pollution and providing for projected growth in population, vehicle traffic, and industry. Adequacy of the strategy must be demonstrated by use of a model of the anticipated air-pollution levels, either a diffusion model or a simple proportional reduction based on the highest measured concentration of pollutant in the region. Current emission levels are determined by a detailed inventory of emission from area sources and point sources.
   3. To provide a system for monitoring air-quality data, to consist of at least a minimum number of devices and sites (see Table III, Chapter 4). Air-quality changes must be reported every three months. Provide emergency action criteria and procedures, including:

      (a) A surveillance systems, including rapid data-processing, to identify the approach of pollutant concentrations toward episode levels (see Table III, Chapter 3)
      (b) Daily forecasts of atmospheric stagnation capable of update each 12 h
      (c) An emergency emission-reduction plan that includes abatement action for each major source (over 100 tons/yr emission).
      (d) Procedures for control and communication of status and action during episodes, including communication with sources and the public (see Figure 1).

   4. To establish a system of controls and permits for stationary and mobile sources of emission consistent with attainment of air-quality standards. These include source inspection, tests on point sources, emission monitoring, enforcement procedure, and means for detecting excessive source emissions.
   5. To provide intergovernmental cooperation to prevent emissions from one region from interfering with the air quality of another.
   6. To require reports on emissions from source operators.
   7. To provide for periodic test and inspection of motor-vehicle sources as required to meet air-quality goals.
   8. To certify new stationary sources for compliance with federal standards or better.
   9. If desired, the state may adopt standards and plans to implement air quality higher than the federal standard.

Table II—(Continued)

III. Source Control Subsystem Objectives

    A. To control emissions from new sources using latest techniques, in conformance with state or federal requirements for certification (see I).
    B. To establish and maintain records from monitoring equipment showing compliance with emission standards. To maintain records and reports on fuels if required by state regulations.
    C. To abate emissions in violation of applicable performance standards within time limits (72 h in some cases) or cease operation, under penalties. Avoid emission of prohibited hazardous pollutants (see I).
    D. To prepare preplanned control strategy for air pollution episode alert in accordance with regional–state regulations. To implement when required to do so by the control activity.
    E. For all sources, to control emission of pollutants in undesirable or illegal amounts by the most economical means meeting requirements, including process changes or plant relocations, if necessary.[22]

state, regional, and local levels; and the sources represent the "controlled plant" and its minor regulatory loops (see Figure 1).

Different time scales are encountered at each level. Federal policy (except in the case of emission standards) is concerned with future decades; local monitoring agencies measure and react to concentrations averaged over minutes, hours, or even a year; but source controls must react at process speeds.

The specific technical objectives of the current national subsystems, defined in part by the 1970 federal law, are detailed in Table II (see also Section 4.2).

5.5. SECOND-GENERATION SYSTEM

The objective of the national air-pollution control system, as outlined in Table I, may appear complex and unwieldy. This is because it is still in the formative stage and is affected by factors of politics, growth, and ignorance. A mature system with funding measured in the billions of dollars that has been shown cost-effective, will have overcome most of these influences. Such a second-generation system will have only four functions:

1. To monitor current pollutants' levels
2. To forecast and predict future trends and steady-state levels
3. To exert effective control and abatement
4. To prevent dangerous "episodes" by quick reaction control.

In the ideal system all the important pollutants and their critical concentrations will be known, mathematical control models and monitoring networks will be installed and operating, and all important sources (including transportation) will be registered or accounted for and capable of having their emissions physically controlled (as by fuel switching, operating or shutdown changes) in accordance with some prearranged strategy determined by the legal control authority.

The second-generation system, then, will consist of individual control agencies acting at each air-quality control region or metropolitan center, each capable of implementing these four functions.

Current-level monitoring is a basic prerequisite for the performance of all other functions in an effective manner. In addition, monitoring satisfies the need for public information and, indirectly, the long-range feedback shown in Figure 1. The bewildering variety of air-pollution receptors and effects complicates the problem of providing an easily understood pollution index. Ideally, this should be expressed as a single number. (As a measure of the difficulty encountered, one recent proposal attempts to combine the effects of six known pollutants and their interactions and reactions with the atmosphere in a single formula.)[23] This topic will be discussed in Section 7.1.

The monitoring system also provides detailed records for legal, administrative, and statistical histories. The latter can improve mathematical models of the airshed transport and diffusion characteristics, aid medical and receptor effects research, and act as a long-term quality-control check.

Forecasting in the real time (see below) provides an early warning to sensitive receptors of potentially dangerous episodes of pollutant concentration. It permits timely control measures to be initiated and, through models (simulators), selection of the most effective control means. Long-term predictive capability, on the other hand, is a tool to aid in decisions on source permits, land use and zoning, and selecting legal emission standards.

Control is the corrective response to short- or long-term predictions of future pollutant concentrations. In the long term, average levels can be held below dangerous values by withholding permits for new sources or placing more stringent restrictions on emissions. The effects of these source modifications can be evaluated with computer-implemented models of the airshed topography and meteorology. Controlled sources may monitor their own compliance with normal and emergency orders by means of stack or ambient monitors, supplemented by patrols of the controlling agencies, providing feedback.

Violation detection is a specific form of control requiring identification of uncooperative sources. Sensors capable of detecting pollutant emissions from remote locations or point sensors may be used in combination with wind direction and other clues. The violator must not only be located and identified, but legal proof of the degree of violation must be obtained. Some major stationary sources are required to maintain recording sensors or facilities for sample probes used by inspectors. Ultimately, effective system operation may demand installation of remote-reading sensors in major sources to permit continuous telemetering of data to a processing and monitoring center.

Quick reaction control of concentrations approaching the level of dangerous episodes can be exerted by demanding the abatement of emissions from major stationary sources or alerting the general public to reduce automobile operation, and similar measures. The communications media can be utilized to alert sensitive receptors in the general public (e.g., to reduce school activities) and take other palliative measures.

5.6 SUMMARY

A comprehensive systems approach toward solving the air-pollution problem, whose potential magnitude is inextricably bound to the growth of our population, energy consumption, and manufacturing wealth, must encompass cost–benefit trade-offs among the broadcast possible mix of emission controls and levels and the carefully monitored use of the atmosphere as a dispersal means and sink. Only by investing in an optimum system can future health and safety be assured without squandering the resources needed elsewhere. The embryonic system is suboptimal from the national view because it is fragmented, insufficiently coordinated, and fails to achieve economies of scale and the control of externalities (the "free" use of air by polluters).[24] The goals or functions of the mature system include identification and prediction of ambient pollutant concentrations and optimal control under both normal and abnormal meteorological conditions. We envision, as do Schwartz and Siegel "an adaptive learning model of [air-pollution control] decision making where the quality of physical, behavioral and decision models is continuously upgraded."[25] Technical scientific, legal, and political constraints exist today, preventing its fruition, but we have only begun to reap the advantages of following the cost–benefit curve.

The optimum system must produce information and predictions about unhealthy air-pollution concentrations in "real time," as it experienced by the human receptor. The technical means by which exotic chemical and meteorological data may be acquired and applied to predictive mathematical "models" of the polluted airspace will be the topic of the following chapters.

REFERENCES

1. H. Koenig, W. H. Cooper, and J. M. Falvey. *Industrialized Ecosystem Design and Management, Part I*, under NSF Grant GI-20. Michigan State Univ. Press: Ypsilanti, Mich., 1971.
2. G. Black. *Air Quality and the Systems Approach*, NTIS publication PB 195797. George Washington Univ. Press: Washington, D.C., 1970.
3. E. Gross. *The National Air Pollution Forecast Program*, U.S. Dept. of Commerce, Environmental Science Services Administrative, Weather Bureau: Washington, D.C., 1970.
4. A. L. Austin and J. E. Brewer. "World Population Growth and Related Technical Problems." *IEEE Spectrum*, **7**, 46 (1970).
5. NAPCA. *Control Techniques for Carbon Monoxide, Nitrogen Oxide, and Hydrocarbon Emissions from Mobile Sources*. Washington, D.C., 1970.
6. "Olefins, Aromatics Face Steady Growth." *Chem. Engr. News*, **48**, 24, (1971).
7. NAPCA. *Control Techniques for Sulfur Oxide Air Pollutants*. Washington, D.C., 1969.
8. E. S. Savas. *IEEE Spectrum*, **6**, 77–81 (1969).
9. HEW. *Air Quality Criteria for Particulate Matter*, Washington, D.C., 1969, pp. 13–14.
10. R. E. Lee, Jr., and S. Gorman. *Cascade Impactor Network*. EPA: Research Triangle Park, N. C., 1972.
11. EPA. *Air Quality Data for Sulfur Dioxide, 1969, 1970 and 1971*, Research Triangle Park, N. C., 1972.
12. H. Neustadter, S. Selk, and J. Brun, Jr. *Statistical Summary and Trend Evaluation of Air Quality Data for Cleveland, Ohio in 1967–1971*. NASA, Lewis Research Center: Cleveland, 1972.
13. *Environ. Sci. Technol.* **6**, 114 (1972).
14. E. L. Katz and T. R. Morgan. "Analysis of Requirements for Air Quality Monitoring Networks." Paper presented to the annual meeting APCA, St. Louis, Mo., June 14–18, 1970.
15. R. Ridker. *Economic Cost of Air Pollution*, Praeger: New York, 1967.
16. L. S. Barret and T. E. Waddell. *The Cost of Air Pollution Damages: A Status Report*. NAPCA: Washington, D.C., 1970.
17. J. E. Yocum and R. O. McCaldin. In A. C. Stern, Ed., *Air Pollution*, 2nd ed., Vol. I. Academic: New York, 1968, p 651.
18. M. Fogel et al. *Comprehensive Economic Cost Study of Air Pollution Control Costs*, Report FR-OU-455. Research Triangle Inst.: Durham, N.C., 1970.
19. CEQ. *Third Annual Report*. Superintendent of Documents: Washington, D.C., 1972, p. 276 (Table 1).

20. V. H. Sussman. *New Priorities in Air Pollution Control. J. Air Poll. Control Assn.*, **21**, 201–203 (1971).
21. *Air Pollution,* 1970, Pt. 5. Hearings before the Senate subcommittee on air and water pollution of the Public Works Committee, U.S. Senate, 91st Congress, 2nd Session on S.3229, S.3406, S.3546, Appendix. U.S. Government Printing Office: Washington, D.C. 1970.
22. Stern, A. C., Ed. *Air Pollution,* 2nd ed., Vol. II. Academic: New York, 1968.
23. L. R. Babcock, Jr. "A Combined Pollution Index." *J. Air Poll. Control Assn.,* **20**, 653–659 (1970).
24. Ref. 2, pp. 47–49, 83.
25. S. Schwartz and G. Siegel. In A. Atkisson and R. Gaines, Eds., *Development of Air Quality Standards,* Merrill: Columbus, 1970, p. 25 et seq.

6

Data Acquisition and Monitoring

In the previous chapter it was concluded that the optimum cost-effective air-pollution control system, causing the least burden on the health and material resources of a society, will require its functional subsystems to perform the following:

1. Monitor current ambient pollution levels
2. Forecast future ambient levels and trends
3. Monitor and control the emissions of man-made polluting sources
4. Provide a quick reaction to the onset of severe episodes of pollution.

Each of these functions require the acquisition and processing of data, much of it on finer temporal and spatial scales than have been accepted in the past.

The kinds of data required and their density depend on the ultimate application. At this point, we wish to consider the sensing process of the "nationwide control system" in the abstract—as an information process. The actual data sensors will be considered in more detail in later chapters. The questions we now wish to consider are the following:

1. What kinds of data are needed? Do we require chemical or meteorological data or both?
2. What spatial density is required of the data? How many data sampling stations are needed per square mile or unit of population?
3. How often shall data be taken? What should be the time constant of the data sensors? How speedily or slowly may it be gathered and put to use?
4. What is the best way to acquire the data? What are the optimum modes and configurations of data sensors and stations, and how shall the data be communicated?
5. How shall the data be processed and used?

Clearly, the answers to these questions will vary according to the particular need. Long-term averaging of pollution levels for monitoring effectiveness of land use or transportation strategy or for historical records does not require very fine-grained data, and it may be collected and examined without undue haste. On the other hand, short-term prediction of rising, rapidly fluctuating, and acutely hazardous levels of pollution in densely populated areas may demand finely spaced data at frequent intervals, as well as the ability to acquire and process it on a very short time scale. In this chapter, we will be largely concerned with the ability to monitor closely and predict pollution levels, since this capability is demanded by our overall strategy (as well as the current situation in most industrialized communities). The more relaxed needs, of course, can be met by the same data-gathering process.

6.1. CHARACTERISTICS OF AIR-POLLUTION DATA

Air-pollution control requires data on the chemical species and particulate matter currently in the ambient, atmosphere; wind, temperature, profiles, or other meteorological factors; information on the locality and intensity of stationary sources or automobile traffic emissions; and synoptic and forecast data on both meteorology and emissions.

6.1.1. Ambient Air-Pollution Data

On the first question alone (What chemical or particulate components are to be measured as indicative of air pollution?), there has been little agreement. In a desire to impose simplicity on what is at best a complex situation, substantial efforts have been put into the development of a single "pollution index" that purports to represent the overall degree of air pollution. Such an index will be discussed in more detail later (see Section 7.1), but it is clear that the general acceptance of such a number would be a great convenience for informing the public and setting into operation the various mechanisms of alert and control.

The simplest index of all would be the acceptance of a single factor, such as a chemical species or a meteorological parameter, that would be proportional to, and represent all of, the elements making up air pollution. Monitoring of that single species, especially if it were relatively easy to measure, would be an economical means to the acquisition of a good portion of the needed data. We have discussed enough of the effects of air contaminants to know that this is not possible in the general sense, however, for restricted situations and times, it may prove to be a feasible approach.

A well-publicized system of air-pollution control based entirely on SO_2 measurements and meteorological data has been placed in operation in the Netherlands Rijnmond industrial complex, between Rotterdam and Euro-

6.1. Characteristics of Air-Pollution Data

poort, which is heavily populated with chemical- and petroleum-refining plants. The assumption on which this system operates is that all significant pollution concentrations rise and fall with the measured SO_2. This was demonstrated, apparently to the satisfaction of Netherlands officials, through an extensive study of the correlation of the variables involved. A similar, SO_2-based system has been used in the city of Chicago. It is clear, however, that the ability to rely on one chemical variable as a pollution index is dependent on a relatively stable set of pollution sources and a traceable means of predicting diffusion of their emissions. The Chicago system could not predict auto-smog concentrations on the basis of SO_2, since they are completely unrelated as to source. During summer weekends, for example, the amount of automobile traffic may well reach a peak, while factory SO_2 emissions are at a minimum.

At the other extreme, Table I lists over 40 species that have been monitored in the atmosphere by U.S. federal agencies. It is clearly impractical and unnecessary to monitor all of these components continuously, everywhere, in the ambient atmosphere, since many would be significantly present only in the presence of specific sources. But many of those listed in Table I and a number of additional components (such as acid mists) are, or will be, the subject of source emission standards and could become ambient standards as well if damage data show the need.

The primary system function is to monitor the ambient air quality in a specific locality. Thus, the answer to the question about what pollutants will be sensed will be determined by both national and local factors. Nationally, the state of knowledge with respect to the effects of pollutants on health and material objects is established by federal criteria. Locally, these criteria will be modified to suit conditions; all pollutants that are the subject of criteria will not be equal hazards everywhere. In the United States, federal criteria have been issued (see Section 4.1) for sulfur oxides, nitrogen oxides, particulates, photochemical oxidants, carbon monoxide, and hydrocarbons.

These criteria not only establish the type of pollutant but the significant concentration level or, more correctly, the concentration and time duration. Current criteria contain a number of statements relating effects to time and concentration (see Table VI, Chapter 3), which can be expressed as points on a two-dimensional plot, as in Figure 1. A line fitted through these points crudely separates the space into "good" and "bad" air-quality regions, representing one form of standard for air quality. A standard of this type is highly nonlinear as a consequence of its mixed origins and is not very amenable to analysis. However, the introduction of time as a dimension of the damage criteria limits the permissible reaction time of the monitoring system instrumentation. This will be discussed in the sequel.

Table I
Atmospheric Pollutants Measured by NAPCA in 1970

Elements	Others
Antimony	Aeroallergens
Arsenic	Asbestos
Barium	β-Radioactivity
Beryllium	Benzene-soluble organic compounds
Bismuth	Benzo(a)pyrene
Boron	Pesticides
Cadmium	Respirable particulates
Chromium	Total suspended particulates
Cobalt	
Copper	*Gases*
Iron	Carbon monoxide
Lead	Methane
Manganese	Nitric oxide
Mercury	Nitrogen dioxide
Molybdenum	Reactive hydrocarbons
Nickel	Sulfur dioxide
Selenium	Total hydrocarbons
Tin	Total oxidants
Titanium	
Vanadium	
Zinc	

Radicals
Ammonium
Fluoride
Nitrate
Sulfate

Source. Ref. 1.

For the present, the criteria have resulted in the National Ambient Air Quality Standards (NAAQS),[3] which have been described (see Chapter 4, Table I). These components can be listed in two groups. The first group represents those elements that the criteria show have of themselves undesirable physiological and material effects, and includes sulfur oxides (SO_2), total suspended particulates (TSP), nitrogen dioxide (NO_2), carbon monoxide (CO), and oxidants (Ox; primarily ozone). Some of this group, notably SO_2 and TSP, also have synergistic effects, as we have noted earlier.

6.1. Characteristics of Air-Pollution Data

The second group includes elements involved in photochemical atmospheric reactions, which, as we have also seen, result in harmful secondary pollutants. This group includes nonmethane hydrocarbons (HC) and NO_2. (It is noted that NO_2 is present on both lists.)

Precursor Components

Because the mechanism of the formation of damaging components from the second group is not as straightforward as is damage from the first group, the criteria and standards, as well as control measures, must be handled in a different way. It will be recalled that the active agents in photochemical smog are the oxidants, largely ozone and the peroxy-or-

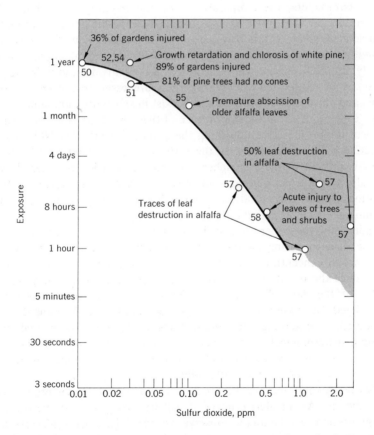

Figure 1 Sulfur dioxide damage to vegetation (courtesy of Academic Press and *IEEE Spectrum*).[2] Numbers refer to data listed in the original reference.

ganic compounds PAN and PBN, which cause eye irritation and extensive plant damage. The significant characteristic of these agents is that they are not emitted by any specific source but are the result of UV energy irradiating a mixture of active hydrocarbons (particularly olefins) and nitrogen oxides. The primary source of these ingredients is automobile exhaust. Others include coal and oil combustion processes, petroleum plants, evaporation of industrial and dry cleaning solvents, and similar industrial sources.

The "manufacture" of oxidants from these ingredients within the atmosphere is much more complex than the emission of SO_2 from, say, a coal furnace. Many reactions—perhaps hundreds—proceed simultaneously and at different rates. These rates and the yields of oxidants are influenced by the concentration of the various hydrocarbons, the meteorological mixing parameters, the temperature, and the intensity of sunlight. The gross results are that nitric oxide is oxidized to the dioxide and ultimately to nitric acid; some NO_2 is recycled to NO and ozone; some hydrocarbons become peroxy compounds, and others, ketones and aldehydes. Additionally, any SO_2 present will be oxidized to the trioxide, and ultimately to sulfate, at an accelerated rate. During these processes, free atomic oxygen, ozone, and various organic free radicals exist in substantial amounts.

The time sequence in the appearance of these products is of the greatest practical importance in establishing the criteria for HC and NO_2 and for predicting pollution episodes. In the early morning, a mixture of nitrogen oxides is present in the atmosphere. Starting at sunrise, nitrogen dioxide gradually forms a peak because of photoreactions. Under further UV irradiation, a peak concentration of oxidants is formed, several hours later, accompanied by a reduction in hydrocarbons.

Empirically, the upper limit of the maximum oxidant concentration that may be reached during the day is known to be a function of the average hydrocarbon concentration in the early morning. But this is a poor indicator, since the maximum potential oxidant concentration is only reached on 1% of the days. The empirical function is only apparent when an unrestricted data base is used; that is, a large number of cities and days are considered, representing all combinations of emission rates and composition, meterological dilution and dispersion, and sunlight intensity, so that these factors are washed out. Otherwise, the data takes on the appearance of an uncorrelated scatter plot.

Clearly, this relationship (Figure 2) is only adequate for the crudest kind of prediction. As an alert criteria, it will be wrong 99% of the time, although its error will be on the conservative side (false alarm). Nevertheless, these data have provided the best available monitoring criteria for O_3 and to some extent NO_2. Consideration of Figure 2 as a "mathematical model"

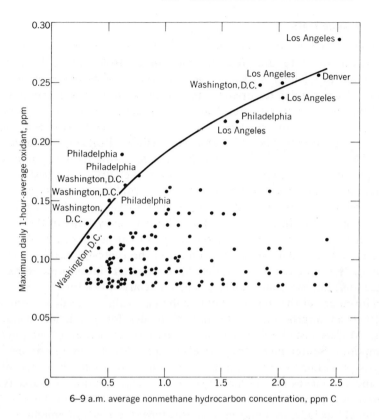

Figure 2 Maximum daily oxidant as a function of morning hydrocarbons (courtesy of *J. Air Poll. Control Assn.* and *IEEE Spectrum*).

for predicting trends and the development of improved photochemical models on a deterministic basis (rather than statistical) will be deferred to a later chapter.

We can conclude this discusion of ambient-pollution data by noting that the present six species and the limits assigned to them by federal criteria (Chapter 4, Table I) are subject to continual reexamination, as they should be under an adaptive, cost-effective system. At this writing, the standards for NO_2, which were based on unreliable past measurements, and those for SO_2 as well, are under reconsideration. Furthermore, any of the components listed in Table I, or several others, may become the subject of NAAQS in the future. One likely candidate is respirable particulates (0.3–1.0 μm), which has probably not been named so far only because suitable instrument technology has not been demonstrated.

6.1.2. Source Data

In addition to the ambient-component data, there are those species that are subject to state and federal emission controls. These have been enumerated in Chapter 4 also. In general, those substances controlled at the source are similar to, or the same as, those measured in the ambient air, although at a much greater degree of concentration. But in addition to the six ambient components, we have in the federal New Stationary Sources Standards (NSPS) and Hazardous Pollutant Standards (NESHAP), at the time of this writing, the following data requirements: H_2SO_4 (acid mist), smoke (opacity), beryllium, mercury, and asbestos.

There are few federal requirements for the continuous measurement of these emissions at present and state regulations vary widely, as we have seen in Chapter 4. In general, air-pollution control systems in the United States have not demanded continuous measurement of source data nor communication of data between source and the central control, as in the Rijnmond system cited. Instead, they have relied on "inventories" for compliance to source emission standards and on "emission factors."

An inventory of the sources of emission in a locality is needed in order to construct an appropriate mathematical model for predicting pollution levels. Models tested to date have all used some degree of simplifying assumptions. Source inventories are also required to design a sampling and analysis program, to serve as a basis for control or zoning laws and planning and as a means for interpreting effects. A classification of source types has been suggested[4] to aid in taking a community inventory (Table II). The data should include, in addition to source type, the kinds of emissions, discharge rates, number of sources, location, and the process raw materials and control used. Sources should also be inventoried as to which ones emit quantities of oxides, CO, organics, and particulates, or specific toxic substances.

Information so obtained should be supplemented with a questionnaire soliciting data on plant size, manufacturing procedures, operating schedules, fuel and solvent usage, and refuse burning. Additional engineering inspection will yield information on process flow diagrams, raw materials, fuel types, emission rates, and gas-cleaning devices. These data can be updated through a computerized registration and permit system, as in Chicago and other major cities.

Emission factors, if not available directly from plant data can be estimated from general tabulations similar to those in Table III or Ref. 5. It is recognized that such information will not be complete or accurate, since emission tables merely represent a statistical average of many installations and operating conditions. In most research to date, approximate data have proved good enough, but this will not hold true in the future.

Table II
Source Types

I. Fuel burning for heat and power
   A. Utilities
   B. Residences (private houses, apartments, etc.)
   C. Industrial
      (1) Manufacturing (per SIC code)
      (2) Commercial (stores, hospitals, hotels, etc.)
      (3) Processing (e.g., laundries, service stations)
II. Incineration
   A. Municipal
   B. Industrial and commercial
   C. Residence
   D. Apartment house
   E. Open refuse burning
III. Transportation
   A. Motor vehicles (gasoline, diesel)
   B. Trucks and buses
   C. Railroad engines
   D. Ships
   E. Aircraft
IV. Industrial and commercial
   A. Manufacturing (e.g., chemical, petroleum, metals)
   B. Agriculture (spraying, dusting)
   C. Commercial (dry cleaning, spray painting)
   D. Miscellaneous (sewage treatment, demolition)

Source. Adapted from Ref. 4 (courtesy of *IEEE Spectrum*).

Source data may be assembled on a gridded map and correlated with demographic and land-use information. The time element is introduced through traffic flow information, temperature (degree-day) data, and similar sources. It is then in a form to be introduced into a computation scheme.

Some use has been made of aerial and satellite photography to identify sources. It is expected that use of IR photography, radiometers, and special sensors for inventory purposes will increase, since they suggest an economical means to mechanization of this task.

U.S. federal regulations now require that states acquire a detailed inventory of SO_2 and TSP emissions from point and area sources in each county (a point source is one that emits more than 100 tons of pollutant under national standards in places over 1 million population or 25 tons in more

Table III

General Emission Factors (Tons/Unit)[a]

Source	SO_x	NO_x	Particulate	CO	HC
Residual oil (1000 hp or more)[b]	0.203[c]	0.052	0.004	negl.	0.0016
Residual oil (less than 1000 hp)	0.203[c]	0.036	0.006	0.001	0.001
Distillate oil (1000 hp or more)	0.024[c]	0.052	0.004	negl.	0.0016
Distillate oil (less than 1000 hp)	0.024[c]	0.036	0.006	0.001	0.001
Anthracite coal (residential)	0.011[c]	0.004	0.10[d]	0.025	0.005
Anthracite coal (commercial, governmental)	0.011[c]	0.004	0.025[d]	0.025	0.005
Bituminous coal (residential, commercial, governmental)	0.019[c]	0.004	0.018[d]	0.025	0.005
Bituminous coal (industrial)	0.019[c]	0.010	0.018[d]	0.002	0.0005
Natural gas (residential, commercial, governmental)	0.0002	0.058	0.010	0.0002	negl.
Natural gas (industrial)	0.0002	0.107	0.009	0.0002	negl.
Gasoline	0.004	0.057	0.005	1.455	0.262
Diesel oil	0.020	0.111	0.055	0.030	0.090
Aircraft	—[e]	—[e]	—[e]	—[e]	—[e]
Open-burning dump	0.0006	0.0003	0.024	0.043	0.040
Municipal incinerator	0.001	0.001	—[f]	0.0004	0.0007
Residential, commercial, governmental, and industrial incineration	0.0002	0.0008	0.010	0.013	0.018
Backyard paper burning	0.0006	0.0003	0.002	—	0.073
Gasoline evaporation	—	—	—	—	0.060
Solvent losses (dry cleaning)	—	—	—	—	1.95

Source. Ref. 5 (courtesy of *IEEE Spectrum*).

[a] Fuel oil, gasoline, and diesel fuel, 1000 gal (3.8 m³): coal, tons; natural gas, 10 ft³ (0.28 m³): refuse, tons; dry cleaning, 1000 people.

[b] 1 horsepower = 746 W.

[c] Dependent on following sulfur content of fuel: residual oil, 2.55%: distillate oil, 0.3%; anthracite coal, 0.6%; bituminous coal, 1.0%.

[d] Dependent on ash content of coal (anthracite coal, 10.0%; bituminous coal, 7.0%), type of firing unit, and type of control.

[e] Dependent on type of aircraft.

[f] Dependent on type of control.

sparsely inhabited places).[6] The data forms are reproduced in the appendix to this chapter.

In the past, much of the data-acquisition activity that has been accomplished has been concerned with measuring or predicting seasonal average rather than episodic air-quality conditions, in order to establish facts for setting air-quality regional boundaries.[7-9] For this purpose, it has been adequate to utilize averages and approximations for sources, meterology, and topography. Local control systems, attempting to predict just these extreme conditions that are excluded by the averaging process, will require more-refined data. In the case of source strength, it is desirable to know the current emission of each major source whenever there is any danger of pollution buildup. It can be envisioned that ultimately, many more sources will be instrumented and many will telemeter information to the data center. In this way, direct feedback of information following control measures can be obtained as an additional benefit.

6.1.3. Other Data

Apart from the aerometric data that monitors the current state of atmospheric pollution in a locality and source data from which control actions may be derived, the next most important need is for meterological data and associated topographical factors. The degree of pollution at a given time and, more important, its trend, whether abating or worsening, is usually related directly to meterological factors, particularly wind and inversions, which determine ventilation or stagnation. Because the measurement and forecast of weather is so important to the forecast of pollution aerometric levels and is integrated completely into the mathematical models that accomplish this task, it will be discussed separately.

In contrast to these immediate, and continual requirements, there is need for data on a much longer time scale, relating to the health of the community being monitored. This kind of information is part of the main feedback loops of the nationwide system block diagram (Figure 1, Chapter 5) which may result in change of national goals, criteria, or standards.

A national program exists that can provide this kind of information, although its scope is not limited to the health effects of the air environment. This program, known as the Neighborhood Environmental Evaluation and Decision System (NEEDS), is a creation of the Bureau of Community Environmental Management, part of the National Institute of Environmental Health Service of HEW. NEEDS was developed and initiated by the bureau as a management tool for city and county governments to assess rapidly and accurately the physical condition of a community, the health status of its residents, and the attitudes, priorities, and aspirations of the community residents, and to use this data to design and carry out the types of community health programs that are needed and desired by the residents.

Assistance in the application of NEEDS technique has been requested by over 50 communities. In 1972, 22 communities, including Indianapolis, Indiana; Norfolk, Virginia; Cleveland, Ohio; and Salt Lake City, Utah, were given use of its services in a survey phase, and in 1973, 9 more were assisted in more advanced stages.

Programs of this nature on a more comprehensive scale, gathering health data on a local level but coordinating it nationally, are a necessity to guide the overall direction of pollution control and to determine the effectiveness and completeness of the clean-air program.

6.1.4. Density of Data

The density of data acquired for the purpose of monitoring air pollution can be expressed in terms of either time or space parameters. The requirement for frequent sampling or "continuous" measurement with instruments of short time constant can impose stringent conditions on the design and cost of these instruments, since we are measuring very small concentrations and correspondingly small signals. Longer periods of sampling between measurement intervals or the integration of continuous reading instruments produces more accuracy or, in general, lower cost for the same performance. Likewise, increasing the number of sampling stations per unit area will improve the statistical accuracy of the resulting readings and improve the resolution of the dynamic picture of trends but, at the same time, will increase the cost of the data-acquisition system in a most direct way. Furthermore, the collection of large quantities of data without a well-conceived plan to process it and utilize the results in an effective way benefits no one.

Temporal Density

A study of the operational need for air-quality monitoring networks having rapid response time was undertaken by MITRE Corporation and reported in 1970.[10] "Rapid response" was defined by MITRE as the ability to report data in less than 6 h. The basis of their findings were interviews of responsible officials in federal and in local–regional air-quality monitoring operational positions. Although the report authors cautioned against gathering too much data, over 75% of the local officials saw a need for the rapid-response information. A lesser number of federal officials opted for rapid rather than delayed response (defined only as "significantly later" than 6 h), but this is clearly justified on the basis that no action could be taken on the shorter time scale by federal officials, as contrasted with local operators.

The uses for rapid response data as specified by the local official, representing six states and seven metropolitan areas, was as follows:

> Episode control 70%
> Source control 30%
> Medical control 15%
> Industrial accident control 15%.*

In general, collection of rapid-response data must be on the basis of actions that can be accomplished in the same time scale and that will substantially affect the air quality of the polluted airshed or respond to the needs of its population. Such actions are usually associated with episode conditions or industrial accidents and can include (1) implementation of preplanned emergency actions for alerts; (2) curtailment of nonessential emissions or automobile traffic; (3) fuel switching (e.g., coal or oil to natural gas); and (4) promulgation of medical alerts to hospitals, schools, and the like.

A justification, if not a need, can also be made on the basis of more economical transmission by the faster mode than by the slower. Although the study was unable to find an extant example, it was indicated that immediate transmission of data to a central control facility could be more economical than on-site storage with subsequent pickup. For example, Hamburg[11] cites a MITRE study as justifying automatic telemetering from a network of 5 or more remote stations (even as few as 3 under special conditions), or 25 sensors. The optimization of data acquisition will be considered in a subsequent section.

Real-Time Data

We see, then, that the need for real-time or rapid-acquisition air-pollution data has been questioned. But it is clear that the delay between the event and the act of doing something about high air-pollution concentrations, even if that action is merely a warning, should depend on how long it takes to harm the receptor of pollution, especially if that receptor is a human being.

Air contaminants taken into the body have a characteristic half-life (during which half the remaining material is eliminated) ranging from less than 20 min for gases such as sulfur dioxide, to 6 months or more for lead and dust. If the local ambient concentration of these substances suddenly rises, perhaps as a result of wind shift or diurnal emission patterns, the

---

* The total exceeds 100% since several indicated more than one of these needs.

concentration in the receptor's critical organs rises exponentially, reaching equilibrium after a time equivalent to three or four half-lives.[12] Fluctuations cycling at rapid intervals, about one-tenth half-life, will have little effect on the concentrations within the body. Thus, the biological effects of a rise in the carbon monoxide level will be maximized at between 4 and 12 h because it takes this long for the carbon hemoglobin levels in the human body to stabilize. However, the effects of sulfur dioxide will be maximized after 1 h or less. This sets an upper bound on the time delay that can be tolerated in acquiring and processing information on dangerous concentration increases. The lower bound may be on the order of 1 or 2 min.

In practice, the Public Health Service continuous air-monitoring program (CAMP), taking data from eight major cities since 1962, has utilized data averaging intervals as short as 5 min, which matches excellently the considerations just stated. Data taken by this agency has been subjected to extended statistical analysis, showing regular distribution patterns [13] that may be used to indicate how frequently real-time data-processing capability may be needed for early warning decisions, considering only the normal fluctuations of emissions and weather conditions. For example, a 1-h average sulfur dioxide concentration of 0.5 ppm (by volume), which if sustained for 24 h would lead to increased mortality,[14] occurred in three out of eight cities between 0.1% and 10% of the time, or about 400 h/yr in Chicago, 40 in Philadelphia, and 9 in St. Louis. In five out of the eight cities, a 1-h average of 0.25 ppm was measured between 0.1% and more than 10% of the time; this concentration adversely affects older sufferers of bronchial disorders.[15] Places with low sulfur dioxide, such as Los Angeles, not appearing on these lists, have similar needs for real-time data to warn of an undesirable rise in the concentration of other pollutants, such as ozone and hydrocarbons. In none of the cities considered can the need for real time data be completely disregarded because of the unpredictable occurrence of episodes, which may not follow the normal statistics of the CAMP data.

Averaging Time

We have pointed out that air-pollution damage is a function of both concentration and time; hence, the average level over a specific time period is the operative parameter. The averaging time of interest may vary widely. For example, damage to crops by oxidant pollution may be significant for exposures of 8 h or more, but odors can be detected if the threshold is exceeded for as short a time as 1 sec. It is always possible to compute averages with data taken at intervals smaller than the averaging period, but the converse is only true if the statistical frequency distribution of the data is of a known and mathematically tractable form.

Furthermore, if the data can be fitted to a known frequency distribution, a knowledge of its parameters can be used to estimate the probability of occurrences for a given period of a particular level. Larsen[16] has analyzed data for six major U.S. cities and shows that many of the common air pollutants (CO, hydrocarbons, NO, NO_2, oxidant, and SO_2) can be described approximately by a mathematical model with the following characteristics: (1) the concentration is log-normally distributed (see Section 1.4.3); and (2) the median (50 percentile) concentration is proportional to the averaging time to an exponent, and plots as a straight line on log paper. This property links all averaging time by the equation:

$$M = M_g S_g^{(0.5) \ln (S_g)}, \tag{1}$$

where M is the arithmetic mean, M_g is the geometric mean, and S_g is the standard geometric deviation.

Larsen has shown that when the latter two parameters have been determined from the data, the maximum concentration expected for any frequency can be calculated as

$$C_{max} = M_g S_g^z,$$

where z is the number of standard deviations of the point from the median. For example, the maximum 1-h average concentration expected once per year (z for 1 h in 8760 h = 3.81 standard deviation), given

$$M_{g\ h} = 0.044 \text{ ppm for 1 h},$$
$$S_g = 2.46$$

is

$$C_{max\ h} = (0.044)(2.46)^{3.81}, \tag{2}$$
$$C_{max\ h} = 1.36 \text{ ppm}. \tag{3}$$

For any other averaging time under one month, the model is a straight line on log paper, or

$$C_{max} = C_{max\ h} (t^b), \tag{4}$$

where b is the slope of the line (obtained from any two calculations) and t is the time. The maximum 1-day average, then, if the slope of the line is calculated at -0.362, would be

$$C_{max\ day} = (1.36)(24)^{-0.362},$$
$$C_{max\ day} = 0.43 \text{ ppm}. \tag{5}$$

Sensor Time Constants

Since the averaging time is the significant parameter that controls the required data density in time, the time delay of data sensors is important.

Some traditional air-monitoring instruments are relatively insensitive and require a long operating period for each independent reading. One that is commonly used has a 20-min time lag, whereas another requires an hour per reading. These instruments would be clearly unsuitable for monitoring a concentration level corresponding to damage within 1 h or less, although they could be used satisfactorily at lower levels that are harmless unless sustained for a day or week. For timely warning of higher concentrations of pollutants such as carbon monoxide or oxidants, it is necessary to have rapid measuring instruments and to provide averaging in the data-processing stage.

The time constant or transient response is of further interest when determining the suitability of instruments to measure sudden step changes in concentration. In the San Francisco Bay Area Pollution Control District (BAPCD), a level of 0.5 ppm of SO_2 is permitted but for not more than 3 min in any 24-h period. The instrument used to monitor this regulation minimally should be able to respond to a "cloud" of SO_2 with a concentration of 0.7 ppm and of such velocity and dimensions as to pass over a monitor in just 3 min. Schnelle[17] has addressed himself to the transient and frequency response of instruments and found that many of those in common use may not meet this type of requirement.

In general, his work showed that the transfer function $G(s)$ of the air-quality monitors tested can be expressed by

$$G(s) = \frac{e^{-Ls}}{(T_1 s + 1)(T_2 s + 1)}, \tag{6}$$

where s is the Laplace transform variable, L is dead time, and T_1 and T_2 are time constants.

Some typical instruments tested by Schnelle produced values of the parameters of Eq. (6) listed in Table IV. A fair approximation of the SO_2 monitors is given by

$$G(s) = \frac{1}{(5s + 1)(50s + 1)}, \tag{7}$$

since the dead time (L) does not affect the shape of the response curve. Applied to the example cited (0.7 ppm and 3 min), the simulated monitor did not respond for the full 3 min necessary to record a violation.

In summary, it is seen that the requirements for data are frequently in real time and of essentially continuous sampling rate and, furthermore, that many of the so-called first-generation instruments based on chemical tests do not meet these requirements. We now consider the necessary data density in space.

Table IV
Instrument Time Constants

Pollutant	Instrument	Dead time (L), sec	Time constants, sec	
			T_1	T_2
SO_2	Beckman 906[a]	20	6.95	65.52
SO_2	Beckman 906-A[a]	18	0.497	52.56
CO	Intertech Uras II[b]	1.18	0.985	1.15
CO	Intertech Uras II[c]	1.15	1.70	1.01
CO	Mine Safety Appliances (MSA) Lira 200[b]	1.82	3.44	3.16
CO	MSA Lira 200[c]	1.41	1.61	4.00

Source. After Ref. 17.
[a] Coulometric.
[b] Nondispersive Infrared (NDIR), 1000 cc/min flowrate.
[c] NDIR = 1500 cc/min.

Spatial Data Density

As suggested, a primary consideration in designing a sensor system is the number and location of sampling stations. Currently, the cost of sensors and the associated installations is the limiting factor.[18] It is seldom that a sufficient number of stations are installed to produce timely data with the desired level of confidence.

West German federal practice, when locating monitoring sites to assess ambient SO_2 levels, calls for 1 station/km², located on a square grid. Other sampling systems have been designed with specific regard for the location and arrangement of existing sources, topography, prevailing meteorological conditions, and population distribution. This has led to some very economical designs, such as the 31 stations (approximately 1 station/mile², or 1 station/2.6 km²) in the locality of Rotterdam, Netherlands, and 1075 total sites for the national survey grid of Great Britain. The flexibility of such an approach is limited, however.

Generally, there is a contrast between two management philosophies that reflects itself in the design of the monitoring network.[19] The area technique (exemplified by New York City's aerometric network), appropriate to a large and flat terrain, averages out spatial variations by uniformily spaced stations and thus obtains net results representative of the area as a whole. The contrasting philosophy is to orient stations to define air quality in the most heavily polluted or populated areas or to provide control feedback from the most prominent sources of pollution. The initial configuration of the Allegheny County (Pittsburgh), Pennsylvania, network represents a maximum surveillance to known sources and reflects the nature of the area topography (a river valley pocket) as well. Obviously, the numbers obtained by averaging the concentrations of these stations represents something entirely different from an area average.

The accuracy of mean values obtained from concentration statistics depends on both the number of readings averaged and the deviation from the mean. Russian studies[20] have been made, for example, concerning the number of measurements required for the determination of the average and maximum concentrations of sulfur dioxide, nitrogen dioxide, carbon monoxide, and soot in the air. They have shown that the calculation of the average SO_2, NO_2, and soot concentrations at a maximum error of 20% requires a minimum of 200 individual measurements, and up to 800 when the dispersion is equal to twice the average concentration. Some 150–200 observations are sufficient for the determination of the average CO concentration, because of the relatively small degree of dispersion of CO.

In the United States, a posteriori analysis of results from random and selected sites has produced the finding that the statistical distribution of the pollutant concentration values from different points in an area fits a normal curve. A consequence of this distribution is a relationship between desired accuracy, averaging period, and the number of measurement stations required. Bryan[21] has shown that the minimum number of stations (N) is related to the allowable departure from the true mean in percent (p) by the expression

$$N = \frac{(CV)^2 t^2}{p^2}, \qquad (8)$$

where t is the tabulated value of Student's t for a given confidence interval and CV is the coefficient of variation in percent.

CV can be computed from Eq. (9) or Eq. (10) if the distribution is either normal or log-normal:

$$\text{Normal:} \qquad CV = \frac{S}{M} \times 100, \qquad (9)$$

$$\text{Lognormal:} \qquad CV = \text{Antilog}\,(S_g \times 100) - 100, \qquad (10)$$

where S is the standard deviation.

Applying this relationship to data from Nashville, Tennessee, it was found that 245 stations, or 4 stations/2.6 km², were required to estimate daily mean sulfur dioxide levels with a 95% assurance of $\pm 20\%$ accuracy.

On the other hand, seasonal variations were estimated equally well with only 1 or 2 stations.

It is clear that these results were completely inconsistent with a projection of 790 stations for the national network (Section 5.2.1). Even if only 1 station/2.6 km^2 is assumed, representing a compromise accuracy of $\pm 40\%$, then New York, Philadelphia, and Chicago would require 1275 stations, and the Los Angeles–Long Beach area, by itself, 4000!

To focus more sharply on the problem, consider again, as a specific example, the continuous automatic air-monitoring system of Allegheny County in Pennsylvania.[22] This system was planned to cover an area of more than 1800 km^2 with 18 stations or double that number if cooperating private industries constructed monitors surrounding major sources. The philosophy behind the Allegheny system, similar to that of the Rijnmond (Rotterdam) system in the Netherlands, is to monitor ground concentrations near large sources with the hope that they will represent maximum exposures for any more distant, populated area and that their magnitude during stagnant weather will serve as a warning in time or roll back emissions at the source. Experiments in New York City, however, indicating that the location of pollution pockets is not constant, expose this concept to some criticism and suggest instead that small-scale measurements are indeed necessary.[23] Experience with the Allegheny system itself also tends to confirm the conclusion that pollution measured at any point is the sum of the contribution of multiple sources and that spot location of ambient sensors cannot pinpoint the emissions from a single source.

However, if Allegheny County were to switch to area grid coverage at 4 stations/2.6 km^2, nearly 3000 stations would be required. To monitor five gases at an average of $2000 for each sensor would total $30 million merely for the sensors. For the 100 cities whose total air cleanup costs were estimated by NAPCA at less than $1 billion/yr, this would entail spending more than three times as much for sensors as the annual cost of the control equipment. This is clearly inconsistent.

Several solutions to this dilemma—not mutually exclusive—exist. First, more versatile sensors than those currently used, capable of time-shared measurement of several different pollutant species, are available and permissible, since an interval of several minutes between readings is consistent with damage criteria. If only two sensors can measure all six pollutants at the same unit cost, the station total is reduced to one-third, or $10 million for the city. Although 10 times the planned expenditure, this amount is certainly more reasonable.

Second, the concept of intelligent placement of sensors to monitor key sources as well as areas cannot be entirely wrong. Less-developed areas and flatter terrain will need fewer monitors and better knowledge of local meteorology and source diffusion will suggest more economical sampling schemes than a uniform grid provides, even if results run short of the Allegheny system's hopes. It is not unreasonble to expect an improvement factor of 2–4 in sampling by good design, reducing the sensor cost per city to within the $2.5–5 million range. Spread over several years, this should be a more equitable burden for the benefits received.

Area is not the only possible basis on which to base sampling designs. An alternate is receptors or population, assuming that the population is not mobile. In point of fact, the *minimum* regional monitoring system mandated by the 1970 Clean Air Act is based on population (see Chapter 4, Table III), although the use of additional stations to achieve area coverage is not ruled out.

Hamburg[25] gives the following rule based on area population:

"With a population less than 100,000 there should be 3 stations; 100,000 to 1,000,000, 4 to 10 stations; over 1,000,000, 11 to 20 stations."

But, in the same reference, he goes on to say that "a better rule is that the number and spatial distribution of stations should reflect the variability of concentration levels throughout the jurisdictional area. This rule could lead to a requirement for more stations in an area where target coverage is sought (like Allegeheny County) than in a larger area where terrain is flat and sources are distributed (like New York)."

To measure this variability he recommends at least one year of data from both mobile and fixed stations (see Russian recommendations, cited earlier in this section[20].

Finally, he lists practical considerations limiting the number and location of remote sites:

1. Physical siting limitations (sharp declivities, no access roads, no power or telephone lines, etc.)
2. Control-agency resources
3. Political and social pressures
4. Full-time accessibility and security against vandalism.

To which we would add yet other system factors that influence the number and location of point sensors. To give representative measurements, sensors should be at least two "building heights" away from windward structures, far enough away from local pollution sources to reduce their contribution to 5% of the total contaminant,[26] and generally be 3–4 m above ground.

An ultimate solution to many of these problems would be the invention of inexpensive, maintenance-free, solid-state sensors (termed by APCO personnel *third generation*).[24] Although the hoped-for price of $1/point sensor has not yet even remotely appeared on the horizon, such a breakthrough would reduce the cost of the monitoring system to essentially that of data transmission and handling, of which more below.

We have assumed in the above that all sensors sample and measure concentrations of pollutants at a single point in space and that statistical techniques will be used to determine the variations in space. Yet another scheme, which in theory will require a lesser number of sensors, involves the use of "long-path" sensors, which respond to the average concentration along a line in space. The advantages and drawbacks of this technique and the state of the art at present will be discussed in the following section.

6.2. NETWORK DESIGN FOR DATA ACQUISITION AND HANDLING

Before leaving the subject of point sensors and network configurations of optimum sampling, it is instructive to consider a detailed proposal for instrumenting the urban center of St. Louis. However, it is suggested that the reader return to this section (6.2.1) after reading Chapter 8.

6.2.1. The RAPS Configuration

The Regional Air Pollution Study (RAPS) of the Stanford Research Institute (SRI) is an experiment designed to prove the worth of mathematical models as means of scientifically managing air-quality improvement.[27] It is based on the principles underlying the 1970 Clean Air Act, which states implicitly and explicitly that emission standards can be derived from air-quality goals through an analytical chain of reasoning that includes a description of the sources and the chemical, physical, and meteorological processes of concentration and dispersion in the atmosphere. The computer-based mathematical model is the central element of this chain that permits the consequences of control, weather, and source-emission events to be predicted. RAPS empirically attempts to find the answers to the same questions we have raised in this chapter: What is adequate data? How timely and how much? How comprehensive a model is required?

The important region to be modeled is the earth's boundary layer, the lowest kilometer of the atmosphere. Surveillance of meteorological and chemical events within this volume and that of the urban region itself—approximately 100 km from the area center (the St. Louis Arch)—must be undertaken on a temporal and spatial scale matching that of the model. This scale may vary from a kilometer or more to the microscale of a city street, approximately 125 m.

The proposed RAPS instrumentation of St. Louis may be considered an ideal case, since its objective is research as well as control, but it stands also as an example of current thinking about system configuration and data acquisition. Furthermore, it is bounded by cost restraints that may place it somewhat on the conservative side: the estimated initial cost of the St. Louis facility given in SRI's report is $4 million, of which $1.6 million is for air-quality instrumentation. This is compatible with the $2.5–5 million estimated for a city the size of Pittsburgh, assuming improved sampling design.

The key question is what is expected of the model? To make strategic control decisions, it is of great significance, economically and otherwise, to determine whether emission must be reduced by 90% or by 95% to achieve a desired air quality. This implies a model accuracy of less than 25% tolerable error in order to make a decision of such precision.

The model is expected to predict the response of the atmosphere to aerodynamic forces and pollution sources and produce results in the form of isotherms, pollution isopleths, and similar displays. In chemical terms, the system is required to monitor the current pollutants—SO_2, CO, $NO-NO_2$, photochemical oxidants, hydrocarbons, and total suspended particulates—but it should have the capabilities of adding those which may be of future interest. These include lead aerosol, fluorides, mercury and other heavy metals, H_2S, asbestos, odors, and polynuclear aromatics.

The scales of interest vary with the details of the model and the forces involved, ultimately leading down to the block-by-block scale of the receptor. In the RAPS proposal, the following area scales are considered of interest: radiation and moisture, 20 km; wind, 5–10 km; and emissions and air quality, 1 km.

Emissions

The objective of this part of the monitoring system is to arrive at an algorithm suitable for input to the computer-based model that will determine the deviation, on an hourly basis, from an emission baseline. In the case of stationary sources, it is necessary to interrogate their management (most likely on the authority of the EPA) to obtain sufficient data to relate their emissions to easily observed operating parameters, such as kilowatts or pounds of steam produced. As many as 100–150 defined "point sources" exist in St. Louis and 1000 or more manufacturing, institutional, and other significant sources, some having a dozen or more stacks or emitting points.

It is also necessary to establish the true emission factors for on-road vehicles. Traffic data can be obtained from traffic-control officials on an hourly basis. Public transportation data (buses, aircraft, trains) can be obtained from schedules and officials.

Topographical and Geographical Data

Important model inputs include the exact location of each stationary source or stack and the location of main traffic thoroughfares (line sources). Secondary feeder streets are considered area sources. Residential emissions are also area sources and may be obtained from surveys, degree-day data, and consumption information from gas utilities and oil and coal dealers.

The topography itself is recorded in the model in terms of deviation from the base altitude. Surface roughness on the scale of centimeters or greater than one meter is an important aerodynamic parameter. Each unit area in the model has associated with a characteristic combination of these two parameter classes.

Monitoring System Configuration

The city area of St. Louis is 158 km^2 and that of the county is 1300 km^2. In order to test the models properly, RAPS proposes to instrument an area enclosed generally by a 100-km radius centered at the Arch. To monitor this area to the model scale would require 30,000 stations! As previously discussed this number is untenable, but it is of interest to note the method by which SRI has achieved a compromise with economic realism.

It is hoped, of course, to achieve "representativeness" of data, minimizing the elements of spatial variability—instrument error, poor siting, and natural variability. To achieve this goal, the general scheme is to have three classes of stations, some of which acquire data at the basic model area scales and the remainder capable of being deployed to obtain microscale information where needed to verify model predictions. In addition, the scheme calls for station density to be greatest where air quality and meteorological data are expected to vary most and be least dense in the "flat areas."

The classes and subclasses of stations and their instrumentation are described in Table V. The A_2 stations acquire data on the surface layer (wind, temperature, humidity, concentration of pollutants) utilizing 30-m towers, while the A_1 stations obtain the same information and additionally the meteorological "forcing functions"—surface energy budget and mesoscale horizontal pressure field. There are 9 class A_1 stations and 8 class A_2; these are deployed about the Arch as a center with a spacing of 40–60 km (see Figure 3). The inner 40-km circle, containing 9 A stations, also encloses 24 class B stations at 12.5-km spacing. These are to give higher spatial resolution data for kinematic analysis of the air and pollutant movements and occupy the region of maximum emission concentration and variability, which is also generally located within 40 km of the Arch.

Table V
St. Louis RAPS (SRI) Instrumentation and Data Stations

	Type and number of stations						Sensor instrumentation[a]
	A_1 (9)	A_2 (8)	B_1 (24)	B_2 (8)	C_1 (16–25)	C_2 (16–25)	

MEASUREMENT (number of sensors per station):

	A_1	A_2	B_1	B_2	C_1	C_2	
CO–CH_4–HC	1	1	1	1	—	—	GC + FID
H_2S–SO_2	1	1	1	1	—	—	GC + FPD
Total S	1	1	1	1	—	—	FPD
O_3	1	1	1	1	—	—	Chemiluminescence[b]
N_2O–NO_x	1	1	1	1	—	—	Chemiluminescence[c]
Visibility	1	1	1	1	—	—	Nephelometer
TSP	2	2	2	2	—	—	Hi-Vol sampler
CO	1	1	1	1	—	—	NDIR
Temperature	1	1	—	—	1	—	—
Wind vector	3	3	1	1	3	1	—
Solar radiation	1	—	—	—	—	—	Pyranometer
Pressure	1	—	—	—	—	—	Transducer
Pressure	1	—	—	—	—	—	Mercury barometer
Net radiation	1	—	—	—	—	—	—
Dewpoint	1	—	—	—	—	—	Hygometer
Precipitation	1	—	—	—	—	—	Gauges

STATION FEATURES:

	A_1	A_2	B_1	B_2	C_1	C_2
30-m tower	×	×			×	
10-m tower			×	×		×
Fixed	×	×	×			
Transportable				×	×	×

Source. EPA Report PB 210017.[27]

[a] GC = automated gas chromatograph
  FID = H_2 flame ionization detector
  FPD = H_2 flame photometric detector
[b] O_3–ethylene reaction.
[c] Or coulometric.
  NDIR = nondispersive infrared analyzer

6.2. Network Design for Data Acquisition and Handling

The remaining *A* stations occupy the 100-km circle of most interest. An additional 8 *B* stations monitor the "poor ventilation" region or the area, downwind of the prevailing southerly and westerly winds.

Altogether, there are 49 of the substantially fixed stations. In order to achieve the necessary microscale definition needed to examine the topographical classes (roughness and height-displacement parameters), RAPS adds 16–25 more stations of type *C*. These may be moved from block to block to examine each parameter class as required.

Additional data is to be obtained from the air. These monitors will include balloon-borne sondes to obtain 1000–1500-m parameters, correlation

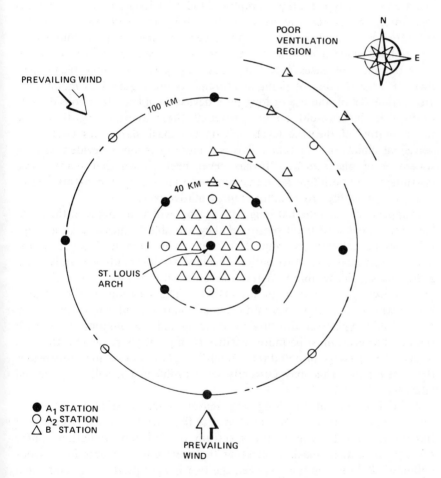

Figure 3 Proposed RAPS deployment of air-quality monitoring stations for St. Louis, Missouri.[27]

spectrometers and "lidar" (see following section) mounted on helicopters to study plume behavior, and profile data swept from aircraft or from ground vehicles searching upward. Added to all these sources will be the standard synoptic weather data regularly obtained in the surrounding region.

6.2.2. Long-Path and Remote Sensors

In the design of future systems to monitor current atmospheric pollution levels, one critical decision will be whether to employ "point" sensors or "long-path" sensors. The former, as implied by the name, measure the concentration of the given pollutant at a single point in the area monitored and are the only type routinely employed today. A long-path (or open-path) instrument measures the average concentration between two points separated by a considerable distance, perhaps one or more kilometers. Lasers have made it possible to measure the concentration of trace quantities of pollutant gases over long paths, using spectroscopic techniques. Since a large number of point measurements are required to monitor an area in fine detail, the use of long-path devices could result in considerable savings in instrument costs, provided their average readings are representative of the true levels and do not mask dangerous pockets or persistent clouds of pollutants. This is a moot question, as evidenced by a statement of one official: "It has never been shown that point-source instruments are optimum for measurement of air quality. Open-path instruments may possibly provide more representative results."[28]

It appears reasonable that long-path sensors will avoid many of the problems associated with point-sensors if they are able to measure an average value of the pollutant concentration over a considerable distance. If the normal statistical area distribution of concentration holds generally, path averaging should be able to determine temporal fluctuations of a cloud of high concentration within an urban area. The location of the original source will be of importance in detecting violations, but long-path sensors may be able to perform this task by scanning and triangulation. Long-path sensors, however, must be more sensitive than point instruments if they are to detect the passage of pollutant "clouds" or plumes of smaller dimensions than their path. This may currently be a problem and will be discussed later.

A brief review of the long-path sensors under current investigation reveals that the most highly developed is the correlation spectrometer of Barringer,[29] which projects the spectrum of a distant radiation source through an optical mask replicating the spectrum of a specific gaseous pollutant. Relative motion between the two images produces a modulated correlation signal, proportional to the amount of pollutant in the space between source and spectroscope. Both NO_2 and SO_2 have been detected

6.2. Network Design for Data Acquisition and Handling

from aircraft by utilizing the UV and visible regions (272–550 nm) and reflected sunlight as a source. Sensitivity to the average concentration of NO_2 across a 1000-m path appears to be of the order of 0.01 ppm. The Barringer instrument is available both as a long-path, or remote, instrument and as an in situ stack monitor.

Hanst and Morreal,[30] followed by Hidalgo,[31] have constructed instruments following the principle of a two beam differential IR spectrometer, designed to operate over paths up to 3 km long, using a tunable CO_2 laser. The laser can be forced to oscillate on any of 20 lines between 9.2 and 10.8 μm by means of a diffraction grating in the cavity or by adding absorbing gas (propylene) to the cavity to shift wavelength. The use of two beams—one tuned to a prominent line of the absorption spectrum of the desired gas and the other to an adjacent, nonabsorbing region—eliminates the effects of atmospheric optical turbulence and scintillation, interference from molecules, and other scattering.

Hidalgo envisions an urban air-pollution monitoring center consisting of a system of lasers emitting radial beams from a center location (see Figure 4). Corner reflectors will return the radiation from distances of 1–2 km to receivers, which determine the differential absorption over the two-way path on both wavelength channels. If a 5% differential absorption can be detected at 0.4862 μm, a sensitivity of 0.03 ppm of ozone should be achieved, with somewhat lower sensitivity for other pollutant gases.

Another technique frequently proposed for long-path measurements involves Raman scattering of a laser beam.[32,33] Raman spectra result from inelastic collisions of monochromatic photons with molecular species during scattering, resulting in bands or lines of shifted wavelength. This wavelength shift is a function of the molecular vibrational modes and permits unique identification of the molecule by comparison with "library" records. The Raman line intensities are a function of concentration, which normally leads to a lack of sensitivity to trace components.[34,35] The combination of the intense monochromatic laser source and a long path tends to overcome this difficulty. The technique devised by Lederman and Widhopf utilized a Q-switched, 1-J ruby laser, a high-gain photomultiplier, and a special absorption chamber, together with narrow-band filters and a spectrograph to isolate the Raman wavelengths and to show the possibility of a kilometer-path analyzer. However, the ranges on plumes reported by Kobayasi and Inaba in field tests did not exceed 160 m.

Microwave absorption spectroscopy has successfully detected formaldehyde, hydrogen cyanide, and a number of other molecules in interstellar space via radio astronomy in recent years[36] and has been proposed by one of its leading investigators, L. E. Snyder, as a detector of atmospheric pollutants. In this technique, receivers and transmitters operating in the gigahertz range would be located at opposite ends of a one-way propagation path. These would be tuned to the characteristic microwave absorption frequencies of common pollutants. Objections have been raised because of interference between molecules at absorption frequencies below that of ammonia.[37]

Still another remote sensing technique that has been tried is that of IR emission spectroscopy. Except for very special situations, such as satellite-borne atmospheric sounders (where a great amount of data processing can be applied), this technique is primarily applicable to hot gases, such as those emitted from stacks, owing to the greater energy emitted. Qualitative spectroscopic detection of the presence of pollutant chemical species in stack gases has been well established by several experimenters.[38] Quantitative measurements are a different matter. A serious problem is the difficulty in determining both the temperature and the concentration of the emitting gases from a remote observation, since both phenomena reflect as brightness levels or strengths of spectral emission lines. Temperature fluctuations and the narrow width of the species-identifying lines in the IR spectrum are complicating factors. Astrophysicists, atmospheric scientists, and rocket engineers have attempted to resolve this dilemma by estimating the composition of the gases and calculating the temperature from spectral emissions at selected frequencies, applying the integral equation governing the emission intensity inversely. The solutions of this calculation, resulting in a temperature profile of the gas layer, are not unique, however. S. H. Chan[39] has recently a possible means of avoiding this computation problem.

A direct experimental approach recently reported by Prengle[40] and his associates at the University of Houston utilized a Michelson interferometer spectrometer (Block Engineering Co.), operating in the 2.8–15 μm range and mounted on a 10-in. Cassegrain telescope. They report an accuracy of $\pm 28\%$ of concentration over the range 0–10,000 ppm at a distance from the top of a stack of 65 m. Their technique included calculation of the plume temperature by adding the total radiation flux over the range 500–40000 cm^{-1}, assuming it to be a gray body, and calibrating the system with an actual laboratory gray body at corresponding distances and temperatures. The results on CO, NO, NO_2, and hydrocarbons, both burned and unburned, to the accuracy stated, must be deemed at least a qualified success.

The effectiveness of long-path sensors as ambient monitors may be evaluated by considering a simple mathematical model that retains only the essential common characteristics, such as highly directional geometry and sensitivity to the average quantity of a specific pollutant in the path. In this model, properties such as cost and reliability are disregarded. Hidalgo has

6.2. Network Design for Data Acquisition and Handling

done this for the laser absorption spectrometer, previously referred to. The instrument is assumed capable of detecting an average concentration of 0.24 ppm of sulfur dioxide, corresponding to a sensitivity of 5% difference in laser energy absorption, over a two-way path of 2300 m, which corresponds to complete area coverage with 1 sensor/2.6 km (see Figure 4).

Since the long-path sensor monitors the average concentration along an extended line and since topography and structures limit the line of sight at desirable monitoring heights (3–4 m), it is not possible, in general, to scan the entire area. With the use of retroreflectors, as suggested by Hidalgo, selected paths can be monitored. If there are gaps in the area coverage, it is legitimate to ask how long it will take to detect the rise in average concentration caused by a cloud of pollutant gas emitted from a source located near one of the monitored lines?

The strength of a moderate SO_2 source—say, a 10,000-lb/h (4536-kg/h) steam boiler using 3%-sulfur coal—is about 1 mole/sec* (1 mole SO_2 equals 64 g). In comparison, a large electric power plant (11,500 Btu/kWh or 12.1 kJ/Wh) using the same fuel will emit 30–40 mole/sec. Hidalgo's calculations show that emissions from either of these sources at a 1000-m distance, propelled directly toward the monitored path by a moderate wind,† would require well over 4 h to build up to a detectable concentration. Even at 100 m, the smaller source would require 15–30 min to be detected. Referring to Figure 4, a sensor covering 2.6 km² with eight paths may have a source at a maximum range of 330 m; this could emit 1 mole/sec for more than 30 min before detection. Under the most unfavorable condition—a dead calm—a pollutant from the same source, transported solely by diffusion, would build up so slowly as to be undetectable within any reasonable time. It is apparent that a choice between an unreasonably large number of expensive point sensors and inadequate or slow long-path instruments is difficult.

More recent work, however, has revived hopes for the laser absorption long-path instrument. The General Electric Company is developing a system, the Infrared Laser Atmospheric Monitoring System (ILAMS), consisting of a spectrally scanning laser that reflects energy from a retroreflector, similar to Hidalgo's, which may be placed 1–10 miles from the source. The CO_2 laser transmitter may be aligned to four separate wavelengths that are scanned in rapid sequence. The wavelengths selected in the intial model were 9.505, 10.532, 10.675, and 10.719 μm, designed to detect ethylene and ammonia in the presence of ozone, H_2O, gasoline-engine exhaust, and atmospheric scintillation.[41]

* Coal emission factor: 38 × (%S) lb SO_2/ton [15.6 × (%S)kg SO_2/metric ton], 87% efficiency, 12,000 Btu/lb (28 kJ/g).
† Transport coefficient equals to 10^5 cm/sec.[31]

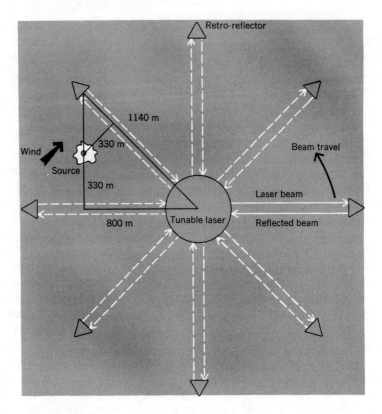

Figure 4 Long-path laser air-pollution monitor (after Ref. 31, courtesy of *IEEE Spectrum*).

This system is theoretically capable of great sensitivity, which has been partially confirmed in laboratory and field tests. After further cleanup of the optical system, including spatial filtering of the laser beam, it is expected to detect the important pollutants (CO, NO, NO_2, SO_2, and O_3) in the 1–15 ppb range over a 1-km (laser to retroreflector) path. An even higher sensitivity (0.1–1.5 ppb) is possible. The response time for Hidalgo model in the example above will be reduced by a ratio of 5 or 50 for the one or two orders of magnitude of sensitivity improvement.[42] (It should be noted that the response time is independent of range of the device.)

Lidar

The laser has also been utilized to detect particulates and visible smoke at a distance in a radarlike mode that has been given the acronym *lidar* (for *li*ght ra*dar*). Lidar was first used in atmospheric scattering studies by Fiocco and Smullin.[43] Systems have been put in the field with peak pulse power up to several hundred megawatts and with pulse widths down to 10^{-9} sec, producing excellent signal to noise ratios (SNR) and good range resolution.[44]

Practical utilization of the lidar to monitor stack plumes at a distance has been accomplished by SRI and others in this country and by Hamilton in England.[45] Scatter from smoke plumes by lidar at a range of 3 km with an atmospheric visibility of 5 km has been reported by latter. Plumes from stacks can be tracked by lidar over long periods and used to predict ground-level concentrations of pollution. Plume-rise can also be measured. An important application has been the measurement of the height of the mixing layer, the adiabatic layer of the lower atmosphere below a stable layer (inversion) in which emitted pollutants are dispersed and mixed. The top of the mixing layer is distinguished by a sharp decrease in lidar backscatter (the significance of these measurements will be detailed in a later section).

Johnson and Collis of SRI, as previously mentioned, utilized a ruby-laser transmitter to monitor the 800-ft stack of a coal-burning power station in Pennsylvania during two periods during 1968. The experiment was primarily for the purpose of obtaining plume cross-section data. Later work, using truck-mounted units, has included mixing-layer depth measurements over the city of San Jose, California.[46-48]

Remote Sensing of the Atmosphere with Sound Waves

It would be inappropriate to conclude a discussion of remote and long-path sensors without at least mentioning the work with sound waves in a radarlike device that has been conducted by McAllister in Australia and by Little and coworkers in the National Center for Atmospheric Research (NCAR) and the Wave Propogation Laboratory of NOAA at Boulder, Colorado.[49] This work is not directly pointed toward measurement of pollution per se but is intended to study the lower atmosphere, up to 450 m, in order to obtain information on profiles of wind speed and turbulence, location and intensity of temperature inversions, and variation of humidity with height. This is just the information required to forecast pollution levels on an urban scale, as will be seen.

The properties of plumes and inversions to 200 m were studied with the original WPL echo sounder developed from standard acoustical components. Excellent detail was obtained, leading the experimenters to believe that this type of device has a practical role to play in urban microscale meterology and pollution monitoring.

Future Prospects

What will be the role of long-path sensors? Clearly, their sensitivity to pollutants must be improved by several orders of magnitude if they are to be sufficiently responsive. Siting difficulties will remain and will require as much ingenuity as is needed for point sensors. To be competitive, the cost of a single long-path instrument, installed and with the geometry shown in Figure 4, must not exceed the installed cost of four point-sensor stations ($15,000-20,000). This assumes that the use of tunable lasers will permit simultaneous monitoring of six pollutants. If these targets can be met, there will certainly be at least a supporting role for these instruments in future monitoring systems.

6.2.3. Automated Data Handling and Transmission

A study of the cost of automating the data collection task for air-pollution monitoring networks, using computer simulation of system models of sophistication ranging from totally manual to completely automated, has supported the belief that full automation may be cost-justified for more than 20-25 sensors even if data is only reported hourly.[50] But it is apparent from the foregoing that accurate and reliable reporting of air quality will require many hundreds of sensors in each large metropolitan area. With large numbers of sensors reporting at frequent intervals, telemetry and data processing will assume increasing importance. Cost and efficiency suggest that digital data communication will be optimum, requiring digitizers and multiplexers at each location. Use of large-scale integrated circuits will keep the cost of providing these functions to a minimum. Communication must be reliable and usually bidirectional in order to provide for polling, calibration, and control commands from the central processor. It is possible that narrow-band wired communication will be preferred in many locations because of its lower cost. Since continuous or real-time monitoring of air-pollution data will not normally require more than six data words per minute from each station, there is every opportunity to time-share communication channels for even greater cost savings.

Some storage for polling, but perhaps not more than a few tens of bits per sensor, will be needed at each remote station. In the event that the sensor is complex and multicomponented, a local digital processor may be used at the remote station to manage it. If so, the processor can easily provide all the communication and storage functions mentioned.

Display and recording at each station should be minimal. For maintenance checks, a single display (digital) could be shared with all local sensors. If recording is needed, it can be provided economically by magnetic tape. Current practices in many existing air-quality monitoring stations are largely manual and rely on graphical records.

At a central location, the data are edited, validated, reduced, and displayed for interpretation. If the data are already in digital form, these functions can be performed automatically by the computer. Editing and validation consist of checking calibrations and recognizing and rejecting errors due to noise, component failures and drift, and statistical abnormalities. The program details will depend on the characteristics of the particular sensors but should present no difficulty. In many current systems, these tasks are still performed manually. They are particularly necessary because of the unreliability of first-generation sensors. Data reduction includes the monitoring of instantaneous values and provisions for an alarm if preestablished limits are exceeded. In addition, data are averaged over various periods ranging from 5 min to 1 yr, peaks and minima noted, and frequency distributions computed. It has been found, even for older networks with only tens of stations, that this data reduction task should be automated if polled more than four to six times per hour.[51] With today's improved and low-cost computers, there should be no question of automating even smaller networks.

Data are displayed by means of conventional printouts, illuminated maps, alarms, and the like. In addition, statistical evaluation and interpretation are conducted, all of which is facilitated by computer routines.

A good outline of the steps that must be taken by a small air-pollution control authority to prepare for data automation and communication with a large computer is given by Dittrich et al.[52] These include such tasks as the development of forms; transcription of information intended for computer input; and the development of codes for station identification, a grid locating system, time, and source classification. The selection of a programming language suitable for the task and the processor is important. The PL-1 language has advantages over FORTRAN, for example, because it permits narrative and numeric information to be mixed.

Data Transmission

In many studies, the cost of transmission of data looms large in the budget of the control agency, especially where large distances are involved. Although the primary output of most aerometric sensors is in analogue form, digital communication from the remote source is becoming the method of choice as the price of integrated circuit A–D conversion modules is rapidly reduced.

Some of the decisions involved in using conventional transmission media—half- or full-duplex, leased or dial-up telephone lines, teletype (narrow-band) channels mentioned earlier, modulator–demodulator (modem) equipment—are summarized in several places,[53] although a detailed discussion is perhaps outside the scope of this book. For the future, cable-television channels, threading through city streets, have been suggested as the media for pollution sensor data, since they could carry practically unlimited amounts of wide-band or high-speed data time-shared with their normal use.[54]

The optimization of air-quality-data communication networks has been analyzed by Maley of the University of Colorado and reported in a recent seminar on remote sensing of the troposphere.[55] Since the spacing of sensors is dependent on the accuracy desired in measuring the spatial characteristics of the atmosphere, he has sought the least bit-rate capacity and length of the transmission line to acquire data from a given density of sensors at minimum cost. The channel bandwidth and modem complexity, and hence their costs, are direct functions of the bit rate, and overall channel length also directly determines cost. (Radio transmission is not considered by Maley because of the number of channels required, which may be so great as to infringe on current assignments.) If the sensor mode is to be controlled from a central station (the sampling rate, sensitivity, or calibration mode, for example), simultaneous two-way transmission (duplex) may be needed, adding to the cost per mile.

Consider an area A monitored by sensors arranged in a grid, each connected to a data center in the middle of the area (Figure 5a). Defining unit *transmission capacity* as that required to transmit one bit per second a

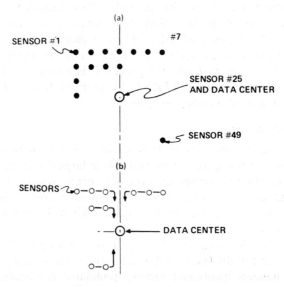

Figure 5 Optimum air-quality data network configurations (after Ref. 55).

distance of one kilometer, the total capacity needed is

$$T = \sum_{i=1}^{N} D_i d_i, \qquad (11)$$

where D_i is the transmission rate (bits/sec) of sensor i, d_i is its distance to the data center, N is the number of sensors in A, and T is the transmission capacity.

Assuming all sensors have the same output $D_i = D$, Maley has shown that

$$T_{\min} = \frac{2}{3} N \left\{\frac{A}{\pi}\right\}^{1/2} D$$

$$= \frac{2}{3} N^{3/2} \left\{\frac{a}{\pi}\right\}^{1/2} D \qquad (12)$$

if N is equal to, or greater than, 6 and

$$a = \text{grid cell area} = \frac{A}{N}.$$

The total length of transmission line (in kilometers) is

$$L = \frac{2}{3} N \left\{\frac{A}{\pi}\right\}^{1/2} = \frac{2}{3} N^{2/3} \left\{\frac{a}{\pi}\right\}^{1/2}. \qquad (13)$$

Since the transmission line could prove too costly for some systems, a configuration is developed that minimizes the length at the expense of a higher data rate (Figure 5b). In this system, duplexed data flow horizontally to the vertical center line and then vertically to the data center. Then the capacity

$$T = \frac{NA^{1/2}}{2} D = \frac{N^{3/2} a^{1/2} D}{2} \qquad (14)$$

and

$$L = (NA)^{1/2} = Na^{1/2}. \qquad (15)$$

Which is the better system depends on the relative costs of bit-rate capacity (bandwidth) and transmission length. In most systems, the latter predominates.

Optimum Network Size

Disregarding the political structure that imposes local, state, and regional levels on data-gathering networks, the optimum size can be analyzed, according to Maley.

Assume that an entire region contains N_1 sensors, and each local system, as before, consists of N sensors with an average data rate D. Then the number of local systems is N_1/N.

For minimum transmission line (Eq. 14),

$$T = \tfrac{1}{2}N^{3/2}a^{1/2}D$$

and the total transmission capacity for all local systems is

$$\frac{N_1}{N}\left(\frac{1}{2}\right)N^{3/2}a^{1/2}D = \frac{N_1}{2}(N)^{1/2}a^{1/2}D. \tag{16}$$

The transmission capacity needed in the entire region of N_1/N local systems, $(Na)^{1/2}$ on each side, producing data at rate ND, to connect the local system to regional data is

$$\frac{1}{2}\left(\frac{N_1}{N}\right)^{3/2}(Na)^{1/2}ND = \frac{1}{2}N_1^{3/2}a^{1/2}D. \tag{17}$$

The total capacity T_t is the sum of Eqs. (16) and (17), or

$$T_t = \frac{1}{2}(N_1^{3/2})a^{1/2}D + \frac{N_1}{2}N^{1/2}a^{1/2}D,$$

$$T_t = \frac{N_1}{2}a^{1/2}D(N^{1/2} + N_1^{1/2}). \tag{18}$$

But T_t is a minimum when $N = 1$; that is, there are *no* local systems.

This is apparently a justification (on the basis of cost) for a single "national" or "world" system for aerometric data. However, it is not totally valid, because of the possible need for local data at the local site for immediate control action.

RAPS Configuration

It is appropriate to conclude this section by comparing these concepts with a description of the data acquisition and handling system proposed by SRI for the St. Louis air-pollution modeling and control facility, discussed in Section 6.2.1.

The proposed facility equipped with sensor instruments reporting at intervals varying from less than a minute to several hours would produce about 1.5 million measurements per day.

Each of the 49 semifixed stations are to be connected by full-duplexed telephone circuits to the central facility for automatic, computer-controlled recording. A digital data terminal at each station interrogates each instrument at (locally) programmed intervals and converts the data on the spot to digital form, storing it in a small magnetic-core memory. At the

6.2. Network Design for Data Acquisition and Handling

same time, as the reading from each instrument is taken, a digital code identifying the instrument is stored adjacently; this code is obtained from a small integrated circuit consisting of a 16-bit shift register and a 16-bit ready-only-memory (ROM).

The station memories are polled by the central computer every 15 min and read out their sensor and I.D. data into its control storage. The computer identifies the instrument and location of each sensor, calibrates the data from updated internal memory, and converts it into engineering units for storage or display. Subsequently, the taped data is transmitted to another computer where the experimental model verification will take place.

The basic principles of this system are as follows:

1. All data are individually and automatically identified, which guarantees its integrity, even at the cost of 24 million bits of tape storage daily.
2. All data are recorded in a "free field format," which permits unlimited expansion in format.
3. Each tape carries a full descriptive header in English.
4. All codes are translated to engineering units at the earliest possible time.
5. Manually inserted data (inserted via teletype) will have the same format as that automatically acquired.

The 17 stations of the system with 30-m towers will report data such as wind components and temperature at 30-sec inervals at three elevations. Pollution measurements will be taken at intervals of 5 min or longer. This implies 500 measurements from each class A tower in 15 min each on 8-bit code, plus another 16 bits to encode the instrument identity (type, serial, position). Thus, the local (station) memory must have at least a 12,000-bit capacity. At 2400 bits/sec into a well-behaved ("conditioned") telephone line a minimum of 5 sec is needed for transmission of the data block. Smaller stations will read out in about 2 sec.

The received data is written on a primary archival tape and also temporarily on disk. A second disk holds a thesaurus (English-language description) of about 2000 active sensors translatable from the I.D. code, as well as the two most recent sets of calibration data for each instrument. (Two sets permit checking of automatic calibration data as received to detect any anomalies or malfunctions.)

At the end of each station transmission, the computer accesses the data block for calibration and engineering-unit conversion, which can be done in the slack period between queries. A secondary tape is made of the converted data. This may be printed or displayed on CRT monitors on request.

6.3. SUMMARY

In this chapter we have considered the nature of the aerometric data that must be collected by the monitoring subsystem, including its density in space and time. The system-sensitive characteristics of current and possible future data sensors have been examined. Finally, we have looked at some real and proposed systems to acquire this data and examined their features in conjunction with ideal considerations.

In the following chapters we will continue to discuss the fundamental uses and interpretations of these data in terms of air quality levels, meteorological trends, and models for the prediction of air quality.

REFERENCES

1. G. B. Morgan, et al. *Science,* **170,** 291 (1970).
2. A. C. Stern, Ed. *Air Pollution,* 2nd ed. Vol. III. Academic: New York, 1968, p. 607, Figure 2; originally published in *Air Quality Criteria for Sulfur Dioxide,* Public Health Service Publ. No. 1619. HEW: Washington, D.C., 1967.
3. *Federal Register,* **36,** Pt. 2, 8187 (April 30, 1971).
4. A. T. Rossano. In Ref. 2, Vol. II, pp. 615–618.
5. R. L. Duprey, *Compilation of Air Pollutant Emission Factors,* USPHS Publication 999-AP-42. NAPCA: Raleigh, N.C., 1965. See also 2nd ed., Publication AP-42, HEW: Research Triangle Park, N.C. 1973.
6. Ref. 3, **36,** 22398–22414 (November 25, 1971).
7. *Report for Consultation on the Washington, D.C., National Capital Interstate Air Quality Control Region.* USPHS, July, 1968.
8. *Report for Consultation on the Metropolitan Denver Air Quality Control Region USPHS, October 1968.* See also *Report for Consultation on the Philadelphia Interstate Air Quality Control Region,* USPHS, October 1968.
9. *Report for Consultation on the Los Angeles Air Quality Control Region,* USPHS, November 1968.
10. E. L. Keitz and T. R. Morgan. *Analysis of Requirements for Air Quality Monitoring Networks,* 63rd Ann. Meeting of Air Pollution Control Assoc. (APCA), St. Louis, Mo., June 14–18, 1970. See also *Environ. Sci. Tech.,* **4,** 723–724 (1970).
11. F. C. Hamburg, *J. Air Poll. Control Assn.,* **21,** 609–613 (October 1971) and Ref. 7. See also H. R. Hickey et al. *Ibid.* 689 (November 1971).
12. B. Saltzman. "Significance of Sampling Time in Air Monitoring." *J. Air Poll. Control Assn.,* **20,** 663 (October 1970).
13. HEW. *Air Quality Criteria for Sulfur Oxides.* NAPCA (PHS): Washington, D.C., 1970, pp. 33–35.
14. Ref. 13, p. 161.

15. Ref. 13, p. 162.
16. R. I. Larsen. Ref. 11, **19**, 24–30 (1969).
17. K. B. Schnelle, Jr., and R. D. Neeley. "Transient and Frequency Response of Air Quality Monitors." 64th meeting of APCA, Atlanta City, N.J., June 27–July 2, 1971.
18. See, for example, S. Greenfield, in *Environ. Sci. Tech.*, **5**, 990–992 (October 1971).
19. Hamburg, Ref. 11, p. 610; also G. B. Morgan, et al., *Science*, **170**, 291 (1971).
20. E. Yu. Bezuglaya. "Statistical Determination of Average and Maximum Pollutant Concentrations" (in Russian). *Tr. Gl. Geofiz Observ.* (Leningrad), **254**, 133–139 (1971).
21. R. J. Bryan, In Ref. 2, Vol. II, p. 442.
22. E. L. Stockton. "Allegheny County's Air Monitoring Program. Paper 69-207, 62nd Ann. Meeting of the Air Pollution Control Assoc., New York City, June 22–26, 1969.
23. E. S. Savas. "Feedback Controls on Urban Air Pollution." *IEEE Spectrum*, **6**, 77–81 (1969).
24. A. E. O'Keeffe. "Needs in Electronic Instrumentation for Air-Pollution Analysis." *IEEE Trans. Geosci. Electron.*, **GE-8**, 145–148 (1970). See also Morgan et al., Ref. 19).
25. Hamburg, Ref. 11, p. 613.
26. Bryan, In Ref. 2, p. 443.
27. Stanford Research Institute. *Regional Air Pollution Study: A Prospectus.* SRI: Menlo Park, Calif., 1972.
28. A. P. Altshuller. "Instruments for Monitoring Air Pollutants." Paper presented at Instrument Society of America meeting, Philadelphia, Pa., May 19–22, 1968; also in R. L. Chapman, Ed. *Environmental Pollution Instrumentation.* Instrument Society of America: Pittsburgh, 1969, pp. 1–6.
29. A. R. Barringer. In *Environmental Pollution Instrumentation*, Ref. 28, pp. 49–67. Also see A. J. Moffat and A. R. Barringer. *Recent Progress in the Remote Detection of Vapours and Gaseous Pollutants* (NASA-CR-127632), Barringer Research Ltd.: Rexdale, Ont., Canada, 1971; and Environmental Measurements, Inc. SO_2 and NO_2 Measurements Metropolitan Los Angeles, Calif. 13–16 July 1971 (PB-212-423). San Francisco, 1971.
30. P. L. Hanst, and J. A. Morreal. *J. Air Poll. Control Assn.*, **18**, no. 11, 754–759 (1968).
31. J. Hidalgo, *Study of the Application of Laser Technology to Atmospheric Contamination Measuremeation*, Final Tech. Report, NASA contract NAS 12-664, Radiation Laboratory, Tulane University, May 1, 1968–April 30, 1969 (Clearinghouse, NBS, Springfield, Va., N69-36282).
32. S. Lederman and Widhopf. "Specie Concentration Measurements Utilizing Raman Scattering of a Laser Beam." Paper 70-224, AIAA 8th Aerospace Sciences Meeting, New York City, Jan. 19–21, 1970.

33. T. Kobayasi, and H. Inaba. "Laser-Raman Radar for Air Pollution Probe." *Proc. IEEE*, **58**, 1568–1571 (1970).
34. E. J. Rosenbaum. In G. L. Clark, Ed., *The Encyclopedia of Spectroscopy*. Reinhold: New York, 1960, pp. 675–678.
35. J. E. Stewart. In Ref. 34, pp. 678–681.
36. G. L. Wick. "Interstellar Molecules: Chemicals in the Sky." *Science*, **170**, 149–150 (1970).
37. *Product Eng.*, Sept. 14, 1970.
38. S. H. Chan, C. C. Lin, and M. J. D. Low. *Env. Sci. Technol.* **7**, 424 (1973). See also M. J. D. Low and F. K. Clancy. *Env. Sci. Technol.* **1**, 73–74 (1967); and J. M. Lepper, Jr. *Study of Infrared Techniques for Monitoring Stack Gases* (APTIC 16067). Dalmo Victor Co.; Belmont Calif., 1965.
39. Chan, Ref. 38, pp. 424–427.
40. H. W. Prengle et al. *Env. Sci. Technol.*, **7**, 417–423 (1973).
41. L. R. Snowman. *Laser Coincidence Absorption Measurements*, General Electric Co., Technical Information Series R72ELS-15, March 1972.
42. R. J. Gillmeister. Private communication June 2, 1972 (a paper comparing point and line monitors will be published by Gillmeister and E. R. Hendrix.)
43. G. Fiocco and L. D. Smullin. *Nature*, **199**, 275 (1963).
44. V. E. Derr, and E. G. Little. *Applied Optics*, **9** 1976–1992 (1970).
45. P. M. Hamilton. *Phil. Trans. Roy. Soc. Lond. A.*, **265**, 153–172 (1969).
46. W. B. Johnson, Jr., and E. E. Uthe. *Lidar Study of Stack Plumes* (HEW-NAPCA). SRI Report 7289. Stanford Research Institute: Menlo Park, Calif., 1969).
47. *New York Times*, December 14, 1970.
48. *Tall Plumes for Tall Stacks*. Stanford Research Institute: Menlo Park, Calif., 1971.
49. *Atmospheric Research*. NCAR, Boulder, Colo., Sept. 18, 1971, p. 3–11. (An excellent bibliography on acoustic probing follows this article.) See also G. C. Little. *Proc. IEEE*, **57** 571–578 (1969); and L. G. McAllister et al. *Ibid.*, pp. 579–587.
50. H. R. Hickey, W. D. Rowe, and F. Skinner. *J. Air Poll. Control Assn.*, **21**, 689 (November 1971).
51. Bryan. In Ref. 2, Vol. II, p. 455.
52. W. W. Dittrich, et al. *J. Air. Poll. Control Assn.*, **21**, 555–558 (1971).
53. B. Boykin, *Res. Dev.* **24** 32–34 (May 1973).
54. T. Johnston. "The Scope of the Wired City." *New York Times*, January 28, 1973.
55. S. W. Maley. In V. E. Derr et al., *Remote Sensing of the Troposphere*. Department of Commerce, NOAA: Boulder, Colo., 1972, Chapter 28.

APPENDIX: (ADAPTED FROM FEDERAL REGISTER, VOL. 36, NO. 228—THURSDAY, NOVEMBER 25, 1971

Point Source Data

(The following information is not required to be submitted with an implementation plan but must be available for inspection by the Administrator, EPA.)

I. General Source Information

A. Establishment name and address.
B. Person to contact on air pollution matters and telephone number.
C. Operating schedule:
    1. Percent of annual production by season.
    2. Days of week normally in operation.
    3. Shifts or hours of day normally in operation.
    4. Number of days per year in operation.
D. Year in which data are recorded.
E. Future activities, if available (e.g., addition of new or expansion of existing facilities, changes in production rate, installation of control equipment, phasing out of equipment, fuel change, etc.).
F. Map or general layout of large complex plants showing locations of various facilities, if available.[1]

II. Fuel Combustion

A. Number of boilers.
B. Type of fuel burning equipment for each boiler.
C. Rated and/or maximum capacity of each boiler, 10^6 B.t.u./hr kcal/hr.
D. Types of fuel burned, quantities, and characteristics:
    1. Type of each fuel used and place of origin.
    2. Maximum and average quantity per hour.
    3. Quantity per year.
    4. Sulfur content (as received), percent.
    5. Ash content (as received), percent.
    6. Heat content (as received), B.t.u. or kcal/unit of measure.
    7. Estimate of future usage, if available.
E. Percent used for space heating and process heat.
F. Air pollution control equipment (existing and proposed):
    1. Type.
    2. Collection efficiency (design and actual), percent.
G. Stack data:
    1. List stacks by boilers served.
    2. Location of stacks by grid coordinates (Universal Transverse Mercator, UTM, or equivalent).[1]
    3. Stack height, feet or meters.
    4. Stack diameter (inside, top), feet or meters.

5. Exit gas temperature, °F. or °C.
6. Exit gas velocity, feet/sec. or meters/sec.
H. Emission data:
1. Based on emission factors.
2. Estimate of emissions by the source.
3. Results of any stack tests conducted.

III. Manufacturing Activities (Process Losses)

A. Process name or description of each product.

B. Quantity of raw materials used and handled for each product, maximum quantity per hour, and average quantity per year.

C. Quantity of each product manufactured, maximum quantity per hour, and average quantity per year.

D. Description of annual, seasonal, monthly, weekly, and daily operating cycle including downtime for maintenance and repairs.

E. Air pollution control equipment in use (existing and proposed):
1. Type.
2. Collection efficiency (design and actual), percent.

F. Stack data:
1. List of stacks by equipment served.
2. Location of stacks by grid location (UTM or equivalent).[1]
3. Stack height, feet or meters.
4. Stack diameter (inside, top), feet or meters.
5. Exit gas temperature, °F, or °C.
6. Exit gas velocity, feet/sec. or meters/sec.

G. Emission data:
1. Based on emission factors
2. Estimate of emissions by the source.
3. Results of any stack tests conducted.

IV. Solid Waste Disposal

A. Amount and description of solid waste generated, quantity per year.

B. Percent of total that is combustible.

C. Method of disposal (on-site or off-site).

D. Description of on-site disposal method, if applicable (incineration, open burning, landfill, etc.) including maximum quantities disposed per hour and average quantities disposed per year and actual operating schedule.
1. Location of the source by a grid system (UTM or equivalent).[1]
2. If method of disposal is by an incinerator, include the following information:
   a. Auxiliary fuel used.
   b. Air pollution control equipment (existing and proposed):
      (1) Type.
      (2) Collection efficiency (actual and design), percent.

c. Stack data:
  (1) List stacks by furnaces served.
  (2) Stack height, feet or meters.
  (3) Stack diameter (inside, top), feet or meters.
  (4) Exit gas temperature, °F, or °C.
  (5) Exit gas velocity, feet/sec. or meters/sec.
  (6) Exit gas moisture content, percent if available.
3. Emission data:
  a. Based on emission factors.
  b. Estimate of emissions by the source.
  c. Results of any stack tests conducted.

Area Source Data[1]

(The following information is not required to be submitted with an implementation plan but must be available for inspection by the Administrator, EPA.)
Grid Coordinate (lower left-hand corner) _____ UTM or equivalent.
Average Stack Height of sources _____ .

I. Fuel Combustion-Stationary Sources

Includes sulfur and ash content of fuels, if applicable).
A. Residential Fuel:
  1. Anthracite Coal (plus type and size of unit)—tons/year or metric tons/year.
  2. Bituminous Coal (plus type and size of unit)—tons/year or metric tons/year.
  3. Distillate Oil (plus type and size of unit)—10^3 gal./year or 10^3 liters/year.
  4. Residual Oil (plus type and size of unit)—10^3 gal./year or 10^3 liters/year.
  5. Natural Gas (plus type and size of unit)—10^6 cu.-ft./year or 10^6 cu.-meters/year.
  6. Wood—tons/years or metric tons/year.
  7. Other—please specify.
B. Commercial and Institutional Fuel:
  1. Anthracite Coal (plus type and size of unit)—tons/year or metric tons/year.
  2. Bituminous Coal (plus type and size of unit)—tons/year or metric tons/year.
  3. Distillate Oil (plus type and size of unit)—10^3 gal./year or 10^3 liters/year.
  4. Residual Oil (plus type and size of unit—10^2 gal./year or 10^3 liters/year.
  5. Natural Gas (plus type and size of unit)—10^6 cu.-ft./year or 10^6 cu.-meters/year.
  6. Wood—tons/year or metric tons/year.
  7. Other—please specify.
C. Industrial Fuel:
  1. Anthracite Coal (plus type and size of unit)—tons/year or metric tons/year.
  2. Bituminous coal (plus type and size of unit)—tons/year or metric tons/year.

[1] Required only when diffusion modeling is utilized.

3. Coke (plus type and size of unit)—tons/year or metric tons/year.
4. Distillate Oil (plus type and size of unit)—10^3 gal./year or 10^3 liters/year.
5. Residual Oil (plus type and size of unit)—10^3 gal./year or 10^3 liters/year.
6. Natural Gas (plus type and size of unit)—10^6 cu.-ft./year 10^6 cu.-meters/year.
7. Wood—tons/year or metric tons/year.
8. Other—please specify.

II. *Process Losses (Hydrocarbons Only)*

A. Surface coating operations, dry cleaning, degreasing operations, etc., unless considered as point sources—appropriate basis for emission estimate.

III. *Solid Waste Disposal*

A. On-site incineration (plus type of unit)—tons/year or metric tons/year.
B. Open burning—tons/year or metric tons/year.
C. Other—please specify.

IV. *Transportation*

A. Gasoline-powered motor vehicles—appropriate basis for emission estimate, including hydrocarbon evaporative losses
B. Diesel-powered motor vehicles—appropriate basis for emission estimate.
C. Off-highway fuel usage—10^3 gal./year or 10^3 liters/year.
D. Aircraft—number of flights per year per type of aircraft.
E. Railroads—10^3 gal. diesel oil/year or 10^3 liters/year.
F. Vessels—10^3 gal. or 10^3 liters of oil/year, tons or metric tons of coal/year, or tons or metric tons of wood/year.
G. Gasoline handling evaporative losses—appropriate basis for hydrocarbon emission estimate from filling tank trucks, service station tanks, and automobile tanks.
H. Other—please specify

V. *Miscellaneous*

A. Forest fires—appropriate basis for emission estimate.
B. Structural fires—appropriate basis for emission estimate.
C. Coal refuse burning—appropriate basis for emission estimate.
D. Agricultural burning—appropriate basis for emission estimate.
E. Other—please specify.

(Pollutant) Emissions Inventory Summary, Tons/yr. (or Metric Tons/yr.) (Example Regions)

_____Air Quality Control Region

Data Representative of Calendar Year_____

Source category	State A					State B [i]		Region [i] total	
	County 1		--------		County N	State region total	Ditto		
	i	iii	i	iii	i	iii	i	iii	iii
I. Fuel combustion—stationary sources:									
A. Residential fuel:									
1. Anthracite coal:									
a. Area sources									
b. Point sources									
2. Bituminous coal:									
a. Area sources									
b. Point sources									
3. Distillate oil:									
a. Area sources									
b. Point sources									
4. Residual oil:									
a. Area sources									
b. Point sources									
5. Natural gas:									
a. Area sources									
b. Point sources									

Source category	State A						State B [i]				
	County 1		----------		County N		State region total		Ditto	Region [i] total	
	ii	*iii*	*ii*	*iii*	*ii*	*iii*	*ii*	*iii*	Ditto	*ii*	*iii*
6. Wood:											
a. Area sources											
b. Point sources											
7. Other (specify):											
a. Area sources											
b. Point sources											
8. Total											
B. Commercial and institutional fuel:											
1. Anthracite coal:											
a. Area sources											
b. Point sources											
2. Bituminous coal:											
a. Area sources											
b. Point sources											
3. Distillate oil:											
a. Area sources											
b. Point sources											
4. Residual oil:											
a. Area sources											
b. Point sources											

5. Natural gas:
    a. Area sources
    b. Point sources
6. Wood:
    a. Area sources
    b. Point sources
7. Other (specify):
    a. Area sources
    b. Point sources
8. Total
C. Industrial fuel:
  1. Anthracite coal:
    a. Area sources
    b. Point sources
  2. Bituminous coal:
    a. Area sources
    b. Point sources
  3. Coke:
    a. Area sources
    b. Point sources
  4. Distillate oil:
    a. Area sources
    b. Point sources
  5. Residual oil:
    a. Area sources
    b. Point sources
  6. Natural gas:
    a. Area sources
    b. Point sources
  7. Process gas:
    a. Area sources
    b. Point sources

Source category	State A						State B [i]			
	County 1		----------		County N		State region total		Region [i] total	
	ii	iii	ii	iii	ii	iii	ii	iii	ii	iii
8. Other (specify):										
a. Area sources										
b. Point sources										
9. Total							Ditto	Ditto	Ditto	Ditto
D. Steam-electric power plant fuel (point sources only):										
1. Anthracite coal										
2. Bituminous coal										
3. Coke										
4. Distillate oil										
5. Residual oil										
6. Natural gas										
7. Process gas										
8. Other (specify)										
9. Total										
E. Total stationary fuel combustion										
II. Process losses:										
A. Area sources [iv]										
B. Point sources										
1. Chemical process industries										

4. Mineral products industries				
5. Petroleum refining and petrochemical operations				
6. Wood processing				
7. Petroleum storage				
C. Total process losses				
III. Solid waste disposal:				
A. Incineration:				
1. On-site:				
a. Area sources				
b. Point sources				
2. Municipal, etc.:				
a. Area sources				
b. Point sources				
B. Open burning:				
1. On-site:				
a. Area sources				
b. Point sources				
2. Dumps:				
a. Area sources				
b. Point sources				
C. Other (specify):				
1. Area sources				
2. Point sources				
D. Total solid waste disposal				
IV. Transportation (area sources only):				
A. Motor vehicles:				
1. Gasoline powered				
2. Diesel powered				

	State A					State B [i]		
Source category	County 1		-------		County N		State region total		Ditto		Region [i] total	
	ii	iii	ii	iii	ii	iii	ii	iii	ii	iii	Ditto	Ditto	ii	iii

Source category:

B. Off-highway fuel usage
C. Aircraft
D. Railroads
E. Vessels
F. Gasoline handling evaporative losses [vi]
G. Other (specify)

H. Total transportation

V. Miscellaneous (area sources only):
  A. Forest fires
  B. Structural fires

C. Coal refuse burning_____
D. Agricultural burning_____
E. Other (specify)_____

F. Total miscellaneous_____

VI. Grand totals_____
   A. Area sources_____
   B. Point sources_____

   C. Total_____

[i] Included only if interstate region.
[ii] "Existing Emissions".
[iii] "Emissions Achieved" with control regulations of implementation plans. Must be submitted in example regions.
[iv] For hydrocarbons only, would include emissions or surface coating operations, dry cleaning, degreasing operations, etc., unless considered point sources.
[v] For hydrocarbons only, would include vehicle evaporative losses.
[vi] For hydrocarbons only, would include losses from filling tank trucks, service station tanks, and automobile tanks.

(Pollutant) Emissions Inventory Summary. Tons/yr. (or Metric Tons/yr.)
(Regions Where Emission Limitations Are Not Developed)

_____ Air Quality Control Region
Data Representative of Calendar Year _____

Source category	State A				State B [i]	
	County 1	....	County N	State region total	Ditto	Regional [i] total
	[ii]	[ii]	[ii]	[ii]	Ditto	[ii]
I. Fuel combustion—Stationary sources:						
A. Area sources [iii]						
B. Point sources						
C. Total						
II. Process losses:						
A. Area sources [iii, iv]						
B. Point sources						
1. Chemical process industries						
2. Food and agricultural industries						
3. Metallurgical industries						
4. Mineral products industries						
5. Petroleum refining and petrochemical operations						

6. Wood processing					
7. Petroleum storage					
C. Total					
III. Solid waste disposal:					
A. Area sources [iii]					
B. Point sources					
C. Total					
IV. Transportation:					
(Area Sources only) [iii, v]					
V. Miscellaneous:					
(Area Sources only) [iii]					
VI. Grand totals:					
A. Area sources [iii]					
B. Point sources					
C. Total					

[i] Included only if interstate region.
[ii] Existing emissions.
[iii] If not available, does not need to be submitted for Priority III regions.
[iv] For hydrocarbons only, would include emissions for surface coating operations, dry cleaning, degreasing operations, etc., unless considered point sources.
[v] For hydrocarbons would include vehicle evaporative losses and losses from filling trucks, service station tanks, and automobile tanks.

Air Quality Data Summary

Pollutant	Sampling site location[1]	Sampling interval (months)	Start date	End date	Number of samples	Maximum 1-hour	Maximum 8 hours	Maximum 24-hours	Annual arith. mean	Annual geo. mean	Geo. std. dev.
Particulate matter	X	X	X	X	X	-----	-----	X	-----	X	X
Sulfur oxides (as SO_2)	X	X	X	X	X	-----	-----	X	X	-----	X
Nitrogen dioxide	X	X	X	X	X	-----	-----	-----	-----	-----	-----
Photo-chemical oxidants	X	X	X	X	X	X	-----	-----	-----	-----	-----
Carbon monoxide	X	X	X	X	X	X	X	-----	-----	-----	-----

X = Date or information required.
[1] UTM Grid coordinate or equivalent.

7

Applications of Air Pollution Data

Processing of raw air-pollution data can serve two useful purposes: to report the current quality of the air and to predict its future state. "Accurate and timely information on status and trends in the environment," according to the CEQ, "is necessary to shape sound public policy and to implement environmental quality programs efficiently."[1] Substantial effort has been expended on behalf of this federal agency to develop means to accomplish the first aim—an index of current quality—but the answer does not come easily. Part of the difficulty, noted in the previous chapter, lies in the number of pollutants and the diversity of their effects. If we consider air-pollution control as a systematic process, the air-quality index is the feedback taken around the entire closed-loop system (see Figure 1, Chapter 5, for example). The problem, as in many complex systems, is one of "multivariable control," to borrow the terminology of control systems analysis. But, in order to obtain a stable operating system, it is not always necessary to control all the variables determining the state of the system, in fact many of the variables may be unobservable (this is assuredly so in the case of most intermediate products in the atmospheric photochemical process). An index that utilizes less than the total number of state variables may be a satisfactory representation of the actual air quality but only if the model on which it is based shows a relationship between the measured and unmeasured (or unobservable) data.

The problem of forecasting is a further projection of that of describing current quality. In this case, the model must be dynamic, not static. As the dynamic forces determining the air quality at any given place and time are primarily meteorological, the pollution forecasting model will be an extension of the weather model. The forecasting model must also differ from the feedback index model in its resolution of space and time. While daily, monthly, and annual averages representing an entire city or district are satisfactory representations of quality, both for public information and broad regulation features, the forecasting model must be concerned with finer detail in order to control individual sources in "real time" and prevent high pollution buildup during stagnant weather conditions.

In this chapter we will concern ourselves with these two topics, the determination of air quality and the meteorological conditions setting the stage for the dynamic prediction model. In the next chapter we will consider the forecasting models themselves.

7.1 INDEXES

During a recent political campaign, the mayor of New York City told his audiences that his administration had succeeded in cleaning the city's air to the extent that the number of "unhealthy" or "unacceptable" days were reduced by two-thirds from 1969 to 1971. Shortly afterward, the *New York Times* reported that the air was actually 7% dirtier than in 1969 and that 58%, rather than 22%, of the previous year's days were "unsatisfactory" from the standpoint of air quality.[2] The mayor was not misleading the public, nor was the newspaper inventing sensational headline: each was using a different index. The earlier index was mainly concerned with sulfur dioxide, which had indeed dropped from 0.8 ppm to 0.4 ppm during the period in question. But particulates, which were actually increasing in concentration during that time, could not be measured accurately by automatic means rapidly enough to be incorporated in the daily index. When the particulate-measuring apparatus was improved, the index was changed and a truer state of quality was uncovered. The improved information also led to further investigations of individual sources of increased particulates.

Although New York abstracted a qualitative judgement from measurements of SO_2 and particulates, during the same period Chicago published a full report. For example, the daily papers on one winter day carried the following information:[3]

"POLLUTION READINGS

Citywide average air pollution levels in the one hour period ended at 9 A.M.: Sulphur dioxide, .07 parts per million parts of air (ppm). Danger level is .11 ppm. Haze, 1.25 coefficient of haze per 1,000 linear feet (COH). Normal reading is .75 COH. Carbon monoxide, 5.3 ppm. Normal level is 10 ppm. Nitrogen dioxide, .07 ppm. Warning level is .13 ppm. Nitric oxide, .02 ppm."

At first glance, this report seems to be a great improvement over the uninformative "satisfactory" rating, since both the hourly average and a normative value is given. But close inspection, even by the relatively informed reader, reveals as much if not more reason for confusion. For example, the 0.11 ppm level for SO_2 labeled "danger" (as an hourly reading) is only 10% greater than the federal secondary standard for 24 h and about one-fifth that for 3 h (see Table I of Chapter 4). The COH hourly "normal" level of 0.75 is comparable to the federal primary annual standard of 75 ug/m^3 and half the 24-h secondary (150), whereas Larsen's relationship (see Chapter 6) would suggest it be larger.

Similarily, the CO "normal," although properly lying between the federal "alert" and standard values of 15 and 9 ppm, respectively, for 8-h averages, is very much less than the federal 1-h standard of 35 ppm. The NO_2 "warning" level is about where it might be expected—below the federal alert (0.15 ppm); however, no normative level at all is given for nitric oxide (NO), for which there is no federal standard. Excessive data coupled with norms that are not simply related to accepted national standards results in as much confusion as too little data.

Figure 1 is a report published in Los Angeles in the same period and is given in terms of peak readings rather than averages. Although these readings may be correlated easily with the LACPD first-stage alert levels, also printed in the report, they do not give a measure of the air quality. The usefulness of this report is much enhanced, however, by the inclusion of a forecast of eye irritation and visibility based primarily on ozone concentration.

What is lacking in these air-quality reports is a weighting of the effects of their components. Between the somewhat arbitrary levels reported in New York City and the detailed analysis of Chicago's air lies an "ideal" index based on agreement as to the severity of the analyzed pollutant's effects. A number of such weightings have been published. One, presented in the context of the total environment, has been derived by H. Reiquam of the Battelle Memorial Institute.[5] Pertinent extracts from this table are listed in Table I. His ratings are calculated by multiplying individual scores based on persistence, range, and complexity scales; for example, a persistence of days results in one point, months in two, years in three, decades in four, and centuries in five. Reiquam's scale is useful in deciding the relative priority of control problems but, being basically subjective, has not gained the general acceptance necessary for air index. Parenthetically, it is interesting to note his high evaluation of CO_2 as a threat, since we have not seen fit as yet to subject this emission to controls (see Chapter 3).

Another approach has been to accept the various factors included in the federal criteria (discussed in Chapters 3 and 6) as a basis for weighting. The simplifying assumption can be made, for example, that all pollutants have the same effect on receptors at the specified air-quality standard and that there is no interaction or synergy between them. The air-quality standards proposed by California in 1969 were used in such a manner by the Bay Area Air Pollution Control District (BAAPCD) as indicated in Table II. These weightings are (very approximately) inversions of the ambient standards and were applied to the pollutant levels and summed to serve, as it was said, "as a public communication tool." A total score of 100 was intended to represent severe pollution.

Smog Report

The Los Angeles County Air Pollution Control District's first-stage smog alerts are based on peak reading: ozone, .50 parts per million; carbon monoxide, 50 ppm; nitrogen oxides (nitric oxide plus nitrogen dioxide), 3 ppm.

The air quality standards of the State Air Resources Board do not involve peak readings, but are for a specific time period: ozone, .10 ppm for 1 hr.; carbon monoxide, 10 ppm for 12 hrs.; nitrogen dioxide, .25 ppm for 1 hr.

FORECAST

Maximum ozone levels today will be .30 ppm in the San Gabriel Valley, .25 ppm in the San Fernando and Pomona-Walnut valleys, and .05 to .15 ppm in all other areas of the Los Angeles Basin. There will be moderate eye irritation in the San Gabriel Valley, and light irritation in all other areas of the basin except coastal. Lowest visibilities will be 1½ miles in the Pomona-Walnut Valley, two miles in the central, coastal and southeast sections, and three to four miles elsewhere.

THURSDAY'S PEAK READINGS

	Ozone	Carbon Monoxide	Nitrogen Oxides
Central Los Angeles	.16	9	.29
Northwest Coastal	.04	6	.14
Southwest Coastal	.05	13	.25
South Coastal	.01	10	.25
Southeast	.18	5	.20
West San Fernando Valley	.32	7	.26
East San Fernando Valley	.29	12	.25
West San Gabriel Valley	.36	10	.36
East San Gabriel Valley	.51	6	.23
Pomona-Walnut Valley	.34	8	.37
Orange County APCD	.15	7	.12

Figure 1 Los Angeles daily air-pollution report.[4]

Table I
Environmental Threat Scores[5]

Pesticides (most severe)	140[a]
Mercury and similar	90
CO_2	75
SO_2	72
Particulates (TSP)	72
Nitrogen oxide (NO_x)	24
Airborne radioactives	20
Oxidants (smog)	12
Hydrocarbons (HC)	10
Carbon monoxide	9
Community noise (least severe)	4

[a] Arbitrary values

7.1.1. Pindex

Babcock[6] undertook a much more comprehensive evaluation based on the same California standards, which he called *Pindex*. He computed tolerance factors based on the hourly values given by, or derived from, the California proposal. He converted the 8-h CO value, for example, to a 1-h standard by assuming that the proportional time–tolerance ratio for SO_2 was linear and universally valid. The visibility standard, roughly incorporated as COH by the BAAPCD, was converted by Babcock to particu-

Table II

	Proposed California ambient air-quality standards (1969)	BAAPCD weighting	"Pindex," ppm	Tolerance factor
Oxidant (Ox)	0.1 ppm–1 h	200	0.1	214
Particulate (TSP)	(visibility 3 mi)	10 (COH)	—	375
NO_x	0.25 ppm–1 h	100	0.25	514
SO_x	0.5 ppm–1 h 0.1 ppm–24 h	—	0.5	1430
CO	20. ppm–8 h	—	32.0	40,000
Hydrocarbons (HC)	—	—	—	19,300

late mass concentration, using the relationship to visibility set forth by Charlson[7] and discussed in Section 3.5.3.

$$\text{Mass concentration} = \frac{1.8 \times 10^6}{\text{Visibility in meters}} \mu g/m^3. \qquad (1)$$

Photochemical reactions were taken into account by assuming that NO_x and hydrocarbons created oxidant (ozone) on a 1:1 molar basis controlled only by solar radiation, yielding the relationship:

$$\text{Oxidant (photochemical)} = 0.0006 \, (SR) \, (\text{Limiting reactant}), \qquad (2)$$

where SR is the incident solar radiation in calories per square centimeter per day (U.S. average is about 375), the constant is a mean conversion coefficient averaged from 10 cities (variation from 0.00024 to 0.00098), and the limiting reactant is the smaller of HC or NO_x (in moles). Synergy between particulates and SO_2 was taken into account by merely doubling the lesser of the two. The resulting index, after applying Eq. (2) and reconverting all to micrograms per cubic meter is

$$\text{Pindex} = \frac{TSP}{375} + \frac{O_x}{214} + \frac{SO_x}{1430} + \frac{NO_x}{514} + \frac{HC}{19,300} + \frac{CO}{40,000}. \qquad (3)$$

Eq. (3) was an early attempt to utilize all the available data but eliminate the confusion associated with gross-weight or concentration reporting and at the same time avoid the biases of Eastern or Western types of pollution. Applied to 10 cities, using data from 1962–1967, Pindex rated Chicago worst with a score of 2.04, followed by Los Angeles and St. Louis, while the cleanest city was San Diego, with a rating of 1.03, followed by Washington and San Francisco, in that order.

7.1.2. Walther's Rankings

Pindex was unfortunate in its choice of denominators, since the California code from which the tolerance factors were derived was made obsolete by the federal standards of 1971. Its assumptions about photochemical conversion and synergy, among others, were based on inadequate knowledge. Nevertheless, the basic concept was accepted and stimulated further work. In 1972, Walther[8] published a rating of pollutants and sources by effect, using tolerance factors derived from the federal standards (Table I, Chapter 4) for a 1-day average where available. For NO_x, the originally proposed 1-day standard was used, and for CO the 1-day toleration was assumed equivalent to that which would result in 2% carboxy hemoglobin (COHb) concentration in the blood.

These factors can be used, as Walther showed, to rank various categories of pollution sources by their overall impact on air quality, even though the "mix" of pollutants was different for each source. Table III shows a comparative ranking of the effects of transportation, fossil fuel (power generation plus space heating), industrial emissions, and solid-waste incineration, using both Pindex and Walther's effects factors.

The mass emissions (percent) are approximately the same for each category for both time periods. However, Pindex (using the proposed California standards) rates fossil-fuel burning as a greater threat than transportation, while the effects factors based on federal ambient standards inverts the order and designates transportation as the primary environmental insult. The marked difference in source evaluation is largely caused by the Pindex emphasis on particulate (TSP), which was derived from a 3 mile visibility standard (see Table II). The high particulate effects rating boosts the rank of both industry and fuel-burning. Walther's interpretation of hydrocarbon rating (as a daily average) has the opposite effect, since transportation was responsible (in 1969) for over 50% of hydrocarbon emissions. Actually, the HC ratings are extrapolated from a 3-h standard, which, in turn, projects an oxidant (ozone) concentration injurious to 5% of economically important vegetation. (The validity of extrapolating a 3-h "precursor" concentration of 160 $\mu g/m^3$ to a 24-h average of 45 $\mu g/m^3$ can be questioned.)

In a more recent paper Babcock revised Pindex to reflect the EPA 1971 standards using a 24-h base, as did Walther.[9] Reapplying the Pindex formula to 1969 mass emissions but without the synergism term for oxides–particulates nor any penalty for hydrocarbons in excess of that creating photo-oxidants, he obtained the same source ranking as the original, in opposition to Walther with the same data (Table III). In short, Pindex absolves the automobile from the onus of being a worse air polluter than fossil fuel because it considers airborne particulates the most potent hazard and hydrocarbons none, other than contributing to nitrogen oxides. It is likely that the inconsistency between these ratings in such fundamental strategic decisions is the result of their inability to interpret in simple numbers the secondary factors of synergism between particulates and SO_x, on the one hand, and the photochemical reaction potential, on the others. Nevertheless, the concept of deriving index numbers from concentration numbers based on the federal standards continues to be useful as a public information vehicle and a check on daily air-quality variations. The indexes currently used by the CEQ to report trends are the Mitre Air-Quality Index, the Oak Ridge Air Quality Index, and the Extreme Value Index.

7.1.3. Mitre Air-Quality Index (MAQI)

MAQI, developed by the Mitre Corporation (an MIT-derived research corporation), is a combined rating based on the *annual* secondary quality

Table III

"Effect" Ranking of Air-Pollution Sources According to Different Indexes

Source	Mass % (1966)	Effect rank % Pindex[6]	Mass % (1969)[a]	Effects rank (1969)	
				Walther[8]	Pindex (rev.)[9]
Transportation	60.7	19	60.2	50.3[b]	55.6[c] 20[a]
Industry	17.7	38	16.5	24.4	21.8 33
Fossil fuel	19.0	41	18.4	19.7	16.8 43
Solid waste disposal	2.6	2	4.9	5.6	5.8 4
Total	100.0	100	100.0	100.0	100.0 100

Pindex:

	California standards		Revised: Federal standards		Walther: Federal standards, adjusted		
	Tolerance (h), µg/m³	Effects factor	Tol. (day), µg/m³ [d]	Effects factor	Tol. (day), µg/m³	Effects factor[b]	Effects factor[c]
CO	40,000	1	7800	1	5600	1	1
HC	19,300	2	—	—	45	125	125
SO$_x$	1,430	28	266	29	260 (365)[c]	21.5	15.3
NO$_x$	540	78	400	20	250	22.4	22.4
TSP	375	107	150	52	150 (260)[c]	37.5	21.5
Ox	—	—	48	162			

[a] Corrected for deletion of "miscellaneous" category, primarily fires.
[b] Federal secondary standards adjusted to 24-h basis (Walther).
[c] Same as (b) but based on primary standards.
[d] Same as (b) (revised Pindex).

standards promulgated by the EPA. It takes into account both the annual averages and the extreme values not to be exceeded more than once per year (Table I, Chapter 4). Like the preceding indexes, it uses the ratio of the pollutant levels to the EPA standard (effects factor), so that an index value greater than 1 means the standard for that pollutant has been exceeded.[10] Because the annual values are compared with actual data spanning the preceding 12-month period, MAQI must always be a retrospective one. Daily changes would be very small and meaningless. MAQI is intended to depict monthly or quarterly changes.[11]

The Mitre index may be applied to as many pollutants as there are standards. It is defined as the root-sum-square (RSS) of the individual pollutant indexes, I_i, where the I_i are computed for some or all of the pollutant components listed in the standards (Table IV). If all are utilized,

$$\text{MAQI} = \sqrt{I_c^2 + I_s^2 + I_p^2 + I_n^2 + I_o^2}, \qquad (4)$$

where I_c represents the components index for carbon monoxide, I_s is that for sulfur dioxide; and subscripts p, n, and o represent TSP, NO$_2$, and oxidant, respectively.

Each component index is the RSS of the ratio of the concentration of that component to its standard applied to the annual longest duration average and to each of the ratios for the shorter averages appearing in the standard if they are greater than 1. For example, the carbon monoxide

Table IV

National Secondary Ambient Air-Quality Standards Used for MAQI[11]

Pollutant	Measurement period	Concentration[a] ppm	Concentration[a] µg/m³
Carbon monoxide (CO)	1 h	35[b]	40,000[b]
	8 h	9[b]	10,000[b]
Sulphur dioxide (SO₂)	3 h	0.5[b,c]	1,300[b,c]
	24 h	0.1[b,c]	260[b,c]
	Annual mean	0.02[c]	60[c]
Total suspended particulates (TSP)	24 h	—	150[b]
	Annual geometric mean	—	60
Nitrogen dioxide (NO₂)	Annual mean	0.05	100
Photochemical oxidants (Ox)	1 h	0.08[b]	160[b]

[a] Secondary standards for NO₂, CO, Ox are the same as primary.
[b] Maximum concentration not to be exceeded more than once per year.
[c] Secondary standard was eliminated in 1973.[41]

index (applied to the 8-h and 1-h ratios) is

$$I_c = \sqrt{\left(\frac{C_{c8}}{S_{c8}}\right)^2 + \delta\left(\frac{C_{c1}}{S_{c1}}\right)^2}, \quad (5)$$

where C_{c8} and C_{c1} are the average concentrations for the latest 8-h and 1-h periods, S_{c8} and S_{c1} are the corresponding standards from Table IV, and δ is 1 if the ratio is ≥ 1 and 0 otherwise. Similarly, for sulfur dioxide,

$$I_s = \sqrt{\left(\frac{C_{sa}}{S_{sa}}\right)^2 + \delta_1\left(\frac{C_{s24}}{S_{s24}}\right)^2 + \delta_2\left(\frac{C_{s3}}{S_{s3}}\right)^2}, \quad (6)$$

where the subscript a refers to the annual average and the remaining numerical subscripts are the hourly averages as before.

Since there are nine standards for the five components, an MAQI value greater than 3 assures that at least one standard has been exceeded, while a value of 1 or less assures that none have been exceeded. Between these two values, the index is ambiguous.

MAQI does not compute any index for hydrocarbons, since "national standards include them only as a guide in devising implementation plans to achieve the oxidant standards."[12] Thus, the controversy introduced by Pindex is averted, but if the oxidant is not included in the MAQI calculation (and the CEQ does not include it), there is no accounting at all for the photochemical factor.

7.1.4. Oak Ridge Air Quality Index (ORAQI)

ORAQI is the index developed by the AEC's Oak Ridge National Laboratory (ONRL) by Thomas, Babcock, and Shults and adapted by the CEQ for their use.[13]

The original version of ORAQI utilizes the concentration levels of the five pollutants of Table IV on a 24-h basis, as did Pindex (see Table III). Hydrocarbons are omitted. A scale is selected so that each pollutant at its standard represents 10 units and the index totals to 100. At a level corresponding to unpolluted "background" (see Chapter 2), the index is 10. The equation relating these parameters is

$$\text{ORAQI} = \left(5.7 \sum_{i=1}^{5} \frac{C_i}{S_i}\right)^{1.37}, \quad (7)$$

where the C_i and S_i values represent the current level and 24-h standard (Table IV), respectively. The 1.37 exponent is more or less arbitrary but is partially justified by exponents correlating SO_2 and particulate levels with episode deaths.[14]

The background levels used by ORAQI were those published by the American Chemical Society[15] in 1969 and so are somewhat dated in view of

7.1. Indexes

the later studies referenced in Chapter 2. The values are as follows:

Ox	0.02	ppm
TSP	37	$\mu g/m^3$
SO_x	0.0002	ppm
NO_2	0.001	ppm
CO	0.1	ppm

Although ORAQI is a somewhat complicated calculation, given that it is intended for radio broadcasters, its application is simplified by use of a monograph (Figure 2). The concentration averages for each component are entered in the right side of each scale and the summing values (left side) totaled. The index is obtained by placing a straight edge through "measured total" and the appropriate point on the "unmeasured pollutant" scale, reading the "air-quality index" scale.

The right-hand scale is so designed as to give an approximately correct index value when fewer than all five pollutants are measured. The corrections are based on the following fractional increases proportional to the annual pollution levels found in six fully monitored U.S. cities:[16]

Ox	26.8%
SO_x	13.8
NO_2	6.3
CO	28.7
TSP	24.4
Total	100.0%

These corrections must be applied with care in areas where one type of pollutant (such as oxidant in Los Angeles) predominate. The designers of ORAQI recognize that "no real substitute exists for a complete set of data."[13]

However, ORAQI offers still one more compromise in the interests of simplication. In areas where "coefficient of haze" (COH) or "soiling index" (SI) are reported instead of TSP in micrograms per cubic meter, the rule-of-thumb conversion is to multiply COH or SI by 100. Since the nature of particulates varies greatly in different localities, use of this conversion tends to invalidate intercity comparisons.[17]

As stated earlier, the CEQ has modified ORAQI for use with three pollutants (SO_x, NO_2, TSP), using the same 24-h normalized standards.

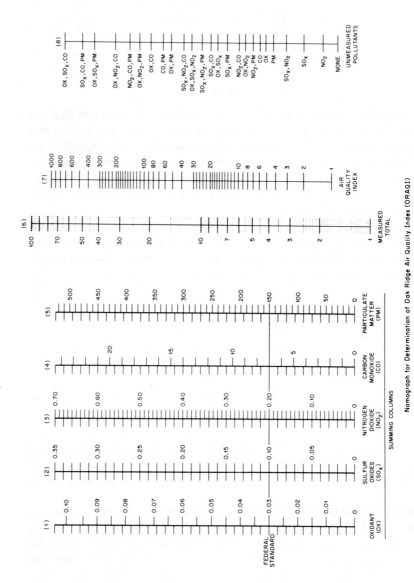

Figure 2 Nomograph for determination of Oak Ridge Air Quality Index (QRAQI) (courtesy of Oak Ridge National Laboratory).[13]

The new formula is

$$\text{ORAQI} = \left(39.02 \sum_{i=1}^{3} \left(\frac{C_i}{S_i}\right)^{0.967}\right). \quad (8)$$

As before, a value of 10 describes the background level and 100 is the equivalent of all pollutant concentrations reaching the federal standards. The suggested rankings for ORAQI readings,[17] in the absence of any meaningful correlation with health effects, is as follows:

ORAQI	Condition
<20	Excellent
20–39	Good
40–59	Fair
60–79	Poor
80–99	Bad
100 and above	Dangerous

7.1.5. Extreme Value Index (EVI)[18]

Whereas ORAQI utilizes normalized 24-h standards and MAQI both annual and short-term, EVI measures only short-term, high-level pollution. It is calculated in a manner similar to MAQI and uses the RSS method:

$$\text{EVI} = \sqrt{E_c^2 + E_s^2 + E_p^2 + E_o^2}, \quad (9)$$

where the subscripts have the same meaning as for MAQI. Since there was no short-term standard for NO_2, this term is omitted. The component EVI's are computed from the RSS of the accumulated extreme values divided by the secondary standard, or for component n,

$$E_n = \sqrt{\sum_j \left(\frac{A_{nj}}{S_{nj}}\right)^2}, \quad (10)$$

where j represents the short-term average; for example, with carbon monoxide, having 8-h and 1-h standards,

$$E_c = \sqrt{\left(\frac{A_{c8}}{S_{c8}}\right)^2 + \left(\frac{A_{c1}}{S_{c1}}\right)^2}.$$

The accumulated extreme values are the sum of the j-h average observations that exceed the secondary standard or

$$A_{nj} = \sum_i \delta_i (C_{nj})_i. \quad (11)$$

If the value of C for the component n and averaging time j does not exceed the standard (S_{nj}), the value of δ is 0; otherwise, it is 1.

An example: The Chicago CAMP station in 1965 measured 1% of the 1-h CO concentrations and 93.4% of the 8-h concentrations as exceeding the secondary standards. From the raw data, the accumulated values were[19]

$$A_{c8} = 16{,}210 \text{ ppm},$$

$$A_{c1} = 2893 \text{ ppm}.$$

The 1- and 8-h standards are given in Table IV. The EVI for Chicago in 1965 is then

$$E_c = \sqrt{\left(\frac{16210}{9}\right)^2 + \left(\frac{2893}{35}\right)^2} = 1803.$$

The remaining component extreme values are computed in the same way and combined by Eq. (9).

EVI is only meaningful if all observations are made for all periods of interest (1, 3, 8, 24 h) during an entire year. Trend analyses based on dif-

Table V
Selected Air-Quality Index Ratings for Some U.S. Cities

	Pindex[6]	ORAQI Basic[9]	CEQ[20]	MAQI[20]	EVI[20]
	1962–1967		1970[a]		
Chicago	2.04	90	—	—	—
Los Angeles	1.91	88	—	—	—
St. Louis	1.47	63	125	4.41	13.15
Philadelphia	1.43	57	150	3.99	13.38
New York	—	—	116	3.48	7.31
Baltimore	—	—	114	4.17	15.15
Detroit	—	—	102	3.39	9.17
Atlanta	—	—	86	2.60	3.44
Birmingham	—	—	76	5.05	16.13
Boston	—	—	76	2.12	1.90
Milwaukee	—	—	70	2.69	6.41
Denver	1.30	55	—	—	—
Cincinnati	1.27	50	—	—	—
Washington	1.18	43	n.a.	1.52	0
San Francisco	1.25	34	—	—	—
San Diego	1.03	31	—	—	—

[a] Based on only one sampling station per city.

fering numbers of observations are misleading. If properly applied, any EVI value greater than 0 indicates that a maximum value "not to be exceeded more than once per year" has been surpassed.

7.1.6. Application of Indexes

Table V is a comparison of the air quality of selected U.S. cities using the various rating methods discussed above. The Pindex is the original version. Looking at the latest (1970) results above, it is seen that none of the ORAQI ratings are better than "poor" and 5 are "dangerous" by the designers' interpretation. Since only five pollutants were used by CEQ to compute the MAQI ratings, any value greater than $\sqrt{5}$ = 2.24 guarantees that at least one standard has been exceeded. Thus, 8 out of the 10 ratings are unambiguously over standard and 2 are dubious. Only one EVI (Washington, D.C.) shows all short-term standards being met. However, it must be noted that all these data are based on only one sampling station per city. Quoting the CEQ, "All . . . data suffer from the problems of too few monitoring sites whose locations are determined by imprecise criteria."[20] Thus, the results shown in Table V may be distorted considerably from the true values for these cities.

7.2. FORECASTING AND TREND PREDICTION[21]

The prediction of future air-pollution levels at any location within a monitoring subsystem's boundaries will be a fundamental function of the nationwide control system. We should be concerned with both short-term fluctuations and long-term trends of air pollution at several levels, varying from the fine scale within cities to that of the entire nation. Furthermore, we should be aware of the global problem, since local systems feeding data into national archives can detect long-term national or global background trends.

At the global level, we must be concerned with the ultimate disposal of all pollutants dumped into the atmosphere. Air parcels move from west to east, generally, making a complete circuit of the globe in about 12 days at 30° latitude but with some north–south meandering rather than along a straight line. Parcels containing pollutants, such as clouds from nuclear or volcanic explosions, have been identified and tracked over such paths for several global circuits.[22]

If the natural purging mechanisms of precipitation, absorption, chemical reaction, and biological processes do not remove all pollutants from the atmosphere in one global circuit, the concentration will ultimately build up to lethal proportions.[23] Evidence of such man-made pollutant buildup over the oceans during the last 40–60 yr has been recently reported.[24]

282 Applications of Air Pollution Data

Such global, long-range predictions are of international interest and beyond the scope of the nationwide systems envisioned here. The necessary global measurements will most likely be made via satellite, and in fact, development of an appropriate instrument (a correlation spectrometer, see previous chapter) is under way. Nevertheless, these considerations illustrate the national and international nature of air-pollution problems and encourage the study of local records over long periods of time to investigate possible trends or cycles.

On the national level, an air parcel traveling in the prevailing direction over the United States might pass over a number of cities, accumulating more pollution than it loses between them over some parts of its path and building up the local average or background levels for the more easterly cities. In one (purely conjectural) path, illustrated in Figure 3, the air mass picks up a heavy concentration on the West Coast, drops nearly to background levels over the desert and Rocky Mountains, builds up to increasing peaks on its way to the East Coast, and then decreases over the Atlantic before building up again over Europe.[23] Such a buildup of average levels, even for a short time, would cause the recipient city to be more vulnerable to high-pollution episodes resulting from its local source and weather fluctuations.

It is obvious that air pollution is no respecter of state or regional boundaries. Even now, something like 7% of the carbon monoxide measured in downtown Washington, D.C., is alleged to come from Baltimore, representing a separate AQCR.[25] It is therefore clear that a national solution to air-pollution problems must include an exchange of data to make air-pollution forecasts feasible.

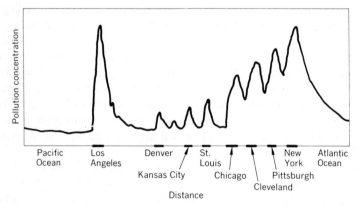

Figure 3 Hypothetical example of pollution distribution across the United States[23] (courtesy of *IEEE Spectrum*).[21]

7.2. Forecasting and Trend Prediction

Short-term forecasting of maximum concentrations that may cause acute pollution symptoms on a citywide or smaller scale is a better-defined objective of the nationwide air-monitoring system and its local activities. In communities where topographical and weather factors favor pollution buildup, increasing industrialization and population make it an imperative function. The primary purpose of such information is to provide early warning for decision and implementation of control measures for existing sources or, in a larger time frame, to serve as the basis for air resources management and land-use planning.

The predictive information subsystem must provide a dynamic model of the local air resource so that its behavior can be simulated, given the following inputs:

1. Current concentration and distribution of pollutants
2. Source strengths and distribution
3. Present or predicted meteorological factors
4. Geographical and topographical theater.

Clearly, the model must be a mathematical simulation capable of being programmed on a computer, so that large quantities of data and alternate measures can be evaluated in real time.

Unless there appears some unique causal relationship or statistical correlation between dependent variables, there is no reason to believe that prediction to a given accuracy will require any fewer input data than current-level measurement. Therefore, everything that we stated earlier about adequacy of current-level monitoring will also apply to the prediction function. The necessary accuracy must be determined on a cost–benefit basis, balancing the cost of acquiring, monitoring, and recording data; the cost of unnecessary control measures; and the risk of failure to control. We have also discussed the means for acquiring source emission data in the previous chapter. It remains to discuss meteorological factors, items 3 and 4 of the predictive subsystem.

Air-Pollution Meteorology

The importance of global air movements in the transportation and dispersal of worldwide pollution is great, but local weather information and air-mass movement are even more vital (as much as source strength and distribution) in predicting pollution concentration maxima. The concept of the self-contained air basin, or airshed, was introduced in Chapters 4 and 5 and its weaknesses, political, demographic, and statistical, have been discussed.[26] Its boundaries are so defined as to result in a minimum of pollution transport. Even in theory, the airshed concept has been challenged on the basis that episodic statisics may differ substantially from those used in AQCR designations.[27] But the prime influence of local meteorology, topography, and source geometry on the concentration and movement of polluted air masses has not been questioned.

Wind velocity determines the mass movement of pollutant clouds. Turbulence causes mixing and rapid diffusion. The lapse rate, or rate of temperature decrease with height, determines the stability or the degree of stagnation of the air mass in the vertical direction (see Figure 4). A parcel of air expanding as it rises, without heat exchange, cools at the (process) adiabatic rate; if the environmental lapse rate is the same, the parcel is hydrostatically in neutral equilibrium. If the environmental lapse is greater (superadiabatic), the air parcel is accelerated upward and is unstable; if lower, the parcel is stable. If the environmental lapse rate is negative (temperature rising with altitude), the stable region is called an *inversion*. An inversion or stable air mass above a city acts like a cap, trapping the pollutants beneath it. The thickness of turbulent air under the cap is called the *mixing depth*. The adiabatic lapse is, roughly, 10°C/km for dry air. Its exact value is affected by temperature, height, and the degree of saturation of the air. For dry air, this condition can be rather easily calculated from

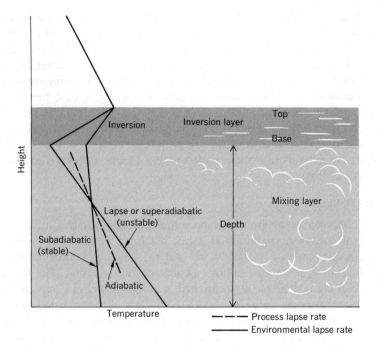

Figure 4 Temperature lapse rate and mixing layer (courtesy of *IEEE Spectrum*).[21]

three elementary equations for gases:

$$p = R\rho T, \qquad (12)$$

which is in the perfect gas law ($PV = nRT$) relating the pressure, p; density, ρ; and temperature T, with the constant, R, for a specific gas:

$$\frac{p}{p_0} = \left(\frac{\rho}{\rho_0}\right)^\gamma, \qquad (13)$$

where γ is the ratio of specific heats, at constant pressure and constant volume, Cp/Cv, and

$$\frac{\partial p}{\partial Z} = -g\rho, \qquad (14)$$

which is the hydrostatic equation for slow air movements, with Z being height and g, the gravitational constant.

Combining the first two and differentiating with respect to Z yields the dry adiabatic vertical temperature lapse:

$$\frac{\partial T}{\partial Z} = \frac{-(\gamma - 1)g}{\gamma R} \qquad (15)$$

For air, γ equals 1.40, R is 287.5 m^2/(sec^2) (°K), and $g = .9.8$ m/sec^2; hence:

$$\frac{\partial T}{\partial Z} = -0.0099°C$$

or about 1°C per 100 m.

If water is present in the atmosphere, the lapse rate of ascending air parcels is a smaller number because of the large amount of latent heat released by condensation. The neutral wet adiabatic lapse rate will vary with water content from about 7°C/km at 0°C down to about 4.5°C/km in the tropics.[28]

Inversions

Any negative temperature gradient of the atmosphere less than the adiabatic lapse rate will impart stability to the air mass; that is, air parcels tending to rise or fall will be opposed by buoyancy forces. Stable air is characterized by horizontal stratification. Even in the presence of wind flow, which is generally made turbulent by irregularities of the earth's surface, a small lapse rate will dampen the eddies, whereas a superadiabatic rate will amplify them.

An inversion of the atmospheric lapse will promote the ultimate in stability. Inversions are of three types, the most common being *nocturnal inversions*. These are caused by loss of ground heat through radiation into space during the early morning hours, which in turn cools the adjacent air, starting at ground level and continuing upward. Such inversions are called *ground based* (Figure 5a). They are often accompanied by radiation fog, formed when air is below its dew point. Generally, ground-level inversions burn off in the morning sunlight and do not contain the pollution.

Elevated inversions are formed by the subsidence of upper level air, which is warmed by compression descending on a layer of cooler air (Figure 5b). Subsidence is typical of an anticyclonic air movement (rotating clockwise in the northern hemisphere), which moves very slowly and may cover an area 1500–2500 miles in diameter. The upper warm layer is very stable and inhibits upward movement of pollution. Inversions of this nature can persist for many days and cause major pollution problems. Rapid cooling by radiation from the top of a closed layer, which exists in the water-vapor bands, can also cause upper-level inversions. Persistent clouds may also form under an inversion layer creating what is called an *anticyclonic gloom*. Upper-air inversions are usually only a few hundred meters high. On the West Coast, subsidence inversions exists semipermanently, particularly over Los Angeles. Several thousand feet of smog may develop under such a formation.

A third type of inversion, which may be even more dangerous, is known as the *urban elevated inversion* and is associated with the so-called urban envelope and heat island effects of cities. The urban envelope is the result of the greater absorption of solar visible radiation by the city's buildings and surfaces and it's re-radiation as heat reflected and trapped between the city walls. Man-made heat from air conditioning or building-space heating adds to the condition. In New York, this may equal the energy received by the sun (on a 32°F mean temperature winter day).[29]

The excess heat causes warm air to rise in the canyons of the central city, recirculating around its periphery but contained within a relatively motionless dome-shaped envelope. Without strong breezes to break up the envelope all emissions tend to stay within the dome.

The exact mechanism of the inversion that tends to seal the urban envelope is not precisely known, but it is related to nocturnal inversion and, like it forms in the early morning. Urban inversions, however, are at an altitude of several hundred meters (in New York City the median height is 310 m). They may reform each night and become very persistent, particularly if accompanied by a subsidence inversion.

Fumigation

Under a stable inversion layer, pollution tends to rise and accumulate at night. The warming effects of morning solar radiation, however, will stir up

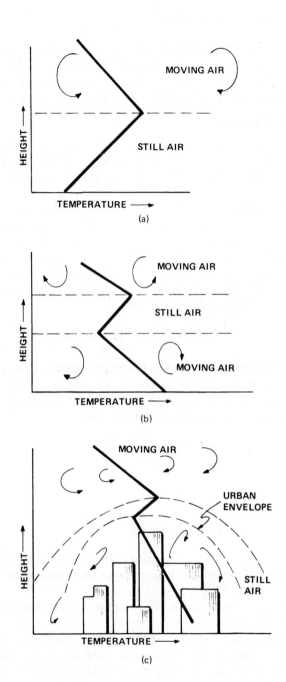

Figure 5 (a) Ground-level or nocturnal inversion, (b) subsidence or elevated inversion, (c) urban elevated inversion.

connection circulation beginning at the surface and working upward. Fumigation is the rapid mixing of the accumulated pollutants within the trapped air mass when the stirred air reaches the inversion layer. The effect is sudden and simultaneous throughout the area under the inversion.

Winds in the Troposphere

Ventilating winds are the prime means of dispersing the products of pollution sources. Winds from 50 m up to about 1000 m are the result of both global Coriolis forces and of convection from solar heating; below this, they are largely caused by frictional forces. In general, wind intensity increases with altitude as the influence of surface roughness wanes up to the height of the planetary boundary layer—½–1 mile. A gradual change in direction of 15–45° results from Coriolis effects. If the lapse rate is adiabatic, the wind strength increases according to a power law. Below 30 ft, logarithmic profiles are observed. Thus, under neutral buoyancy and with turbulent flow, the mean wind flow at height z in centimeters can be expressed (following Prandtl) as[30]

$$\bar{u}(z) = u_* k^{-1} \ln \frac{z}{z_0} \text{ cm/sec,} \tag{16}$$

where z_0 is a characteristic roughness height in centimeters, u_* a friction velocity related to horizontal shear stress and density, and k is the von Karman number (0.4). For practical application, it is sufficient to use the drag coefficient defined for a specific height, $u(h)$:

$$C_D = \left(\frac{u_*}{\bar{u}(h)}\right)^2. \tag{17}$$

Then

$$\bar{u}(z) = 2.5\sqrt{C_D}\, u(h) \ln \frac{z}{z_0}. \tag{18}$$

Values of z_0 and C_D for various ground surfaces and for a height of 2 m, $u(2)$, are given by Portman and Ryznar (after Deacon) and are reproduced in Table VI.

The conditions for wind flow under unstable (lapse) conditions are much more complex and less quantifiable. Some information is given in Ref. 30. Surface roughness and resulting increased turbulence enhances the diffusion properties of surface winds. Horizontal wind variations are less than vertical, but temporal changes may be as much as a factor of 1000 from day to day.

7.2. Forecasting and Trend Prediction

Table VI
Characteristic Roughness Height (z_0) and Drag Coefficient (C_D) for Wind at 2 m[30]

Surface	z_0 (cm)	C_D
Smooth mud flats	0.001	0.0010
Desert	0.03	0.0020
Snow on short grass	0.005	0.0012
Snow on natural prairie	0.10	0.0028
Grass, mown (1.5–3.0 cm)	0.2–0.7	0.0034–0.005
Grass, mown 4.5 cm and $\bar{u}(2)$ = 2–8 m/sec	2.4–1.7	0.0085–0.007
Grass, long (60–70 cm) and $\bar{u}(2)$ = 1.5–6.2 m/sec	9.0–3.7	0.016–0.010

Topography affects winds greatly. Cold air flowing down a mountainside results in a *katabatic* wind, especially when the cooling is accelerated by radiation from the top of a fog layer.[31] A katabatic wind will bring pollution down a mountain slope and cause an inversion a few meters off the ground, effectively trapping it. An *anabatic* wind, on the other hand, blows upward from a valley in a thick layer, resulting from the warming of morning sunshine. It may be trapped by an anticyclonic inversion or by a snowline.[32]

Topography can change wind speed and direction rapidly by channeling it through valleys and city-street canyons or by separation of flow (see below). The effect of heat sinks or sources on the ground is significant. Rough city profiles are good kinetic energy sinks and slow down winds. Other energy changes are caused by freezing, melting, evaporation, or man-made sources such as warm air rising (thermals) from power plants. A windblown parcel passing over a ridge increases in speed because of constriction and drops as it expands over a valley. Surface features such as houses cause wind-profile instabilities up to three times their height. Sea–land and mountain–valley winds are another topographical local deviation caused by differential heating.

Eddies are defined as motions larger in scale than the molecular mean free path and may vary in size up to thousands of meters. Eddies are random and can only be described by their statistical distributions of size and energy (spectra). Sources of eddies include thermal convection and flow past objects (see below). In meteorology, eddies are classified by their effect on wind direction and fluctuation.

Separation

Wind flowing over a solid body (as a hill or a stack) will increase in velocity to a point of minimum pressure, according to Bernoulli's relationship, and then decrease in velocity on the lee side until rising pressure causes flow to stop. At this point (the line of zero velocity), fluid flows from both directions and the flow is separated from the body to form a turbulent wake. In this wake, eddies circulate in an area of low pressure. If there is a sharp edge near the line of zero velocity, the separation will become initiated at that point, which is then known as *salient edge*. At some later time, the flow may rejoin the body.

The wake eddies may become detached and flow downstream, causing gustiness. Eddies may also be trapped in valleys or at the base and top of steep cliffs, containing pollution. Effluent from chimneys or buildings may also become trapped in the wake; it is then known as *downwash*. The sharp corner of a building, acting as a salient edge, may also trap effluent in the wake.

Plumes

The diffusion of pollutants emitted from point sources such as industrial stacks, under the influence of wind and turbulent eddy forces, is one of the most significant micrometeorological phenomena for air-quality control. In a highly stable atmosphere, diffusion takes place by molecular transport according to Fick's law, which states that the rate of outward transport (diffusion) is proportional to the concentration gradient and to the boundary area. In an unstable atmosphere, the mechanism of diffusion is eddy transport. If the eddies are small compared to the plume, the form of the diffusion is the same as the Fickian type, but it occurs much more rapidly.

In the following, an elementary explanation will be given for the shape and growth of plumes (adopted from Scorer).[33] More complex formulations will be considered in the next chapter.

Figure 6a shows a mass of pollutant emitting from point Q, traveling downwind and simultaneously spreading outward. At a distant x from the source the concentration within a small volume of unit length and radius R is proportional to I/R^2 for $r < R$ and is 0 for $r > R$. Hence, the rate of diffusion outward is proportional to the area of the plume boundary of radius R and unit length and the concentration gradient normal to that area:

$$\frac{dR}{dt} = K'A \frac{dc}{dR} = -K \frac{R}{R^2} = -KR^{-1}. \tag{19}$$

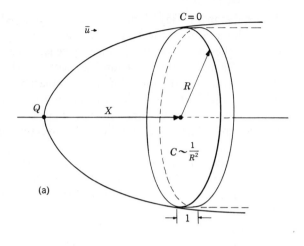

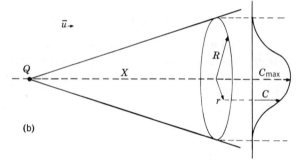

Figure 6 Plume formation from a point source.

Integrating, we obtain

$$t = KR^2 + \text{const. and } R \sim t^{1/2}. \qquad (20)$$

But also

$$x = \bar{u}t, \qquad (21)$$

so that the shape of the envelope traced by $R(x)$ in time is a paraboloid. Since the diffusion coefficient K is not usually uniform in all directions, because of changes in turbulence with height, a true paraboloid shape is not achieved.

The eddies having the most influence on the shape of the plume near the source are those of small size consistent with R, having characteristic averaging time of 2 min. Larger eddies are perceived as wind direction changes. But at larger distances—say, 2 miles or more—large eddies of a characteristic time of about 20 min affect the plume shape, while the parabolic spreading effect of the smaller eddies has nearly ceased. The plume is contained within a boundary set by the predominant eddies, which at any given distance from the source are those of size comparable to plume width. If we assume that the strength of eddies varies uniformly with their size, the plume will grow uniformly within a cone-shaped locus. In the usual case of eddies caused by structures and thermal convection, this is approximately so.

The distribution of concentration across a transverse diameter of the paraboloid is nearly Gaussian. Within the cone through which the plume meanders, similar distributions can be assumed (Figure 6b). Assume a cosine-shaped distribution

$$C = \frac{1}{2}\left(1 + \cos\frac{\pi r}{R}\right)$$

$$= C_{max} \quad \text{for } r = 0; \quad 0 \text{ for } r \geq R, \tag{20}$$

which is close to the Gaussian curve and easier to manipulate. The total mass emitted from the source of strength Q in grams per second must equal that transported across a vertical section of the plume of area S by the average wind $\bar{u}$:

$$Q = \int \bar{u} C \, dS \tag{21}$$

or

$$Q = \int_0^R \bar{u} C_{max} \frac{1}{2}\left(1 + \cos\frac{\pi r}{R}\right) 2\pi \, dr, \tag{22}$$

which, when solved, yields the relationships

$$C_{max} = Q/\pi R^2 \left(\frac{1}{2} - \frac{2}{\pi^2}\right) \tag{23}$$

and

$$C = \frac{1}{2} C_{max} \left(1 + \cos\frac{\pi r}{R}\right). \tag{24}$$

In the presence of an inversion lid, the plume is confined to a layer of constant height. If there are no large eddies, the concentration will be proportional to R^{-1}, not R^{-2} as in Eq. (23). That is, the dilution downwind will vary as $x^{-1/2}$, since $R \sim x^{1/2}$ for a parabola. The effect of some other

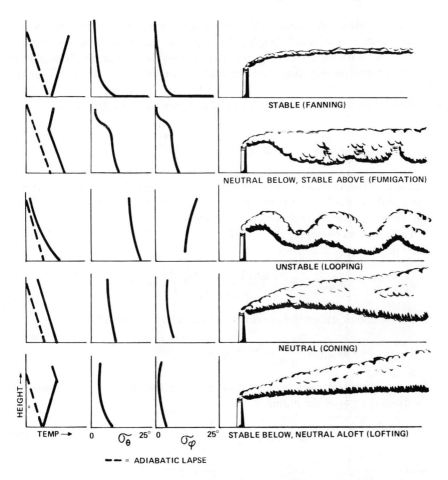

Figure 7 Types of plume behavior under various conditions of stability. Broken lines represent the dry adiabatic lapse rate; and full lines, existing lapse rates.

lapse conditions on the shape of plumes is seen in Figure 7. (See also Ref. 34.)

Penetration Through Inversion

Breakup of an inversion by mechanical means is impractical at this time. According to Scorer, the work required to mix the air over 1 km² below an inversion of 5°C at 500 m into twice this depth so as to reduce the pollution concentration by half is equivalent to the combustion of 25 tons of fuel oil at 100% efficiency, or 100 times this at the usual efficiency of stirring processes. However, since most inversions are only the order of two or three times normal stack height, it is possible that a hot plume rising upward will retain sufficient heat energy to penetrate the barrier. It will do so if the plume temperature is in excess of the inversion at the inversion height.

The mean central upward velocity of a plume rising in still air is[35]

$$W_{max} = 4.7 \left(\frac{F}{Z}\right)^{1/3} \tag{25}$$

where Z is the altitude above the stack top; F is $g(B) \cdot$ (volume emitted per sec); and B is $\Delta T/T$ (ΔT being the excess temperature of emitted stack gas above T, the absolute ambient temperature).

Taking as an example $\Delta T = 100°C$, $T = 300°K$, $Z = 125$ m, volume $= 1500$ m³/sec, $g = 9.8$ m/sec²;

$$W_{max} = 4.7 \left(\frac{9.8(100)(1500)}{(300)(125)}\right)^{1/3} = 16 \text{ m/sec}.$$

For $Z = 250$ m, twice the previous altitude above the stack,

$$W_{max} = \frac{16}{2^{1/3}} = 12.7 \text{ m/sec}.$$

For a plume bent over by a wind u, Scorer's formula for the upward velocity of the top of the plume is

$$W_{top} = 0.75 \left(\frac{F}{Zu}\right)^{1/2} \tag{26}$$

and the maximum velocity in a bent-over plume

$$W_{max} = 2W_{top}.$$

For wind $= 4$m/sec and other conditions equal to the first example:

$$W_{max} = 1.50 \left(\frac{4900}{(125)(4)}\right)^{1/2} = 4.6 \text{ m/sec}.$$

For $Z = 250$ m,

$$W_{max} = \frac{4.6}{\sqrt{2}} = 3.25 \text{ m/sec}.$$

The typical velocities of eddies below an inversion cover a range of ½–1 m/sec, so that the penetration is the same as if the air were calm.

The maximum buoyancy ($B_{max} = \Delta T/T$) for a cone-shaped plume is

$$B_{max} = 11 \frac{F^{2/3}}{gZ^{5/3}}, \tag{27}$$

so that

$$B_{max} = \frac{11(4900)^{2/3}}{(9.8)(125)^{5/3}} - 0.13$$

and

$$\Delta T_{max} = 0.13 \times 300°F = 39°C.$$

The mean excess temperature ΔT is $1/3 \Delta T_{max} = 10°C$ for the plume at $Z = 125$ m above the stack top, and for $Z = 250$ m, $\Delta T = 10/2^{5/3} = 3°C$. For the bent-over plume ($\bar{B}$ = average buoyancy),

$$\bar{B} = \frac{2F}{gZ^2 u},$$

and for the same conditions as before ($Z = 125$ m),

$$\bar{B} = \frac{2(4900)}{9.8(125)^2(4)} = 0.016$$

or

$$\Delta T_{mean} = 0.016 \times 300 = 4.8°C.$$

For $Z = 250$ m,

$$\Delta T_{mean} = \frac{4.8}{2^2} \approx 1°C.$$

Since the typical inversion is 3°C to 10°C, the plumes in the cases calculated will penetrate the 125-m inversions and will be marginal or not penetrate 250-m inversions. If penetration occurs, winds above the inversion will disperse the pollution rapidly.

To assure penetration with a moderately high inversion typical of the region, it is necessary to build fairly high stacks. Since these become very costly over a few hundred feet and nearly impractical over 300 m, alternate means to add energy to plumes have been considered. One hopeful concept is the vortex ring generator. The vortex ring tends to hold its shape and momentum, and achieves two or three times the normal plume height before dispersing.[36] However, this solution remains for future research.

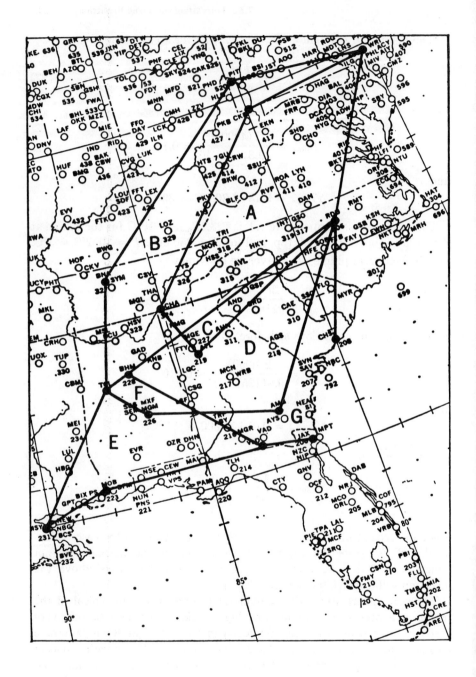

Advisory No. 122 (National Meteorological Center Advisory of High Air Pollution Potential, 28 July to 1 August 1970, 1300 E.S.T.).
    28 July: begin areas A, C
    29 July: continue areas A, C
             begin areas B, D, F
    30 July: continue areas C, D, F
             begin areas E, G,
             end areas A, B
    31 July: continue areas E, F,
             end areas C, D, G
    1 August: end areas E, F

ESSA Meteorological Support Unit statements of local high air pollution potential, 27 July to 30 July 1970.
    27 July: Washington, D.C. (and Baltimore, Md..), begin 1130 E.D.T.
    28 July: Washington, D.C. (and Baltimore, Md.), continue;
             Philadelphia, begin 1300 E.D.T.
    29 July: Washington, D.C. (and Baltimore, Md.), continue;
             Philadelphia, continue;
             New York City, local statement on dispersion conditions (not local high air pollution potential) were issued at 1100 and 1500 E.D.T.
    30 July: Washington, D.C. (and Baltimore, Md.), end 1700 E.D.T.;
             Philadelphia, end 1200 E.D.T.

Figure 8 Advisory of high air-pollution potential, National Meteorological Center[40] (courtesy of *IEEE Spectrum*).

7.3. METEOROLOGICAL MEASUREMENTS[21]

Surface wind direction can be obtained by means of vanes of various types equipped with electronic direction sensors, and the velocity of winds aloft measured by pilot balloons or "tetroons" and tracked by radar or theodolites. Smoke rockets and colored or fluorescent tracers are also used. Wind speed is measured with a variety of anemometers: cup, propeller, windmill, and hot-wire types (thermocouple or thermistor). Speed aloft can also be obtained from monitors placed on television towers.

Turbulence can be measured using rapid-response vanes and instrument computers that determine the standard deviation of the direction fluctuation, known as *sigma*. A two-directional vane or bivane will measure sigma in the vertical, as well as the horizontal, plane. Other methods used include gust accelerometers, bladed wheels similar to restrained cup anemometers, and various airborne devices.

Lapse rate may be measured from thermocouple-instrumented towers, aircraft, wire sondes, and similar means. Near the ground, it may be measured by a thermal radiometer, a two-sided differential thermopile instrument that determines the net cooling or heating load at the surface. The vertical temperature structure can also be explored with microwave or IR radiometer techniques. Lidar is capable of measuring the height of the mixing layer through reflection from particulates concentrated below the interface. This height is an indirect measure of lapse rate. Conversely, the mixing depth, or the height of the mixing layer (a significant parameter in current diffusion models), may be calculated by comparing the dry adiabatic lapse rate, measured from the surface temperature, with the morning environmental sounding, noting the altitude where the two temperatures are equal.

Solar radiation is important in photochemical reactions, which produce oxidants as secondary pollutants. A number of instruments (actinometers) have been designed to measure this parameter.

Hydrometers used for pollution meteorology are of standard design, as are precipitation instruments, rain gauges, and the like. Precipitation is the most important mode of purging the air of pollutants.

Recommended instrumentation for a moderate-size city may consist of several television towers with wind vanes and totalizing cup anemometers at 10, 20, 40, and 80 m, topped by a gust accelerometer, bivane, or three-axis anemometer. Sigma computers, rain gauges, and actinometers are required. A wide horizontal deployment of sensors is recommended in order to monitor valley or shoreline winds. Investigations of low-level inversion prevalence using wire sonde temperature probes is also suggested.[37]

The system installed by the city of Chicago, described by Cramer,[38] obtains meteorological data from 8 remote stations but has a capacity for 59, each handling five channels of information. Wind direction and speed are digitized at 5-sec intervals, electronically stored, averaged over a 15-min period, and recorded. The stored averages are polled at intervals and telemetered to the central data station for printout and display. In more recent installations, on-line analogue and digital computers are being used for assessment of pollution conditions and diffusion analysis. An interesting example of direct computer control of a pollutant source through meteorological measurements was installed in the Enrico Fermi atomic power plant at Monroe, Michigan. A valve in the stack operated by an analogue computer accepted wind-speed and direction data, computing the mean speed and the direction standard deviation (sigma), and comparing them with preset dispersion criteria. If they were not met, the stack gases were diverted to temporary storage until good dispersion was restored.[39]

Supplementing the measure of pollutant dispersion that can be deduced from local meteorological data are those of a synoptic nature. The Environmental Science Services Administration (ESSA) in cooperation with

EPA provides forecasts over large areas whenever persistent high-pressure air masses lead to stagnant conditions conducive to air-pollution episodes. Figure 8 provides an example of one of these "high air-pollution-potential advisories." The extent of this stagnant area is cited as a reason for "nationwide forecasting and coordinated control strategy."[40]

In the next chapter we shall discuss the formidable problems of merging these data into a model describing the complexities of gas dynamics, airmass transport, chemical reactivity, and urban topography, and thus be able to generate a detailed, short-term pollution forecast that will permit meaningful control actions to be taken on the time scale of sensitive human receptors.

REFERENCES

1. CEQ. *Environmental Quality* (Third Annual Report). Supt. of Documents: Washington, D.C., 1972, p. 3.
2. *New York Times*, May 20, 1972.
3. *Chicago Daily News*, February 10, 1972.
4. *Los Angeles Times*, July 14, 1972.
5. H. Reiquam. *138th Meeting, American Association for the Advancement of Science*, Philadelphia, December 1971.
6. L. R. Babcock, Jr. *J. Air Poll. Control Assn.*, **20**, 653–659 (1970).
7. R. J. Charlson and N. C. Alquist. *J. Air Poll. Control Assn.*, **7**, 17 (1967).
8. E. G. Walther. *J. Air Poll. Control Assn.*, **22**, 352–355 (1972).
9. L. R. Babcock and N. L. Nagda. "Indices of Air Quality," *138th Meeting, American Association for the Advancement of Science*, Philadelphia, December 27, 1971.
10. Ref. 1, pp. 7–8.
11. Ref. 1, pp. 34–39.
12. Mitre Corp. *National Environmental Indices: Air Quality and Outdoor Recreation* (MTR-6159), April 1972.
13. W. A. Thomas, L. R. Babcock, Jr., and W. D. Shults, *Oak Ridge Air Quality Index* (ORNL-NSF-EP-8), Oak Ridge National Laboratory, Tenn., September 1971.
14. Ref. 6, p. 9.
15. American Chemical Society. *Cleaning Our Environment: The Chemical Basis for Action*. Washington, D.C., 1969, p. 24.
16. NAPCA. *Air Quality Data 1966*. Durham, N.C. 1968.
17. Ref. 13, p. 5.
18. Ref. 1, p. 43.
19. Ref. 1, pp. 39–43.

20. Ref. 1, pp. 10, 11, 44.
21. Adapted from R. Bibbero. *IEEE SPECTRUM,* **8,** 73–81 (November 1971), by permission.
22. R. Newell. "The Global Circulation of Atmospheric Pollutants." *Scientific American.* January 1971, pp. 33–34.
23. C. B. Ludwig et al. *Study of Air Pollution Detection by Remote Sensors* Rept. CR-1380. NASA: Washington, D.C., 1965, p. 1-1–1-2.
24. *New York Times,* October 18, 1970.
25. *Report for Consultation on the Washington, D.C., National Capitol Interstate Air Quality Control Region.* USPHS: Washington, D.C., 1968.
26. *Report for Consultation on the Los Angeles Air Quality Control Region.* USPHS: Washington, D.C., 1968.
27. A. C. Stern. Cited in *Chem. Eng. News,* **48** (June 1970).
28. R. Scorer. *Air Pollution.* Pergamon: London, 1968, p. 14.
29. E. Ferrand. *Sci. Technol.,* **90,** 10 (June 1969).
30. R. Machol, Ed. *System Engineering Handbook.* McGraw-Hill: New York 1965, pp. 5–12.
31. Ref. 28, p. 70.
32. Ref. 28, p. 76.
33. Ref. 28, pp. 29–39.
34. E. V. Somers. In W. Strauss, Ed., *Air Pollution Control,* Pt. 1. Wiley: New York, 1971, p. 25,
35. Ref. 28,p. 60.
36. P. R. Sticksel and S. D. Ban. *Research Outlook,* Vol. II. Battelle Memorial Institute: Columbus, Ohio, 1970, p. 29.
37. E. W. Hewson. In A. L. Stern, ed., *Air Pollution,* 2nd ed., Vol. III. Academic: New York, 1968, pp. 329–387 (this is a complete survey of instrumentation requirements).
38. H. E. Cramer. In R. L. Chapman, Ed. *Environmental Pollution Instrumentation.* Instrument Society of America: Pittsburgh, 1969, p. 1.4.
39. R. C. Wanta. In Ref. 37, Vol. I, pp. 187–223.
40. G. B. Morgan et al. "Air Pollution Surveillance Systems." *Science,* **170,** 289–296 (October 16, 1970).
41. Federal Register, **38,** 25678–25681 (September 14, 1973).

8

Mathematical Models of Air Pollution

Mathematical modeling of the atmosphere's capacity to diffuse and transport effluents from specified air-pollution sources is the crucial element in any system attempting to achieve control by the most cost-effective means. As stated by H. Moses,[1] a mathematical model "may be likened to a transfer function where the input consists of both the combination of weather conditions and the total emission from sources of pollution, and the output is the level of pollutant concentration observed in time and space."

The symbolic relationship between the proposed mathematical air-pollution control model, its inputs, and its outputs is diagramed in Figure 1. The model is needed in order to synthesize all of the known or assumed causal information into a meaningful pollution forecast. Knowledge of the status of receptors—those people and things affected by air pollution—permits prediction and display of the effects.

Whenever the refinement is justified, a mathematical model may also be used to facilitate the optimizing of control actions, through simulation, in order to evaluate the effects of alternate means of source control. Introduction of cost factors permits this evaluation to be on a cost-effective basis. If the model is not hopelessly complex, the decision process may be mechanized through linear programming routines. Decision simulation may proceed in real time or decisions may be predetermined off-line.

To act in these capacities, the model must be implemented in real time; that is, a solution must be obtained sufficiently in advance of the predicted effects to take corrective action. In the case of real-time simulation for control decisions, even more lead time is called for. Considering the complex causes and relationships that generate air pollution, a computerized model is strongly suggested.

Figure 2 states in more detail the input–output relationships of the model and suggests, qualitatively, its internal structure. The model may be entirely or partly empirical, based on statistical correlations (regression),

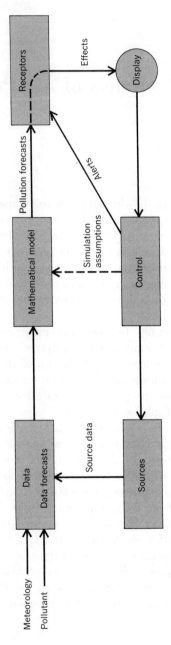

Figure 1 Air-pollution control model (courtesy of *IEEE Spectrum*).

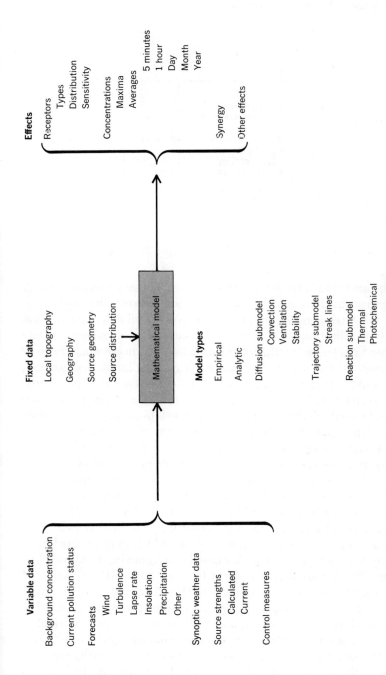

Figure 2 The input-output of relationships of a mathematical model (courtesy of *IEEE Spectrum*).

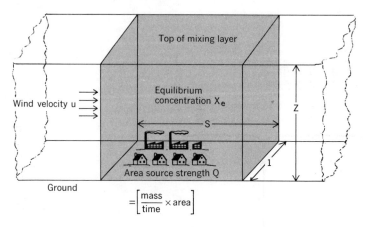

Figure 3 Area-source "box" diffusion model (courtesy of *IEEE Spectrum*).

or it may be analytical and derived from explicit physical relationships. A simple empirical model might be the correlation between degree-days (a measure of mean temperature for heating purposes) and SO_2 concentration averages. Generalization of this kind of relationship to places or time periods other than those for which the correlation was made would be very suspect because of the many other factors (e.g., ventilation, topography, fuel) that could influence the ambient sulfur concentration. This is true even if the correlation for a particular community, season, and averaging period is quite good. Nevertheless, such relationships at a given time may be the only kind that are available, as has been true for photochemical oxidant prediction. Such statistical relationships, if properly validated for each locality and intended use, comprise a valuable predictive tool.

An elementary analytical and physical model representing gaseous diffusion from an area source (which may be a city with many closely spaced houses or buildings burning fuel) is shown in Figure 3. The "box" model is considered to enclose the city, bounded by the ground as its base and the height of an assumed mixing layer Z (which may be fixed by a temperature inversion) as its top. Within the box, mixing is assumed to be thorough. The box itself is considered to be of unit width and oriented so that its length S lies in the direction of the wind, which passes freely through its ends with a average velocity u. The ventilation rate, defined as the volume of air passing through a unit width of the box, is then equal to uZ. If Q is the area source strength, or the mass emission rate per unit area, then QS is the rate for a unit width, corresponding to the ventilation rate. With some mathematical manipulation, it can be shown that the equilibrium concentration X_e of a gaseous pollutant in the box equals the emission–

ventilation ratio, or

$$X_e = \frac{QS}{uZ} \quad \text{(mass per unit volume)},$$

using any consistent units.[2,3] Also, 90% of the equilibrium concentration is reached within a certain time, 2.3 S/u. For particulates, which settle with time, these expressions must be modified accordingly.

The box or prism model may be used for rough-order-of-magnitude assessments of a city's pollution concentration, using approximate numbers for mixing heights, wind vector, and source strengths. On the other hand, it is far from answering the primary question asked of a mathematical model: What will be the pollutant concentrations at any point in the air quality region, given all the data on sources and meteorological conditions? In theory, a complete answer to this question can be given only by continuously tracking the pollutants emitted from individual sources and computing the concentration of each species at every point as they are transported by the wind, spread by diffusion, mixed by turbulence, and reflected or channeled by surfaces such as the ground or buildings. The basic consideration of mass continuity in fluid dynamics leads to complex vector equations that describe the time-varying changes in the concentration field. The problem is much like that of trying to describe the changing intensity of a soluble dye at every point after it is dropped into a swirling, turbulent brook, except that in our case we must also account for the chemical decay of each species with time and their reactions to each other.

8.1. QUANTITATIVE BASIS OF AIR-POLLUTION MODELS

It must be understood that there is no complete theory of the dispersion of pollutants in the atmosphere. Instead, there are many theories, partly empirical, and little comprehensive data by which to choose. Nevertheless, there are many useful formulas, some mentioned in the previous chapter, a good many of which generate insight into the problem, if not precision. The modeling of air-pollution dispersion is akin to weather meteorology and forecasting and, along with these counterparts, is very much an evolving science.

In order to perceive the present state of this art, an examination of its quantitative basis in fluid dynamics and diffusion theory is useful. In this endeavor we will be guided by a comprehensive treatise, *Meteorology and Atomic Energy—1968*, edited by D. H. Slade for ESSA.[4] Much of the theory and test data involved for protection against radioactive emissions from nuclear plants are equally applicable to the more mundane effluents from industrial and domestic sources.

8.1.1. Diffusion Theory

Most pertinent atmospheric diffusion theory is an attempt to apply on a macroscale an analogy to the molecular process of heat and momentum transfer. The initial concept is the response of a viscous (Newtonian) fluid or gas to a shearing stress, such as that caused by a difference in horizontal wind velocity at two different heights. The shear stress, τ, in units of force per unit area, is given by

$$\tau = \mu \frac{d\bar{u}}{dz}, \tag{1}$$

where $\bar{u}$ is the mean horizontal wind, z is height, and μ is the dynamic viscosity.

Viscosity is a molecular property of "internal friction" and can be considered a momentum conductivity, analogous to thermal conductivity. Eq. (1) is descriptive of laminar flow, in which the streamlines (lying in the direction of flow at all points and time) remain distinct from each other and do not cross.

Above a certain velocity, the streamlines become unstable and break into random cross-flow eddies or turbulence. The critical parameter is the Reynolds number (Re), defined as the dimensionless ratio of inertial to viscous forces. A Reynolds number pertinent to the atmosphere can be expressed as

$$\text{Re} = \frac{LV}{\mu \rho^{-1}}, \tag{2}$$

where L is a characteristic length, V is a flow velocity, and μ/ρ is called the kinematic viscosity, ν.

A Reynolds number greater than 10^3 is characteristic of turbulent flow. If the "characteristic length" is chosen as the height above ground, since ν is on the order of 0.15 cm^2/sec (stoke), a typical velocity of 100 cm/sec at 10 m height is equivalent to

$$\text{Re} = \frac{(1000)(100)}{0.15} = 5 \times 10^5,$$

so that the atmosphere is normally turbulent.

The diffusion of momentum in turbulent flow is three or more orders of magnitude greater than diffusion in laminar flow, where it is governed only by the molecular viscosity, μ. If gaseous and aerosol concentrations are considered to move with the atmosphere (nonbuoyant and having no mass), their diffusion follows that of the air momentum.

Boussinesq in 1877 introduced the concept of turbulent viscosity to account for this greater diffusion. Eq. (1) can be replaced by

$$\tau = (\mu + A)\frac{d\bar{u}}{dz}, \tag{3}$$

where A is the *Austauch*, or exchange coefficient, defining the kinematic eddy-viscosity coefficient,

$$K_M = \frac{A}{\rho}. \tag{4}$$

An eddy can now be defined, somewhat more precisely, as a "material wind structure having the ability to transfer air properties across [mean] flow" in a manner similar, but much greater, than molecular transfer. Since K_M is a function of z, eddies are larger, or more energetic, at greater heights above ground.

Because K_M is much larger than ν, near the ground the tangential shearing stress is

$$\tau_0 = (\rho K_M)\frac{d\bar{u}}{dz}, \tag{5}$$

where τ_0 is a function of z but constant for z near 0.

Then $\bar{u}(z) = f(\nu, z, \rho, \tau_0)$, and to obtain the vertical wind profile, we must resort to dimensional analysis. Applying the Buckingham π theorem:

$$f\left(\frac{d\bar{u}}{dz}, \frac{z}{V}, \frac{z\nu}{\nu}\right) = 0$$

$$\frac{d\bar{u}}{dz} = \frac{V*}{z} f_2\left(\frac{zV*}{\nu}\right) \tag{6}$$

where $V^{2}* = \tau_0/\rho$ and $V*$ is called the *frictional velocity*.

The function f_2 in Eq. (6) represents the contribution of laminar flow (ν), and since $V*$ is observed to be in the order of 100 cm/sec, the laminar atmospheric sublayer (where $Re < 10^2$) is about 1 mm thick. Hence, f_2 can be neglected for any surface roughness greater than 1 mm (grass, pebbles, etc.). Thus,

$$\frac{d\bar{u}}{dz} = \frac{V*}{kz}, \tag{7}$$

where k is a "universal constant" (von Karman's) experimentally shown to equal 0.4. Integrating Eq. (7) and setting the boundary condition $\bar{u} = 0$ for $z = z_0$ yields

$$\bar{u}(z) = \frac{V_*}{k} \ln \frac{z}{z_0}, \qquad (8)$$

where $z \geq z_0$, the "roughness length." This is the log wind profile stated in Eq. (16) of the previous chapter, where experimental values of z_0 were given (Table VI). Values of V_* (m/sec) corresponding to similar z_0 values (cm) are as follows:

	z_0	V_*
Smooth mud, ice	0.001	0.16
Smooth sea	0.02	0.21
Snow or smooth lawn (1 cm)	0.1	0.27
Lawn, 5-cm grass	1–2	0.43

for $\bar{u}$ (2 m) equal to 5 m/sec.

8.1.2. Buoyancy Forces

In the previous chapter the dry adiabatic lapse rate, Eq. (15), approximately 1°C/100 m, was shown to be the temperature profile that would impart no buoyancy forces to an ascending or descending air parcel. An ambient profile differing from the adiabatic will signify temperature differences between the moving parcel and displaced air, resulting in acceleration or decelerating forces. The previous discussion has not accounted for these forces; hence, K_M, the eddy-viscosity coefficient (Eq. 4), relates solely to mechanically induced turbulence.

Since adiabatic ambient conditions normally exist only briefly, for a few moments near dawn and dusk, net forces enhancing turbulence (superadiabatic) are present during the day and conversely at night.

The buoyancy force on a parcel of volume V is equal to the difference between the weight of the parcel W_P and that of the displaced atmosphere W_A, or

$$F = W_A - W_P = gV(\rho_A - \rho_P), \qquad (9)$$

where positive F is an upward force. Dividing Eq. (9) by the mass yields the parcel acceleration

$$a = g \frac{\rho_A - \rho_P}{\rho_P}, \qquad (10)$$

and since $P_A = P_P$, from the equation of state ($\rho = \rho RT$)

$$a = g \frac{T_P - T_A}{T_A} = \frac{dw}{dt},$$

where T is in degrees Kelvin and w is the vertical upward velocity acquired by the parcel.

Where γ, the diabatic lapse rate existing in the atmosphere, equals dt/dz; Γ is the dry adiabatic lapse; and Δz is the height change,

$$\frac{dw}{dt} = g\frac{(\gamma - \Gamma)\Delta z}{T_A}. \tag{11}$$

Hence, for *neutral stability*,

$$\gamma = \Gamma \quad (T_P = T_A).$$

For the *unstable* case (vertical motions amplified by buoyancy),

$$\gamma > \Gamma \quad (T_P > T_A).$$

For the *stable* case (vertical motions damped),

$$\gamma < \Gamma \quad (T_P < T_A).$$

A dimensionless number somewhat analogous to Re can be used to account for the effect of diabatic lapse on the turbulent energy and vertical wind shear profile $(d\bar{u}/dz)$. The Richardson number (Ri) is defined as the rate of consumption of turbulent energy by bouyancy forces, divided by the rate of production of turbulent energy by wind shear.

From dimensional considerations it can be shown[5] that

$$\text{Ri} = \frac{g}{T_A}\frac{(\gamma - \Gamma)}{(d\bar{u}/dz)^2}. \tag{12}$$

Eddies also transport heat flux under diabatic conditions, being at a different temperature than the surrounding mass. The heat flux

$$H = \rho C_p K_H (\gamma - \Gamma), \tag{13}$$

where C_p is the specific heat at constant pressure and K_H is the eddy-heat conductivity coefficient.

Combining Eqs. (13), (12), and (5), we obtain

$$\text{Ri} = \frac{gHK_M}{T_A C_p K_H \tau_0 (d\bar{u}/dz)} \tag{14}$$

or, as an alternate "flux form" of Ri,

$$R_f = \text{Ri}\frac{K_H}{K_M} = \frac{gH}{T_A C_p K_H \tau_0 (d\bar{u}/dz)}. \tag{15}$$

8.1.3. Gradient Transport Theory: Fickian Diffusion

The classical conduction equation describing the diffusion of a property from a region of high concentration to one of low concentration, analogous to Ohm's law, states that the rate of diffusion is proportional to the gradient of the concentration, or (in one dimension)

$$\frac{dq}{dt} = K \frac{\partial^2 q}{\partial x^2}. \tag{16}$$

If the coefficient of proportionality, K, which can be identified with the eddy-diffusion coefficient in the case we have been discussing is constant, Eq. (16) is identical to Fick's law and the diffusion is Fickian.

In three dimensions, Eq. (16) can be written as

$$\frac{dq}{dt} = \frac{\partial}{\partial x}\left(K_x \frac{\partial q}{\partial x}\right) + \frac{\partial}{\partial y}\left(K_y \frac{\partial q}{\partial y}\right) + \frac{\partial}{\partial z}\left(K_z \frac{\partial q}{\partial z}\right). \tag{17}$$

The K permit description of the flux of a passive scaler quantity such as smoke concentration, where flux is defined as $K_x \partial q/\partial x$, and so on. The K values are assumed to be determined solely by the atmospheric properties and the concentration flux does not affect these dynamics. If the turbulence is largely mechanical, $K \approx K_M$, and if thermal, $K \approx K_H$.

In practice, since K cannot be well specified because of the difficulty in measuring heat flux [see Eq. (13)], the K are specified by reference to observed diffusion data.

Equation (17) is complex but mathematically tractable and leads to many interesting solutions when integrated with respect to the proper variable and initial conditions. Integration with respect to t yields the equation of diffusion from a continuous point source, and with respect to the Y axis, that of a crosswind infinite line source (e.g., a heavily traveled highway), etc.

For example, Eq. (16) may be applied to the instantaneous release from a point source of total strength Q, a "puff," or explosion, (where Q = mass and q = mass per unit volume). The boundary conditions are as follows:

(1) $\quad\quad\quad q \to 0 \quad$ as $\quad t \to \infty$

(2) $\quad\quad\quad -\infty < x < +\infty$

(3) $\quad\quad\quad q \to 0 \quad$ as $\quad t \to 0 \quad$ except at $\quad x = 0$

and the continuity condition is

$$\int_{-\infty}^{+\infty} q\, dx = Q.$$

The fundamental solution

$$\frac{q}{Q} = \frac{1}{at^{1/2}} \exp\left(-\frac{b_x^2}{t}\right) \tag{18}$$

is a symmetrical cloud growing from an initial state of $q = 0$ (except at $x = 0$) with a Gaussian distribution of concentration q through the axis at any time $t > 0$.

Applying the continuity conditions, it can be shown that

$$\frac{q}{Q} = \frac{1}{(4\pi Kt)^{1/2}} \exp\left(-\frac{x^2}{4Kt}\right). \tag{19}$$

Comparing Eq. (19) with the standard form of the Gaussian density function (normal curve) for zero mean,

$$Y = \frac{1}{\sigma(2\pi)^{1/2}} \exp\left(-\frac{x^2}{2\sigma^2}\right), \tag{19a}$$

it is seen that

$$\sigma^2 = 2Kt, \tag{19b}$$

where σ (the standard deviation) is the value of x where the concentration q falls to 0.6066 of its value at the cloud center.

Equation (19) corresponds to a no-wind condition or to that of coordinates moving with a constant mean wind in the x-direction ($\bar{u}$) and zero horizontal crosswind ($\bar{v}$) or vertical ($\bar{w}$) components.

For the nonisotropic case in three dimensions ($K_x \neq K_y \neq K_z$),

$$\frac{q(x, y, z, t)}{Q} = [(4\pi t)^3 K_x K_y K_z]^{-1/2} \exp\left[-\frac{1}{4t}\left(\frac{x^2}{K_x} + \frac{y^2}{K_y} + \frac{z^2}{K_z}\right)\right]. \tag{20}$$

8.1.4. K-Theory

The assumption in Eq. (16) and subsequent is that the eddy-diffusivity coefficients K are constant. But this cannot be so because of the variations of mean wind shear and heat flux in time and space. We have seen that both K_M and K_H vary with height, for example. K-theory is an attempt to account for these variations in K.

Consider, for example, the case of an infinite crosswind line source in the steady state ($dq/dt = 0$). In Eq. (17), the flux variation in y at ground level,

$$\frac{\partial}{\partial y}\left(K_y \frac{\partial q}{\partial y}\right),$$

is also zero because of the infinite length of the source.

For a mean wind $\bar{u}$ on the x-axis, it is a reasonable assumption that

$$\bar{u}\frac{\partial q}{\partial x} \gg \frac{\partial}{\partial x}\left(K_x \frac{\partial q}{\partial x}\right);$$

that is, the x-transport due to mean wind flow greatly exceeds eddy-transport flux. For reasons of continuity,

$$\bar{u}\frac{\partial q}{\partial x} = \frac{\partial}{\partial x}\left(K_z \frac{\partial q}{\partial z}\right), \tag{21}$$

for which a solution is[6]

$$\frac{q}{Q} \approx \frac{1}{(2\pi K_z x \bar{u})^{1/2}} \exp\left(-\frac{\bar{u}z^2}{4K_z x}\right) \tag{21a}$$

The evaluation of Eq. (21a) then depends on finding an expression of K_z.

This case has been solved by assuming a power law for the wind

$$\bar{u} = V_* r' \left(\frac{z}{z_0}\right)^{\alpha'} \tag{22}$$

where r' and α' are chosen to give the best fit to the log wind profile and V_*, z_0, retain their previous meaning for surface roughness effects. This expression has been verified experimentally up to 1000m from a line source.

It can then be shown that

$$K_z = kV_* z_0 \left(\frac{z}{z_0}\right)^{\beta} \tag{23}$$

$$\bar{u}(z) = V_* r' \left(\frac{z}{z_0}\right)^{\alpha^*}, \tag{24}$$

where β and α^* are obtained from the observed diabatic wind profiles.

Slade[4] reports solutions to the elevated crosswind line source and work on the continuous point source. K-theory at best, however, is semiempirical, since the basic Fickian equation depending on constant K cannot be valid. It has been said that the theory is "useful in practice but not in principle."

8.1.5. Statistical Theories of Turbulent Diffusion

A statistical–kinematical approach to diffusion has the advantage that no knowledge of the causes of turbulence, dynamical or thermodynamical, are needed. This approach leads to the well-known Sutton equations, which are the best known form of diffusion models. In his method, the variations of concentration with time within a cloud can be forecast from a knowledge of the turbulence statistical spectrum.

If we consider the emission of particles from a source with discrete random right or left lateral movements in each time interval (Slade gives the example of a professor randomly handing out pennies to students in the first row of a class, each of whom randomly hands them right or left to the student in the next rearmost row, etc.), we find that for a large number of lateral steps (m) and emissions or time intervals (n), we arrive at the binomial or Bernoulli distribution

$$P(m, n) = \left(\frac{2}{\pi n}\right)^{1/2} \exp\left(-\frac{m^2}{2n}\right)$$

for the probability density (or concentration). For very large values of n and m, we approach the Gaussian continuous distribution, Eq. (19a). A major point of difference in this approach, however, is that we are following the individual particle and observing the resulting standard deviation from the moving particle frame of reference. This is termed the Lagrangian frame as distinguished from the observations of turbulent flux from a fixed point (Eulerian frame), as in the Fickian theory.

The random-particle, or "drunkard's walk," view of dispersion is highly valid on the molecular scale and leads directly to the observed Brownian movement. But since large-scale turbulence tends to be highly autocorrelated rather than random, a much greater time scale is needed to observe a Gaussian distribution.

The standard deviation of the crosswind concentration distribution was derived by G. I. Taylor for the case of continuous motion[7] in a flow that possesses statistically homogeneous turbulence (which, however, cannot be the case in the real atmosphere near the surface). The crosswind distance y traveled by a particle because of turbulent crosswind fluctuations v' during time t is

$$y(t) = \int_0^t v'(t_1) \, dt_1, \tag{25}$$

where approximately $t = x/\bar{u}$.

The second moment or variance (mean square diffusion) along the y-axis is

$$\bar{y}^2 = 2\bar{v}'^2 \int_0^t \int_0^{t_1} R(\xi) \, d\xi \, dt_1. \tag{26}$$

The variable $R(\xi)$ is known as the one-point Lagrangian velocity correlation coefficient because it is normalized so that $R(0) = 1$, and it refers to the particle velocity rather than that measured at a fixed point. (The distinction between these velocities is roughly that measured by tracking a free-floating neutral balloon compared with that measured at a fixed point by anemometer.) $R(\xi)$ is accordingly defined as

$$R(\xi) = \frac{\overline{v'(t)v'(t+\xi)}}{\bar{v}'^2}. \tag{27}$$

Then, for a short time scale (t small),

$$R(t) \approx 1 \quad \text{and} \quad \bar{y}^2(t) \approx \bar{v}'^2 t^2. \tag{27a}$$

But for large t, R approaches 0 and

$$\bar{v}'^2 \lim_{t \to \infty} \int_0^t R(t_1)\, dt_1 = K_1,$$

where K_1 is a constant, so that

$$\bar{y}^2(t) \approx 2K_1 t. \tag{28}$$

K_1 can be recognized as closely allied to the original eddy-diffusivity coefficient of Fickian theory, since $\bar{y}^2$ is identical to σ^2 in Eq. (19b).

8.1.6. Sutton Diffusion Model

O. G. Sutton proposed an extension of Taylor's model to represent average plume diffusion,[8] hypothesizing that $R(\xi)$ depended on the turbulence $\bar{v}'^2$, the viscosity ν, and on ξ, only. His formula

$$R(\xi) = \left(\frac{\nu}{\nu + \bar{v}'^2 \xi}\right)^n, \quad \text{where } 0 < n < 1, \tag{29}$$

can be combined with Eq. (26) to reach, with some approximation,

$$\bar{y}^2(t) = \frac{\nu^n}{(1-n)(2-n)\bar{v}'^2} (\bar{v}'^2 t)^{(2-n)}. \tag{30}$$

Defining the constant

$$C_y^2 = \frac{4\nu^n}{(1-n)(2-n)\bar{u}^n} \left(\frac{\bar{v}'^2}{\bar{u}^2}\right)^{(1-n)}, \tag{31}$$

he obtained for the variance of the plume concentration in the y-axis

$$\sigma_y^2 = \bar{y}^2 = \tfrac{1}{2} C_y^2 (\bar{u} t)^{(2-n)}. \tag{32}$$

Furthermore, Sutton accounted for atmospheric flow where turbulent-viscosity effects greatly exceed those of molecular viscosity by introducing the "macroviscosity,"

$$N = V_* z_0, \tag{33}$$

to replace ν.

Also, in the vertical direction, Sutton defined

$$C_z^2 = \frac{4\nu^n}{(1-n)(2-n)\bar{u}^n} \left(\frac{\bar{w}'^2}{\bar{u}^2}\right)^{(1-n)} \tag{34}$$

and
$$\sigma_z^2 = \bar{z}^2 = \tfrac{1}{2}C_z^2(\bar{u}t)^{(2-n)}. \tag{35}$$

Originally, it was thought that the parameter n, which determines the time rate of diffusion growth [in Eqs. (32) and (35)] could be defined by the wind-profile power law expression

$$\frac{\bar{u}_1}{\bar{u}_2} = \left(\frac{z_1}{z_2}\right)\frac{n}{2-n}. \tag{36}$$

But this expression with finite n is neither valid for a homogeneous turbulence field per the original Taylor assumption, nor are the n physically related.

In practice, Sutton's formulas have been widely used with good success for ranges exceeding a few kilometers but with n_y and n_z determined empirically and as separate parameters. Hence, the Sutton equations may be regarded as semiempirical.

8.1.7. Relative Diffusion

The Lagrangian statistical basis of diffusion according to Taylor refers to the random motions of a single particle. The expressions so derived describe the average spreading of a plume from a fixed axis oriented with the wind. This can be observed in a time-exposure photograph of a plume averaging the density over about 5 min. It can be shown that the plume variance $\bar{y}^2$ in Eq. (26) is proportional to

$$\frac{\sin^2(\pi n t)}{(\pi n t)^2}, \tag{37}$$

where n is the frequency of the Fourier components of an eddy-energy spectrum. Expression (37) will be recognized as a low-pass filter that causes the small values of n to be most effective in diffusion for large t. In other

Table I

Horizontal Eddy Diffusivity at Various Scales[9]

L (cm)	K_y (cm²/sec)	Source
5×10^{-2}	1.7×10^{-1}	Molecular diffusion
1.5×10^3	3.2×10^3	Low-level wind shear
5×10^4	6×10^4	Pilot balloons, 100–800 m
5×10^6	5×10^8	Volcanic ash
1×10^8	1×10^{11}	Cyclonic storms

words, the steady-state component of wind is most effective in the long term. The result with respect to plume shape is the "conical plume" described by Scorer and mentioned in the last chapter.

On the other hand, an instantaneous snapshot of a plume shows a meandering centerline from which the plume also spreads. This diffusion is described by the joint Lagrangian statistics of two particles spreading relative to each other. This relative diffusion, it was discovered by Richardson,[9] is affected by eddies of approximately the same size as the instantaneous plume width. He published the diffusivity values as a function of scale shown in Table I and suggested the empirical formula for diffusivity

$$K = 0.2(L)^{4/3}. \tag{38}$$

8.2. APPLICATION OF MODELS

In the preceding, it was shown that both the Fickian and statistical models of diffusion lead to a Gaussian (normal) probability distribution of particle density (or concentration), under the assumption of homogeneous conditions and long diffusion times. Because of the central limit theorem of statistics, which describes a very general class of random phenomena whose distribution functions may be approximated by the normal distribution function,[10] this conclusion may also hold without these restrictions.

Starting with an *instantaneous point source* or *puff* of Q g of material, the concentration in the cloud after time t, relative to a fixed point of coordinates $x = y = z + 0$ and with $r^2 = (x - \bar{u}t)^2 + y^2 + z^2$ is

$$\chi(x, y, z, t) = \frac{Q}{(2\pi\sigma_y^2)^{3/2}} \exp\left(-\frac{r^2}{\sigma_y^2}\right), \tag{39}$$

where it is assumed that $x = \bar{u}t$; $\bar{u}$ is the mean wind; and, for *isotropic* conditions, $\sigma_y = \sigma_x = \sigma_z$.

Eq. (39) can be combined with any of the diffusion coefficients discussed above:

$$\sigma_y = (2Kt)^{1/2} \qquad \text{(Fickian)} \qquad (19b)$$

$$\sigma_y = \bar{v}'^2 t \qquad \text{(Taylor)} \qquad (27a)$$

$$\sigma_y = \frac{1}{\sqrt{2}} C_y (\bar{u}t)^{\frac{2-n}{2}} \qquad \text{(Sutton)} \qquad (32)$$

and the corresponding expressions in z or x. Of these, Eq. (27a) is good for a short time and Eq. (19b) for a very long period (continential scale).

For the *nonisotropic puff* case, assuming diffusion is independent in all three axis,

$$\chi(x, y, z) = \frac{Q}{(2\pi)^{3/2}(\sigma_x\sigma_y\sigma_z)} \exp\left[-\frac{(x-\bar{u}t)^2}{2\sigma_x^2} - \frac{y^2}{2\sigma_y^2} - \frac{z^2}{2\sigma_z^2}\right]. \quad (40)$$

A *continuous point source* will give rise to a *plume*, which may be described as the time integral of the puff model or a linear superposition of an infinite number of overlapping puffs. Neglecting diffusion along the x-axis compared to the transport of the wind $\bar{u}$, this formulation yields an average concentration:

$$\bar{\chi}(x, y, z) = \frac{Q'}{2\pi\sigma_x\sigma_z\bar{u}} \exp\left[-\frac{y^2}{2\sigma_y^2} - \frac{z^2}{2\sigma_z^2}\right], \quad (41)$$

where Q' is the source strength in grams per second and σ_x, σ_z are functions of x, the downwind distance (see Figure 4).

The effect of a ground plane on a continuous source at height h if no absorption of material by the earth is postulated must be equivalent to total reflection. It can be computed by assuming an "image source" symmetrically located beneath the ground plane. Then,

$$\bar{\chi}(x, y, z) = \frac{Q'}{2\pi\sigma_y\sigma_z\bar{u}} \exp\left(-\frac{y^2}{2\sigma_y^2}\right)$$

$$\times \left\{\exp\left[-\frac{(z-h)^2}{2\sigma_z^2}\right] + \exp\left[-\frac{(z+h)^2}{2\sigma_z^2}\right]\right\}. \quad (42)$$

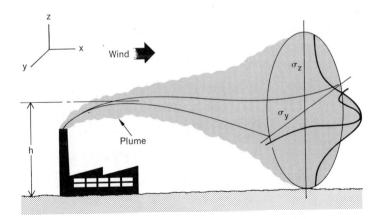

Figure 4 Gaussian diffusion from a point source (courtesy of *IEEE Spectrum*).

For the same conditions but with the receptor at the ground plane ($z = 0$),

$$\bar{\chi}(x, y, 0) = \frac{Q'}{\pi \sigma_y \sigma_z \bar{u}} \exp\left[-\left(\frac{y^2}{2\sigma_y^2} + \frac{h^2}{2\sigma_z^2}\right)\right], \quad (43)$$

which is the most commonly used form of the Gaussian equations.

A number of expressions have been developed for single and multiple reflections from inversion traps as well as the ground plane.[11]

A *fluctuating*, or *meandering*, plume may be described by a Gaussian distribution of the plume center from a fixed axis (mean center line), with variance $\bar{D}^2$ ($= \bar{D}_x^2 = \bar{D}_z^2$). The mean concentration

$$\bar{\chi} = \frac{Q'}{2\pi \bar{u}(\bar{Y}^2 + \bar{D}^2)} \exp\left[-\frac{r^2}{2(\bar{Y}^2 + \bar{D}^2)}\right], \quad (44)$$

where $r^2 = y^2 + z^2$ and $\bar{Y}^2$ is the relative diffusion parameter (mean-square relative spreading coefficient for a two-particle system).

A number of expressions have been derived for *line sources*, which may represent a heavily traveled highway (continuous source) or an aircraft on takeoff (instantaneous line source). Some are reproduced below from a recent collection.[12]

For a *finite line at height h*, length y_1 to y_2, at right angles to wind direction (x), the ground concentration is

$$\bar{\chi}(x, y, 0) = \frac{2q}{(2\pi)^{1/2} \sigma_z \bar{u}} \exp\left[-\frac{1}{2}\left(\frac{h}{\sigma_z}\right)^2\right] \int_{P_1}^{P_2} \frac{1}{(2\pi)^{1/2}} \exp\left(-\frac{P^2}{2}\right) dp, \quad (45)$$

where $P_1 = y_1/\sigma_y$, $P_2 = y_2/\sigma_y$ and q is the emission rate per unit length [g/(sec)(m)].

For a *finite line source on the ground* (road),

$$\bar{\chi}(x, y, 0) = \frac{2q}{(2\pi)^{1/2} \sigma_z \bar{u}} \int_{P_1}^{P_2} \frac{1}{(2\pi)^{1/2}} \exp\left(-\frac{P^2}{2}\right) dp. \quad (46)$$

For an *infinite line at height h*,

$$\bar{\chi}(x, y, 0) = \frac{2q}{(2\pi)^{1/2} \sigma_z u} \exp\left[-\frac{1}{2}\left(\frac{h}{\sigma_z}\right)^2\right]. \quad (47)$$

For an *infinite line at ground level*,

$$\bar{\chi}(x, y, 0) = \frac{2q}{(2\pi)^{1/2} \sigma_z \bar{u}}. \quad (48)$$

If the *wind* is blowing *at angle* ϕ ($<45°$) *to an infinite line*,

$$\bar{\chi}(x, y, 0) = \frac{\{\text{Eq. (47) or (48)}\}}{\sin \phi}.$$

Area sources can be treated in two ways. The area can be divided by a square grid and each source square considered as a point source at the grid center. This method was used by Pooler.[13] Or the grid areas may be approximated, as done by Turner,[14] by normally distributed line sources oriented crosswind and passing through each area center, with the grid distance equal to 4σ. Downwind from an area source one may consider the plume as originating from a "virtual" point source upwind of the source square,[15] which will result in the assumed line distribution.

Gaussian plume models are source-oriented; that is, to compute the concentration at a receptor point, sum the individual contributions of the point sources in the vicinity (superposition principle). It is permissible, also, to invert the problem; that is, all sources may be considered combined at the receptor location and the concentration at each actual source point computed (after reversing the wind direction). The sum of all these inverted concentrations is the receptor concentration. But the plume approach is less well adapted to the converse problem. It is not easy to determine at the receptor the effect of an area source. Furthermore, the Gaussian equations are completely undefined for a time-varying source that is the common real-world situation.

Table II

Pasquill Turbulence Types[18,19] [a]

Surface wind speed, m/sec	Daytime insolation			Nighttime conditions	
	Strong	Moderate	Slight	Thin overcast or $\geq 4/8$ cloudiness[b]	$\leq 3/8$ Cloudiness
<2	A	A–B	B		
2	A–B	B	C	E	F
4	B	B–C	C	D	E
6	C	C–D	D	D	D
>6	C	D	D	D	D

[a] A, extremely unstable; B, moderately unstable; C, slightly unstable; D, neutral (heavy overcast, day or night); E, slightly stable; F, moderately stable.
[b] Fraction of sky above local apparent horizon covered by clouds.

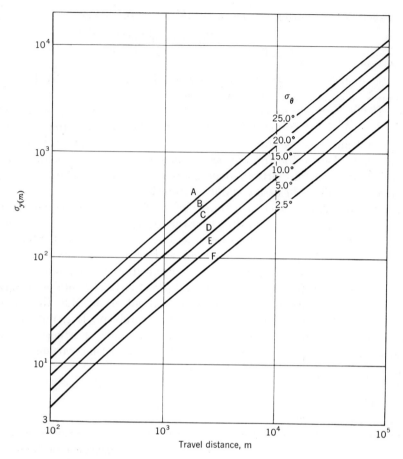

Figure 5 Horizontal diffusion coefficient as a function of travel distance for Pasquill's turbulence types.

8.2.1. Diffusion Coefficient Estimates

Since the numerical values of the diffusion (or spreading) coefficients σ_x and σ_z, are critical to the application of the Gaussian forms and since there is no universal method of achieving them, their evaluation has been discussed at length in the literature. Somers[16] undertakes a recent critique; classically, however, the method of Pasquill[17] is followed with frequent warnings to limit it to specific conditions and similar caveats. Pasquill defines six categories of atmospheric stability (others have used four to seven), defined in readily measured parameters, such as surface wind, cloudiness (at night), or solar radiation (daytime). The spreading coefficients are defined for a given sampling interval and the lower few hundred meters of the atmosphere.

Pasquill's stability conditions are defined in Table II. They can also be related to the standard deviation of horizontal wind direction, σ_θ, which can be measured by a simple recording anemometer:

Pasquill category	σ_θ
A	25.0°
B	20.0
C	15.0
D	10.0
E	5.0
F	2.5

Similar classes have also been defined for gustiness.[16]

For a given stability class, the diffusion coefficient can be stated as a function of downwind distance x in meters, using the data of Pasquill as evaluated by Gifford.[20] Figures 5 and 6 reproduce these data. It should be noted that these coefficients represent an atmosphere without an inversion lid. The conditions are 10–60 min sampling period and a short averaging time.

8.2.2. Gaussian Formulas for Special Meteorological Conditions

The plume behavior represented by different lapse conditions (Figure 7 of Chapter 7) can be quantified in terms of concentration by Eqs. (41)–(43) and their variations, together with the diffusion coefficients defined above.

Fanning represents a slow vertical diffusion during stable conditions, which is defined by Eq. (43) for a receptor on the ground with σ for stable conditions. Here, $\sigma_y \gg \sigma_z$, for example with condition F at 1000 m downwind, $\sigma_y = 39$ m, and $\sigma_z = 14$ m.

Fumigation is the rapid downward mixing of material trapped aloft during a stable inversion at h_i when the latter breaks up. Equation (43) is integrated over z and the resulting material distributed uniformly through the layer of height h_i

$$\bar{\chi} = \frac{Q'}{(2\pi)^{1/2} \bar{u} h_i \sigma_y} \exp\left(-\frac{y^2}{2\sigma_y^2}\right). \qquad (49)$$

The height h_i may represent the bottom of a persistent inversion layer, the top of the planetary boundary layer, or the tops of ridges surrounding a valley. Briggs[21] has expanded this case to include the reflecting properties of

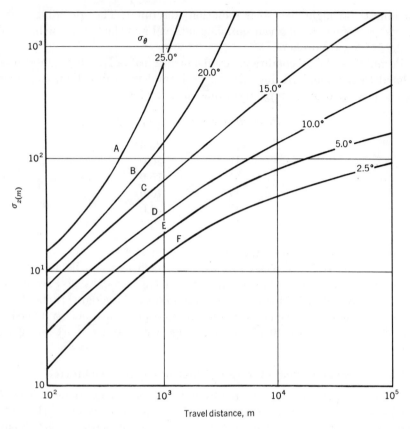

Figure 6 Vertical diffusion coefficient as a function of travel distance for Pasquill's turbulence types.

the ground and inversion base:

$$\tilde{\chi} = \frac{Q'}{\pi \sigma_y \sigma_z \bar{u}} \left\{ \exp\left(\frac{-h^2}{2\sigma_z^2}\right) + \exp\left[\frac{-(2h_i - h)^2}{2\sigma_z^2}\right] + \exp\left[-\frac{(-2h_i - h)^2}{2\sigma_z^2}\right] \right\}.$$
(49a)

Looping corresponds to a Pasquill type A condition and can be estimated from Eq. (43). The maximum concentration $\tilde{\chi}_{max}$ corresponds to a ground-based source (as the loop touches the ground), $y = 0$, $h = 0$, downwind distance $x' = (x^2 + h^2)^{1/2}$,

$$\tilde{\chi}_{max} = \frac{Q'}{\pi \sigma_y(x') \sigma_z(x') \bar{u}}.$$
(49b)

Coning occurs in a neutral or slightly stable atmosphere; hence, Eq. (43) is used with a Pasquill D or E condition.

Lofting describes the condition of stable below and neutral aloft; hence, it is a favorable state for pollution dispersion (see the referenced figure) but is also a possible precursor of fumigation. Briggs suggests treating the inversion base as $z = 0$ to obtain concentration along the plume centerline.

Long-period average: The "wind rose" describes the percentage of time the mean wind comes from any direction of the compass and its strength. To obtain an average concentration over a long period compared with the mean-wind averaging interval, Eq. (47) is employed, which not only describes a continuous infinite line source, as a freeway, but also the cross-wind integrated concentration from a continuous source. Multiply this integral by the percentage of time the wind flows toward the given section (f) and divide by the width of the sector at the distance of interest. If there are n sectors, each of $2\pi/n$ radians, the sector width at x is $2\pi x/n$. Then,

$$\bar{\chi}_{\text{long-term av.}} = \left(\frac{2}{\pi}\right)^{1/2} \cdot \frac{0.01 f Q'}{\sigma_z \bar{u}(2\pi x/n)} \exp\left(\frac{-h^2}{2\sigma_z^2}\right) \quad (50)$$

Maximum concentration from an *elevated source* is a very difficult situation to compute, and no simple formulation exists. If we assume a vertical or slightly unstable condition, $\sigma_x \approx \sigma_z$. Differentiating Eq. (43) and setting equal to 0 yields the maximum:

$$\bar{\chi}_{\max} = \frac{2Q'}{\pi h^2 e \bar{u}}, \quad (51)$$

which occurs where $h^2 = 2\sigma_z^2$ and can be solved for x, since $\sigma_z = f(x)$. For conditions where horizontal and vertical cloud growths are proportional,

$$\bar{\chi}_{\max} = \frac{2Q'\sigma_z}{\pi h^2 e \bar{u} \sigma_y} \quad (52)$$

(also at $h^2 = 2\sigma_z^2$) or

$$\bar{\chi}_{\max} = \frac{2^{1/2} Q'}{\pi h e \bar{u} (\sigma_y)_{\max}} \quad (53)$$

where $(\sigma_y)_{\max}$ is that applying to the maximum concentration distance.

Isopleths, or the loci of points where the concentration from a source drops to $p\%$ of its value on a plume axis, are commonly plotted. These points are

$$y_p = \left(2\sigma_y^2 \ln \frac{100}{p}\right)^{1/2}, \tag{54}$$

$$z_p = \left(2\sigma_z^2 \ln \frac{100}{p}\right)^{1/2}. \tag{55}$$

8.2.3. Height-of-Rise Formulas

The Gaussian formulas given to this point do not take into account the buoyancy and momentum of the plume rising from a stack and assume that the gas emitted is neutral with respect to the ambient air.

In actuality, hot stack gases rise buoyantly so that the plume appears to be emitted from an "effective height" some distance above the actual stack height. Some results of this theory were cited in the previous chapter.

Over 20 formulas for height of rise published between 1950 and 1968 have been reviewed by Briggs,[21] and research has continued since that time.[22] Many of these formulas have been empirical, and none are completely accepted, because of the experimental difficulties involved. Two of the more successful expressions are the Holland (or Oak Ridge) formula as modified by Stümke and the Bosanquet expression. The former gives good results for moderate-size plants but fails for large ones (by a factor of 3). The Bosanquet formula fails at large distance. The Holland-Stümke expression is

$$\Delta h = 1.5 \frac{w}{\bar{u}} d + 65.0 \frac{d^{1/2}}{\bar{u}} \left(\frac{\Delta T}{T_s}\right)^{1/4} \tag{56}$$

(in MKS units), where w is the stack gas efflux velocity; $\bar{u}$, the mean horizontal wind component; d, the inside stack diameter; ΔT, the excess of stack gas temperature, T_s, over the ambient, T, or $T_s - T$. The $\frac{1}{4}$-power relationship is also cited by Somers.[16] Bosanquet's formula, revised in 1957, gives a rise trajectory, not only the final rise. For the formulation, which is complex, the original reference is recommended.[23]

By application of dimensional analysis to the 12 possible combinations of windy and calm, stable, neutral, and transitional rise; momentum- and buoyancy-dominated plumes, Briggs arrives at the following formulas (it is recommended that the one be used that gives the minimum rise):

Transitional rise:

$$\Delta h = 2.0 F^{1/3} \bar{u}^{-1} x^{2/3} \tag{57}$$

Final, neutral:

$$\Delta h = 400 \frac{F}{\bar{u}^3} + 3r \frac{w}{\bar{u}} \tag{58}$$

Final, stable with wind:

$$\Delta h = 2.6 \left(\frac{F}{\bar{u}S}\right)^{1/3} \tag{59}$$

Final, stable, calm:

$$\Delta h = 5.1 F^{1/4} S^{-3/8} \tag{60}$$

where

$$S = \frac{g}{T} \frac{\partial \Theta}{\partial z},$$

$\frac{\partial \Theta}{\partial z} = \frac{\partial T}{\partial z} + 9.8°C/km =$ "potential temperature gradient", g is the gravitational constant, T and T_s are the ambient and stack temperatures (absolute), r is the inside stack radius, M and M_s are average molecular weights of air and stack gas, Cp, Cps are the specific heats at constant pressure of air and stack gas, and the other symbols are defined as before. For $M_s \approx M$ and $Cp \approx Cps$,

$$F \approx \frac{\Delta T}{T_s} gwr^2 = 3.8 \times 10^{-5} Q_h \text{ m}^4/\text{sec}^3,$$

where Q_h is the rate of heat emission of the stack in calories per second.

The objective of increasing plume rise is to reduce ground concentration and penetrate inversions, if possible, as noted in the last chapter. Plume rise can effectively increase stack height and can be considered in an economic trade-off with the high cost of tall stacks. The following considerations have been noted:

1. Increasing stack gas efflux velocity has a rise effect approximating $1.5 \, d(w/\bar{u})$ [see Eq. (56)] and is generally not economical. A TVA stack equipped with a nozzle doubling w was noted by Briggs to have no significant effect on rise during high-pollution conditions. However, downwash is eliminated if $w > \bar{u}$.
2. Raising the heat efflux $(T_s - T)$ produces a $\frac{1}{4}$ power effect on rise [Eqs. (56) and (60)] and is generally uneconomical because of the added heat cost.

3. Combining pollutants in one stack is economical because the buoyancies are additive and a single, tall stack may be less costly than several smaller ones.

The plume rise models may be combined with the Gaussian diffusion formulas by judicious choice to calculate the effect on concentration. Only a few of the many cases are presented.

As an average condition, consider the windy neutral case. The maximum Gaussian ground concentration is [Eq. (52)]

$$\bar{\chi} = \frac{2Q'\sigma_z}{\pi h_s^2 e \bar{u} \sigma_y}$$

where h_s is stack height. For neutral conditions, assume

$$\frac{\sigma_z}{\sigma_y} \approx 0.5.$$

A conservative neutral buoyant plume rise, from Eq. (58), is (neglecting momentum effects of efflux velocity)

$$\Delta h = 400 \frac{F}{\bar{u}^3}.$$

Combining,

$$\bar{\chi} = \frac{Q'}{\pi (h_s + 400F/\bar{u}^3)^2 e \bar{u}}. \tag{61}$$

Differentiating with respect to u and setting equal to 0 yields

$$\bar{u}_c^3 = 2000 \frac{F}{h_s} \quad \text{or} \quad \bar{u}_c = 12.6 \left(\frac{F}{h_s}\right)^{1/3} \tag{62}$$

for the critical wind speed that gives the highest maximum ground concentration.

Substituting in Eq. (61), this is

$$\bar{\chi}_{\max} = \frac{0.0065 Q'}{F^{1/3} h_s^{1/3}}. \tag{63}$$

For a nonbuoyant plume, using the Holland–Stümke expression, Eq. (56), with $T_s = T_A$, $\Delta h = 1.5\, d(w/\bar{u})$, Briggs obtains

$$\bar{u}_c = 3 \frac{wr}{h_s},$$

$$\bar{\chi}_{\max} = 0.01 \frac{Q'}{wrh_s}. \tag{64}$$

For a large plant, $\bar{u}_c$ may be as high as 40 mph.

For the worst condition, that of no wind followed by fumigation, Briggs suggests on dimensional grounds:

$$\bar{\chi}_{max} = b \frac{Q'}{F^{1/2}S^{1/4}(h_s + 5.1F^{1/4}S^{-3/8})}, \quad (65)$$

where he has estimated b to be 0.05–0.06 from TVA observations.

For a valley of width W and a night drainage wind $\bar{u}$, "box model" considerations lead to

$$\bar{\chi}_{max} = \frac{Q'}{\bar{u}W[h_s + 2.6(F/\bar{u}S)^{1/3}]}. \quad (66)$$

8.2.4. Particle Deposition

Where buoyancy and momentum effects elevate the plumes from stacks and reduce ground concentration, the settling of particles has the opposite effect. The Gaussian equations must be modified to reflect a downward tilt.

Stokes's equation gives the fall velocity of particles under gravitational force:

$$v_g = \frac{2r^2 g \rho}{9\mu}, \quad (67)$$

where r is the particle radius; ρ, their density; μ, the atmospheric viscosity; and g the gravitational acceleration. Shape, altitude, and particle size modify this result. For a fall velocity of 1–100 cm/sec, assuming no wind shear, the Sutton form of Gaussian equation for ground concentration may be modified as follows:[24]

$$\bar{\chi}_{(x,y,0)} = \frac{2Q'}{\pi C_y C_z \bar{u} x^{2-n}} \exp\left\{-x^{n-2}\left[\frac{y^2}{C_y^2} + \frac{(h - xv_g/\bar{u})^2}{C_z}\right]\right\}, \quad (68)$$

where n is the Sutton stability parameter. If the particles are deposited as they touch the ground, the deposition pattern is

$$W = v_g \bar{\chi}_{(x,y,0)}, \quad (69)$$

where W is the amount removed per unit time and area.

Materials, including gases or vapors, may be deposited from plumes by a variety of mechanisms other than Stokes's, such as chemical, electrostatic, and velocity impaction. In this case, where the mechanism may be unknown, a "dry deposition velocity" ($= v_d$) is defined experimentally, where

$$W = v_d \bar{\chi}_{(x,y,0)}. \quad (70)$$

Considering the source to be depleted by dry deposition, the Sutton equation (for a ground-level source) is

$$W = v_d \bar{\chi} = \frac{2Q'v_d}{\pi \bar{u} C_y C_z x^{2-n}} \exp\left(\frac{-4v_d x^{n/2}}{n\bar{u}\pi^{1/2}C_z}\right) \exp\left(\frac{-y^2}{C_y^2 x^{2-n}}\right). \quad (71)$$

More complex situations are worked out graphically in Ref. 24.

8.2.5. Flow Around Structures

The concept of wake separation during wind flow over topographical discontinuities such as hills, cliffs, and buildings was introduced in the Chapter 7. Topographical and man-made urban features cause a high degree of aerodynamic distortion, which seriously modifies the ideal predictions of the models discussed earlier. At the same time, there is little or no analytical theory to account for building effects. Current technique is to make measurements on models, using wind tunnels,[25] or other fluids, such as water.[26] Providing certain criteria for similarity are met, flow fields measured in this way may be extrapolated to other, full-scale structures. It has been shown in the references that the Reynolds number is not a consideration for similarity of stack plume use, providing flow is turbulent. However, at least four other similarity criteria must be obeyed, including the Froude number (ratio of inertia to buoyant forces) and, for large distances, the Rossby number (ratio of advective to Coriolis accelerations).

Flow over a flat plate of height L causes separation of the wake streamline for about $17\,L$ downwind. The wake boundary is a paraboloid, beneath which is a turbulent cavity. Measurements made in wind tunnels lead to an expression for longitudinal turbulence:

$$\frac{\sigma_\mu}{V} = A\left(\frac{X}{L}\right)^{-2/3}, \quad (72)$$

where V is the background flow in the tunnel, X is the downwind distance, and A is a constant equal to 0.25 on the X-axis passing through the center of the plate and varies for other plate-center–wake-boundary radii. The plane through the plate center may be considered a ground plane, and because of ground image effects, the plate simulates a building $L/2$ in height. In actual buildings, turbulence will show large deviations from this number when the wind does not blows perpendicular to the plate.

To apply similarity theory to the pollutant concentration field around structures, a concentration coefficient K_c can be defined as the ratio of actual to reference concentrations for any point in the field:

$$K_c = \frac{\bar{\chi}L^2V}{Q'}. \quad (73)$$

For a uniform and homogeneous turbulent flow field such as a tall stack (far from the ground), Eq. (73) can be applied to the Sutton expression for $\bar{\chi}/Q'$, with L equal to the source height and V equal to the uniform wind velocity.

For fields close to buildings no analytic expression is available, so that K_c must be determined by experiment:[27]

1. $\bar{\chi}$ is measured while a tracer contaminant at is released rate Q'.
2. A reference area $A = L^2$ is selected relevant to the building size. This may be the side area projected on a plane perpendicular to the wind. A reference wind velocity V is chosen at an appropriate height.
3. $\bar{\chi}$ is divided by Q'/AV and isopleths of constant K_c drawn around the structure.
4. These isopleths can be applied to another, similar structure, providing appropriate source location, orientation to wind, L (or A) and V factors are chosen. The field differs only in scale for any similar configuration (having the same hydrodynamic parameters), so that the same K_c plots can be used.

The most recent studies of flow around structures have been conducted within the environment of city streets. For example, it has been reported that the concentration of carbon monoxide can vary by a factor of two from one side of the street to the other. At high levels of city traffic this may be an important health consideration.

Chang, et al.[28] considered diffusion in a simplified city street model, a rectangular trough excited by a transverse wind. In both laboratory and computer simulations, the wind was found to separate from the edge of the upstream roof and form a recirculation cavity flow in the depression at street level. Djuric and Thomas[29] were able to simulate the three-dimensional flow around buildings of a downtown Houston area for up to 36 min of real time, using a computer-implemented numerical model.

Johnson et al.[30] conducted field tests in San Jose, California, using a street model they developed for carbon monoxide, and obtained good results; correlation coefficients of measured and model-predicted values were between 0.6 and 0.7, and 80% of the observed values (ranging up to 16 ppm) were within 3 ppm of the model predictions. Ludwig and Dabberdt[31] tested Johnson's model in the environment of St. Louis, as a subsystem of a larger urban diffusion model, precursor of that proposed for RAPS (see Section 6.2.1). Two adjacent downtown street canyons were instrumented to obtain CO measurements at 30 points and wind measurements up to 130 meters at 8 points. The canyon height-width ratio parameter was 1.5, compared to 0.7 for the San Jose model. This resulted in the formation of a single-cell helical distribution of CO within the street canyon. Between the simulation and actual measurement, errors of 3–4 ppm were found.

However, these could be reduced by 1 ppm using linear regression "calibration." The model assumptions were as follows:

1. CO is inert (conserved during residence time).
2. Mean wind varies only in the vertical.
3. Net horizontal flux is negligible compared to mean wind transport.
4. No vertical transport through topographical volume is considered.

The basic mesoscale model gives the concentration of CO at roof level. It is a combination Gaussian plume and box model. The sources are traffic arteries, stored as line segments in a computer memory, represented by the geographic location of their endpoints. The emissions, Q, from each line segment can be represented by the hourly traffic data times a factor

$$E = \alpha S^{-\beta}, \tag{74}$$

where E is the grams CO emitted per vehicle mile and α and β are functions of car emission controls and age mix (for California, $\alpha = 700$ in 1971 and will decrease to 160 in 1972–1974, 16 in 1975–1979, and 8 after 1980; $\beta = 0.75$). Here S is the average traffic speed in mph.

Within the first km of the source line the Gaussian model is 45° wide and divided into five radial segments. Five additional segments $22\frac{1}{2}°$ wide extend to greater distance and are represented by a box model for each segment (i):

$$C_i = \frac{Q_{Ai}(\Delta x)}{uh}, \tag{75}$$

where Q_{Ai} is the emission from the ith line segment, Δx is its length, and u and h are the wind speed and mixing height. These parameters are computed for each segment, using temperature profiles and local meteorological data plus lidar. Turbulence for diffusion coefficients is determined from the Pasquill σ_θ criteria.

The CO concentration level from the mesoscale model is augmented below roof level by a modified box model of the street canyon. For a cross-street wind, the CO added to roof-level value to the building fronts facing the roof-level wind is

$$\Delta C_W = \frac{0.1 KNS^{-0.75}}{W(u + 0.5)} \text{ ppm}, \tag{76}$$

where N is the average volume (cars per hour), W is the street width in meters, u is the wind speed in meters per second above roof level, K is a constant, and 0.1 and 0.75 are factors depending on the auto-emission control mix. There is a linear variation with height:

$$\Delta C_W = \{\text{Eq. (76)}\} \cdot \frac{(H - z)}{H}, \tag{77}$$

where z is the height and H is canyon height.

On the opposite side of the street (downwind-facing buildings),

$$\Delta C_L = \frac{0.1KNS^{-0.75}}{u + 0.5} \cdot \frac{1}{[(x^2 + z^2)^{1/2} + z]}, \tag{78}$$

where X is the horizontal distance to the nearest traffic lane.

When the wind is parallel to the street, the average value for both sides of the street is used

$$\Delta C_I = 0.5\{\text{Eq. (77)} + \text{Eq. (78)}\}. \tag{79}$$

8.3. EXPERIMENTAL VERIFICATION OF PLUME MODELS

Two of the more extensive series of tests held to verify the Gaussian models were those at Hanford and at Cape Kennedy and Vandenberg Air Force Base, both reported by Slade.[4] The Hanford 1959–1962 tests (project Green Glow 30) included 66 diffusion experiments using ZnS fluorescent tracer, with 833 sample locations 200–25,600 m distant from a 1.5-m source height. The generalized Gaussian equation, Eq. (32), was used for stable conditions, in terms of the concentration time integral or exposure ($\psi = $ mass $\times$ time per unit volume).

The ground level exposure (stable) is

$$\frac{\psi}{Q_0} = \frac{Q_x/Q_0}{\pi \sigma_y \sigma_z \bar{u}} \exp\left[-\left(\frac{y^2}{2\sigma_y^2} + \frac{h^2}{2\sigma_z^2}\right)\right], \tag{80}$$

where $\bar{u}$ is the average wind speed at emission height h, Q_0 is the amount released, and Q_x/Q_0 is the depletion factor due to deposition velocity.

In Eq. 80

$$\sigma_y = At - \frac{A^2}{2(\sigma_\theta \bar{u}^2)}\left\{1 - \exp\left[-2\frac{(\sigma_\theta \bar{u})^2 t}{A}\right]\right\}, \tag{81}$$

where t is the travel time and $A = 13 + 232\sigma_\theta \bar{u}$, the experimental constant for the turbulence scale.

Also,

$$\sigma_z = a[1 - \exp(-k^2 t^2)] + bt; \tag{82}$$

Table III

Sutton Parameters Used at Hanford[32]

	Release level	Wind, m/sec	Unstable	Neutral
C_y	Ground	1.0	0.35	0.21
		5.0	0.30	0.15
		10.0	0.28	0.14
	Elevated	1.0	0.30	0.15
		5.0	0.26	0.12
		10.0	0.24	0.11
C_z	Ground	1.0	0.35	0.17
		5.0	0.30	0.14
		10.0	0.28	0.13
	Elevated	1.0	0.30	0.15
		5.0	0.26	0.12
		10.0	0.24	0.11
n			0.20	0.25

where a, b, k^2 are stability functions; namely:

		Moderately stable	Strongly stable
a	(m²)	97	34
b	(m²/sec)	0.33	0.025
k^2	sec⁻²	2.5×10^{-4}	8.8×10^{-4}

For neutral and unstable conditions, Hanford used the Sutton form for the exposure:

$$\frac{\psi}{Q_0} = \frac{2 Q_x/Q_0}{\pi C_y C_z \bar{u}^{2-n}} \exp\left[-\frac{1}{x^{2-n}}\left(\frac{y^2}{C_y} + \frac{h^2}{C_z^2}\right)\right]. \quad (83)$$

The Sutton coefficients C_y, C_z, and n are listed in Table III for the conditions at Hanford (use for other terrain and meteorological conditions is subject to verification at the local site). Depletion factors Q_x/Q_0 are the ratio of apparent released material (Q_x) to that actually released (Q_0), as observed at distance x from the source. The formula differs with stability conditions:

Stable:

$$\frac{Q_x}{Q_0} = \exp\left[-\left(\frac{2}{\pi}\right)^{1/2}\left(\frac{v_d}{\bar{u}_0}\right)\left(\frac{\bar{u}_0}{\bar{u}}\right)\bar{u} \cdot \int_0^t \frac{\exp(-h^2/2\sigma_z^2)}{\sigma_z} dt\right] \quad (84)$$

Neutral and unstable:

$$\frac{Q_x}{Q_0} = \exp\left[-\frac{2}{C_z\pi^{1/2}}\left(\frac{v_d}{\bar{u}_0}\right)\left(\frac{\bar{u}_0}{\bar{u}}\right) \cdot \int_0^x x^{(n-2)/2} \exp\left(\frac{h^2}{C_z^2 x^{2-n}}\right) dx, \quad (85)$$

where v_d is the deposition velocity, $\bar{u}_0$ is the wind speed at the surface, and $\bar{u}$ is the wind speed at emission height. Table IV lists the deposition factors used with Eqs. (84) and (85).

It is clear that the complexity of these relationships demands the use of a computer for numerical solutions on a routine basis.

The 1961–1962 Cape Kennedy experiments (Project Ocean Breeze) and those at Vandenberg Air Force Base (Project Dry Gulch), a total of 185 experiments, offered a test of varied terrain. That at Kennedy is flat with 5-m-high vegetation, while the Vandenberg terrain is more complex, including a valley and elevation variations of 60 m. The Kennedy tests covered a distance of 1.2–4.8 km from the source. The purpose of these tests was to provide input to an automatic computer-controlled meteorological data-acquisition and data-processing system. A statistical machine analysis was used to process half the data, and the other half used for verification. Hence, the expression derived empirically was not a true Gaussian expression, although the exponential form was retained.

The normalized peak concentration in seconds per cubic meter is

$$\frac{\tilde{\chi}_p}{Q'} = 0.00211 x^{-1.96} \sigma_\Theta^{-0.506} (\Delta t + 10)^{4.33}, \quad (86)$$

where σ_Θ is the standard deviation of the azimuthal wind direction in

Table IV
Deposition Factors Used at Hanford[32]

Atmospheric condition	Deposition coefficient ($vd/\bar{u}_0$)		Wind speed ratio, u_0/u 60 m[a]
	Fine particles ($\times 10^{-4}$)	Halogens ($\times 10^{-3}$)	
Very stable	1.5	2.4	0.17
Moderately stable	2.2	3.4	0.35
Neutral	3.0	4.6	0.50
Unstable	6.0	8.0	0.70

[a] For other levels, interpolate shear factor by assuming log or power function for wind profile.

degrees (15 sec average running over 30-min observation period), x is the downwind travel distance in meters, and $\Delta t = T_{54 \text{ ft}} - T_{6 \text{ ft}}$.

With independent data on these sites, the prediction was equal or better than a factor of 2 in 72% of the cases and by a factor of 4 for 97%.

8.4. PHOTOCHEMICAL MODELS

The models discussed to this point include the box type, in which the concentration of pollutants is homogeneously distributed throughout the volume, and integrated puff or plume Gaussian distributions. Although these latter models can account for a wide range of source geometries and boundary conditions, it is only with the greatest difficulty that noninert characteristics, such as chemical reaction, depletion, and deposition, can be accommodated. With the increased order of complexity demanded by consideration of the important photochemical and other atmospheric reactions, the plume models become unable to describe many urban situations realistically.

Current efforts toward solution of this problem require a return to the fundamental approaches of continuity and mass conservation, of which the Fickian diffusion equation was a special case. The mass-conservation equations have been stated by Roth in a recent review[33] to require the following:

1. Terms describing pollutant transport by wind and dispersion by turbulent air motions
2. Source terms describing influx of new pollutants
3. Sink terms to account for removal of materials
4. Chemical reaction terms.

The two types of models based on the continuity-equation solutions are the trajectory approach, where a (hypothetical) moving column of air is advected by the wind and is followed by a moving coordinate system (Lagrangian), and the grid approach, having fixed (Eulerian) coordinates. In the latter, the airshed is divided into three-dimensional cells, each of which may be on the order of a mile or two per side and 100 ft high. The trajectory coordinates offer a natural means of following the chemical reactions, since they can be envisioned as defining a relatively static "batch" reaction chamber into which pollutants are injected, and in which the chemical reactions take place.[34] On the other hand, the physical existence of distinct air parcels is a dubious concept. Experiments have been described to prove the validity of this approach by following parcels with balloons, tracers, and instrumented helicopters and ground units to measure chemical parameters.[35]

The fixed coordinate grid, on the other hand, leads to a better definition of the diffusion parameters (especially as they are influenced by local conditions) and of point sources. But errors may arise in following the chemistry while moving the reaction calculations discontinuously from cell to cell. Both models suffer from poor resolution caused by "smearing" point emissions over a relatively large grid area.[34]

A complete mathematical description of the equations of continuity of a chemically reacting system in vector form is[36]

$$\frac{\partial c_i}{\partial t} + \nabla \cdot (\vec{u} c_i) = -\nabla \cdot \vec{q}_i + R_i + S_i, \tag{87}$$

where c_i is the time-averaged concentration of ith chemical species, $\vec{u}$ is the vector wind velocity, R_i is the rate of chemical production or loss of i, S_i is the emission rate of i from sources, and $\vec{q}_i$ is the mass flux of i due to turbulent diffusion and is approximately equal to $-K \nabla c_i$ (K being the turbulent diffusion coefficient and molecular diffusion being neglected).

This equation can be expanded into the normal three-axis coordinate form, when it becomes[37]

$$\frac{\partial c_i}{\partial t} + u \frac{\partial c_i}{\partial x} + v \frac{\partial c_i}{\partial y} + w \frac{\partial c_i}{\partial z}$$

$$= \frac{\partial}{\partial x}\left(K_x \frac{\partial c_i}{\partial x}\right) + \frac{\partial}{\partial y}\left(K_y \frac{\partial c_i}{\partial y}\right) + \frac{\partial}{\partial z}\left(K_z \frac{\partial c_i}{\partial z}\right) + R_i(c_1, c_2, c_3, \ldots, c_p) + S_i \tag{88}$$

where $i = 1, 2, \ldots, p$ and the limits on z are the terrain elevation $h(x, y)$ and the inversion base height $H(x, y, t)$. (In practice, diffusion terms in the x- and y-directions are neglected).

We now have p nonlinear partial differential equations in x, y, z, t space. The equations are nonlinear because the R_i terms, representing the reactions between all reacting species present, are in general nonlinear. Furthermore, all of the p equations are coupled (through the R_i terms), so that they must be solved simultaneously. In order to keep the time for machine solution of these equations to an acceptable value, as few chemical species as possible should be chosen. Yet, as we have seen in the first chapter, an enormous number of reactions actually occur in the atmosphere. The problem is one of judicious choice of a kinetic submodel.

8.4.1. The Chemical-Reaction Submodel

A large number of kinetic mechanisms have been proposed to describe the rates of chemical reactions in the atmosphere. These have varied from

Table V

Atmospheric Photochemical Reactions for Simulation Model

Step[a]	
1	$NO_2 + h\nu \rightarrow NO + O$
2	$O + O_2 + M \rightarrow O_3 + M$
3	$O_3 + NO \rightarrow NO_2 + O_2$
4	$O_3 + 2\,NO_2 \xrightarrow{H_2O} 2\,HNO_3 + O_2$ (composite)
5	$NO + NO_2 \xrightarrow{H_2O} 2\,HNO_2$
6	$HNO_2 + h\nu \rightarrow OH\cdot + NO$
7	$CO + OH\cdot \xrightarrow{O_2} CO_2 + HO_2\cdot$
8	$HO_2\cdot + NO_2 \rightarrow HNO_2 + O_2$
9	$HC + O \rightarrow \alpha\,RO_2\cdot$
10	$HC + O_3 \rightarrow \beta\,RO_2\cdot + \gamma\,RCHO$
11	$HC + OH\cdot \rightarrow \delta\,RO_2\cdot + \epsilon\,RCHO$
12	$RO_2\cdot + NO \rightarrow NO_2 + \theta\,OH\cdot$
13	$RO_2\cdot + NO_2 \rightarrow PAN$
14	$HO_2\cdot + NO \rightarrow NO_2 + OH\cdot$

Source. Ref. 37, Appendix B.

[a] $OH\cdot$, $RO_2\cdot$, $HO_2\cdot$ are radicals; $\gamma, \beta, \ldots, \delta$ represent stoichiometric proportions which are functions of R, the particular organic species considered; R is an organic species (e.g., an olefin) that must be specified; M is any molecule; and $h\nu$ is the solar-derived photon energy.

highly simplified (having fewer than 10 reactions) to complex models (60 or more). Hecht[38] has recently described 42 reactions known or suspected to take place in polluted atmospheres. For the purposes of implementing Eq. (88), in a model design to predict CO, NO_x, O_3, and HC, Hecht and Seinfeld determined that mechanisms consisting of less than 10 steps that had previously been incorporated with urban models (such as Eschenroeder's in 1964) were insufficiently realistic. Further, those having over 25 reactions are impractical for numerical solution, owing to the cumulative error in reaction rates that can accrue. This is seen by considering the expression for R_i, the algebraic sum of rates of production of i over all reactions in which i participates:

$$R_i = \sum_{j=1}^{n} a_{ij} k_j c_{j1} c_{j2}, \qquad (89)$$

where a_{ij} is -1 if species i is a reactant, $+1$ if it is a product, and 0 if not present; c_{j1} and c_{j2} are the reactant concentrations in reaction j; and kj is the specific reaction rate constant (or "rate constant") for j.

A total of 14 steps is selected for Hecht's model (Table V).

For each reaction, the reaction rate can be expressed as a function of the rate constant and concentrations.[39] For example, in the ninth step of Table V, the rate of HC formation can be written

$$r_9 = \frac{d(\text{HC})}{dt} (9) = k_9 (\text{HC})(0), \qquad (90)$$

where the parentheses indicate concentration (molecules per unit volume).

Combining the individual rate expressions, such as Eq. (90), we can obtain a set of four differential equations that are functions of the four species:

HC: $\quad \dfrac{d(\text{HC})}{dt} = -r_9 - r_{10} - r_{11} \qquad (91)$

NO: $\quad \dfrac{d(\text{NO})}{dt} = r_1 - r_3 - r_5 + r_6 - r_{12} - r_{14} \qquad (92)$

NO$_2$: $\quad \dfrac{d(\text{NO}_2)}{dt} = -r_1 + r_3 - 2r_4 - r_5 - r_8 + r_{12} - r_{13} + r_{14} \qquad (93)$

O$_3$: $\quad \dfrac{d(\text{O}_3)}{dt} = r_2 - r_3 - r_4 - r_{10}. \qquad (94)$

We could also write, if desired, for aldehydes and PA,

$$\frac{d(\text{RCHO})}{dt} = r_{10} + \epsilon r_{11},$$

$$\frac{d(\text{PAN})}{dt} = r_{13}$$

and similiar expressions.

We may also write steady-state equations as a function of the concentration of the following:

O: $\quad r_1 - r_2 - r_9 = 0 \qquad (95)$

HNO$_2$: $\quad 2r_5 - r_6 + r_8 = 0 \qquad (96)$

OH·: $\quad r_6 - r_7 - r_{11} + \theta r_{12} + r_{14} = 0 \qquad (97)$

RO$_2$·: $\quad \alpha r_9 + \beta r_{10} + \delta r_{11} - r_{12} - r_{13} = 0 \qquad (98)$

HO$_2$·: $\quad r_7 - r_8 - r_{14} = 0. \qquad (99)$

Given the initial concentrations of NO, NO_2, CO, O_3, and HC the four differential equations (91)–(94) may be numerically integrated to predict the concentration versus time behavior of the four species. These equations are nonlinear and their integration involves a number of technical difficulties described in the reference. Furthermore, the rate constants for each step in Table V must be determined from the literature or estimated (all were not known at the time of the study), and the solar input, $h\nu$ (steps 1 and 6), must be determined. Nevertheless, the kinetic mechanism has been successfully validated by predicting the results of smog chamber experiments using propylene and a number of other hydrocarbons in varying ratio with NO_x.

8.5. IMPLEMENTATION OF URBAN MODELS

The practical implementation of any of the models discussed in an urban situation is, of course, the critical test. But before it can be applied, one must determine just what is expected of the model. To some extent this is a function of what the model can do, so that an iterative growth pattern may be anticipated in the present course of development.

Model tasks may be divided into those of long and short time scales. Among the long-term model objectives may be included:[40]

1. Assessment of the effectiveness of alternate control strategies
2. Cost-effective studies, in terms of overall economic impact
3. Land planning, including location of projected power plants and freeways
4. Impact of major changes of energy use and life styles on air quality.

On a short time scale ("real time") the uses of models may include:

5. Prediction of air-pollution episodes and selection of short-term strategies to prevent them
6. Identification of individual sources by predicting their emission pattern, or "signature"
7. Spatial and temporal interpolation of air quality between monitoring stations (e.g., small-scale prediction).

In both long- and short-time-scale uses, the question of model computation time becomes important. In the real-time control tasks, such as 5 and 6, the requirement is obvious. In the longer-time problems it is still an economic requirement to minimize machine time.

8.5.1. Urban Box Models

The scheme that represents an entire city area as a single "box" capped by an inversion lid has much to recommend it from the standpoint of computational simplicity, but having no fine scale resolution, it is extremely limited in usefulness. Stern[41] describes two versions: one in which there is no wind and in which the concentration builds up continuously ($=Qt/Z$; see Figure 3, where Q is the emission rate per area, t is elapsed time, and Z is the inversion height) and the slightly more complex ventilated model, in which concentration equals QS/uZ. Lettau[42] has treated the box model much more completely and has shown its utility for computing the response junction of cities to a direction and pollution release variations. The adverse effects of "urban sprawl," for example, can be demonstrated by showing the relationship of city size to pollution buildup.

One recent nondiffusion model experiment was that of Leahey,[43] who developed an advective box model to predict SO_2 concentration within the New York City "heat island." A rather complete inventory of sources in the 30 × 40-mile area about Manhattan was available and assumed to be released according to the temperature (degree-day). Allowance was made for stack height of large sources, including inversion penetration. The mixing-depth data was computed from a thermodynamic submodel based on artificial heat release. Mean-wind data was taken from three balloon stations. Concentration was then calculated for each square mile of a grid oriented with respect to the wind and compared with 400 observations taken on five separate mornings. The results yielded a correlation coefficient of 0.83 for the computed and observed points and a standard error of 0.13 ppm in an observed range of 0.19–0.52 ppm.

8.5.2. Gaussian Urban Models

At this point, we should examine the results of some of the applications of the Gaussian diffusion equations. Much success has been achieved using the Sutton equation and its variants to locate and design stacks for industrial power plants. Diffusion models have also been used routinely by NAPCA (now APCO) to determine the average distribution of pollutants in urban areas and to establish AQCR boundaries. Nine separate tests of models conducted on a more rigorous scale in various cities have been reported in detail in papers by Wanta[42] and Seinfeld.[44]

An even more detailed survey of urban models was published by H. Moses of Argonne Laboratories,[45] covering the entire history from Frenkiel's pioneer work on Los Angeles in 1956 to that of Hilst, Badgley, et al. in Connecticut, published in 1967.

In most cases so reported, the Gaussian equation or some simplified version of it comprised the kernel of the basic model. On the basis of source and meteorological data, future concentrations were predicted for the pollutants SO_2, NO_x, and CO (or CO_2 in some cases). The only chemical reaction considered was the decay of SO_2, usually as an exponential factor.[46]

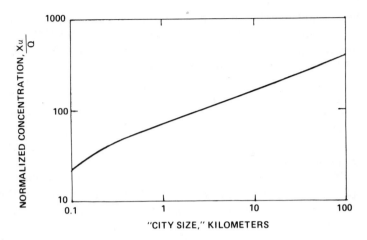

Figure 7 Relationship of normalized pollutant concentration and "city size" [after Miller and Holzworth; *Federal Register*, 36, 22406 (1971)].

Generally, the resolution of the model in space measured a kilometer or more and, in terms of time, from 1–2 h to a month. Under these conditions, Pooler[47] in 1961 was able to predict half of the monthly averages of SO_2 at 123 stations in Nashville, Tennessee, within a factor of 1.25. In 1964, Clarke[61] refined the time scale to 2 h and was able to predict 24-h averages of NO_x at 14 of 19 stations in Cincinnati, Ohio, to within 0.02 ppm. Turner[48] reduced this error by half 58% of the time, and in 1967 Koogler[62] was able to report 90% of 8-h SO_2 averages correctly within 0.01 ppm, employing the same model. Perhaps the most extensive tests of the Gaussian diffusion model have been conducted by Miller and Holzworth,[49] in three different cities, with 2-h-average predictions at an accuracy comparable to Koogler's.

Considering the coarseness of the input data and the resolution obtained, these results appear quite promising, although the models are severely limited by their inability to account for the time variation of source strengths and to other than simple sources and inert contaminants. Nevertheless, they have a legal standing in the United States through the 1970 Clean Air Act. The *Federal Register* (Vol. 36, November 25, 1971) specifies, for example, a simplified version of Miller and Holzworth's referenced work as one official means for a state to estimate air quality of a region. Figure 7 is a graphical representation of the relationship between normalized concentration and "city size" for an "area" model, according to these authors. The city size is defined as one-half the square root of the urban area measured in kilometers, while the estimated concentration (X) is in micrograms per cubic meter, the wind speed through the mixing layer (u) is in meters per second, and the emission density (Q) is in micrograms per square meter per second. For example, the Standard Metropolitan Statistical Area (SMSA) of Chicago is 2500 km², giving an urban size of 25 km; hence, if $Q = 17.8$ and $u = 7.3$ (1967 data for SO_2),

$$\frac{Xu}{Q} = 230 \quad \text{(from graph)}$$

$$X = \frac{(230)(17.8)}{7.3} = 561 \text{ } \mu g/m^3.$$

In the event that all emissions in a region are from a single source, the corresponding model[50] is (for 1-h average ground-level concentration):

$$X_1 = \frac{Q'}{(2\pi)^{1/2} \sigma_y L u} \text{ g/m}^3,$$

where Q' is the emission rate in grams per second and L is the height of the mixing layer in meters. σ_y (the plume standard deviation in meters) and u, the wind speed, are assumed to be constant for a 6-h period. (The maximum 24-h concentration is presumed to be one-fourth of this.)

8.5.3. Mass Conservation Urban Models

Lamb[51] in 1968 returned to the basic diffusion equation, a special form of Eq. (88), and applied it to compute CO concentration at 1200 grid points (as close to each other as 200 m) for a single day in Los Angeles. Lamb's model utilized simple chemical reaction rates and included absorption of components by the ground. His point, line, and area sources were variable in space and time. Stability (inversion height) and the K values of Eq. (88) were considered constant. In the numerical solution of Lamb's integral equations, the sources were considered to emit a puff of pollutant at each time step: these puffs were advected by the x, y components of surface wind computed at each grid point and were followed until fully dispersed.[44] The effects of all these dispersed emissions were then totaled to obtain concentration as a function of time and location.

The model's predictions did not correlate perfectly with measurements at various stations. Its faults have been ascribed to lack of a vertical wind component, giving concentrations too high at the convergence of trajectories and results too low during the afternoon, suggesting an influx of sources from outside of Los Angeles. In order to improve this model substantially, it is also necessary to include nonlinear chemical-reaction terms. This, precisely, is the model we have described in Section 8.4.1.

Table VI

Computation Parameters for Selected Large-Scale Mathematical Models

Reference	Problem size	Model type	Computation resolution	Source inputs
Turner[48] 1964	24-h concentration forecast, Nashville 17 × 10-mile area	Gaussian	99 grid pts. 2-h time step	—
Roberts, Croke, Kennedy[52] 1970 (Argonne)	Short-term prediction (2–6 h averages), Chicago 25 × 10-mile area	Integrated puff (Gaussian), time-dependent	1-mile grid ~200 pts 1-h time steps	50 point sources area inventory
Rote, Gudenas 1971[53]	(same)	Gaussian plume modified for low wind speeds (≤ 1 mph)	(same)	(same)
Shieh et al.[54] (NYU)	50 × 50 miles² (NYC) SO_2, 48-h forecast	Statistical, Gaussian, non-isotropic, time-dependent	2000 pts 0.2 mile and 1.0-mile grids 2-h steps	0.5–5.0 mi grids 565 area sources 148 pt sources
Walden Research[55]	Metro. Boston, CO and particulates	Gaussian	221 receptors	500 source locations
Roth, Reynolds, Roberts, Seinfeld[37,55]	CO in L.A. basin (model designed for 4 species). 11-h forecasts	Mass conservation grid, photochemical model (4 partial diff. eq's, see text)	2-mile grid 25 × 25 × 10 vertical (625 locations) 25-min time steps	Initial conc. at each location
Pacific Environmental[55]	Trajectory of air parcel, 3 h O_3, NO_2, NO, CO	Mass cons., chem. kinetics	15 min	Initial conc.
Center for Environment and Man[55]	8000-pt grid	Mass cons.	8000 pts 150 time steps	3 changes emission data
Systems, Science and Software[58]	CO in L.A. basin, 18-h prediction 60 × 45 miles × 1200 ft	Mass cons.	22 × 16 × 4 cells high (~2.8 mile grid)	(same as Roth et al.)
Grumman Aerospace[55]	Kennedy Airport 24-h avg. CO	Gaussian	250-pt grid	—

Meteorological inputs	Machine and language	Computing time; cost/run	Memory, kilo-bytes	Validation
—	IBM 7090	2 min	—	32 station 58% ± .01 ppm
Interpolated mixing ht.	IBM 360-75	0.5–0.75 min	—	8 AQ monitoring stations −3 to +16% error
(same)	(same)	Less than above	—	Similar
Wind, temperature vertical soundings	CDC 6600	3-min met. 19-min diffusion	—	Two 5-day tests
—	360–65 PL-1	$395	202	0.85–0.90 corr. coeff. (particulates only)
Hrly ground wind, inversion maps (25K data entries/ 12 hr day)	360	1 min/h (est 1 h/8 h for 6 species) $500	315	1 day CO (1969) 10–20% error (1–1.5 ppm)
Winds	360–50	4–5 min (cpu) 5–20 (run) $40–$60	86	2-day test in L.A.
3 changes met. data	UNIVAC 1108	4.5 min	40	—
(same as Roth et al.)	360–75	10 min (cpu) 14 min (run) $150	480	0.73 corr. coeff. 0.67 ppm mean error (2-day test)
—	360–67	45 min $300	—	—

Table VI

Continued

Reference	Problem size	Model type	Computation resolution	Source inputs
General Research[58]	L.A. basin, 11 species, trajectory between two pts for 3 h	Mass. cons.	5 vertical, 10-min time steps	—
Computer Sciences[55]	CO, 1200 pts for 18 h	Gaussian	1200 pts	Traffic data
Intercomp[55]	SO_2, 10 h 10 × 10-miles × 4000 ft	Mass cons.	15–30-min average 2300-pt grid (3-dimensional)	—
Mt. Auburn Research Assn.[55]	(a) Trajectory— low turbulence	Gaussian	754 parcels	—
	(b) Trajectory— high turbulence	(same)	115 parcels	—

8.5.4. Data-Processing Requirements for Large-Scale Models

Since the ultimate objective of most urban models is to predict the onset of high-air-pollution episodes in time to avert them through appropriate control measures (at least 6 h of lead time in a typical large city), the ability to compute forecasts rapidly becomes most important. Fortunately, our computing power has advanced rapidly, perhaps more rapidly than our ability to formulate useful models and gather data, but it is still necessary to be very economical in the choice of scale and computing variables. For example, it has been estimated that a full solution of the Navier–Stokes equations for boundary-layer turbulence with random initial conditions might require as much as 10^{13} bits of storage in a typical problem.[26]

Computation of rather coarse-grid Gaussian models has not introduced any difficulties in this respect. Turner, in 1964, was able to make 24-h forecasts of inert pollution concentrations in Nashville, Tennessee, in 2 min of computing time, using a 7090 machine. But we have seen that this kind of model is not an adequate solution for urban smog prediction. Table VI lists some recent experiences with larger and more powerful models and processes. With the photochemical model described in Section 8.4, medium to large processors currently available will produce forecasts

Meteorological inputs	Machine and language	Computing time; cost/run	Memory, kilo-bytes	Validation
—	CDC 6400	3.8 min (cpu) 4.1 (run) $30	270	CO $\pm$ 20%
Met data	CDC 6500	60 min (cpu) 120 (run) $700	450	1 day within factor of 3
—	CDC/ UNIVAC	3 h (cpu) $75–110	375	Satisfactory
—	360–75	3 min	—	—
—	—	21 min	480	—

at the rate of 1–7.5 min of computer time per hour of prediction, depending on the number of pollutants, with acceptable errors up to 12 h predicted time or more.

The memory requirements are rather severe. Many of the models currently being run demand a capacity of over 3×10^6 bits, which is definitely in the large computer class. A recent survey by Darling of transportation-oriented pollution models[55] indicated that most programs were in the 1000–2500 line and larger class (Fortan IV). Nearly half the programs being run at the time of the survey required in excess of 200 kilobytes of storage.

One of the present authors[56] as well as others[57] have suggested that large parallel-processing arrays such as Illiac IV[58] may be used to solve photochemical models, so that processing time will cease to be a problem. One writer has estimated that these machines can produce about one day of forecast using a complex model per hour of computing time.[59] Thus, one Illiac IV would be sufficient to serve at least 20 cities on a time-shared basis.[60]

REFERENCES

1. H. Moses. *Mathemetical Urban Air Pollution Models*. Argonne National Laboratory: Chicago, 1969.
2. R. C. Wanta. In A. C. Stern, Ed., *Air Pollution*, 2nd ed., Vol. I. New York: Academic: 1968, pp. 216–217 and pp. 220–223.
3. M. E. Smith. *International Symposium: Chemical Reactions Lower Atmosphere*, Advance Papers. Stanford Research Institute: San Francisco, Calif., 1969, pp. 273–286.
4. D. H. Slade, Ed. *Meteorology and Atomic Energy—1968*. Environmental Science Services Administration: Silver Springs, Md., 1968.
5. Ref. 4, p. 75.
6. E. V. Somers. In W. Strauss, Ed., *Air Pollution Control*, Pt. 1. Wiley: New York, 1971, p. 6.
7. G. I. Taylor. *Proc. London Math. Soc.*, **2**, 196–202 (1921).
8. O. G. Sutton. *Proc. Royal Soc. (London), Ser. A.*, **135**, 143–165 (1932). See also *Micrometeorology*. McGraw-Hill: New York, 1953.
9. L. F. Richardson. *Proc. Royal Soc. (London), Ser. A.*, **110**, 709–737 (1926). See also Ref. 4, p. 93.
10. E. Parzen. *Modern Probability Theory and Its Application*. Wiley: New York, 1960, p. 238.
11. Ref. 6, pp. 9–12.
12. E. M. Darling, Jr. *Computer Modeling of Transportation-Generated Air Pollution* (DOT-TSC-OST-72-20). U.S. Dept. of Transportation: Washington, D.C., 1972.
13. F. Pooler, Jr. *Int. J. Air Water Poll.*, **4**, 199–211 (1961).
14. D. B. Turner. *J. Appl. Meteorol.*, **3**, 83–91 (1964).
15. Ref. 1, p. 22.
16. Ref. 6, pp. 19–23.
17. F. Pasquill. *Meteorol. Maq.*, **90**, 33–49 (1961).
18. F. Pasquill. *Atmospheric Diffusion*. Van Nostrand: New York, 1969.
19. Ref. 4, p. 406.
20. F. A. Gifford, Jr. *Nucl. Safety.* **2**, 47–51 (1961).
21. G. A. Briggs. In Ref. 4, Chapter 5.
22. J. C. Weil and D. P. Hoult. *Effective Stack Heights for Tall Stacks*. MIT Fluid Mechanics Laboratory: Cambridge, Mass., 1971.
23. C. H. Bosanquet. *J. Inst. Fuel*, **30**, 326 (1957).
24. I. Van der Hoven. In Ref. 4, p. 203 (Eq. 5.38).
25. T. A. Hewett et al. *Laboratory Experiments of Smokestack Plumes in a Stable Atmosphere*. MIT Fluid Mechanics Laboratory: Cambridge, Mass., 1970.

26. W. H. Snyder. *Boundary-Layer Meterol.*, **3,** 113–134 (1972).
27. J. Halitsky. In Ref. 4, pp. 221–255.
28. P. C. Chang et al. "Turbulent Diffusion in a City Street." *New Mexico State Univ., Las Cruces Proc. Symp. Air Pollut., Turbul. Diffus.*, Las Cruces, N. Mex., 1971, pp. 137–144.
29. D. Djuric and J. C. Thomas. "A Numerical Study of a Gaseous Air Pollutant in the Vicinity of Tall Buildings." In Ref. 28, pp. 27–34.
30. W. B. Johnson et al. *Field Study for Initial Evaluation of an Urban Diffusion Model for Carbon Monoxide.* Stanford Research Institute: Menlo Park, Calif., 1971.
31. F. L. Ludwig and W. F. Dabberdt. *Evaluation of the APRAC-1A Urban Diffusion Model for Carbon Dioxide,* Stanford Research Institute: Menlo Park, Calif., 1972.
32. Ref. 4, pp. 140–142.
33. M. C. Dodge. *Workshop on Mathematical Modeling of Photochemical Smog: Summary of Proceedings, Oct. 30–31, 1972,* National Environmental Res. Center, U.S. Environmental Protection Agency, Research Triangle Park, N.C., Jan. 1973, p. 5.
34. R. C. Sklarew. In Ref. 33, p. 35–6.
35. W. A. Perkins. In Ref. 33 (refers to the 1973 Los Angeles smog season atmospheric measurement program, EPA).
36. R. A. Papetti and F. R. Gilmore. *Endeavor,* **30,** 107–111 (1971).
37. P. M. Roth, S. D. Reynolds, P. J. W. Roberts, and J. H. Seinfeld. *Development of a Simulation Model for Estimating Ground Level Concentrations of Photochemical Pollutants,* Final Report 71SA1-21. Air Pollution Control Office, EPA: Durham, N.C., 1971.
38. T. A. Hecht. In Ref. 33, pp. 10–11 and Appendix A.
39. J. H. Perry et al. *Chemical Engineers' Handbook,* 4th ed., McGraw-Hill: New York, 1963, pp. 4–3.
40. R. Papetti. In Ref. 33, p. 37.
41. A. C. Stern. *Clean Air,* **4,** 31–32 (1970).
42. A. C. Stern, Ed. *Proc. of Sympos. on Multiple-Source Urban Diffusion Models,* U.S. Environmental Protection Agency: Research Triangle Park, N.C., 1970, pp. 2-1–2-6.
43. D. M. Leahey. *J. Air Poll. Control Assn.*, **22,** 548–550 (July 1972).
44. J. H. Seinfeld. "Mathemetical Models of Air Quality Control Regions." In A. Atkisson and R. S. Gaines, Eds., *Development of Air Quality Standards.* Merrill: Columbus, Ohio, 1970, pp. 169–195.
45. Ref. 1.
46. H. G. Fortak. In Ref. 42, pp. **9-1–9-33**.
47. Ref. 13.

48. Ref. 14.
49. M. E. Miller and G. C. Holzworth, *J. Air Poll. Control Assn.*, **17**, 46–50 (Jan. 1967).
50. D. B. Turner. *Workbook of Atmosphere Dispersion Estimates*, PHS publication 999-4P-26. HEW: Cincinnati, 1960.
51. R. G. Lamb. "An Air Pollution Model of Los Angeles." M. S. thesis. University of California, Los Angeles, 1968.
52. Ref. 42, p. 6–1 ff.
53. D. M. Rote and J. W. Gudenas. "A Steady State Dispersion Model Suitable for Air Pollution Episodes." *64th Ann. Meeting Air Poll. Control Soc.* Atlantic City, N.J., June 27–July 2, 1971.
54. L. J. Shiek, B. Davidson, and J. Friend. In Ref. 42, p. 10–1 ff.
55. E. M. Darling, Jr. *Computer Modeling of Transportation-Generated Air Pollution* (DOT-TSC-OST-72-20). U.S. Dept. of Transportation, Office of Supt. Dev. and Tech.: Washington, D.C., 1972.
56. R. J. Bibbero. *IEEE Spectrum*, **8**, 51 (December 1970).
57. K. T. Yen. "Meteorological Air Pollution Monitoring." *64th Ann. Meeting Air Poll. Control Soc.*, Atlantic City, N.J., June 27–July 2, 1971, p. 22.
58. W. J. Bouknight et al. *Proc. IEEE*, **60**, 369–388 (1972).
59. S. Schwartz and G. B. Siegel. "Models for and Constraints on Decision Making." In A. Atkisson and R. S. Gaines, Eds. *Development of Air Quality Standards*. Merrill, 1970, Columbus, p. 40.
60. D. M. Stern. Personal communication, 1970.
61. J. F. Clarke, *J. Air Poll. Control Soc.*, **14**, 327 (1964).
62. J. B. Koogler, R. S. Sholtes, A. L. Danis, and C. I. Harding, *J. Air Poll. Control Soc.*, **17**, 211 (1967).

9

Community Air-Quality Management: Control, Monitoring, and Abatement

Given a formidable degree of authority, and a potential to understand, measure, and predict the consequences of air pollution—which, although still incomplete, far exceeds that of previous generations—it is still up to the community to make the best use of this armamentarium. In a sense, this chapter has yet to be written, since the emissions of atmospheric pollutants, especially sulfur and nitrogen oxides, show continual increase, even in the most recent official tabulations.[1] Nevertheless, we will attempt at least to sketch an outline of the air-pollution management means currently used and available to communities in the United States.

The desirable order of applying air-pollution controls might be as follows:

1. Prior controls
2. Posterior controls
3. Monitoring and abatement
4. Emergency controls.

By "prior controls," we mean those which go into effect before a pollution source exists, while "posterior" reflects the regulations by which a source is permitted to operate. Monitoring, of course, refers to the surveillance of both ambient concentrations and source emissions in order to detect and avert any adverse trends in air quality. Only in the last resort, preferably never, should we need to apply emergency controls to abate an "episode" of high-concentration pollution. The word *control* is used in two senses, meaning both the legal emission standards and ordinances reflecting the social power over pollution-emitting sources and the physical devices that actually reduce emission. We shall discuss mainly the concepts and means of implementing the legal standards (see Chapter 4), including measuring devices at the sources to see that these controls are enforced.

Although the physical controls are of immense technical and monetary interest to industry, representing perhaps 90% of the funds that must be expended to clean the air at this time, they represent little that is not well known. Much of the expenditures for air-pollution control go for such mundane devices as bag filters, cyclones, gas scrubbers, and absorbers. Electronic precipitators for removing particles, especially, are immensely expensive, as mentioned earlier, and of greater technical interest, but even these have been used since the beginning of this century. Tall stacks, of course, do not reduce total emissions, but as we saw in the last chapter [Eq. (52)], they will reduce the concentration of ground pollution by the square of their height; however, they are no less expensive, costing from $3000 to $7500/m. The greatest area for innovation may be actual changes in the pollution-causing processes themselves. If materials can be used more efficiently or recycled, the end result will be a saving rather than an expense to the manufacturer.[2,3]

Legal control may be exerted prior to, or following, the establishment of a man-made source of air pollution. Physical controls and tests to prove the ability to meet emission standards may be required as a prerequisite for construction or operation. Once a source is permitted to come into existence, its emissions must be brought into line with the regional air-quality standards by continual monitoring and inspection. If it is found wanting in this respect, additional controls or change in process may be required.

9.1. PRIOR CONTROLS

Ideally, air-resource management starts with community planning. All aspects of land-use planning—transportation; zoning for industry, commerce, or residential use; waste disposal; park and open-space reservation—should be considered in planning to meet quality standards for the air, and conversely. A survey of the local background of contaminants should be made prior to land development. Such factors as future growth of population, increased automobile use, location of new electric power plants, and introduction of new mass-transportation methods have obvious impact on future air quality. The effect of these changes as well as the expected values of pollution with current land usage can be estimated, employing mathematical diffusion models.

Given a single source with a constant rate of emission and stack height, there is a combination of wind direction and velocity, atmospheric stability, and distance for which the ground concentration of pollutant will be greatest, as shown in Chapter 8. This concentration can be set equal to the desired air quality in order to obtain an emission standard for this stack. In the case of several large and many small sources, the problem of allocating the total emission may become very complex, but this is precisely the basic problem of fairly dividing up use of the public's air resource. (Croke and Roberts,[4] for example have suggested emission standards based on land area owned rather than by individual stack or industrial plant.) A computer program that has for inputs the yearly inventory of source strengths and a long-range analysis of meteorological variables, with the capability of computing the time and place of maximum pollution, has been suggested.[5] The program would then print out the contribution of each source to this "worst pollution day" and serve as a basis for allocation.

The means by which the sources in a region can control their emissions to predetermined maximums may be optimized from a cost standpoint by the use of linear programming models and similar economic models discussed earlier. Studies of this nature will lead to a rational basis for land-use and zoning decisions and for rules governing construction permits and licenses. Where local interests conflict with quality standards based on public health and welfare, an educational program or intervention by a higher political entity, through the imposition of state or federal emission standards, may be required. Likewise, these may be necessary in the event of pollution transport between local regions.

Emission standards and zoning regulations, established by local ordnance or statute, imply the existence of a local control activity to implement them. Engineering analysis of plans, with special attention to emission points, stack heights, process-flow sheets and quantities, and proposed control methods and monitoring instruments or access, form the rationale for approval of construction permits. At this time, the data base for a source-emission inventory is established. A data bank, which may or may not be mechanized, depending on the needs of the system, is a necessary part of the activity.

9.2. POSTERIOR CONTROL OF STATIONARY SOURCES

Control can only be exerted over stationary air-pollution sources, once they are established, with the specific legal and financial authority necessary to create and support an enforcement agency to act at the local level. The CEQ, in its 1970 annual report to the U. S. Congress,[6] indicated that most of the state programs, as well as those at the local and regional level, were inadequate in this respect. There were 144 local agencies receiving federal grants, but only 44% were adequately funded for a minimal program. The 1970 Clean Air Act requires repair of this situation.

Prior to August 1971, there was in effect only one federal emission standard—that established for new automobiles—but performance standards for certain categories of new stationary plants are now in effect.[7] By 1970, all states had air-pollution legislation, but only 42 actually had any kind of regulation to control emission. For example, only 6 states regulated emissions of on-the-road vehicles.[8] This matter is still a question for debate.

As a result of the 1970 act, the legal basis for regulation and control of sources has accelerated. Patterns of source-control regulations have been established, and it is believed that new regulations will continue to follow these trends.

9.2.1. Particulates

Regulations for the control of particulates, for example, have been based on four different concepts: opacity, concentration, process weight, and emission potential. The Ringelmann standards for visible emissions have already been discussed (Chapter 4).

Other particulate regulations specify the concentration (mass per unit volume) of effluent gas. These are based on a "model smoke ordinance" developed by the ASME in 1948. Typically, the limits range from 0.2–0.3 gr/standard ft^3 (at 60°F and 1 atm), depending on the definition of particulate, the sampling method, and the gas composition.

Another rule governs the emission of dust as a function of the weight of material processed in order to circumvent attempts to avoid the concentration rule by diluting the gas stream. The process-weight concept,

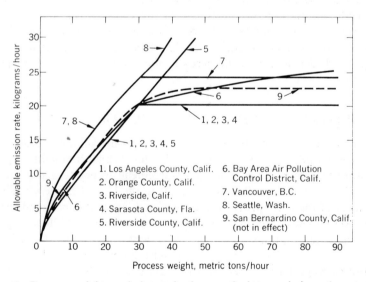

Figure 1 Process–weight emission rule for particulate emissions (courtesy of *Chem. Eng.*).[9]

9.2. Posterior Control of Stationary Sources

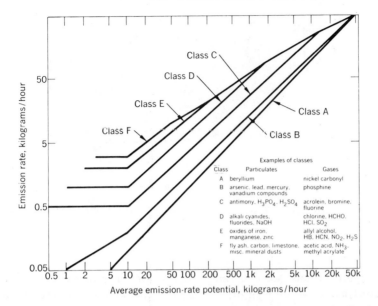

Figure 2 "Emission potential" standard (courtesy of *Chem. Eng.*).[9]

which is also a feature of the state regulations discussed in Chapter 4, is demonstrated in Figure 1. It can be seen that permissible emissions under this rule can be increased by using two or more small units, rather than one large device.

The concept of control of both gases and particulates by "emission potential," depicted in Figure 2 has been adopted by New York State, among others. The abscissa of each curve represents the potential rate at which contaminants would be emitted if there were no gas-cleaning devices. The rules increase in stringency depending on the toxicity class of the material; thus, about 15% of iron oxide potential can be emitted but only 1% of beryllium.[9]

It is clear that this type of standard, recognizing the effect on the receptor, is most rational. Some regulations go even further by distinguishing betweeen fine and coarse particles, which have different physiological and physical effects. It is possible that the development of a convenient, accurate, objective instrument that would correlate with the more important receptor effects would create a new standard means of measuring particulates and replace the visual tests.

9.2.2. Gases

Most standards for the emissions of gases from stationary sources have been directed against SO_2, although other gases and vapors, including fluo-

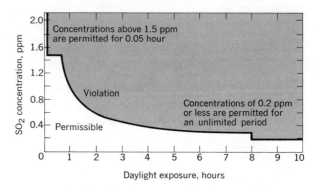

Figure 3 Ambient ground-level limits on gaseous emissions (courtesy of *Chem. Eng.*).[9]

rides, hydrocarbons, solvents, and other sulfur compounds, are also controlled (Chapter 4). In California, the rules give an operator the option to monitor ambient ground-level concentrations, rather than to adhere to a fixed stack concentration limit. The ambient limits (Figure 3) must be monitored by at least three continuous SO_2 analyzers and one recording wind station.[9]

Some emission standards are combined with a design standard, such as stack height or adjusted height (corrected for temperature of flue gas). Stack height criteria result from the diffusion models discussed earlier. Other regulations control emissions indirectly by specifying fuel standards; for example, the volatile content of coal, the olefin content of gasoline, and the sulfur content of heating fuels are utilized in some of the state regulations of Chapter 4.

9.2.3. Source Testing and Monitoring

If, in the long term, the application of air-pollution controls to old and new sources is the response of the national system to intolerable ambient levels, then source testing and monitoring represent the feedback that closes the control loop. Source emissions are tested by the plant operator or by a control agency for many reasons: survey; licensing or inspection; checking the efficiency of collectors; monitoring process malfunctions or accidents; compliance with legal standards; including federal "new source" performance standards; or for the record. Such monitoring is now usually accomplished in the stack, by roof monitor, or at other effluent locations. Control agencies may, in addition, monitor or test sources with in-stack or remote sensors to answer complaints or to detect violations. In connection with alleged offences, the method must meet local or federal legal criteria.

9.2. Posterior Control of Stationary Sources

Source testing must take into account all the technical problems of obtaining a valid and representative sample. Pertinent factors include cyclic or random fluctuations of the effluent, both in quantity and composition, and physical or chemical instability of the sample. Each problem must be solved in the context of the particular process, plant, and installation. The temperature and the dewpoint of flue gases are significant. In general, particulates pose greater sampling problems than gases because of agglomeration and variations in particle size.

Flowrates, as well as composition, directly determine a plant's contaminant emissions. Flow measurement by means of Pitot tubes in ducts and stacks must cope with plugging by particulates. Locations of sample or duct traverse points must follow good practice for these instruments. Because of the dirty, hot, or corrosive nature of effluents, special sampling equipment must often be designed. Probes and nozzles may have to be heated to prevent condensation of the sample before reaching a collector. In sampling particulates 3 μm or greater in size, care must be taken to sample "isokinetically"—that is, to match the velocity of the sampling nozzle with that of the gas stream. Otherwise, an imbalance of heavy of light particles enters the nozzle.

Accurate sampling of a large duct, ensuring the proper number of sampling points, changing nozzle sizes for isokinetic conditions, flow metering, and so forth, can take 10 or more man-hours per trial. The samples so collected must be tested in the laboratory, which involves additional time and expense. Therefore, approximate or automated methods are indicated.

Continuous analysis is required to meet more stringent regulations. The time constants of analysis equipment such as the SO_2 conductimetric monitor are short enough to gather real-time data for the guidance of operators. Photoelectric monitors installed in the stack give the same kind of information for particulates. On the other hand, some monitoring equipment, such as tape samplers, requires a definite sample period and average over this time lag. Monitoring devices relying directly on pollution effects may have very long time constants: examples are lead sulfate candles, lead acetate tiles, rubber strips, paint, and metal. These may take months or years to measure contamination.

In general, the stack-mounted analysis equipment in current use is similar to air-quality monitoring equipment, except that the former measures higher concentration ranges. Wet-chemical analyzers including both photometric and conductimetric systems are used for SO_2 and H_2S. Flame photometric techniques are also used. Carbon monoxide is monitored by nondispersive IR instruments; particulates, by photoelectric and IR opacity meters and tape samplers. Recorders, and sometimes level alarms, are usually incorporated in these instruments.

A novel method of stack sampling for SO_2, the correlation spectrometer, was mentioned in Section 6.2.2 as a remote sensor. In this application the instrument measures the intensity of the UV spectrum of stack gases sampled by a slotted tube in the stack, comparing it with a photographic mask of the SO_2 spectrum. The photocell monitor is calibrated to read in parts per million and is protected from the flue gases by means of an air curtain. Chemical analysis equipment is discussed in more detail in the next two chapters.

Although modern photoelectric and spectroscopic devices may solve many of the problems of source monitoring, equipment capable of reading stack-emission compositions at a distance is even more attractive. In this way not only are installation and sampling problems obviated, but the possibility of time sharing as well as portable application for law-enforcing agencies is introduced. Remote-reading stack monitors are by no means a new idea. The Ringelmann smoke chart is a remote comparison device, and several other portable visual smoke guides and viewing devices are available.

To replace visual particulate measures, the pulsed-laser instrument (lidar) has been mentioned; it enjoys popularity as a research instrument but is not yet in common use.

Active, single-ended spectrographic devices also are being studied. These include IR backscatter instruments; another concept monitors emission spectrums from hot stack gases by means of a spectrophotometer; Raman spectroscopy is being investigated for stack monitoring, as noted in Chapter 6. The difficulty with these remote systems at present—apart from the need to prove their feasibility, consistence, and acceptability as legal standards—is their cost and complexity.

9.3. CURRENT AMBIENT MONITORING SYSTEMS

Supplementing land-use and source control is ambient monitoring, used both as a check and a predictive device. As Chapter 4 described, the 1970 Clean Air Act requires some form of monitoring in each air-quality control region; however, the mandatory requirements are minimal. Many communities have been able to install monitoring systems of some sophistication and have gained thereby valuable experience.

An overview of some continuous aerometric networks currently being used and developed by state and local air-pollution control activities will give some idea of how far we are along the road toward a nationwide monitoring and control function. This survey will disregard the National Air Sampling Network (NASN) and the Continuous Air Monitoring Program (CAMP) of the federal government, despite the fact that these are nationwide, because these projects are intermittent or limited

9.3. Current Ambient Monitoring Systems

Several typical state and local continuous monitoring networks are described in Table I, with the Rijnmond (Rotterdam) network developed for the Netherlands government included for comparison. Network characteristics can be conveniently broken down into four groups for purposes of this tabulation. These are the network mission, the sampling design, the data-handling methods, and the control action decision. The actual performance of control activities such as fuel switching are not part of the aerometric network, but the compliance of sources on a legal or voluntary basis is, of course, essential to the entire undertaking. The network may also take on supplementary functions, such as storage of source-emission inventories and of data in standardized format (SAROAD) for nationwide exchange. The objectives of the networks are reasonably consistent. All have as one goal monitoring for unhealthy concentrations, which is mandated by the requirement to establish an alert procedure in order to qualify for federal funds. (Most network funds—for example, two-thirds of Philadelphia's total—are derived from the Federal Clean Air Act.) There is a strong tendency to spread out the objectives to longer-ranger items, such as criteria or mathematical-model development, to help justify expensive data processors.

In measurement and sampling, there are two schools of thought: one utilizes SO_2 as a "tracer," or index of general pollution, whereas the other finds it necessary to monitor individual chemical variables. Those relying on SO_2 are either less concerned with automobile-derived photochemical smog or anticipate a strong correlation between pollution components. This difference is also reflected in the number of meteorological variables considered necessary.

The most striking inconsistency lies between the design of the sampling systems, particularly the station density, and the criteria for "representativeness" of the samples. Studies of SO_2 pollution in several areas of the world have agreed that the spacing between sampling stations should not exceed 0.8 km to obtain reasonably accurate estimates of the daily average. We have noted earlier that one standard, that of the West German Federal Republic, approaches this measure by requiring 1 station/km² for pollution surveys. The actual densities tabulated for U.S. networks are at least two orders of magnitude below this criterion. Attempts have been made to justify this parsimony theoretically by appeal to diffusion models and by placing sensors near large sources. It has been shown in the Allegheny County system, as we have seen, that this does not work.[10] In Rijnmond, however, where the topography and source parameters appear less complicated, it may prove possible to obtain accurate results with about one-tenth the theoretical coverage. If this proves to be the case, it will serve as economic justification for greater application of diffusion modeling in support of sampling network design.

Table I

Continuous Aerometric Methods

System	New York State	Pennsylvania (Planned)	New Jersey	Delaware	Allegheny Co. (Pittsburgh)
Mission					
Coverage	450 linear miles	(10 air basins)	(State)	(State)	430 miles^2
Population ($\times 10^6$)	—	—	—	—	1.6
Major system objectives	Monitor-alert-criteria	Monitor-alert-criteria. Control-model development	Monitor-alert-criteria	—	Monitor-control-historic data
Sample System Design Variables:					
Chemical	8	7	10	7	7
Meteorological	8	5	7	2	4
Poll interval (min)	15	1	—	3	3
Basic averaging time	15 min	1 hour	15 min	15 min	5 min
Number of stations	35 (50)^{a,23}	1 (25)	18 + 3 mobile	4	7 (18)
Sample density (stations/mile^2)	—	—	—	—	0.02
Site design (primary criteria)	Geography-pop. levels	Diffusion model-10 air basins	—	—	Topography-large sources
Data transmission	Digital-dial-up line	Digital-leased voice line	analog	—	Digital-voice line
Data Handling					
Data processor	B-3500	Spectra 70 + control computer	Spectra 70-45 + PDP-8	—	IBM 1801
Display	CRT + TTY	Graphics map + printout	Printout	Printout	Printout (reads red over std)
Episode prediction	—	(Diffusion models)	—	—	COH-met, regression model
Control optimizing	—	—	—	—	—
Control Decision					
Alert criteria	SO$_2$-CO-particulate levels	—	—	—	Meteorology forecast + SO$_2$-CO-particulate levels
Control procedure	—	—	—	—	Source abatement
Aux.					
Emission inventory	—	(Computerized)	—	—	(Computerized)
Data-exchange format	—	—	—	—	SAROAD
References	12	13	14	15	10, 16

^{a} Parentheses signify plans or goals.

Chicago	Los Angeles County	New York City	Philadelphia	Rijnmond (Rotterdam)
250 miles²	1035 km²	200 miles²	125 miles²	125 miles²
6.2	6.9	11.4	—	—
Monitor–alert–control	Monitor–alert–control	Monitor–alert–criteria. History	Monitor–alert	Alert–control
1 (SO_2)	6	3(5)	6	1 (SO_2)
2	4	3	(total)	2
15	Demand or 1 h	5–30	5	1
15 min	1 h	5 min–4 h	—	1 h
8	12	10(30)	6	31
0.03	0.003	0.05	0.05	0.25
Topography–industry–pop.	—	Pop. density–geography	—	Diffusion model–sources
Digital–TTY line	Digital–wire + microwave	Analogue (PWM)–voice line	Analogue–voice line	Analogue (PFM)–phone (120 Hz)
—	—	PDP-8	—	Phillips P-9201
Printer (CRT)	Real-time graphic hourly printout	TTY printer	—	Printer
(2–24-h forecast diffusion model)	Analytical diffusion model	—	—	Statistical (mean deviation)
(Optimizing short and long-term model)	—	—	—	—
SO_2 (0.30 ppm), 48-h stagnation	O_3–CO–SO_2–NO_x	Meteorology forecast + SO_2–CO particulate levels	—	Meteorology forecast + SO_2
Fuel change or shutdown	3-stage alert (burning, traffic)	3-stage alert	—	Source curtailment
Large sources (computerized)	—	—	—	Major sources known
—	—	SAROAD	—	NA
17, 18	11	19–22	—	23, 24

The data-handling systems should show few surprises, since the current numbers of sensors do not require any of the novel techniques discussed in Section 6.2.3., for example. In view of the debate on manual or analogue versus digital data transmissions, it is noteworthy that most networks have chosen or have switched to the digital technique.[11] The use of general purpose computers for data processing is practically universal. In several systems, elaborate diffusion models are planned, and in at least one case, these are for short-term pollution forecasting and feedforward, real-time control of abatement procedures.

9.4. SHORT-TERM AND EPISODE CONTROL

We have seen in Section 4.2.2. that federal law requires AQCR plans to provide adequate authority to deal promptly with emergency air-pollution conditions, and in Section 4.3.12 (Table XVI) we have noted some alert concentrations and procedures. The governors of states have the power to take action in the event of dangerous air-pollution episodes. Alert plans call for action to be taken upon observing high concentrations of certain pollutants or combinations for as little as an hour or less. The designation of greatly increased numbers of AQCR and implementation of their approved plans implies that a large number of areas will have the capability to respond within hours or less to transient, but dangerous, air-pollution levels.

In the present stage of development of the national air-pollution control system, however, the pace is generally much more leisurely. All regions have not implemented alert plans. The need for rapid transmission of air-monitoring data to a data center is minimized or even doubted by some officials, as has been discussed. There have been, fortunately, few episodes in this hemisphere where widespread fatalities and illness could clearly be laid to air pollution, so that no "clear and present danger" is generally feared.

Nevertheless, the factors referenced in Section 3.6, for example, including increased population, greater industrial activity, and more automobiles, are pointing to higher levels of contamination concentration over a longer-term period, while the criteria—especially those based on health effect—move toward recognition of lower levels of tolerance.[25] (A temporary respite, perhaps through 1980, may be obtained by stringent enforcement of federal automobile and stationary-source controls and may thereby allow more time to achieve more-basic means of restraint.) Consequently, potentially dangerous episodes should become increasingly prevalent through both increased base concentration levels and statistical fluctuations, as well as by more rigorous definitions. It follows that the mature national air-pollution control system must have a fully developed

Table II
Quick Reaction Control System

1. ALERT DECISION
    A. Resulting from predictions of probable dangerous pollutant levels, a decision is made to exert control over emission sources. (Note: Existing systems (e.g., Los Angeles) alert on the basis of current rather than predicted levels, plus meteorological forecasting of pollution potentials in the New York–New Jersey area; see Table I.)
    B. Requirements for decision function are the following:
        1. Prediction model
        2. Law and doctrine
        3. Display of current and predicted levels, including meteorology and geographic display
        4. Mobilized command and control organization

2. TRADE-OFF DECISION
    A. Objective is to determine the least costly effective control action
    B. Control actions available by law to the control team are as follows:
        1. Point source abatement or shutdown
        2. Stationary area source abatement (heating, process combustion, open burning, etc.)
        3. Mobile source abatement (e.g., auto traffic diversion or reduction)
    C. Determine the minimum cost objective according to the cost-effectiveness model or doctrine.

3. COMMAND AND CONTROL
    A. Objective: to exert control over emissions
    B. Requirements:
        1. Data central and status display
        2. Personnel who are authorized to take competent action
        3. A communications net to controlled sources; also a public information net to control area emissions, give out information, etc.

4. FEEDBACK
    A. Quality-monitoring and trend-prediction.
    B. Violation detection: patrol with mobile remote sensors
    C. Source instrumentation and telemetering

capability to predict, detect, and react to dangerous air-pollution concentrations, acting within a time scale of hours or fractions of an hour.

In addition to this capability, which will be designed to combat the rare, natural coincidence of source-strength fluctuation and unfavorable weather, the high-speed response of a monitoring system will be effective against industrial accidents and violations of control regulations.

The quick-reaction system envisioned will depend very much on high-speed data-processing equipment and refined mathematical models that will permit prediction and prevention of dangerous concentrations, not merely their detection after the fact. Furthermore, the system will depend on a more widespread network of monitoring instrumentation than is currently deployed and may utilize continuous central monitoring sources through permanently installed telemetering and measuring instruments as remote-acting stack monitors.

Equally important, a quick-reaction system must have a fully developed command and control structure in as real a sense as that of a military air-defense system. Information must not only be collected but displayed without delay to personnel capable of making rapid decisions (with computer aid) on the course of action that will avert the predicted emergency. It is understood, then, that although some portions of the envisioned system may now exist in the more highly developed air-quality control systems, both here and abroad, all the necessary components do not and cannot exist until further research and development are carried out. An outline of the functions of such an ideal system is presented in Table II. These can be compared with the characteristics of some existing systems described in the previous section.

Not the least of the problems that must be solved before adequate numbers of monitoring stations can be employed in control of air quality is that of chemical instrumentation. There are few chemical or physical mechanisms that respond uniquely to a single chemical species in the proportion of 1 ppm or better without interference from other and more abundant constituents. It is not surprising that most chemical instruments of this sensitivity are designed for laboratory use rather than for the continuous and unattended requirements for air monitors. The cost of constructing and operating these instruments and keeping them reliable and calibrated is a serious inhibiting factor in implementing aerometric networks. In the following chapters we will discuss some of the important characteristics of existing air-quality-measuring instruments and the problems involved in developing new and improved devices.

REFERENCES

1. J. H. Lavender et al. *Nationwide Air Pollutant Emission Trends, 1940–70* (AP-115). EPA: Research Triangle Park, N.C., 1973.
2. S. J. Bailey. "Control Practice in the Electric Power Industry," *Control Eng.*, **18,** 42–44 (Sept. 1971).
3. M. R. Gent and J. W. Lamont. "Minimum-Emission Dispatch," *Proc. 7th Power Industry Computer Applications Conf.*, Boston, Mass., May 24–26, 1971.
4. E. M. Croke and J. J. Roberts. "Air Resource Management and Regional Planning." *Bull. Atom. Scientists,* **Feb. 1971,** pp. 8–12.
5. A. C. Stern, Ed. *Air Pollution,* 2nd ed., Vol. III. Academic: New York, 1968, p. 620.
6. CEQ. *Environmental Quality, First Annual Report.* Superintendent of Documents. U.S. Government Printing Office: Washington, D.C., 1970, pp. 83–88.
7. *Federal Register,* **36,** 5931, (March 31, 1971). See also Chapter 4.
8. S. E. Degler. *State Air Pollution Control Laws,* Rev. ed. Bureau of National Affairs: Washington, D.C., 1970.
9. G. E. Yocum. *Chem. Eng.,* **69** (July 23, 1962).
10. E. L. Stockton. "Experience with a Computer Oriented Air Monitoring Program." *J. Air Poll. Control Assn.,* **20,** 456–460 (July 1970).
11. J. Mills. "Continuous Monitoring." *Chem. Eng.,* **77,** 217–220 (April 27, 1970).
12. D. C. Hunter. "The Air Quality Monitoring Program in New York State." Paper presented at Air Pollution Control Assn. meeting, New York City, June 22–26, 1969.
13. B. Brodivicz, V. Sussman, and G. Murdock. "Pennsylvania's Computerized Air Monitoring System. *J. Air Poll. Control Assn.,* **19,** 484–489 (July 1969).
14. P. C. Wolf. "Carbon Monoxide Measurement and Monitoring in Urban Air." *Environ. Sci. Technol.,* **5,** 212–217 (March 1971).
15. P. E. Wilkins. *Monitoring for Compliance.* Monitor Labs, Inc.: San Diego, Calif., 1970, pp. 10–11.
16. B. Bloom, Allegheny County Air Pollution Control Board. Personal communication.
17. W. J. Stanley. *A Real-Time Air Pollution Monitoring Program.* Dept. of Air Pollution Control: Chicago, undated.
18. H. E. Cramer. "Meterological Instrumentation for All Pollution Applications." In R. L. Chapman, Ed., *Environmental Pollution Instrumentation.* Instrument Society of America: Pittsburgh, 1969, pp. 15–16.
19. A. N. Heller and E. F. Ferrand. *The Aerometric Network of the City of New York.* EPA, Dept. of Air Resources: New York, undated.

20. S. Klein, "New York City Steps Up War on Foul Air." *Machine Design,* Dec. 19, 1968, p. 38.
21. *New York Times,* May 5, 1970.
22. M. Eisenbud. "Environmental Protection in the City of New York." *Science,* **170,** 706–712 (Nov. 13, 1970).
23. F. Cabot. "So Goes SO_2." *Ind. Res.,* 70–72 (Sept. 1970).
24. "A New Approach to the Prediction and Control of Air Pollution." Philips Gloeilampenfabrieken: Endhoven, Netherlands, 1960.
25. G. Black. "Air Quality and the Systems Approach." Staff discussion paper 104, Program of Policy Studies in Science and Technology, NASA Grant NGL 09-010-030. George Washington Univ. Press: Washington, D.C., 1970, pp. 17–18, 33, 39.
26. W. Forester. *Environ. Sci. Technol.,* **8,** 498–505 (June 1974).

10

Gas- and Vapor-Monitoring Instruments

10.1. SYSTEM REQUIREMENTS OF POLLUTANT MONITORS

10.1.1. Considerations in the Use of Instruments

Monitors for Ambient, Stack, and Mobile Sources

In other sections of this book we have discussed the contribution of various stationary and mobile sources to the total pollutant load of the atmosphere. The sum of all these contributions, the chemical reactions in the atmosphere, and the residence times therein, gives rise to a dynamic resultant: the ambient concentrations of the various pollutants. These pollutants and their reaction products, of course, have been considerably diluted compared to the concentration originally present at the emission source. Thus, analyzers of very high sensitivity are required in order to estimate ambient atmospheric pollutant concentrations. Even higher sensitivities are required to measure the background concentrations. The analytical information will then tell us how far above background any particular ambient is and how effective the abatement procedures are in achieving any desired level of air quality.

The systems approach not only requires information on the resultant but also on the inputs. For this reason, it is necessary to estimate concentrations and total emissions of pollutants at their original sources. At the source end, the concentrations will be relatively high and the problem of estimation to any given precision will not be so severe as estimation of background or ambient concentrations. However, this advantage is more than offset by the very severe conditions existing at many industrial emission sources, such as high temperatures, excessive particulate matter, corrosive gases, and excessive moisture. These conditions lead to extremely difficult analytical and sample treatment problems. While conditions at the exhaust pipe of an automobile are less severe, they still constitute a challenge in sample treatment.

Table I

Typical Concentrations of Pollutants in Samples of Interest in Air Pollution

Pollutant	Background	Urban ambient	Stack effluents	Auto emission
CO	0.1 ppm	5–10 ppm	2000–10,000 ppm	1–4%
SO_2	0.2 ppb	0.02–2 ppm	500–3500 ppm	50–100 ppm
NO_x	0.2–5 ppb	0.2–1.0 ppm	1500–2500 ppm	1500 ppm
O_3	10 ppb	0.1–0.5 ppm	—	—
Suspended particulates	10 $\mu g/m^3$	60 $\mu g/m^3$	35×10^6 $\mu g/m^3$	
Methane	1.5 ppm	1–10 ppm		
Other hydrocarbons	<1 ppb	1–100 ppb		1000 ppm

Sources. CEQ, *Environmental Quality*, Third Annual Report, August 1972 (available from Supt. of Documents, U.S. Gov't. Printing Office, Washington, D.C. 20402, Stock Number 4111-0011); and Ref. 9.

In Table I, data are given on the concentrations of some pollutants of interest in urban ambient atmosphere as well as in emissions from stationary and mobile sources. For comparison, the natural background is also given. It is seen that as we go from background to urban atmosphere, the concentrations increase roughly by a factor of 1000; in the next step, from urban ambient to emission sources, we see another factor of about 1000. Thus, the determination of any one pollutant by a specific technique would require that the technique be capable of giving results of the requisite precision and accuracy over a concentration range of 10^6 as well as overcoming the problems of interferences and sample treatment. Few techniques or instruments are capable of analyses over a dynamic range of this magnitude. Thus, we find that instruments, techniques, and general analytical approaches are designed for application to specified concentration ranges as represented by background, ambient, and emission concentrations.

Until quite recently automatic instruments were not available for making real-time measurements in the ambient atmosphere for such pollutants as sulfur dioxide and oxides of nitrogen, precisely because the concentrations of these materials were so low. In order to bring the material of interest into analytical range, large volumes of air were bubbled through an aqueous solution in order to absorb and concentrate the material to be measured. The solution from the aqueous scrubber was then taken to the laboratory and subjected to classical colorimetric procedures to estimate the pollutant of interest. It is obvious that this approach could only give results averaged over the sampling time and the sampling volume and could not indicate short-lived episodes of high concentration that were a possible danger to health and safety. Thus, these methods give a smeared-out account of past history and can only be used to control air quality in a very crude fashion.

For a number of the important pollutants, automatic instruments have recently been developed that can make real-time measurements at the concentrations present in the ambient atmosphere. These will be described in more detail below. At this point, we should mention that the precision and accuracy of many of the instruments have not been thoroughly investigated and that there is a great deal of doubt as to the significance of the data generated by these instruments. It is expected, however, that with the development of adequate methods for standardization and calibration and their application to instruments, reliable information will become increasingly available. The subject of standardization and calibration of analytical instruments and methods is discussed below (Section 10.4).

Point, Remote, and Long-Path Measurements

Analytical data pertaining to the atmosphere can be obtained in a number of ways and the manner of making the measurement has a marked influence on the significance of the results. It is frequently required that one know the concentration of a specific pollutant at a specific point in space. The most obvious example is the concentration of solvents in a paint factory or at work stations or the concentration of carbon monoxide in a vehicular tunnel or busy thoroughfare. Measurements made at a specific point in space are termed, in this discussion, *point measurements*. They are usually accomplished by installing automatic instruments at the point of interest and making real-time measurements of the concentration of the pollutant of interest as a function of elapsed time. Thus, the pollutant is continuously monitored and concentrations that exceed limits consistent with health or safety may be noted and appropriate action taken. The main idea in point measurements is that they are not space-averaged; however, they may be time-averaged. For example, the scrubber type of analysis described above may involve sampling the atmosphere at a specific point in space over a long period of time, and thus, an averaged concentration is obtained rather than a real-time concentration. It can be seen that time-averaged and space-averaged data have completely different significance from real-time non-space-averaged measurements at a point.

It is convenient to discuss another manner of making a measurement that we shall term *remote*. By this term we mean that the analytical instrument is placed at an appreciable distance from the portion of the atmosphere we wish to examine. For example, it may not be convenient or economical to monitor the emissions from a smokestack at the smokestack. It may be more feasible, especially from an inspection viewpoint, to use optical means 100–500 yd from the smokestack to estimate the concentration of pollutants in the plume. It must be understood that a remote measurement is not necessarily a space- or time-averaged measurement but is a separation of the analytical instrument from the analytical sample over a long distance because of a convenience, jurisdictional, or inspection problem. It is clear that the most frequent application for remote measurements will be for high concentration of pollutants such as in stack emissions. We will discuss instruments capable of making these kinds of measurements below.

In the *long-path* mode, a measurement of a pollutant is made, usually by optical means, over a long sample distance and thus is a space-averaged measurement; it can be done in real time or be time-averaged also. This method of measurement is frequently of value in estimating the ambient background of a pollutant in the countryside or of estimating the overall air quality of an urban area without the necessity for making many point measurements in order to obtain an overall picture of the air quality.

The term *in situ* is sometimes used to refer to measurements that do not involve removal of the sample from the point of origin. For example, if we are interested in estimating the SO_2 content of a stack emission precisely at the stack outlet, we may do so in a number of ways. We can have an instrument right at the outlet or we can send a signal from some distance away and note changes in the return signal. These are in situ measurements. Another method is to remove a sample at the outlet and send it by pipe to an instrument at the base of the stack. This is a *sampled*, not an in situ, measurement.

The significance of point, remote, and long-path measurements in the general scheme of air-pollution analysis may be clarified somewhat by considering a taxonomy of air-pollution instrumentation as in Figure 1.

Sampling and Sample-Handling Problems

As in the case of other analytical determinations, the significance of air-pollution analysis results is heavily dependent on the nature and treatment of the sample presented for analysis. In the ordinary analytical situation, which deals with solids and liquids, we are greatly concerned with obtaining a "representative" sample. Since we cannot, in many cases, analyze the whole sample (e.g., a carload of coal), we select a portion of it that "represents" the whole. More concretely, if we are interested in the composition averaged over the whole of a massive sample, the smaller sample taken for analysis is said to be representative if it has that averaged composition. In the usual case, we do not know the averaged composition (that

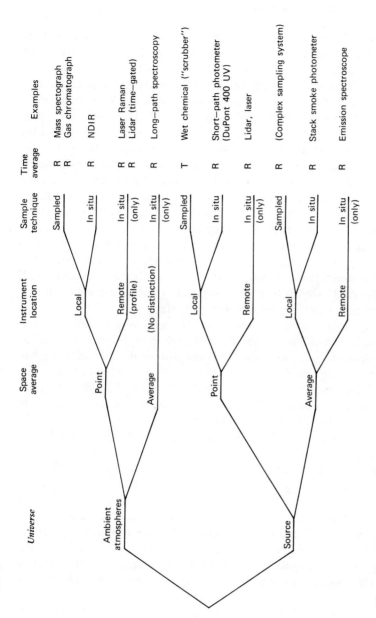

Figure 1 Taxonomy of air-pollution instrumentation: R signifies real time, and T, time-averaged.

is the purpose of doing the analysis), but we wish to be assured that the sample taken for analysis has that averaged composition, whatever it may be. In the language of statistics, we wish to insure that the sample chosen for analysis is a random sample of the entire universe or population of interest. A random sample is obtained in such a way that each value in the population has an equal chance of being selected. The result of a truly random selection of portions of the whole population for inclusion in the sample is a sample that represents the whole population (approaches the averaged composition of the whole population). For this reason, fairly elaborate schemes have been developed for obtaining representative samples of coal, metallic ores, milk, ice cream, and so on. In addition, a statistical theory has been developed to support and justify many sampling schemes. The test of any practical or theoretical sampling scheme lies in experiment; that is, a series of "representative" samples when subjected to analysis must give concordance of results within the desired precision and accuracy. The assumption in this test is that the analytical procedure itself has been previously shown to have at least the same or a higher order of precision and accuracy than the sampling procedure.

Frequently, it is desired to estimate the composition, not of some massive sample, but of a small well-defined portion of the universe that has special significance. Some examples are the nature of dirt particles in a ball bearing, the composition of corrosion products on an electrical contact, and particulate contamination in a drug or foodstuff. We find similar situations in the air-pollution field, especially in specifying the composition of particulate matter present in the atmosphere.

In the field of air-pollution measurements we are most frequently dealing with the sample of special significance, but within this framework, sampling problems involving "representativeness" arise. For example, in analysis of emission from a stack or an automobile exhaust, we are interested in the representative composition of the emission even though only a small portion of the total emission is withdrawn for analysis. This is a special problem with particulate analysis of stack emissions, where, in order to obtain a representative sample, the latter must be taken isokinetically so that segregation by particle size does not occur during sampling. The concept of isokinetic sampling is discussed in more detail in Chapter 11.

There is another problem in air-pollution analysis that is of special importance as compared to ordinary analysis: sample deterioration due to nonequilibrium conditions of the air sample itself and untoward effects of the analytical procedure itself on this metastable condition. From a

10.1. System Requirements of Pollutant Monitors

thermodynamic viewpoint, sulfur dioxide, nitric oxide, nitrogen dioxide, and ozone are not compatible with each other. Yet at sufficiently low concentrations these materials can coexist and persist in an air mass for long periods of time. The object of an analytical measurement is to estimate the concentration of the component of interest in the sample as it is at the time of the analysis. Great care must be taken that the sampling technique and its associated hardware do not change the amount of the component of interest sufficiently to create a significant disparity between the "true" composition at the sample inlet and the final analytical readout. Some examples of this kind of problem will illustrate the point:

1. Ozone is frequently decomposed by catalytic interaction at the surface of sampling lines. Thus, designers of instrumentation should choose materials of construction that minimize catalytic decomposition of ozone and make the sampling lines as short as possible.
2. Sulfur dioxide is strongly absorbed on many kinds of surfaces, especially metallic surfaces. The moisture content of the air sample will strongly affect this absorption.
3. In the neutral iodide method for ozone determination, nitrogen dioxide will be counted as ozone, while sulfur dioxide will cause low results.

The above discussion has referred primarily to ambient atmosphere analysis at a point. The usual situation for this type of analysis is that a pump draws a sample of the ambient into the instrument where it undergoes estimation with or without chemical modification. In the case of long-path or remote type of analysis, there is usually no modification, because the analytic means involves optical interaction with the molecules of interest with low probability of chemical reaction. Optical analysis in the IR region involves sufficiently low energy radiation that changes in chemical composition may be ignored. Radiation in the UV region is of sufficient energy that chemical changes may be initiated; the analytical method should be tested to insure that any reaction initiated by UV that may occur does not affect the results significantly. The removal of UV radiation may also lead to errors, as indicated by Butcher and Ruff.[1] The concentrations of NO, and NO_2, and O_3 in air depend on solar radiation intensity. When the sample enters an analytical instrument, it is no longer irradiated. The overall effect is to reduce O_3 and NO concentrations and increase NO_2, while the sum ($NO + NO_2$) remains unchanged. The relative errors will be large at low concentrations.

10.1.2. Analytical Considerations

Range of Measurements

Measurements of pollutants in air involve essentially four major classes of samples:

1. *Background.* The atmosphere unaffected by the activities of mankind
2. *Ambient.* The atmosphere in cities and other population centers as influenced by the activities of men
3. *Stack effluents.* Gaseous effluents from industrial operations such as combustion of fossil fuels and other processes
4. *Automobile exhausts.* Gaseous effluents from automobiles.

Typical data on these classes of samples for the pollutants of major interest have been given in Table I. We noted that from an analytical viewpoint the range of over a millionfold will give rise to severe analytical problems and that it is probably unreasonable to expect any one instrument or one technique to be adequate for all applications. Thus, for background analysis, concentration and trapping techniques involving long averaging times and large volumes will be required and will give adequate data because the background concentrations are not time- or space-dependent. For city ambient, long averaging times do not give adequate information about dangerous peaks (see Section 10.1.3). Therefore, instruments of great sensitivity and rapid response are required. These are the instruments that need the greatest improvement. Finally, in the determination of pollutants in combustion sources, the analytical problem is mainly one of sample treatment for high moisture and high particulate content. The quantitation problem is relatively easy once the sample has been suitably prepared.

Accuracy and Reproducibility

In the measurement of any physical quantity three major components are involved: the system on which the measurement is made, the instruments used, and the human observer.[2] Since no measurement is free of error, it follows that the source of the errors resides in the components brought together to create the measurement and their manner of interaction. We have already discussed one aspect of sample system definition above—namely, sampling errors and sample treatment, which in chemical determinations are frequently the largest source of error. Scientists are aware of observer bias and the long history of efforts to minimize this source of error. Instrumental sources of error arise from poor calibration, random fluctuations in instrument behavior, and deterioration in instrument performance with time. These will be dealt with briefly below, where instruments and techniques are described in more detail. In

10.1. System Requirements of Pollutant Monitors

this section, we confine ourselves to a brief consideration of the concepts related to errors in measurements. For a fuller treatment of errors and statistical treatment of data, standard texts in the field should be consulted.[3-6]

Following Sandell,[2] we define the error of a measured value as the numerical difference between it and the true value:

$$E = M - T, \qquad (1)$$

where E is the absolute error in M, in the same units as M; M is the measured or observed quantity; and T is the true value of the quantity being measured. If the quantity is truly an experimental one, then T can never be known exactly. The best that can be done is to estimate T between sufficiently narrow limits using various techniques and applying appropriate statistical treatment to the different estimates of T. Then, if the uncertainty in T is small compared to E, we can, as a practical matter, speak of the error in the measurement M, using T as the standard.

With these considerations as a basis, we are in a position to discuss accuracy and precision. The error is an inverse (not reciprocal) measure of the accuracy: the smaller the error, the greater the accuracy. While the error is a measure of the discrepancy of M from T, the accuracy is a measure of the concordance between M and T. We can appreciate this more easily by considering the relative error, $(E/T) \times 100$, which expresses the error in M as a percentage of T. The accuracy, A, of M may now be expressed as

$$A(M) = 100 - 100\frac{E}{T} = 100\left(1 - \frac{E}{T}\right). \qquad (2)$$

It is seen that as $E \to 0$, $A(M) \to 100$. A value of $A(M) = 100$ would express complete concordance, or 100% accuracy. We may express the error, E, in the last equation in absolute terms without regard to sign:

$$A(M) = 100\left(1 - \frac{|E|}{T}\right). \qquad (3)$$

This expression gives a numerical measure of the accuracy. Values of $A(M)$ then range from 100 on down, with 100 representing the highest accuracy attainable and values less than 100, lesser values of accuracy.

By precision, we mean the concordance of a series of measurements among themselves. While we can express the accuracy of a single measurement, we need a number of values of M in order to evaluate the precision (reproducibility, repeatability). We speak of high precision when the measured values agree with each other closely; we speak of low precision when differences between the measured values are large. The precision is evaluated by calculating the mean of all the measurements and then the deviation of each measurement from the mean. There is an inverse relation between the size of the deviations and the precision: the smaller the deviations, the higher the precision. The mean of a series of measurements is one indicator of the central tendency of the series, while the deviations are used to estimate the dispersion of the series. Using the same approach as above, we may write for the precision of a single measurement, $P(M)$,

$$P(M) = 100\left(1 - \frac{|D|}{\bar{M}}\right), \qquad (4)$$

where
$$|D| = |M - \bar{M}| = \text{deviation from the mean without regard to sign,} \qquad (5)$$

and
$$\bar{M} = \text{mean of the measurements, } M$$

$$= \frac{\sum_{i=1}^{n} M_i}{n} \qquad (6)$$

where n is the number of measurements. Here again, as $|D| \to 0$, $P(M) \to 100$, or perfect precision. More usually, the precision is expressed as a deviation around the mean.

The Advisory Board of Analytical Chemistry recommends the use of confidence limits for a measure of the precision of a mean:[7]

$$\text{Confidence limits of the mean} = \bar{M} \pm \frac{ts}{\sqrt{n}}, \qquad (7)$$

where t is the t-table value at the stated confidence level and s is the standard deviation of the group of measurements represented by $\bar{M}$. The confidence limits of the mean, $M - (ts/\sqrt{n})$ and $M + (ts/\sqrt{n})$ "are the limits around the measured mean within which the mean value for an infinite number of measurements can be expected to be found with the stated level of probability."[7] If the chosen confidence level is 95% (probability = 19/20), this means that the "true" mean ($\bar{M}$, as $n \to \infty$) has 19 chances out of 20 of being within the confidence limits calculated on the basis of n measurements and 1 chance out of 20 that it will be outside these limits. The calculation of s and tables of t at various confidence levels are given in Refs. 3–6.

The most useful measure of precision is the standard deviation of the results expressed in absolute terms (s) or more frequently as a percentage of the mean [s(rel)]. It is fairly clear that the concepts of accuracy and reproducibility are intertwined. However, as indicated above, they are dis-

10.1. System Requirements of Pollutant Monitors

tinct and tell much about the usefulness of an analytical method or instrument. We shall illustrate with examples.

Example A. A sample is known to contain 100.0 units of X. A series of analyses gives the following results:

	X, units	Deviation from Mean
	98.8	+0.1
	99.0	+0.3
	98.5	−0.2
	98.5	−0.2
	98.7	0.0
Σ	493.5	0.0
$\bar{M}$	98.7	—
s	—	0.2
s(rel)	—	0.2%

The method is said to be accurate to within 1.3% with a systematic bias on the low side; the reproducibility is 0.2% at the level of 100 units. Expressing this result in accordance with the previous discussion, we may say that the accuracy is 98.7% and that the 95% confidence limits are 98.7 ± 0.3.

Example B. The sample in Example A is analyzed by another method and gives these results:

	X, units	Deviation from Mean
	101.0	1.0
	99.5	−0.5
	99.0	−1.0
	100.5	+0.5
	100.0	0
Σ	500.0	0
$\bar{M}$	100.0	—
s		0.8
s(rel)		0.8%

In this case there is no systematic bias in the method, but the reproducibility is quite poor, compared to Example A. The 95% confidence limits are 100.0 ± 1.0. In this case, although we do not have sufficient confidence in any single measurement and must resort to a number of measurements to obtain reliable information, the accuracy is better than the reproducibility.

We conclude from these examples, that a method may have high reproducibility but be quite inaccurate. The inaccuracy is nearly always caused by some systematic bias in the method. On the other hand, there may be no systematic bias, but the method may be poorly reproducible. For practical purposes, the accuracy of the method is limited by its reproducibility. In nearly all cases we are interested in good reproducibility, since even if we obtain an inaccurate result, the bias may be corrected for if we have confidence that the procedure is repeatable within narrow limits. Poor reproducibility places a great deal of doubt on any one result and means that the method is at least unreliable to the extent of the nonreproducibility. The difference between precision and accuracy is illustrated in Figure 2: the difference between the "true" value and the mean of the measured values is a measure of the accuracy; the dispersion of the measured values around the mean is a measure of the precision.

It is important that all instrumental techniques as well as laboratory methods include procedures for testing the accuracy of the procedure, and this can best be accomplished by the use of calibration standards (see Section 10.4). Previous discussion in this book suggests that an accuracy of 5% is required for adequate system control. This will place more or less severe requirements on the analytical technique, depending on the range as we have previously indicated.

With respect to instrument reproducibility, it is advisable that this be in the neighborhood of 1% of the full scale. The reason for the tighter requirement on reproducibility is to insure that the entire analytical system, sample, sampling system, analytical module, and readout give a result within the accuracy limit required. If the reproducibility limit is relaxed we get into the difficulty illustrated by the above examples.

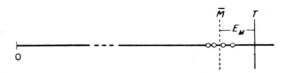

Figure 2 Precision and accuracy: The arithmetic mean of 4 measured values (circles) is $\bar{M}$; the true value is T. Deviations from $\bar{M}$ are a measure of the reproducibility (i.e., precision.) The error of the mean is E_M. A number of other measurements would give a new mean, differing somewhat from $\bar{M}$. The greater the number, the more closely the mean would approach the true mean, which is free from indeterminate errors. All the values shown are low because of one or more determinate errors, the sum of which has the value $\bar{M}_\infty - T$, where $\bar{M}_\infty$ results from a very large number of measurements. E_M is a rough approximation of the determinate error.[2]

Readouts and Reporting Forms

In the case of manual laboratory procedures, the readouts will be in the form of handwritten or typewritten reports giving the analytical data in the most convenient form. Until recently, the reporting form has been in terms of parts per million of the gaseous pollutant in the atmosphere. This, of course, is a form independent of temperature and gives the moles per million moles or the volume per million volumes of the pollutant in the atmosphere, regardless of changes in temperature and pressure. This is probably the most convenient and useful reporting form and is most consistent with effects of the pollutant on health and safety. More recently, the EPA has suggested micrograms per cubic meter as the reporting form of choice. For particulate matter this form is not temperature- or pressure-dependent, but for gaseous components it is. The conversion of one to the other may be accomplished by means of the following formulas for gases or vapors whose molecular weight is known, recalled from Section 1.4.3:

$$\text{microgram per cubic meter} = \frac{PM}{RT} \times 1000 \times (\text{ppm}) \quad (8)$$

$$\text{ppm} = \frac{RT}{PM} \times \frac{1}{1000} \times (\text{microgram per cubic meter}). \quad (9)$$

The EPA has recently defined the standard conditions as 760 mmHg and 21.1°C (70°F). Under these conditions the above formulas reduce to

$$\text{micrograms per cubic meter} = (\text{ppm}) \times M \times 41.4, \quad (10)$$

$$\text{ppm} = \frac{(\text{micrograms per cubic meter}) \times 0.0241}{M}. \quad (11)$$

In Eqs. (8)–(11), P is pressure (atm), M is molecular weight (g/mole), R is the gas constant (0.0821 l-atm/mole-°K), and T is the absolute temperature (°K).

Leithe[8] has given factors for interconversion of these quantities for molecular weights from 1 to 200 and for 0° and 20°C at 760 mmHg (see Section 1.4.3). These will be found useful, since the atmospheric pressure deviates very little from 760 mmHg. It is unfortunate that the EPA has defined a standard temperature different from the usual European or American standard temperatures: 0°, 20°, or 25°C. However, the above formulas will provide means for calculating the factor at any temperature or pressure.

Interfacing and Telemetry

Since systems evaluation and control can only be achieved on the basis of analytical data on the concentrations of pollutants in the atmosphere, it is mandatory that the analytical data be available at some centralized facility for logging, calculation, statistical evaluation, and action along the lines discussed in other portions of this book. For this purpose, it is necessary that the analytical instruments have outputs that can interface suitably with hardware for transmission of the analytical data to the central facility. The transmission of signals by telemetry equipment is also of concern. One efficient way for transmission of unambiguous signals is by digital pulses with suitable safety features built into the pulse code. The problems of signal transmission are discussed in more detail in Section 6.2.3.

10.1.3. Instrument Characteristics

Manual Versus Automatic Techniques

As previously described, much of the data gathered to date has been derived by laboratory analysis of samples of air scrubbed through aqueous solutions. Manual methods thus imply, at the very minimum, time-averaging of the sample as well as errors and inaccuracies inherent in manual methods. With respect to analysis of the components of particulate matter, manual techniques are combined with instrumental methods in order to determine a wide variety of elements and compounds in the particles. These are still classified as manual techniques because of the intervention of analysts in achieving the final analytical result, as contrasted with automatic instruments, which present a result without human intervention. It seems clear that in order to obtain real-time response over the air-pollution system and to predict conditions dangerous to life and health and give adequate warning, automatic analytical instruments will be required that operate in real time or very close to real time.

It is not always true that on-line monitors are to be preferred over manual methods. Lodge[9] has indicated that the choice depends on the motivation for obtaining the data and has discussed the trade-offs involved. The three main motivations mentioned are exploration, case study, and monitoring. In general, automatic instrumentation is not warranted for exploration or case study, because the capital and maintenance costs are high and not justifiable for short-term data acquisition. The use of continuous monitors is probably most easily justified when the purpose is to warn of dangerous concentrations. In other cases, cost–benefit analysis is recommended. At the present time, many of the commercially available instruments display poor design, poor construction, and/or poor performance. Even the best instruments require periodic maintenance, which must be considered part of the total cost. As instrument performance improves, it can be expected that cost–benefit analysis will favor on-line automatic instrumentation in more and more cases.

Field Instrument Specifications

It is useful to list the requirements for the ideal instrument to be used in the field with a minimum of human intervention. This "wish list" is given in Table II. It is obvious that even if we could quantitatively define all of the factors mentioned, no single instrument could meet all of the criteria listed at a reasonable cost. Therefore, it is of value to convert these ideal requirements to quantitative requirements that are presently feasible and that comply with the system requirements that we have been discussing in this volume. In order to define some of these realistic requirements, we discuss the factors listed in Table II in somewhat more detail.

While it may be of advantage to have one instrument to measure all of the pollutants of interest, it is not mandatory that this be so. In some circumstances, it may be of great advantage to have only a single-parameter instrument available. One or two parameters may provide all the information necessary to define adequately the air quality of a region or of a small volume of space. Far more important from an economic and manufacturing point of view is to have instruments that operate on the same principle but that, with slight modification, may be applied for analyses of

Table II

The Ideal Field-Analytical Instrument

1. The instrument shall completely analyze the sample for all components of interest to any desired degree of accuracy and reproducibility.
2. The instrument shall respond to changes in sample composition with no time lag.
3. The instrument shall give information continuously.
4. The process of bringing the sample to the instrument and giving it any necessary premeasurement treatment shall not affect the analytical results with respect to the original sample or subsequent samples.
5. The analytical measuring process shall not affect future samples or the environment.
6. The instrument shall show no drift.
7. The instrument shall be self-calibrating.
8. The instrument shall be simple, easy, and inexpensive to install, operate, and maintain.
9. The instrument shall not be affected in its operation or in the quality of its results by ambient conditions or change in ambient conditions.
10. The output shall be available for a variety of purposes such as recording, alarm, control, telemetry transmission, and computer interfacing.
11. The instrument shall be inexpensive.

the different pollutants. For example, optical instruments may be employed for the determination of sulfur dioxide, nitric oxide, and nitrogen dioxide by slight modification of the source, of the filter, or of the detector, while the basic instrument remains unchanged. In a similar way, a gas-chromatographic instrument may be modified to analyze for different constituents in the atmosphere. Multicomponent instruments tend to be more complex, more expensive, and more difficult to set up and maintain than single component instruments, although this is not true in all cases. Thus, the multicomponent requirement is frequently in conflict with the last item on the list—that is, expense—and a suitable trade-off must be made depending on the situation. In the usual case, it is advisable to sacrifice multicomponent operation and complexity in favor of ruggedness, reliability, easy maintainability, and long-term uninterrupted operation.

No real instrument responds instantaneously to changes in sample composition, if only because it takes time to bring the new sample to the analytical device and because of the finite time it takes to operate on and measure the sample. Even in the case of optical instruments that respond to changes almost instantaneously, there will be a lag due to flushing the sample lines and the optical cell with the new sample. The best that one can hope to do is to minimize this lag, but one cannot eliminate it all together. As indicated in the previous discussion we would want the total instrument response to be of the order of 2 min or less when a step-change in composition is presented to the instrument. We discuss dynamic behavior in somewhat more detail below.

The need for continuous information on a real-time basis has been discussed in a number of places in this text. While this is not a universal requirement, especially in the rural, or "background," type of environments, it is mandatory in those cases where the pollution load is high and where the probability is high of exceeding legislated limits imposed because of health or safety. In this situation, a short response time on the part of the sysem for abatement or alarm purposes is required. A continuous analysis with very a short response time gives the maximum of information about the air quality of a particular environment. One need not use all of this information, but one is on the safe side if it is available. Of course, this must be traded off with respect to the cost of obtaining information of this type.

Point 4 in the list is self-explanatory and need not detain us any further here, except to point out that this is the sine qua non of all the sampling devices. As with the measuring technique itself, nothing in the sampling technique or in the measuring technique shall affect the significance of the results or the quality of the samples that will be presented to the analyzer in the future. We have discussed sampling requirements in greater detail in Section 10.1.1.

10.1. System Requirements of Pollutant Monitors

A drift-free instrument is very hard to come by and is probably not worthwhile to achieve economically. As a matter of fact, it may be worthwhile to combine low cost with a certain amount of drift along with a self-calibrating feature, which will bring the instrument scale within the limits of precision required for the analysis. If a calibration is done frequently enough, then indeed the instrument may show rather high drifts as long as means are provided to bring the instrument back to scale within the drift deviations normally presented to the instrument. In view of the very low precision and accuracy of many of the present-day analytical procedures, a requirement for self-calibration should be imposed on all instruments used for monitoring air pollutants and estimating air quality. If the calibration can be done rapidly enough, then it should be done often enough within the limits required for continuous information, to insure that the instrument is always within calibration.

There are a number of trade-offs involved in point 8. The ideal instrument, of course, requires no maintenance, but in order to approach this ideal, costs become prohibitive. It would appear to be a more economical instrument design philosophy to build a certain amount of routine maintenance into the instrument in the interest of lowering of the manufacturing cost and improving reliability of operation. Thus, an instrument that requires weekly maintenance would cost less than one requiring monthly maintenance, but all instruments would be designed for routine maintenance in order to push the operating time well over 99% and increase the reliability of the analytical data. Cost studies that show how effective this approach is are not readily available, but experience in the instrumentation industry indicates that costs to build a maintenance-free and 100% up-time instrument are prohibitive. At the same time, experience in the petroleum-refining industry demonstrates that scheduled continuing maintenance can achieve the on-time goal of over 99%.

The requirement for ruggedness and reliability in the face of environmental changes in the field is, of course, quite obvious and is the chief bugaboo of the instrument designer of field equipment. Here again, it may be of advantage from an overall cost and economic viewpoint to build highly reliable and inexpensive analyzers designed to operate in an indoor environment and then recreate this environment in the field by use of housings, constant-temperature ovens, and so on. It would appear that this approach would result in an overall increase in operating reliability and minimize manpower input.

The problem of providing outputs for a variety of purposes has been solved, and should new requirements become important, the problems would appear to be relatively easy to solve. There are a wide variety of digital to analogue and analogue to digital converters, amplifiers, signal conditioners, and boosters, so that any reasonable signal available from an analytical instrument can be put into a form that is useful for any particular purpose.

Expense is always a relative term and bears directly on the philosophy we have been discussing in this book. Our viewpoint is that effectiveness of hardware has to be considered in the light of the results desired from the use of the hardware. When a particular result must be achieved, then cost considerations take second place. We see this philosophy put into practice especially in the development and use of military hardware. With respect to air-pollution problems, the trade-off is between an absolutely clean environment at infinite expense versus a contaminated environment whose level of contamination will be below the threshold that causes economic loss, illness, or other loss to society. In addition, we must consider the total number of instruments that will have to be employed in the network in order to get adequate data from a systems viewpoint. From this point of view it would be of great advantage to have analytical instruments that meet all the other criteria that we have been discussing and cost about $500. This, of course, has not been achieved to date, most instruments costing $2000 or more, especially the more complicated optical correlation and absorption instruments.

Dynamic Behavior: Real-Time Versus Integrating Instruments

In many instruments, instantaneous or very short response-time measurements are obtained; other devices provide an integrated measurement over various lengths of time. Instantaneous recording instruments and those with very short sampling and response time—say, 2 min or less—produce data from which integrations over any desirable time period may be computed. However, it is not possible to reverse the procedure; that is, it is not possible to recover data about a fixed point in time from averaged or integrated measurements. The averaging time is not significant when one is concerned with the normal constituents of the atmosphere, such as oxygen and nitrogen, since the concentrations of these materials over short periods will deviate very little from the long-time average. However, the pollutants of interest to us are introduced from point or area sources in high concentrations and are slowly diffused into the atmosphere. Thus, at any point in space the concentration of a pollutant of interest will vary quite markedly with time, depending on meteorological conditions. If we are dealing with hazardous pollutants, averaged data, which may fall well below the threshold of danger, are of little use in assessing danger to health or safety in a specific region should high concentrations be developing because of unusual emissions from the source or unusual meteorological conditions. A brief discussion of this point given by Tebbens[10] shows that averaged data completely hides large concentration peaks.

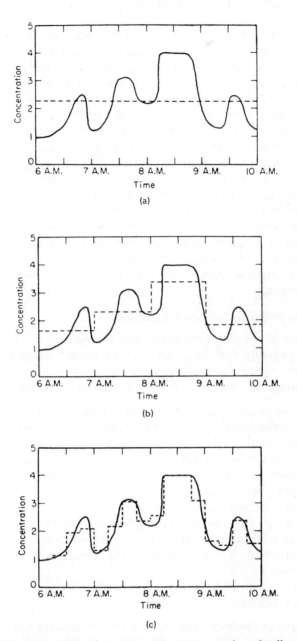

Figure 3 Effect of sampling time on apparent concentration of pollutant compared to actual concentration. Simulated.[23] Solid line is instantaneous concentration; dashed line is time-averaged concentration. (a) 4-h sampling period, (b) 1-h sampling period, (c) 15-min sampling period (from *Air Pollution,* Vol. 2, Pt. 6, 2nd ed., Academic Press, 1968).

The idea is strikingly illustrated in Figure 3,[23] which shows how increasing the sampling time smears out detail and loses large concentration peaks. Conversely, as the sampling time is reduced, the apparent concentration may be made to approach actual concentrations more and more closely. In the next section, however, we discuss difficulties that may arise when we are dealing with short-lived peaks (i.e., peaks lasting only a few minutes).

We can conclude that in an urban environment, we would wish to have instantaneous data, or data as close to instantaneous as possible, on all the pollutants of interest that contribute to the total air quality of the region. At the very least, we would want to match the sampling and instrument response times to the duration and intensity of peaks of pollutants found to be hazardous to health and safety. This might require an intensive research investigation so that automatic instrumentation most nearly matching real-world requirements could be designed for most economical monitoring. The alternative is the creation of instrumentation with extremely short sampling and response times (1 min and less). The latter might very well be unduly expensive.

Real-Time Response and Legal Requirements

No instrument responds instantaneously to changes taking place in the real world. In addition to the "dead time" involved in bringing a changing sample from outside the instrument to the point within the instrument where it is measured, there are a number of other instrumental delays:

1. Purging the analytical part of the instrument with new sample. Complete replacement of old sample with new is time-dependent and usually exponential in character. Thus, a sudden change in pollutant level will not be recorded when the change takes place. The output will show a gradual increase to the new level after a delay, characteristic of the instrument.
2. For wet methods there is chemical-reaction time, involving addition of reagents, mixing, color development time, and the like.
3. For optical methods, the optical measurements themselves are extremely rapid, but purging of the optical cell takes time, as described.
4. Read-out delays due to dynamic behavior of sensors, recorders, amplifiers, and signal conditioners.

An example of the importance of dynamic response of real-time instruments (see Section 6.1.4) is given by Schnelle and Neeley.[11,12] In the San Francisco Bay Area Air Pollution Control District (BAAPCD), a relatively high level of 0.5 ppm of SO_2 is permitted, but only for a maximum of 3 min in any one 24-h period. The question is: If the SO_2 level were to go from

10.1. System Requirements of Pollutant Monitors

below this limit to just above 0.5 ppm for, say, 3.1 min, would the instruments presently available record this event and thus signal a violation?

To respond properly in real time so as to give information related to the time and precision of the legal regulations described above, instruments must not only be accurately and precisely calibrated, but they must have sufficiently rapid transient response to follow rapid fluctuations in ambient concentration of the components of interest (COI). For example, if it takes 5 min for the instrument to respond to a change that lasts only 3 min, then the change will not be seen either in its true magnitude or in its true time duration (Figure 4). A smeared-out signal shorter in duration than 3 min and significantly less than the maximum concentration will result. This type of instrumental response is described by Schnelle and Neeley.[11, 12]

They set up an apparatus that enabled them to make a rapid switch between two streams of known, but different, concentration feeding into an instrument. Permeation (Section 10.4) was used to establish known SO_2 concentrations, while CO of known concentrations was obtained from carefully analyzed cylinders. A pulse of known duration was created by switching from one stream to the other and then back again after a measured time; a square pulse was assumed. The instruments were calibrated with known concentrations of the pollutant in air flowing under steady-state conditions. The steady-state instrument responses were found to be linear with concentration. From response of the analytical instrument to step changes as well as to pulsed concentration changes, it was possible to calculate the instrument time constants, including lag time, using well-known techniques,[13-17] and to arrive at a transfer function that fit instrument behavior quite closely. The most appropriate was found to have the form of a second-order transfer function with a lag time:

$$G(S) = \frac{e^{-Ls}}{(T_1 s + 1)(T_2 s + 1)} \qquad (12)$$

(from Chapter 6, Eq. 6), where $G(S)$ is the transfer function, s is the Laplace transform variable, L is dead time, and T_1, T_2 are time constants.

The results for a number of air pollution monitors were given in Table IV of Chapter 6. An analogue simulation study was then conducted to illustrate the response of a typical SO_2 monitor to a violation of the San Francisco BAAPCD regulation mentioned above, using the time constants $T_1 = 5$ sec and $T_2 = 50$ sec. Dead time was not used, because it does not affect the shape of the response curve on open loop test (see again Section 6.1.4) but just displaces the curve on the time axis. The result is shown in Figure 4. It can be seen that the instrument failed to indicate a violation of the regulation. Thus, either the regulations concerning the time duration of a pollutant concentration must be rewritten to conform to the behavior of

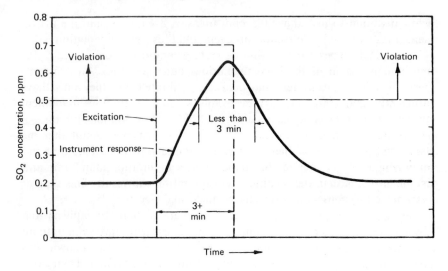

Figure 4 Instrument response to violation of BAAPCD regulation for SO_2. Simulated.[11] Dashed line: violation; solid line: instrument response (Courtesy of J. Air. Pollut. Cont. Assoc.)

instruments presently available or the regulation must include a requirement on the instrument to be used, specifying the dynamic response including the time constants.

The table showed that L, the dead time, is significantly lowered when the flow rate is increased. This is consistent with observations that reducing the volume swept out by the gas sample and absorbing liquids, as well as increasing the sample flowrate, produces a more rapid response. Large holding time and poor mixing are caused by large analytical cell volumes and low flowrates. These observations show the way to improved instrument design from the point of view of dynamic behavior. Small cells and large flowrates, however, also place an additional burden on detector sensitivity and response characteristics.

Mage and Noghrey,[18] using the same approach as Schnelle and Neeley to determine the constants for the transfer function of an analytical instrument, have gone one step further. Writing the transfer function in finite-difference form they employed the convolution integral in order to obtain the input function (true concentration of pollutant as a function of real time) from the output function (recorded concentration as a function of real time that includes the sum of the lag time and the time constants for

10.1. System Requirements of Pollutant Monitors

instrument response). To quote these authors:

"Once the transfer function is obtained, one can write a computer program which will give the output of the instrument for any arbitrary input. By doing so, it is then possible to obtain a feel for the true nature of the input by examination of the recorded output of the instrument. When the data acquisition system is part of an air pollution alert system for a community, a computerized data transmission system can make these corrections which will aid the calling of alerts at the earliest possible moment."

Thus, where air-quality standards are based on the maximum time that an air pollutant can exceed a level deemed injurious to health, safety, or property, the techniques discussed by these authors can be used to determine true pollutant concentrations. This information will not be available in real time, because one must wait for the instrument to record the output information before the latter can be operated on to compute the true input. This delay will be of the order of magnitude of the sum of the lag time plus the instrument time constants. Although not ideal, this is better than relying on false information, as indicated by Schnelle and Neeley.[11,12] In addition, since the standard has been set with a factor of safety in it, sufficient warning time should be available. In those cases

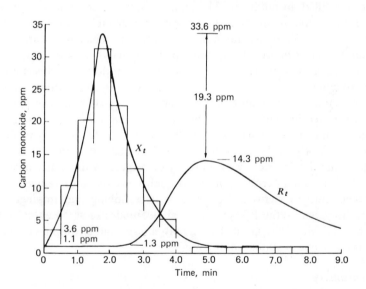

Figure 5 Re-creation of true input concentration, $X(t)$, from instrument output curve, $R(t)$.[18] (Courtesy of J. Air Pollut. Cont. Assoc.)

where warning time is critical, it may be necessary to revise standards downward in order to take into account delays and distortions due to the dynamic behavior of analytical instruments. Figure 5 illustrates the results of the method of Mage and Noghrey[18] in converting instrument output, $R(t)$, to the true input function, $X(t)$. There are two striking things to note in this figure. The first is that the true peak concentration is 2.3 times the recorder concentration, confirming the results of Schnelle and Neeley. The second is that there are substantial time delays in awareness of the true situation. First, the recorded peak lags the true peak by 3 min. Second, even if the computer calculation is instantaneous, knowledge of when the true peak occurred and its value is delayed by 5–7 min.

10.2. INSTRUMENTS AND METHODS REQUIRING SEPARATION BEFORE MEASUREMENT

10.2.1. Aqueous Scrubbing Followed by Classical Wet Analysis

Impingers

The classical technique for analysis of pollutants in the atmosphere involves passing a known volume of air through an aqueous solution in an apparatus called an impinger. The aqueous solution contains reagents that absorb the compound of interest from the atmosphere. After passing the required volume of sample, the aqueous solution is subjected to analysis for the component of interest by various standard analytical procedures. These will be described in more detail in succeeding sections. At this point it is useful to point out that these classical procedures, even though they are time-consuming and require large input of trained manpower, are still being employed in many areas. A number of them have been automated as in the Technicon Autoanalyzer.[19,20] Standards and calibration techniques issued as recently as 1971 by the EPA[21,22] are based on these techniques. Much of the data presently available with respect to the concentration of air pollutants and determination of air quality are based on aqueous scrubbing techniques. Extensive descriptions of scrubbing and impinger techniques are given by the EPA[21,22] as well as Hendrickson,[23] Leithe,[8] Katz,[24] and the ASTM.[25,26] A number of supply houses make equipment required for this technique available commercially.[27-31]

Colorimetry

In a colorimetric analysis, a solution containing a component to be analyzed is treated with a reagent so as to produce a colored substance. The amount of colored material is proportional to the amount of the

10.2. Instruments and Methods Requiring Separation Before Measurement

component of interest originally present; the former is measured by a number of different techniques. In the classical colorimetric technique, the intensity of the unknown color is compared by eye with solutions of known concentration of the colored material in a colorimeter.[32] More recently, other optical techniques have been used that are less dependent on human subjective evaluation. Advantage is taken of Beer's law to obtain more accurate and reproducible results without human intervention and forms the basis for most automatic instrumentation in which a colored material is measured in solution.

Colorimetric monitoring instruments are in widespread use for the continuous measurement of sulfur dioxide in the ambient atmosphere. The most widely used of the colorimetric methods is the West–Gaeke procedure.[33] The method is applicable to sulfur dioxide concentrations in the range 0.005–5 ppm. The air to be analyzed is passed through an aqueous scrubbing solution of $0.1M$ sodium tetrachloromercurate. The sulfur dioxide in the atmosphere quantitatively reacts with the mercuric salt to form a nonvolatile, stable, sulfatomercuric compound. After the requisite amount of air has been passed through the scrubbing solution, the latter is treated with acid-bleached pararosaniline and formaldehyde to form a red-purple color because of the sulfur dioxide. The amount of the color is determined with a spectrophotometer or may be measured by eye with a colorimeter. The color system obeys Beer's law up to about 0.6 μl SO_2/ml of absorbing solution. The method is not subject to interference from other acidic or basic gases or solids such as SO_3, H_2SO_4, NH_3, or CaO. It is recommended, however, that the color be developed within a week after sample collection and that the concentrations of ozone and NO_2 in the sample should be less than that of SO_2. The chemical apparatus and detailed analytical procedure are described in the literature.[33, 34] While originally designed for laboratory manual analysis, the West–Gaeke procedure has also been modified for continuous automatic analysis in commercial instrumentation. The countercurrent principle of absorption of SO_2 from the atmosphere into the aqueous solution is used and the aqueous solution continually analyzed for dissolved SO_2 by the colorimetric procedure described above.[19, 20]

Sulfur dioxide may also be determined colorimetrically by measuring the decolorization of an aqueous starch–iodine solution, which has a deep blue color. On passing the sample containing SO_2 through a scrubber containing starch–iodine solution, the iodine is reduced by the sulfur dioxide, and thus, a decrease in the intensity of the starch–iodine color results. A number of other colorimetric methods for sulfur dioxide have been described in the literature, but they are not much in use. Among these, we may mention the fuchsin–formaldehyde method, the Stratman method, and the barium chloranilate method. In the fuchsin–formaldehyde method, SO_2 is scrubbed from the air sample using a solution of $0.1N$ sodium hydroxide containing 5% glycerol. If SO_2 is present, a red-violet color is developed on addition of a reagent containing basic fuchsin, sulfuric acid, and formaldehyde. The absorption maximum is at 570 nm. In the Stratman method SO_2 is absorbed from the atmosphere on silica gel and then reduced with hydrogen to H_2S on a platinum contact catalyst at 700–900°C. The H_2S is passed into a solution containing 2% ammonium molybdate in $0.4N$ sulfuric acid; a blue-violet molybdenum complex is formed, which is estimated colorimetrically. Finally, the barium chloranilate method involves the reaction of solid barium chloranilate with sulfate ion at pH 4 in 50% ethyl alcohol. The highly colored acid-chloranilate ion is liberated; it has an absorption peak at 530 nm. The SO_2 in the atmosphere is caught and converted to sulfate with a 0.5% aqueous H_2O_2 solution in a scrubber. Details of these methods may be found in the literature.[24, 34]

The most widely used colorimetric method for nitrogen dioxide (NO_2) is the so-called Saltzman method.[38] This method is based on the Griess–Ilosvay reaction, in which a pink-colored dye complex is formed between sulfanilic acid, nitrite ion, and α-naphthylamine in an acid medium. Saltzman substituted N-(1-naphthyl)ethylenediamine for the α-naphthylamine used by Griess. The absorbing solution consists of the sulfanilic acid and the diamine dissolved in an acetic acid medium. Nitric oxide may be determined by the same procedure if the sample is first passed through an acid permanganate bubbler, which oxidizes the NO to NO_2. Nitric oxide is then calculated as the difference between total NO_x and the nitrogen dioxide. The colorimetric aspects of the procedure are very similar to those described for SO_2. Simultaneous determination of NO and NO_2 may be made by dividing the same sample of air into two portions: one portion is passed into the color-developing reagent directly and one is passed through permanganate before color is developed.

As in most absorption methods, there is a question of absorption efficiency and this question has not been definitely settled. In addition, there is also a question about the stoichiometric factor to be used in calculating the percentage of nitrogen dioxide in a sample from the color developed by a sodium nitrite standard. Lyshkow[39] has described a modified Griess–Saltzman reagent that results in more-rapid color development and is a more efficient absorber of NO_2. The Saltzman method is sensitive in the parts per million range.

A method for NO_x determination that has also been widely used and was originally recommended by the EPA[21, 22] is the Jacobs–Hochheiser procedure.[40, 41] In this method the air sample is drawn through a $0.1N$ sodium hydroxide scrubber. NO_2 in the sample reacts in the alkali solution to

10.2. Instruments and Methods Requiring Separation Before Measurement

produce sodium nitrate ($NaNO_3$) and sodium nitrite ($NaNO_2$). The latter is determined by using the classical diazotization and coupling procedure to form a deeply colored azo dye. The intensity of the color is measured colorimetrically using standards made up with pure sodium nitrite. The color-forming reagent consists of phosphoric acid, sulfanilamide, and N-1-(naphthyl)ethylenediamine dihydrochloride. Problems arise with this procedure also, because of poor collection efficiency, variable stoichiometry, and poor correlation with the Saltzman procedure. More recently, it has been found that collection efficiency, assumed to be 35%, varies with concentration,[42] falling off with increasing concentration. Thus, if a constant factor of 35% is used, low concentrations are overestimated, leading to conclusions that NO_2 concentrations exceed legal limits when they do not.

As a result of these difficulties with the Jacobs–Hochheiser procedure, the EPA has withdrawn this technique as a reference method and proposed the substitution of one of three alternates, the Saltzman, chemiluminescence, and the arsenite methods.[43] The last is similar to the Jacobs–Hochheiser, except that the air is bubbled through a sodium hydroxide–sodium arsenite solution rather than merely through the hydroxide, to form a stable solution of sodium nitrite. The collection efficiency of the arsenite bubbler is taken as 85%. The range of the analysis is 0.02–2 µg (0.003 to 0.4 ppm with 200 cm³/min sampling rate for 24 h).

For the colorimetric determination of NO_x in stack effluents (5 ppm to several thousand ppm), a frequently used method is the phenoldisulfonic acid procedure.[26] The gas sample is admitted to an evacuated flask containing a solution of H_2O_2 in sulfuric acid. The oxides of nitrogen are converted to nitric acid, and the resultant nitrate ion react with phenoldisulfonic acid. A yellow color is produced, which is measured colorimetrically.

Ozone and oxidants are determined colorimetrically by passing the gas through a solution of neutral, buffered potassium iodide.[44] The solution used is 1% KI buffered at pH 6.8. In the presence of oxidizing agents the iodide is oxidized to iodine giving the solution a pale yellow color whose intensity is measured colorimetrically. Although fritted bubblers are much more efficient absorbers than straight tubes, it is not recommended that they be used for determination of ozone and other oxidants, because on passing through the frit the ozone is decomposed and low results are obtained. In the alkaline KI method, the reagent is 1% KI in $1N$ sodium hydroxide. A stable product is formed that can be stored several days before final determination, in contrast to the neutral KI method, which must be completed within 30–60 min. The color in the alkaline method is developed by addition of phosphoric-sulfamic acid reagent, which liberates the iodine. The phenolphthalein method of Haagen-Smit and Brunelle[34] depends on the development of the characteristic pink color when reduced phenolphthalein is oxidized by hydrogen peroxide or other oxidants in the presence of copper sulfate. The method is subject to many interferences. Both the neutral and alkaline methods have been studied and improved by Saltzman and co-workers[45-47] and form the basis for an automatic instrument commercially available.

Conductimetry

A sample of air containing sulfur dioxide bubbled through a solution 2×10^{-5} normal in sulfuric acid and 3×10^{-3} molar in hydrogen peroxide undergoes a change in electrolytic conductivity because of the formation of sulfuric acid by oxidation of the SO_2:

$$H_2O_2 + SO_2 \rightarrow H_2SO_4 \rightarrow H^+ + HSO_4^-. \tag{13}$$

This change in conductivity is the basis of a number of laboratory methods for determining SO_2 in the air.[48] The principle is also employed in several automatic analytical instruments.[49-53] The advantage of using the above described solution is that the determination of sulfur dioxide is not interfered with by carbon dioxide normally present in the atmosphere. Carbon dioxide may also be determined by this same method. In this case, the scrubbing solution is $0.005N$ sodium hydroxide containing 0.3% of n-butyl alcohol. The reduction in conductivity resulting from the conversion of sodium hydroxide to sodium carbonate is a measure of the carbon dioxide in the air:

$$CO_2 + OH^- \rightarrow HCO_3^-. \tag{14}$$

The hydroxyl ion has an ionic conductivity about 4.5 times that of the bicarbonate ion,[54] so that in the dilute NaOH solution the loss of hydroxyl ions by the above reaction is easy to detect as a loss in conductivity. Other acid gases do not interfere seriously with the method, because they are present in such small amounts compared to carbon dioxide.

Potentiometry

Potentiometric methods involve the insertion of a pair of electrodes into an aqueous solution and the measurement of the voltage between them. The potential of one electrode, the indicator, depends on the concentration of an ion of interest in the solution; the potential of the second electrode, the reference, is constant. The measured EMF follows the Nernst equation,[55] which may be written

$$E \text{ (cell)} = k + b \log A(i), \tag{15}$$

10.2. Instruments and Methods Requiring Separation Before Measurement

where E (cell) is the measured EMF in volts, k is a constant that depends on the nature of the electrodes and the temperature (volts), b is a constant that depends on the temperature and the nature of the ion being measured (volts), and $A(i)$ is the activity of the ion of interest, a function of the concentration of the ion being measured.

This technique is not often employed in measuring pollutants in the atmosphere, because suitable specific electrodes have not yet been developed for the pollutants of major interest. In the case of the determination of fluorides present in the atmosphere, a procedure has been developed in which the gaseous and particulate fluorides are quantitatively collected by filtration and chemisorption on cellulose acetate fibers impregnated with sodium formate.[56] The interference due to the presence of iron and aluminum as well as the pH and ionic strength are controlled with sodium citrate buffer, which is used to extract all the fluoride from the filter. The final measurement is made with the fluoride-specific ion electrode.[57] Chlorides may be determined in a similar manner, using a chloride electrode. Methods have also been developed for the potentiometric measurement of NO_2 and NO in the parts per million range in gaseous mixtures. DiMartini[58] describes a procedure in which NO and NO_2 are converted in the gas phase to N_2O_5 by reaction with ozone. The N_2O_5 is hydrolyzed to nitric acid, and the resultant nitrate ion is measured with the Orion model 92-07 nitrate-specific ion electrode.[57] An apparatus is described that performs these operations automatically and continuously on ambient air. Driscoll et al.,[59] investigated the use of the nitrate electrode for measuring NO_x in stack effluents. The scrubbing solution first used was the same as for the phenol disulphonic acid (PDS) method (i.e., H_2O_2 in H_2SO_4 solution). The sulfate was found to interfere in the nitrate measurement at the 50:1 level. Removal of sulfate by ion-exchange methods and precipitation with barium salts proved to be unsatisfactory; when the scrubbing solution was adjusted to be below the 50:1 level, excellent agreement was obtained with the PDS method.

Titrimetry

Titrimetric methods involve the absorption of a component of interest into an aqueous solution, followed by addition of a suitable aqueous reagent to a well-defined endpoint. The volume of the reagent added is a measure of the component of interest absorbed into the solution. Frequently, the component itself is not titrated, but a reaction is carried out to create another material that can be titrated. We will give examples of these below.

Sulfur dioxide may be determined in the laboratory by scrubbing the atmosphere through an aqueous solution of hydrogen peroxide and then titrating the resultant sulfuric acid with dilute sodium hydroxide standard solution. This method is said to be applicable in the range of 0.01–10 ppm.[35] Obviously great volumes of air must be passed through the scrubbing solution in order to obtain sufficient acid for titration.

Sulfuric acid mist in the atmosphere may be determined by filtering the sample through filter paper, dispersing the paper in neutral distilled water, and titrating the liberated acid with sodium hydroxide.[36]

Carbon monoxide may be determined by the reaction of the air sample with iodine pentoxide (I_2O_5). This reaction may be written

$$5CO + I_2O_5 \rightarrow 5CO_2 + I_2. \tag{16}$$

The reaction is carried out by passing the sample through a column of the solid iodine pentoxide in granular form, keeping the column at about 150°C. The liberated iodine vapor may be caught in a solution and titrated with standard thiosulfate or measured colorimetrically.[37]

10.2.2. Filtration with Cloth or Paper

The classic determination of an air pollutant by this technique is that of hydrogen sulfide using a paper or cloth tape impregnated with lead acetate. This may be set up for laboratory determination. Automatic instruments are also available on the market that employ this same technique. Great sensitivities may be obtained, depending on the amount of sample passed through the tape. It is clear that this is a time-averaged technique and that the time required must be extended, the lower the concentration to be measured. While tape spots are normally measured photometrically in the laboratory hours or days after the spot has been created, some automatic analyzers make measurements every 15 min.[60] The tape samplers designed for tape measurements in the laboratory suffer from the disadvantage that the lead sulfide spot fades appreciably in the presence of air, depending on the time, the temperature, and the amount of light permitted to fall on the spot.[61, 62] In addition, the lead sulfide spot is bleached by ozone, sulfur dioxide, and nitrogen oxides in the concentrations normally present in the ambient.[61, 62] In order to overcome the limitations of lead acetate as a reagent for H_2S determination, Pare[62] investigated the use of mercuric chloride for black spot formation. He found mercuric sulfide spots to be stable in the presence of air and light for at least three days and to be unaffected by O_3, SO_2, and NO_x.

10.2.3. Gas Chromatography

Principle of Operation

The analytical technique of gas chromatography (GC) has achieved wide application in many areas, and it is not surprising that it has been found useful for analysis of air pollutants. It is, therefore, appropriate that this method be discussed in some detail, not only because of its widespread use but because of the possibility of more intensive application in the field of air-pollution analysis. The basic physical phenomenon on which the gas-chromatographic method depends is that of *partition* of a solute between two phases, one of which is stationary and one of which is mobile.[63]

In order to make the idea of partitioning clear, consider Figure 6, which shows a closed vessel containing a liquid phase (say, dibutylphthalate) and a gas phase (for example, air). If an organic vapor like methane is injected into the air space, then the methane will dissolve in the liquid, but some will remain in the gaseous phase until an equilibrium is reached. The equilibrium is described by the partition coefficient, $K(T)$, which depends only on the temperature. $K(T)$ is defined as follows:

$$K(T) = \frac{\text{Concentration in the gas phase}}{\text{Concentration in the liquid phase}} = \frac{c(a)}{c(l)}. \tag{17}$$

If some more of the same solute is injected, then obviously more will dissolve in the liquid phase and the amounts in the liquid and gas phases will then adjust themselves until they satisfy the equilibrium constant relationship indicated in Eq. (17). If another solute is injected into the same space—say, ethane—it too will dissolve in the liquid phase and achieve its own equilibrium between the gas and liquid phases characterized by an identical type of equilibrium constant but of a different value because its solubility in the liquid is different. In the usual case, the solutes of higher molecular weight will tend to be more soluble than those of lower molecular weight.

If, after equilibrium has been achieved for the system just described, the entire gas phase in the first container is removed to another container with the same amount of pure solvent, at the same temperature, then the respective solutes will dissolve in the liquid as before until an equilibrium is achieved that satisfies both equilibrium constant equations. However, because some of the solutes have remained behind, dissolved in the liquid in he first container, the actual amounts of solutes in the gas phase will obviously be less than in the original container. Furthermore, that solute, ethane, which is more soluble in the liquid will be present in a relatively smaller amount in the second container. If this process is repeated with successive containers, the net effect will be a relative enrichment of methane down the series of containers relative to ethane, as compared to their ratio at the start of the experiment.

Let us go back and look at what happens to the first container when its gas phase has been removed and replaced by pure air. Since the equilibrium constant for both solutes must be satisfied, portions of both of the solutes

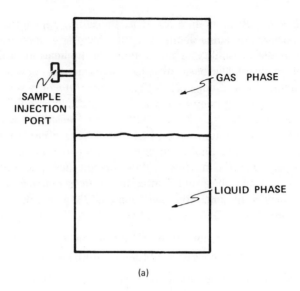

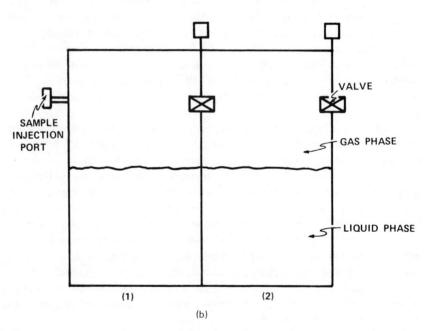

Figure 6 The concept of partitioning: (a) equilibrium partitioning (distribution) of gaseous solute between gas and liquid phases; (b) transfer of equilibrium gas mixture from container 1 to container 2.

10.2. Instruments and Methods Requiring Separation Before Measurement

will leave the liquid phase and reenter the gas phase so as to reestablish the respective equilibria. Thus, if the contents of the gas phase in each container is moved to the next container and if those in the first are replaced with air, ultimately all of the solute will leave the liquid in the first container. The net effect, however, of the entire procedure is to retard selectively the motion of the ethane relative to the methane, so that if the process is continued for a sufficient number of containers, ultimately the two hydrocarbons will be completely separated from each other, or more precisely, the separation can be carried out as far as one pleases, limited only by the sensitivity of the means used to measure the separation.

If a sample contains more than two components, these same principles would apply, and as the gas phase is moved down the series of equilibrium containers, the solutes will move at a speed inversely with their solubility in the liquid phase. In an actual gas-chromatographic apparatus, the containers are replaced by a long tube called a *column*. The liquid is held on solid granular material called the *solid support*, and the transfer of the gas phase is done continuously by means of a carrier gas. If the liquid phase is present in a thin enough layer, then equilibrium is achieved very rapidly and it is possible to carry out separations of multicomponent systems in 5 min to about 1 h. If the column is made extremely small in both diameter and length, it is even possible to carry out a 5–10-component separation in seconds.[64-66] For this to be achieved, the particle size of the solid support must be very small and uniform, and great precautions must be taken to eliminate dead spaces in the apparatus that would permit dilution or mixing after separation has been achieved.

In Figure 7, the basic components of a gas-chromatographic analyzer are indicated schematically. These may be listed as follows:

1. *Carrier-gas source.* The carrier is a gas, usually helium, that transports the solutes in the vapor phase down the column. Helium is chosen because the detector at the end of the column is usually a thermal-conductivity detector, which shows the greatest sensitivity when the thermal conductivities of the components gases are most unlike. Next to hydrogen, helium has the greatest thermal conductivity.
2. *Injection port.* This, in the laboratory, consists of a T in the column covered by a rubber septum. The sample is injected by means of a hypodermic syringe into the carrier-gas stream as rapidly as possible so as to constitute a slug in the carrier-gas stream. In automatic instruments, a special sampling valve accomplishes the same purpose.
3. *Column.* This is a metal or glass tube filled with the supporting ma-

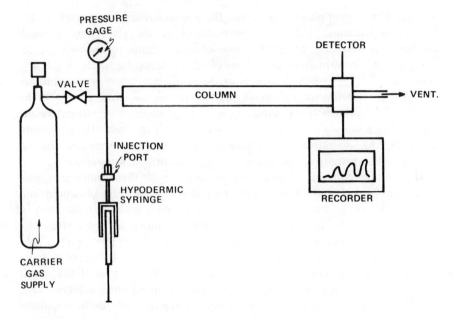

Figure 7 Schematic of a gas chromatograph.

terial coated with a suitable partitioning liquid as described above. In the case of capillary columns, a thin layer of liquid coats the column walls.

4. *Detector.* The sum total of a chromatographic process is to convert a multicomponent system into a series of binary systems, each consisting of one component of the sample and the carrier gas. Since binary systems can be easily quantitated by suitable calibration, it is not necessary that the detector be a specific one for the components leaving the column. The measurement of some general property will suffice to indicate as well as quantitate deviation from purity of the carrier gas. The most widely used chromatographic detector is the thermal-conductivity detector, which can be made extremely sensitive. Other detectors presently in use are the flame-ionization detector, which is also a nonspecific detector and, less frequently, the flame-photometric detector, which can be specifically sensitized for sulfur and phosphorus compounds.

5. *Readout.* The signal from the detector is usually traced on a strip-chart recorder.

10.2. Instruments and Methods Requiring Separation Before Measurement

It is clear from the above discussion that one need not be confined to the use of liquid stationary phases. Solid absorbants may be used that show a differential absorption of the components of a sample. In this case, we speak of absorption rather than partition coefficients. When a liquid stationary phase is used we speak of *gas–liquid partition chromatography* (GLPC). When an uncoated solid absorbant material is used, we speak of *gas solid chromatography* (GSC). The fixed gases, such as oxygen, nitrogen, and carbon monoxide, may be separated using a solid absorbant, while hydrocarbons and similar materials usually require liquid stationary phases. It is not necessary to coat the liquid on a granular material if a thin enough layer can be prepared on the column tubing wall of small radius. This has been done in the case of the so-called capillary columns. Capillaries are required in order that the diffusion distance from the gas phase into the liquid phase be sufficiently small so that equilibrium is achieved within reasonable times.

Applications

There are a number of problems in applying the gas-chromatographic technique to the determination of the pollutants of interest in the atmosphere. The biggest problem is the very low concentrations that are found in the ambient. If a reasonably sized gaseous sample is injected into the carrier-gas stream of a chromatographic apparatus, the amount of the pollutant of interest is so small as to be practically undetectable by ordinary GC detectors. It is not practical to inject larger samples so as to have enough of the pollutant to be measureable at the detector. In addition, most of the pollutants of interest are extremely reactive materials, so that they will not pass through the column and appear at the detector. For example, SO_2 is an acidic material and is strongly absorbed by many liquids and solid supports used in gas chromatography as well as on the walls of the tubing employed. Ozone is very reactive and is destroyed catalytically at many surfaces. Nitric oxide (NO) and nitrogen dioxide (NO_2) are extremely reactive materials and do not survive very long in a gas-chromatographic environment. Thus, in order to employ gas-chromatographic techniques usefully in this field, special materials and extremely sensitive specific detectors must be used in order to obtain results of value.

Stevens, O'Keeffe, and their co-workers have addressed themselves to the problem of gas chromatography of pollutants in air and devised a chromatographic column support and stationary partitioning liquid that would separate a number of sulfur compounds found in the atmosphere.[67-69] They have employed the flame-photometric detector (FPD) specifically for the detection and measurement of the sulfur-containing effluents from their column. A Teflon column was employed in order to minimize absorption of the constituents of interest on the walls. The solid support and liquid partitioning material, while able to separate the sample components, were specially selected to do that job, without interacting with them chemically. More recent work has been reported using improved solid support materials and partitioning liquids for shorter analysis times.[70]

Gas chromatography has also been applied to the determination of light hydrocarbons, total hydrocarbons, carbon monoxide, and carbon dioxide in air. Villalobos and Chapman[71] describe a gas-chromatographic apparatus for the separation of quantitation of these compounds employing a flame-ionization detector (FID). Since the FID is not sensitive to CO or CO_2, means for conversion of these compounds to methane by hydrogenation were developed so that they could also be quantitated with the FID. Sensitivities in the low parts per million level were obtained.

Trapping and Desorption Followed by Gas Chromatography

As indicated in previous sections, great modifications of the ordinary gas-chromatographic apparatus are required to carry out analysis for air pollutants. In order to overcome some of these problems a freeze-out technique has been employed, which concentrates the air pollutants from a large volume of air into a small volume. This is accomplished by passing the sample through a charcoal trap maintained at liquid-nitrogen temperatures. Programmed increases in temperature may then be employed to release classes of components from the trap into the chromatograph, or rapid heating can be used to send all of the condensed components into an instrument of ordinary sensitivity for separation and quantitation by standard techniques. While not very widely employed in the field at the present time, the technique is extremely useful for laboratory investigation and may be important for the future in analyzing components that may be of significance in the sub-part per billion range.

Subtraction Techniques in Gas Chromatography

Freeze-out and desorption are part of a class of techniques that may be called subtraction techniques. In this case, the technique is used to "subtract" large volumes of air from the sample entering the chromatograph. Other techniques in this group may be used for enhancing the separation of unresolved components, masking unwanted components, or removing unwanted components from the sample or from the chromatographic record. An extensive and systematic discussion of subtraction in chromatography has been written by Szonntagh.[72]

10.3. INSTRUMENTS AND TECHNIQUES THAT MEASURE THE SAMPLE DIRECTLY

10.3.1. Optical Measurements

Absorption Spectrophotometry

Principle of Absorption Measurements. Whenever a beam of polychromatic radiation passes through a transparent medium, some attenuation of the beam energy occurs (Figure 8). If the medium is a fluid, some loss occurs at each of the phase boundaries encountered by the beam because of differences of refractive index. Other losses occur because of scattering by particles or by thermal fluctuations in the medium. These losses are minor compared to the attenuation due to interaction (absorption) of specific frequencies in the radiation with specific molecules or ions in the medium.[73, 74] The nature of the interaction changes with the frequency,[75] as illustrated in Figure 9.

The attenuation due to absorption is best summarized by some statement of Beer's law.[73-78] In differential form, one statement is: Successive increments in the number of absorbers attenuate equal fractions of the incident radiant energy, or

$$-\frac{dP}{P} = k'dN \qquad (18)$$

where P is the radiant power of the beam = radiant energy passing through a given area per second, k' is a proportionality constant, N is the number of

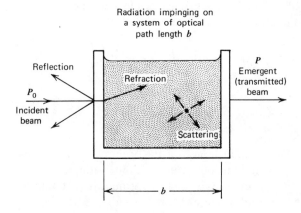

Figure 8 Attenuation of an optical beam.

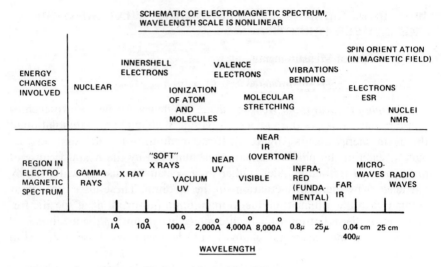

Figure 9 Interaction of radiation with matter.

absorbers in the beam. The minus sign indicates that P decreases as N increases. The integration may be carried out by assuming:

1. The area of the beam is constant, and the length of the beam path in the absorbing medium is fixed. These two statements imply a fixed volume of the medium with a constant number of absorbers in the beam path.
2. The absorbers act independently of each other.
3. The radiation is monochromatic:

$$-\int_{P_0}^{P} \frac{dP}{P} = k' \int_{0}^{N} dN; \quad \ln \frac{P}{P_0} = -k'N \tag{19}$$

$$\frac{P}{P_0} = e^{-k'N} = 10^{-k''N} \tag{20}$$

where

P_0 = radiant power of the beam before attenuation,
   = radiant power incident on a sample,
P = radiant power of the beam after attenuation,
   = radiant power transmitted by a sample.

The number of absorbers in the path, N, is not a particularly useful

10.3. Instruments and Techniques That Measure the Sample Directly

quantity. Equation (20) may be transformed to a more useful form as follows:

N = number of absorbers in the volume, V, swept by the beam,

= (area of beam) × (path length of beam) × (number of absorbers per unit volume)

$$= B \times b \times n, \qquad (21)$$

n = (concentration of absorbers per unit volume) × (Avogadro's number)

$$= cN_0, \qquad (22)$$

where c is expressed in moles per liter. Putting this result into Eq. (21):

$$N = B \times b \times c \times N_0, \qquad (23)$$

$$k''N = (k''BN_0) \times (b) \times (c) = abc. \qquad (24)$$

Equation (20) then becomes

$$\frac{P}{P_0} = 10^{-abc} = T = \text{transmittance.} \qquad (25)$$

This form of Beer's law is particularly useful for liquid and solid samples, where temperature effects on volume are small; that is, the number of absorbers in the beam path are not strongly influenced by temperature. In the usual case, c is expressed as moles per liter, b is in centimeters, and, since the exponent must be dimensionless, a is in liters per mole per centimeter. The absorptivity, a, depends on the nature of the absorber, the nature of the medium, and the wavelength of the radiation. For a given medium it is an intensive property of the absorber. In the case of gases, medium effects are usually negligible and a is a property of the absorbing molecule itself.

If logarithms are taken on both sides of Eq. (25), we have

$$-\log T = abc = A = \text{absorbance,} \qquad (26)$$

$$c = \frac{A}{ab} = \frac{\log 1/T}{ab} = \frac{\log P_0/P}{ab}. \qquad (27)$$

Thus, the concentration is linear with absorbance, provided a and b remain constant. The test of constancy is the so-called Beer's law plot of A versus c. If the plot is linear passing through the origin, the system is said to obey Beer's law in the concentration range studied. For unit path length, the slope of the Beer's law plot is a.

In the case of gases the number of absorbers in the beam path is strongly influenced by temperature. Thus, it may be more convenient to write Beer's law in a form more suitable for gases. This may be most easily accomplished by assuming that the gas of interest and air are ideal gases obeying Dalton's law of partial pressure:

$$p_i V = M_i RT, \tag{28}$$

$$c_i = \frac{M_i}{V} = \frac{p_i}{RT}, \tag{29}$$

where p_i is the partial pressure of the component of interest and M_i is the number of moles of COI. Then Eq. (26) becomes

$$A = \frac{abp_i}{RT}. \tag{30}$$

The partial pressure is rarely used as a read-out form; more frequently employed is the relative pressure, $p_i / \sum_i p_i$. For ideal gases, the mole fraction, x_i, may be written

$$x_i = \frac{p_i}{\sum_i p_i} = \frac{M_i}{\sum_i M_i} = \frac{v_i}{V} \tag{31}$$

where

v_i = volume of component i at $p(t)$ and T,

$p(t)$ = total pressure = $\sum_i p_i$

$$\therefore p_i = x_i \sum_i p_i = x_i p(t) \tag{32}$$

$$A = \frac{abx_i p(t)}{RT} \tag{33}$$

If one wishes to refer all absorbance measurements to standard pressure and temperature then, Eq. (33) becomes

$$A = \frac{a'bx_i}{R} \times \frac{p(t, x)}{p(t, s)} \times \frac{T(s)}{T(x)} \tag{34}$$

where

a' = a modified absorptivity,

$p(t, x)$ = total pressure of the real sample,

$p(t, s)$ = standard total pressure,

10.3. Instruments and Techniques That Measure the Sample Directly

$T(s)$ = standard temperature,

$T(x)$ = temperature of the real sample.

If we wish to use the readout form of parts per million, then Eq. (34) becomes

$$A = \frac{a'b\,(\text{ppm})_i}{R \times 10^6} \times \frac{p(t, x)}{p(t, s)} \times \frac{T(s)}{T(x)}. \tag{35}$$

Equation (35) may be simplified to

$$A = a''b(\text{ppm})_i \times \frac{p(t, x)}{p(t, s)} \times \frac{T(s)}{T(x)}, \tag{36}$$

where a'' is an absorptivity that includes all of the constants, the parts per million reporting form as well as the appropriate dimension for b.

It is clear from Eq. (31) that x_i as well as $(\text{ppm})_i$ is independent of temperature and pressure. However, A is not; it depends on the number of absorbers in the beam path of length b and area B. The terms in p and T in Eq. (35) correct A for the change in the number of absorbers because of pressure and temperature effects.

Absorption measurements may be made in any region of the spectrum. For the pollutants of general interest (SO_2, CO, CO_2, O_3, HC, NO, and NO_2) measurements are generally made in the UV or IR region. In the usual case, a wavelength band is chosen where the COI shows a maximum in absorbance for a given concentration, since the greatest sensitivity is achieved in this way. If a wavelength can be found where this is the case and no other component shows absorbance, the ideal analytical situation obtains. In many cases, however, other components show some absorbance at the wavelength of maximum absorbance for the COI (i.e., there is an analytical interference). This situation can be dealt with in a number of ways. In one way, another wavelength is chosen where none of the other components absorb at some sacrifice in sensitivity for the COI. In another way, maximum sensitivity for the COI is retained by measuring its absorbance at the wavelength of its maximum absorbance and correcting for interferences by making further absorbance measurements. This procedure will now be described in detail.

Suppose that, at the wavelength of greatest absorbance of the component of interest (COI), component 1, another component, 2, shows a significant absorbance (Figure 10). Since the absorbances at any wavelength are additive, then at this wavelength, α,

$$A(\alpha) = [a(1, \alpha) \cdot c(1) + a(2, \alpha) \cdot c(2)]b. \tag{37}$$

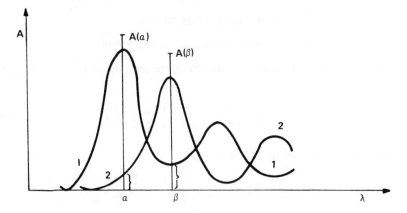

Figure 10 Spectrophotometric measurements: mutual interference.

One then measures A at wavelength β, at which component 2 shows large absorbance with some interference from 1 (Figure 10). Then, at β,

$$A(\beta) = [a(1, \beta)\cdot c(1) + a(2, \beta)\cdot c(2)]b. \qquad (38)$$

In Eqs. (37) and (38), for a term such as $a(2, \beta)$, the number refers to the component and the Greek letter to the wavelength. A, a, b, and c have the significance previously given, in Eqs. (21)–(26). All of the quantities in Eqs. (37) and (38) are experimentally measurable except for $c(1)$ and $c(2)$. We thus have two linear simultaneous equations in two unknowns, $c(1)$ and $c(2)$, and these can be solved for $c(1)$ and $c(2)$. This principle can be extended to as many absorbance measurements as necessary to take care of the effect of all significant interferences on measurement of absorbance of component 1. We will need n absorbance measurements for the n substances (including the COI) that absorb at α. This will generate n simultaneous equations, which can be solved rapidly by use of a computer.

If a wavelength for 2 can be found at which 1 shows zero absorbance (Figure 11), then Eq. (38) becomes

$$A(\beta) = a(2, \beta)\cdot c(2)b; \qquad a(1, \beta) = 0. \qquad (39)$$

Equation (37) becomes

$$A(\alpha) = a(1, \alpha)\cdot c(1)b + \frac{a(2, \alpha)A(\beta)}{a(2, \beta)}. \qquad (40)$$

In Eq. (40), the terms $A(\alpha)$ and $A(\beta)$ are measured and all the other terms except $c(1)$ are constants. Thus, $c(1)$ can be calculated. The advantage of

10.3. Instruments and Techniques That Measure the Sample Directly

this expedient is that while two absorbance measurements are required, only one equation is dealt with.

Another technique that has been found useful is to make the second absorbance measurement at a wavelength different than the wavelength for the COI. However, this wavelength, γ, is chosen to conform to the following criteria:

1. Component 2 has the same absorptivity at γ as at α; that is,

$$a(2, \alpha) = a(2, \gamma); \quad \text{therefore,} \quad A(2, \alpha) = A(2, \gamma). \quad (41)$$

2. Component 1 has zero absorbance at γ; that is,

$$a(1, \gamma) = 0. \quad (42)$$

This situation is illustrated in Figure 12. Under these circumstances equation (37) becomes

$$A(\alpha) = a(1, \alpha) \cdot c(1)b + A(2, \gamma)$$
$$= a(1, \alpha) \cdot c(1)b + A(2, \alpha). \quad (43)$$

In Eq. (43), $A(\alpha)$ and $A(2, \gamma)$ are measured and the other terms, except $c(1)$, are constants; thus, $c(1)$ can be calculated. In both Eq. (40) and Eq. (43) two absorbance measurements are made. In both cases the second wavelength is chosen, so that $a(1, \beta \text{ or } \gamma) = 0$. In the first case, the correction term, $A(\beta)$, must itself be corrected by the appropriate ratio of $a(2)$ terms. In the second case, the correction term, $A(\gamma)$, is subtracted from $A(\alpha)$ directly.

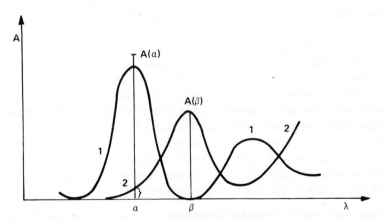

Figure 11 Spectrophotometric measurements: $a(1, \beta) = 0$.

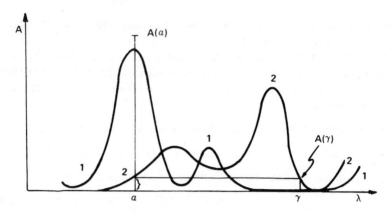

Figure 12 Spectrophotometric measurements: $a(2, \alpha) = a(2, \gamma); a(1, \gamma) = 0$.

Short-Path Point Instruments. A short-path instrument is defined as one in which the radiation source, sample cell, and radiation detector are all contained in one package. The sample of air from a specific region in space is drawn into the sample cell and the attenuation of the radiant power continuously measured photometrically. The path length may range from several centimeters to several meters. Path lengths in the meter range are usually achieved by use of reflecting mirrors at the ends of a 1- or 2-ft cell such that the radiation passes back and forth within the cell to achieve the desired path length. Examination of Beer's law [Eqs. (25) and (26)] shows that for a given sample, the absorbance can be increased by increasing the path length. This is frequently the means taken to obtain significant amounts of absorption, since the concentration of the absorbers in the air is so low. The radiation regions usually employed are the UV and the IR, since the pollutants of interest generally show little or no absorption in the visible. Short-path point instruments are rarely employed for ambient monitoring (background and urban), because the concentrations are too low. However, they are frequently employed for source monitoring (stacks, automobiles, ambients in factories and mines), since concentrations are sufficiently great for useful absorption measurements. Short-path instruments for this application are commercially available from several instrument makers. Problems involved in the application of optical instruments to emission monitoring include presence of particulate matter, excessive moisture, high temperatures, and corrosive sample. These are discussed in detail in a number of papers in the literature.[79-83]

Long-Path Space-Averaging Instruments. In optical instruments of this kind, the radiation source is separated from the radiation detector by large

10.3. Instruments and Techniques That Measure the Sample Directly

distances, tens to hundreds of meters. Attenuation of the radiation, as indicated in Eq. (20) is a function of the number of absorbers in the beam path. If the path length is known and fixed, then the absorbance may be written

$$A = a \int_0^b c \, db. \qquad (44)$$

In the usual case, c is not known as a function of b. If c is replaced by c(av.), where (av) is the mean concentration of the absorbers over the path $0 \to b$, then Eq. (44) becomes

$$A = ac(\text{av.})b \qquad (45)$$

and the absorbance measurement gives c(av.), but no information about the concentration at any point along the path.

Because ordinary radiation tends to spread over long distances, laser beams are frequently used so that the area of the beam remains essentially constant over the path length.

Long-path optical measurements are useful for assessing ambient concentrations (background and urban) where estimates of air quality over a large region are desired.

Microwave Instruments. In rotational microwave spectroscopy, a polar molecule absorbs energy at specific frequencies in the microwave region ($\lambda \sim 1$ cm, $\nu \sim 10^{10}$ Hz) when it is a freely rotating body. Often referred to as *molecular rotational resonance* (MRR) *absorption,* the frequencies are determined only by the energy separating two rotational levels of the entire molecule. Thus, the absorption spectrum in this region for a given compound is unique to that compound and may be used for qualitative and quantitative analysis in the same way as absorption in other spectral regions. The technique is at present in the laboratory stage and is applied to compounds such as aldehydes and alcohols. Specificity is said to be quite high owing to the sharpness of the spectral peaks. These are typically about 0.5 MHz wide and the peak frequency can easily be measured to better than ± 0.05 MHz (10^{-6} cm^{-1}) compared to IR bands, which are generally measured to ± 1 cm^{-1}. Although extremely high specificity with the microwave technique appears to be feasible, sensitivity is quite low. For example, one maker[84] states that 0.01 millitorr of SO_2 can be detected in a mixture of sulfur-containing compounds with cell pressures ranging from 20 to 100 millitorr. This is 1 part in 10,000 (assuming the higher pressure), whereas sensitivities in the parts per billion range are required for ambient SO_2 monitoring Thus, the technique seems to be suitable for emission monitoring. However, component concentration techniques are being developed so that pollutants at ambient concentrations can also be measured. At the present time, MRR spectroscopy appears to be mainly a research tool. However, it may very well turn out to be an important monitoring technique for specific pollutants, since spectral resolution, as indicated above, is extremely high, and the signal is linear up to 100% concentration. The limitations that have to be overcome are (1) low resolution at atmospheric pressure; (2) very low pressure operation for high resolution in the Doppler-limited regime; and (3) inability to detect molecules that lack a dipole moment in the ground state (e.g., CO_2, ethylene, benzene).

Second Derivative Instruments. Williams and Hager[85] have discussed the theory and construction of a derivative spectrometer. Starting with Beer's law, Eq. (25), these authors show that the second derivative, $d^2P(\lambda)/d\lambda^2$, is proportional to bc, where $P(\lambda)$ is the transmitted radiant power at a particular wavelength, λ; c is the concentration of absorbers; and b is the path length. The use of the second derivative of the transmitted signal is said to be especially effective in picking out the signals of very small amounts of pollutants from very large amounts of the background atmosphere signal. Thus, sensitivities in the parts per billion range are claimed for sulfur dioxide, nitric oxide, nitrogen dioxide, and ozone. These authors describe the construction of a laboratory instrument that operates on the transmitted signal and prints out the second derivative as a function of wavelength. This instrument has been embodied in a commercial form.[86]

The wavelength region chosen for analysis is in the UV from 190–400 nm. This region is chosen because it is essentially free from absorption lines from the major constituents of air such as oxygen, nitrogen, and carbon dioxide. In addition, there is no interference of water in this region and precautions need not be taken to remove the water, as is necessary with IR instruments. Finally, light sources and detectors in the UV region are readily available and the latter are much more sensitive than similar detectors in the IR. Since $a(\lambda)$ for gases is small, long path lengths are required; lengths of the order of 10 m are achieved by using cells with a folded light path (i.e., mirrors on the end walls).

The second derivative (d^2) spectrometer is said to increase the sensitivity of a transmission spectrometer by 2–3 orders of magnitude. Absorption changes as small as 0.05% are readily resolved. It should be understood that in order to obtain a usable second derivative signal, it is required to scan a suitably wide wavelength region and this takes time. Thus, it is not possible to obtain a continuous readout of the desired concentration. However, it is probably easy to get data with a high degree of frequency. The problem of interference from gases with the same absorption bands is overcome by a suitable choice of the wavelength region for the component of interest. The presence of interference is detected with the d^2 spectrometer by a change in the normal shape of the absorption band being scanned or by a change in the normal relationship of peak heights of the d^2 signature of the gas. The manufacturer claims the minimum detectable concentration for four important pollutants as follows: sulfur dioxide, 3 ppb; nitric oxide, 5 ppb; nitrogen dioxide, 25 ppb; and ozone, 25 ppb.

Real-time analysis is claimed, even though a scanning mode is required. The scanning time for each pollutant is 72 sec and the four pollutants mentioned can be measured independently in a total scan and reset time of 7 min. The analyzer can be programmed to measure a single pollutant at frequencies as high as 20 times per hour. The instrument is presently being evaluated for in-field applications; few data are available on its behavior with respect to precision, accuracy, and sampling problems in real-world situations.

Laser Sources. In the UV, visible, and IR regions of the spectrum a combination of a broad-band light source and a monochromator is used to obtain radiation of desired bandwidth at various wavelengths. Frequently optical filters are used instead of a monochromator. This approach to obtain sufficiently monochromatic radiation has the drawback that the power available in a given bandwidth is quite low. UV and visible radiation sources are of high energy, and the available detectors for these wavelength regions are quite sensitive. If high-resolution absorption measurements are to be made in these regions, little difficulty with signal–noise (S–N) problems are encountered. In air-pollutant analysis, IR spectroscopy is an important technique, since each of the major gaseous pollutants contains at least one strong absorption band in the 3–15 μm range. However, as one goes into the IR region, the radiant power per unit of bandwidth becomes smaller; as one attempts to obtain higher-resolution absorption measurements, the signals become weaker and are ultimately swallowed up in noise. The expedient used in scanning of IR spectra, for example, is to program an increasing opening of the monochromator slits as the wavelength increases so that sufficient radiant power to overcome detector noise becomes available. The price one pays for this is poor resolution and, in the case of quantitative absorption measurements, a loss of sensitivity. Even more important, interference from adjacent absorption peaks of other species present in the sample is encountered. Thus, if high-energy, high-resolution IR sources or high-sensitivity IR detectors were available, they would be extremely useful for quantitative photometry of air pollutants in this region. The development of laser technology in recent years appears to be meeting the need for high-energy and high-sensitivity devices in the IR region. We proceed to discuss some of these developments, which were introduced in Section 6.2.2.

The conditions for an ideal high-sensitivity, high-selectivity quantitation by optical absorption photometry may be stated as follows:

1. The source of radiation shall be highly monochromatic and of high intensity.
2. The COI shall have an absorption band at the wavelength of the monochromatic source; the absorptivity shall be sufficiently high for precise quantitation.
3. None of the other species present in the sample shall absorb at the source wavelength (i.e., $a_i = 0$ for all $i \neq$ COI).

Laser sources are characterized by highly monochromatic radiation of high power. High monochromaticity increases the probability that Beer's law will be obeyed and interferences will be avoided; high power reduces the requirement for high-sensitivity detectors. The likelihood, however, that the COI will have an absorption band at precisely the wavelength of the laser emission line and that no other species will absorb at this wavelength is rather small. Thus, it would be of extremely great advantage to match the emission line to an appropriate COI absorption band by shifting the emission line to that band. This has been accomplished by means of the tunable laser.

Tunable lasers have been described in a number of review papers that discuss their construction, operation, and application to air-pollutant measurements in the rock-salt region of the IR.[87-92] Fixed-frequency lasers have been extensively used for air-pollution work.[93] By selecting appropriate laser lines that overlap absorption lines of gaseous pollutants, concentrations of the latter can be measured by direct absorption measurement. However, the match is seldom ideal, and for SO_2, no entirely satisfactory lines are available. Even when a good match is found, difficulty is still encountered because of absorption by components of the normal atmosphere or by other pollutants. Thus, the recent development of a variety of tunable lasers provides the analyst with a powerful analytical technique of high sensitivity and excellent selectivity.

There are six main types of tunable lasers presently available or in the stage of active development. These are listed in Table III, along with their range of tunability and power output (continuous wave and pulsed). It is seen that they cover a wide range of wavelengths from the UV to the IR. Of these, the semiconductor diode lasers are of special interest because of the variety and ease of tuning methods available. Before assembly, the tuning range can be adjusted by adjusting the composition of the ternary compounds (i.e., adjusting x in $Pb_{1-x}Sn_xTe$, for example). This changes the band gap and thus the main emission frequency of the laser. After assembly, the semiconductor laser may be tuned by applying pressure, by

10.3. Instruments and Techniques That Measure the Sample Directly

Table III
Tunable Lasers[a]

Type	Range, μm	Power output, W		Method of tuning
		Single mode, cw	Pulsed	
Organic dye lasers	0.34–1.2	0.05	10^7	Optical resonators
Parametric oscillators	0.5–3.75	0.003	10^5	Temperature, crystal orientation
Semiconductor diode lasers	0.63–34	0.001	10^2	Composition, current, temperature
Spin-flip raman lasers (CO)	5.3–6.2	1	10^3	Magnetic field
(CO$_2$)	9.2–14			
Bulk semiconductor lasers (optically pumped)	0.32–34[b]	0.001	—	
High-pressure gas lasers (electron-beam-pumped)				
(CO)	4.8–8.5[b]	—	—	
(CO$_2$)	9.1–11.3[b]	—	—	

[a] Data from Table I and text of Ref. 74.
[b] Predicted.

applying an external magnetic field, or by varying the laser temperature.[88] In addition to these advantages, the assembly and electronics involved in semiconductor diode lasers are very much simpler than those associated with spin-flip and parametric-oscillator lasers. A major disadvantage is that they must be operated at liquid-helium temperatures for continuous operation, with pulsed operation possible at liquid-nitrogen temperature. The diode laser provides effectively infinite resolution for detection of pollutant gases, as demonstrated by a heterodyne experiment with a CO_2 laser.[88] The line width of a specific composition of a PbSnTe diode laser at 10.6 μm was 54 kHz (1.80×10^{-6} cm^{-1}). By comparison, the narrowest Doppler-broadened linewidths of even the heavest molecular pollutants are tens of megahertz at room temperature. Figure 13 compares the spectrum obtained with a high-quality laboratory grating spectrometer and the resolution achieved in a small portion of this spectrum by a diode laser scan.[88] Figure 13 summarizes resolution data with respect to gas spectra, as well as instrumentation.[89]

Because of the enormous increase in resolution made possible with tunable laser sources, the possibility of finding absorption lines for

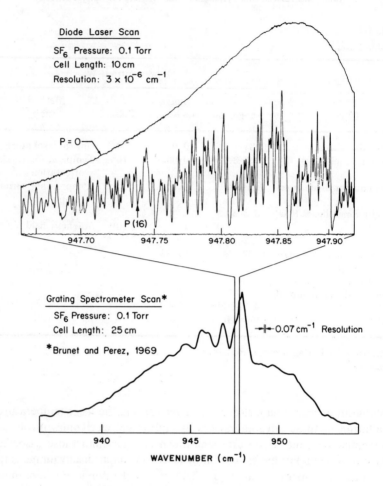

INFRARED SPECTRUM OF ν_3 BAND OF SF_6

Figure 13 Spectra of sulfur hexafluoride. Grating-spectrometer scan of the p_2 band (bottom), taken in a 25-cm cell at a pressure of 0.1 torr SF_6 has 0.07-cm^{-1} resolution. Diode-laser scan (top: SF_6 pressure 0.1 torr, cell length 10 cm) is of band segment near the P(16) CO_2 laser line. The diode laser is capable of resolving frequencies to 3×10^{-6} cm^{-1}. Grating-spectrometer scan from H. Brunet and M. Perez, *J. Mol. Spec.*, **29,** 472 (1969). (Courtesy of *Physics Today*.)

10.3. Instruments and Techniques That Measure the Sample Directly

pollutants of interest that are free from interference of other gases present is also greatly increased. The estimate made by Hinkley and Kelley[88] in this regard is of great interest. If we consider an air sample at atmospheric pressure and assume that 25 molecular species are present that absorb in the "window" region (transparent to CO_2 and H_2O) between 8 and 13 μm, the lines are broadened to about 6×10^9 Hz (0.2 cm^{-1}). Most of the structure in the P, R, and Q branches is lost; even so, 100 resolvable lines per molecule remain, assuming the peaks are separated by one linewidth. If a line is resolvable when an interval of three linewidths ($\sim 18 \times 10^9$ Hz) is not occupied by an absorption line of another molecule and if a uniform probability distribution of lines is assumed, there should be an average of five resolvable lines per molecular species (see Figure 13).

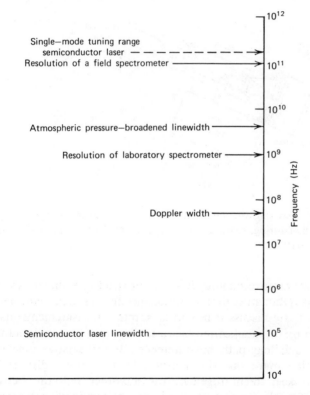

Figure 14 Comparison of resolution requirements of atmospheric-pressure-broadened and Doppler-broadened gas-absorption lines with resolution of various spectrometers (copyrighted by IEEE; reprinted with permission from *IEEE Trans. Geosc. Electro.*).[89]

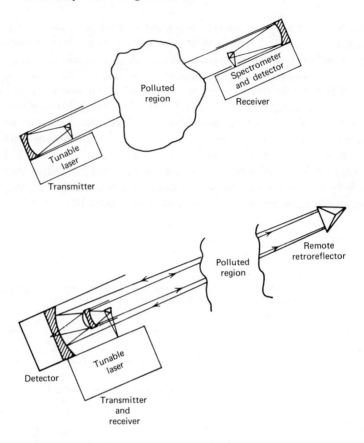

Figure 15 Two possible long-path absorption schemes using laser radiation for pollutant monitoring; schematic (copyrighted by IEEE; reprinted, with permission from *Proc. IEEE*).[92]

Although laser technology has been applied to pollutant measurement in other ways (Raman scattering, resonance fluorescense, and thermal radiation, which are discussed below), absorption measurements may have the widest range of application because they can be used both for point sampling and long-path measurements. For the latter, one must either separate the source and the detector by the sample distance or, if one wishes to keep them together, use a corner reflector (retro-reflector) (Figure 15). The important questions for long-path measurements are, What is the maximum path length that is feasible for pollutant measurements in the atmosphere and what sensitivities are possible for the various pollutants? Kildal and Byer[92] have calculated the minimum detectable

10.3. Instruments and Techniques That Measure the Sample Directly

concentrations for a number of pollutants using a photomultiplier or an indium antimonide detector over a path length of 100 m (328 ft) for two changes of intensity, 5% and 0.1%. This data is given in Table IV. It is seen that even for a transmission change of 5%, the minimum detectable amount for gaseous pollutants of interest is in the fractional parts per million range. The sensitivity of absorption measurements increases with pathlength limited only by atmospheric absorption, atmospheric scattering (Rayleigh and Mie), and background radiation. Even so, ranges of up to 50 km (30 miles) are possible when working in the IR. For point measurements Kildal and Byer[92] recommend folded path cells with gold-coated mirrors. The latter have a reflectivity of 98.3% in the IR, and beam intensity is reduced by a factor of 0.2 after 100 bounces. Thus, a low-power laser can provide enough intensity for a good S–N ratio over an effective path of 10–30 m.

Although this section has been concerned primarily with laser sources, this is a convenient point to mention an instrument using a spin-flip tunable Raman laser source combined with a novel detection principle—namely, optoacoustic spectroscopy. The latter is of special interest because it is a highly sensitive general detector that may be used for a wide variety of pollutants; specificity is achieved by means of a tunable laser as described above. Kreuzer and Patel[94] describe the optoacoustic method of detection. It involves the use of a cell containing the COI through which chopped IR radiation passes. If the IR radiation is tuned to a specific absorption line of the COI, periodic heating of the sample occurs, which is a function of the COI alone. The resulting pressure fluctuations are detected by a capacitor microphone in the cell. The sensitivity is limited by the Brownian motion of the microphone diaphragm; powers as small as 10^{-9} W can be detected. The spin-flip Raman laser source used by Kreuzer and Patel consisted of a liquid-nitrogen-cooled CO pump laser emitting at 1888.31 cm^{-1}. This radiation was focused into an indium antimonide (InSb) crystal $2 \times 2 \times 4$ mm held in a magnetic field. The radiation emitted by the InSb is de-

Table IV

Minimum Detectable Concentration in Parts per Million for Laser Absorption Measurement over a Path Length of 100 m (328 ft) at Two Levels of Intensity Change

Pollutant λ, μm	CO 4.7	CO 2.35	Na 0.5896	Hg 0.2537	C_6H_6 0.2500	NO_2 0.4000	SO_2 0.2900
5%	0.1	20	5×10^{-7}	4×10^{-6}	0.1	0.7	0.6
0.1%	0.002	0.4	9×10^{-9}	7×10^{-8}	0.003	0.01	0.01

Source. Ref. 92, Table V.

termined by the magnetic field according to

$$\omega_s = \omega_0 - g\,\mu_B B, \tag{46}$$

where ω_s is the Raman laser radiation from InSb, ω_0 is the pumping laser radiation from CO laser, g is the gyromagnetic ratio for electrons in InSb, μ_B is the Bohr magneton, and B = the magnetic field.

A dielectric-coated filter was used to reject the pump radiation allowing only the tunable spin-flip Raman radiation to pass. By varying the magnetic field, monochromatic high-intensity laser power in the range 1843–1788 cm^{-1} was generated. Nitric oxide has several strong absorption lines in the 1810–1825 cm^{-1} range, coinciding with the tunable range of the source. Nitric oxide concentrations as small as 0.01 ppm were detected operating at a pressure of 300 mmHg. Determinations of NO in automobile exhaust gas in the range of 50 ppm were made as well as in ambient air in the 0.1 ppm range. The technique for determining NO required only 1 cc of sample, with a sensitivity of 0.01 ppm NO and a 4-sec integration time. The instrument is presently undergoing commercial development.[95, 96]

Laser Techniques Other than Absorption

Laser sources and techniques based on lasers have also been applied to optical measurements of pollutants based on phenomena other than absorption of radiation. These are[89] Raman backscattering, resonance fluorescence (backscattering), and emission of thermal radiation. As Kildal and Byer[92] point out, the Raman and resonance backscattering schemes require optical and data-reduction equipment of great complexity and relatively high cost. Absorption schemes are less complex and less costly and require minimal laser power. While absorption schemes lack depth resolution, they are the only approach sensitive enough to measure dispersed pollutants. However, for the measurement of stack plumes and other pollutant emission sources, the other schemes mentioned offer advantages such as ability to measure high pollutant concentrations with depth resolution. We give below a brief resume of the basic physical phenomena involved in each of the techniques and the special advantages and applications of each.

Raman Backscattering. If a substance is irradiated with monochromatic light, absorption occurs if the energy of the photon, $h\nu$, matches the difference between two energy levels of the molecule—that is, if ΔE (molecule) = $h\nu$ (radiation). If there is no match, interaction does not occur and the light passes through; the material is transparent to that wavelength. In the usual case there must be a match. In spite of this, however, there is a small, but finite, probability, even though the exciting radiation is

10.3. Instruments and Techniques That Measure the Sample Directly

not in an absorption band of the molecule, that the molecule will absorb part of the energy of the photon, releasing the remainder. This is the Raman effect. The absorbed energy exactly matches the energy gap between two vibrational or rotational states of the molecule. The difference in frequency between the exciting radiation and the scattered radiation corresponds to a rotational or vibrational frequency of the molecule. In the laboratory, radiation in the visible or UV is passed through the sample and the scattered radiation observed at 90° to the exciting radiation. In addition to the exciting frequency, a series of relatively weak lines on the low-frequency side of the exciting line is observed and a similar, but still weaker, series on the high-frequency side.[97] The frequency difference between the exciting line (Rayleigh line) and a displaced line (Raman line) $\Delta \nu$, corresponds to a molecular vibration or rotation as indicated above and expressed as

$$\Delta \nu = |\nu_E - \nu_S| = \nu \text{(molecule)}, \tag{47}$$

where ν is the frequency, E is the exciting radiation, and S is the scattered radiation. Since the probability of this type of event is very small, the intensity of the displaced lines is also small. The advantages of Raman observations lie in the use of high-energy sources in the UV and visible as well as the availability of extremely high sensitivity detectors in these regions. While the energy differences correspond to IR frequencies, the experimental work is carried out in a convenient high-energy region of the spectrum.

In general, a molecular vibration will produce a Raman line when the vibration results in a change of polarizability of the molecule. The polarizability of a molecule is a measure of the ease of displacement, within the molecule, of positive and negative charges with respect to each other under the influence of an electrical field. IR absorption occurs for those molecular transitions involving a change in dipole moment. A Raman line corresponding to the origin of a vibrational band often appears where none is found in the direct IR vibration–roation absorption spectrum. The two methods (IR absorption and Raman scattering) are thus complementary; when vibrations are active in both spectra, the agreement is excellent. The Raman shifts of the important pollutant molecules are shown in Figure 16.

The Raman backscattering scheme for remote detection of pollutants depends on the Raman effect. Kobayasi and Inaba[98] describe their instrument as a laser–Raman radar, a form of lidar (see Section 6.2.2). The term backscattering indicates the radiation detected returns to the instrument (Figure 17). Only a single wavelength laser is required, and it need not be tuned. However, the backscattered radiation includes Raman lines from all the species present in the sample. Since the total width of a Raman line is typi-

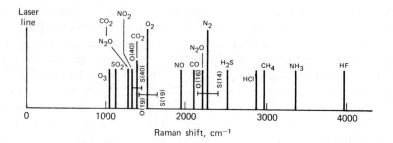

Figure 16 Raman shifts of some pollutant molecules relative to exciting laser frequency (copyrighted by IEEE; reprinted, with permission, from *Proc. IEEE*).[92]

cally over 100 cm^{-1}, interference, especially from N_2 and O_2, is a problem. The latter may be minimized by limiting detection to the narrow Q branch of a Raman band at some cost in the effective scattering cross section. The reduction for CO_2, for example, amounts to only 10%.[92]

Equations for scattering cross section as well as backscattered intensity are given by Kildal and Byer,[92] and a schematic of their instrument is shown in Figure 17. The returning radiation is collected by a large collecting mirror receiving telescope. The radiation is then passed through a spectrometer and filters in order to reject backscattered Rayleigh and Mie radiation. The Raman signal intensity is proportional to the Raman cross section of the species, the atmospheric transmittance, the laser energy per pulse, and the species concentration. Since the cross section varies inversely as the fourth power of the exciting wavelength, it is advantageous to operate in the UV. A quantitative evaluation of all of these factors is given by Kildal and Byer[92] including, in addition to the above, noise in the detector, distance to the sample, rejection of Rayleigh line, and area of the receiving mirror. With a 1-m^2 mirror and (S/N) = 70, the detectability is about 10 ppm at 100 m with a depth resolution of 10 m. Beyond 1 km, the schme is severely limited because of small Raman signal and loss of laser-exciting radiation by atmospheric scattering. The practical range of the Raman scheme is thus less than 1 km. It is not sufficiently sensitive for measurement of ambient concentrations (0.01–0.1 ppm) but can be used to monitor emissions from smokestacks and other sources. The absolute concentration of each pollutant is determined by comparing the backscattered intensity with the N_2 and O_2 Raman lines.

The advantages of Raman backscattering for remote pollutant detection may be listed as follows:

- Only a single wavelength, nontunable laser is required.

10.3. Instruments and Techniques That Measure the Sample Directly 421

- The laser source and detector optics are in the same location.
- Pollutant concentrations can be measured with good depth resolution.

The disadvantages may be summarized:

- Not sufficiently sensitive for ambient concentrations
- Ineffective at distances over 1 km
- Requirement for high-powered laser with possible eye safety hazards

Resonance Fluorescence Backscattering. In fluorescence, the exciting radiation is absorbed in a well-defined spectral region, causing a molecular transition. Radiation is simultaneously emitted at a lower frequency accompanied by an intermediate transition within the molecule. The molecule

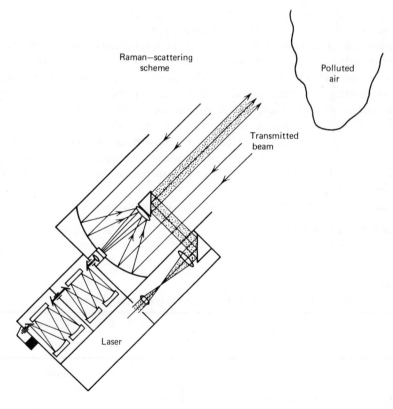

Figure 17 Raman-scattering scheme for pollutant monitoring showing transmitting laser, receiving optics, and spectrometer (copyrighted by IEEE; reprinted, with permission, from *Proc. IEEE*).[92]

ultimately achieves the ground state by nonradiative processes. The absorption region and the emitted frequency are specific to each species; great selectivity and sensitivity are possible. Thus, with a tunable laser source it is possible to excite selectively various pollutants. Excitation of IR vibrational transitions and excitation of atomic and molecular electronic transitions are two possibilities. The excited species emits radiation in all directions. As in the Raman scheme, laser source and detector are at the same location. Detection of the emitted radiation back at the source optics (thus, the designation backscattering) permits remote qualitative and quantitative evaluation of pollutants.

Many of the factors that affect signal intensity at the detector for the Raman scheme enter into the signal intensity for resonance fluorescence. These are also discussed quantitatively by Kildal and Byer.[92] The information most readily available from resonance backscattering from vibrational transitions is the distance to the beginning of the polluted region. Since there is no straightforward relationship between intensity of backscattered signal and pollutant concentration, quantitative estimation of the latter is difficult. The minimum detectable concentration depends on the distance and the thickness of the polluted layer, increasing with distance and decreasing with thickness. Sensitivity can be increased by enlarging the receiving mirror area. Atmospheric attenuation limits the range to a few kilometers if IR excitation is used. Minimum detectable concentrations for selected materials by resonance fluorescence are given in Table V.

The advantages of the resonance fluorescence scheme are its excellent selectivity and moderate sensitivity at small distances and the ease of estimating distance to polluted layers.

The major disadvantage is the difficulty in quantitating concentrations.

Emission of Thermal Radiation. When materials are heated, they emit radiation dependent on the temperature and the nature of the species

Table V

Minimum Detectable Concentrations at 100 m by Resonance Fluorescence

Species	λ, μm	$c_m{}^a$ ppm	Depth resolution, m
CO	4.7	2	10^2–10^4
Hg	0.2537	10^{-6}	26
C_6H_6	0.23–0.27	10^{-3}	—

Source. Ref. 73, Table I.

[a] c_m = minimum detectable concentration at 100 m.

present. Hinkley and Kelley[88] have proposed monitoring the thermally excited radiation from a smokestack or jet engine exhaust. If a PbSnTe-diode laser is used as a tunable local oscillator, IR heterodyne detection techniques may be used to detect emission lines from a smokestack effluent. The technique is similar to that used by radio astronomers to detect weak emission from distant sources. These authors claim, on the basis of a number of assumptions, that it should be possible to detect SO_2 in smokestack plumes in excess of 10 ppm. The temperature of the plume must be known and may be estimated by (1) heterodyne detection of one of the stack gases at two wavelengths; (2) multicolor broadband radiometric measurements; and (3) direct measurement at the top of the stack. Heterodyne detection has been demonstrated for hot ethylene (600°K) at 10 μm, using a CO_2 laser as a local oscillator and a CuGe photoconductor at 4.2°K as a detector.[89]

The advantages of using direct thermal emission measurements are that a high-power laser radiation source not required and well-known heterodyne detection techniques of weak signals are applicable.

The disadvantages would appear to be poor sensitivity and poor selectivity.

Correlation Spectroscopy

In correlation spectroscopy (CS), an incoming spectral signature (transmission spectrum of a sample) is continuously cross-correlated in real time against a replica of the transmission spectrum of the COI. The COI spectrum is stored within the spectrometer either optically (photographic film, glass plate, or slitted mask) or magnetically. The technique may be used from the far UV through the visible into the IR regions of the spectrum.

In Figure 18[99] the transmission spectrum of iodine vapor in the region 500–600 nm is shown, a representative example of the type of spectrum stored in the correlation spectrometer. Figure 19[99] shows one type of correlation spectrometer, an Ebert dispersive spectrophotometer. The incoming radiation is dispersed by the grating and emerges through the exit slit (not shown) to a photodetector. The grating is rotated so that successive regions of the spectrum fall on the exit slit. The incoming radiation is now made to vibrate periodically through a limited wavelength excursion corresponding to the peak-to-peak distance of the iodine spectrum by means of the tuning-fork–refractor-plate arrangement. If a mask replica of the iodine spectrum is placed at the exit slit, the incoming spectrum will be moved periodically into and out of correlation with the mask spectrum. At one point in time, peaks of the sample spectrum will be lined up with mask spectrum peaks leading to maximum light transmission; at another time, peaks will

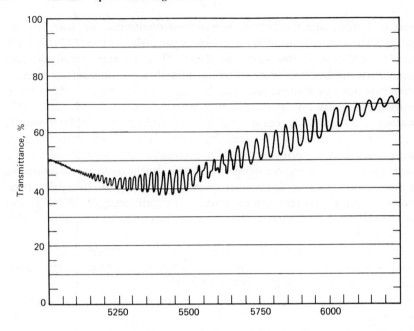

Figure 18 Absorption spectrum of iodine (copyrighted by the American Chemical Society; reprinted, with permission, from *Anal. Chem.*).[99]

be lined up with valleys, leading to minimum light transmission; at intermediate times the transmission will correspond to the degree of positive correlation. A beat signal will result and is detected by the phototube. Phase-sensitive synchronous detection techniques can be used for signal measurement, employing the drive signal for the vibration as the reference signal.

The basic physical process on which the operation of the correlation spectrometer depends is attenuation of radiation by an absorbing medium and is described by the Beer–Lambert law (see Section 10.3.1):

$$P = P_0 e^{-Ebc}, \qquad (48)$$

where E is the absorption coefficient, at a fixed wavelength and the other symbols have the same meaning as in Eqs. (20), (21), and (22).

If at two different wavelengths λ_1 and λ_2, $E_1 \neq E_2$ but P_0 remains the same, we have

$$P_1 = P_0 e^{-E_1 bc} \quad \text{and} \quad P_2 = P_0 e^{-E_2 bc} \qquad (49)$$

or

$$\frac{P_1}{P_2} = e^{-(E_1 - E_2)bc}.$$

10.3. Instruments and Techniques That Measure the Sample Directly

In general, however, the exit slits have a finite width, so that P_0, E_1, and E_2 are not constant. Thus, a more general expression is[99]

$$\frac{P_1}{P_2} = e^{-(E_1-E_2)bc} \cdot \frac{\int_{\lambda_1}^{\lambda_1+\Delta\lambda} N_\lambda \, d\lambda}{\int_{\lambda_2}^{\lambda_2+\Delta\lambda} N_\lambda \, d\lambda} \qquad (50)$$

where P is the radiant power passing through the exit slit in watts and N_λ is the intensity of radiation without absorbers at wavelength λ.

One would expect that if the slits in the mask are properly chosen, the effect of spectral distribution of the light source and attenuation due to gases other than the COI would be minimized. The reason for this is that the mask is chosen to correlate exactly with the spectrum of the COI. Spectral distribution of the light source as well as the transmission spectrum of other absorbers will correlate randomly with the COI spectrum, so that the effect of these is canceled out with respect to the mask spectrum as the sample spectrum moves relative to the mask. The greater the concentration of the COI, the greater the amplitude of the beat signal. This can be seen by considering the situation when the COI is absent. In this case, the random background will cancel out and give a steady signal. As more and more COI is added, greater and greater correlation and anticorrelation with the mask occurs, giving a bigger beat signal.

If N_λ is constant throughout the wavelength of interest, $E(\lambda)$ is a periodic function over the same wavelength; c is constant over the path, b,

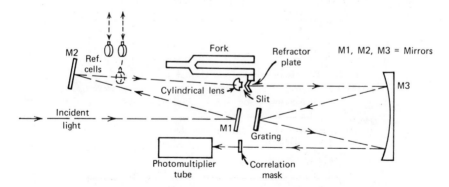

Figure 19 Dispersive system for gaseous pollutant detecting using spectrum correlation filter (copyrighted by the American Chemical Society; reprinted, with permission, from *Anal. Chem.*).[99]

which also remains constant. Under these circumstances, it can be shown[99] that the detector response is

$$R = \alpha J(1 - e^{-(E_2-E_1)bc}), \qquad (51)$$

where R is the instrument response in volts, α is the load impedance of the phototube in ohms, and J is the phototube response in amperes. R will thus be a function of bc and can be calibrated against known bc products in the laboratory. For a remote-sensing instrument using scattered sky illumination, only data on cb is obtained; if b can be estimated, then a space-averaged value of c can be obtained.

On the other hand, the instrument can be designed for ambient monitoring by using a cell of fixed optical pathlength. Under these conditions, c can be determined precisely, because b is known and the light source is controlled. Figure 20 shows a SO_2 ambient monitor. Relative motion of the two spectra is achieved by vibrating the grating with the torque motor. In a long-path monitor, the radiation source is separated from the optics and detector by a known distance (up to 1000 m = 0.62 mile). As indicated above, the long-path instrument gives an average of the COI concentration over the path length used and is obviously more representative of the COI concentration over this path than point monitors. However, as pointed out in previous discussions, it cannot detect local wide fluctuations from the average. The performance of SO_2 and NO_2 long-path instruments is summarized in Table VI.

The advantages of correlation or mask spectroscopy are good selectivity against interference, very high sensitivity, and real-time readout of results.

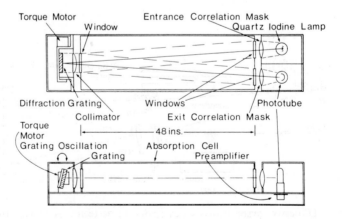

Figure 20 Schematic of SO_2 ambient monitor (copyrighted by the American Chemical Society; reprinted, with permission, from *Anal. Chem*).[99]

10.3. Instruments and Techniques That Measure the Sample Directly

Table VI

Performance of Long-Path Correlation Spectrometers for SO_2 and NO_2

Species	SO_2	NO_2
Spectral bandwidth, A	2800–3150	4130–4500
Minimum detectable conc. in 1000 m	2 ppb	1 ppb
Linearity error (9–2500 ppb in 300 m)	<2%	<3%
Speed of response F.S.	40 sec	20 sec

Source. Ref. 82, Table I.

These advantages become possible under the following conditions: (1) For good selectivity, a region of the spectrum of the COI is chosen such that a series of evenly spaced absorption maxima and minima are found. The spacing of the peaks in this region must be sufficiently different from those of interfering spectra that correlation between the COI and the interferences is at a minimum or absent altogether. Correlation to interferences can be minimized by maximizing the number of peaks in the mask for the COI. (2) For good sensitivity, it is only necessary, having met the first condition, that a sufficiently long light path and sufficiently powerful radiation sources and sensitive detectors be used to achieve any desired sensitivity. Parts per billion sensitivity for most air pollutants has been achieved with presently available technology over light paths of about ½ mile. In addition, the sensitivity is directly proportional to the number of peaks employed in the mask.

A general discussion of correlation spectroscopy has given by Davies,[99] Williams and Kolitz,[100] and Harney et al.[101] Basic papers in the development of the technique for astronomical work have been written by Bottema et al.[102-104] Strong[105, 106] uses the technique in the IR. Kay[107] describes an instrument applying the technique to ambient SO_2 detection. Design, development, and application of commercial instruments have been discussed by Barringer et al.[108-110]

10.3.2. Excitation Methods: Chemiluminescence

Introduction

An excellent brief review of measurements based on gas-phase chemiluminescence has been prepared by Fontijn, Golomb, and Hodgeson.[111] A discussion on chemiluminescent reactions as applied to air-pollutant measurements has also been given by Stevens and Hodgeson.[112] This paper is especially valuable because commercial instruments are listed.

Chemiluminescence is the emission of radiation from chemi-excited species. The latter are created as a direct result of the formation of new chemical bonds. Excited species are those having energy in excess of that possessed by the same species in the ground or equilibrium state; they are usually short-lived. Species formed by collisional excitation, dissociative reactions, charged species impact, or fluorescence are not considered under chemiluminescence. This discussion is confined to chemiluminescent reactions for quantitation of air pollutants even though they have great utility in a number of research areas.

The majority of air-pollutant detectors are based on homogeneous gas-phase chemiluminescence. The detectors may be classified into two types. The first, the ambient temperature detector, employs the chemiluminescent reaction between the small molecule of interest, X, in air and a second reactant species, R, added in excess to a flow reactor. The intensity of the emitted radiation is directly proportional to the product of the reactant concentrations:

$$I = k\,[R][X]. \tag{52}$$

Since R is maintained in great excess, [R] is essentially constant so that

$$I = k'\,[X]. \tag{53}$$

In the other type of detector, chemiluminescence resulting from reactions between atomic or molecular fragments produced from primary molecules introduced into a flame is observed in the secondary combustion zone. Detectors based on flame excitation are generally less specific than those based on low-temperature chemical reactions and are more applicable to classes of compounds (e.g., phosphorus or sulfur compounds).

Chemiluminescence as the basis for quantitation of air pollutants has the advantages of high sensitivity, specificity, and simplicity. With the use of high-gain, low dark-current photomultiplier tubes, extremely low levels of radiation can be detected. The advantage over radiation absorption methods for trace analysis of pollutants is that a positive quantity, emission of radiation, is measured; in absorption the difference of two large photometric quantities is measured. The chief disadvantages are that not all pollutants of interest are amenable to measurement by this technique and calibration with known concentrations is required.

Low-Temperature Excitation

Homogeneous Gas-Phases Reaction. In 1965, Nederbragt et al.[113] described the chemiluminescent reaction between ozone (O_3) and ethylene (C_2H_4) at atmospheric pressure. A broad emission band was observed in the

435-nm region. Prototype detectors based on the Nederbragt reaction were not built until 1970. The technique has been so successful in field and laboratory measurements of O_3 that it has been designated the reference method for routine O_3 measurements required by recent EPA air-quality standards.[114] Commercial versions of the instrument are available from a number of manufacturers.

Ozone may also be quantitated by reaction with nitric oxide (NO):

$$NO + O_3 \rightarrow NO_2 + O_2 + h\nu \qquad (\lambda \geq 600 \text{ nm}). \qquad (54)$$

Results essentially equivalent to those obtained with the Nederbragt technique are obtained. This reaction is presently used almost exclusively for measurement of NO and NO_2. The latter is prepared for quantitation by conversion to NO:

$$NO_2 \xrightarrow{\text{energy}} NO + \tfrac{1}{2} O_2. \qquad (55)$$

The reaction with NO is carried out by mixing the sample and excess ozone in a vessel maintained at a pressure of 2 mmHg. The limit of detection is 2 ppb, and a linear response is obtained up to 1000 ppm of NO. The technique is extensively applied for measurement of NO_x in automotive exhaust gases. A recent paper describes an instrument operating at atmospheric pressure with a range of 1–1000 ppm.[115]

The reaction for conversion of NO_2 to NO may be carried out by heating the sample in a stainless-steel tube kept at 625°C; NH_3 may interfere, since it is oxidized to NO at these temperatures. The ammonia (NH_3) may be removed with acid scrubbers without affecting NO_2. When a converter is used, NO_2 may be estimated by difference from measurements of NO_x (with conversion) and NO (no conversion). An apparatus has recently been described for the rapid simultaneous determination of NO, NO_2, and NH_3 in mobile source emissions,[117] using the thermal converter technique for NO_2 and oxidation of NH_3. Measurement of NO_2 may be accomplished directly by photolysis of the NO_2 creating O atoms; excess NO is added to react with the O atoms and the emitted radiation measured. The response is linear below 1 ppm, and the limit of detection is 1 ppb. Ammonia, in the range 1–10 ppb, has been measured using a chemiluminescent technique.[116]

Gas–Solid Reaction. Regener[118, 119] described the first practical chemiluminescent detector; it was developed for quantitation of ozone in air and was based on the heterogeneous reaction of O_3 with rhodamine-B dye impregnated on activated silica gel. The method is specific for O_3 with a limit of detection of 1 ppb. Instruments based on this principle are commercially available. The main shortcoming is the variable and slowly decaying sensitivity of the rhodamine-B surface. Thus, frequent calibration is required with samples of known O_3 concentration. In a recent paper, Hodgeson, et al.[120] discuss a laboratory investigation of the deficiencies, operating problems, and unknowns in the Regener procedure. On the basis of their study, design criteria for improved O_3 monitors were given. They studied the response characteristics of the chemiluminescent surface (sensitivity to ozone detection, surface conditioning, agent effects, and activation and decay of the chemiluminescent reaction), the effect of moisture, and changes in flowrate, temperature, and pressure.

At the present time, no other gas–solid chemiluminescent reaction is employed for air-pollutant detection and quantitation.

High-Temperature Excitation: Flame-Photometric Detector (FPD)

When sulfur compounds are burned in a hydrogen-rich flame, a strong blue chemiluminescence is emitted. The emitting molecule is S_2^* (the asterisk indicates an excited species). Since two atoms are involved, the intensity of chemiluminesence is porportional to the square of sulfur compound concentration in the flame, for compounds containing only one S atom. The original FPD was described by Draeger.[121] Crider[122] in 1965 and Brody and Chaney[123] in 1966 described an FPD sensitive to phosphorus (P) and sulfur compounds. When an optical filter with a passband at 394 nm ± 5 nm was used, a highly specific response to sulfur compounds was obtained. The FPD of Brody and Chaney has been incorporated into a commercial sulfur dioxide monitor that has been extensively applied in ambient monitoring. Although the FPD, with an optical filter at 394 nm is sensitive to sulfur compounds besides SO_2, laboratory work by Stevens et al.[124] as well as extensive field investigations[125] showed that for ambient monitoring, better than 90% of the total sulfur measured by the FPD is caused by SO_2. In those cases where determinations of the specific sulfur compounds is required, the gas chromatograph is used in conjunction with the FPD. The gas chromatograph has been described above (Section 10.2.3).

Since it has already been mentioned, it is appropriate to point out that for phosphorus compounds, the FPD is operated with an optical filter centered around 526 nm. Gaseous phosphorus compounds have not as yet been identified as pollutants in the atmosphere. The FPD for phosphorus compound determination has been mainly used for determination of pesticide pollutants in water.

10.3.3. Excitation Methods: Flame-Ionization Detector (FID)

If organic molecules are introduced into an oxygen-rich hydrogen flame, fragments are produced that carry a charge. The increase in ion current between the burner tube (a stainless-steel hypodermic needle) and an electrode held above the flame is proportional to the hydrocarbons in the sample. Usually, the burner is negatively charged and the electrode positively charged, so that the negative ions are collected at the electrometer electrode.[126] FID sensitivity is nearly uniform for organic compounds composed of carbon and hydrogen only (hydrocarbons). A number of carbon compounds give little or no signal: COS, CS_2, CO, CO_2, and $HCOOH$; the same can be said for the fixed gases, oxides of nitrogen, H_2S, SO_2, H_2O, and NH_3. Thus, the FID sees primarily C–H bonds, although its ability to do this may be seriously reduced if atoms of O, N, P, S, or halogen are present in the structure (e.g., formic acid).

10.3.4. Mass Spectroscopy

In mass spectroscopy, a portion of the sample to be analyzed is introduced into a chamber maintained at relatively high vacuum of 10^{-5}–10^{-6} mmHg. Ionization of the sample in this chamber is achieved by means of high-velocity electrons emitted by a hot cathode and accelerated by high positive voltage. High-energy electrons impinging on molecules of the sample knock out electrons in the outer shells, producing positive ions and more electrons. The positive ions thus produced are accelerated through a slit by means of plates maintained at high negative potential. In this manner, they are introduced between the poles of a magnetic field. The field acting on the moving charge particle deflects the particle into a circular orbit whose radius depends on the mass–charge (M–C) ratio of the particular particle. Thus, various particles appear at the downstream side of their travel in a spectrum determined by the M–C ratio; the number of particles at each M–C ratio is measured by an ion collector and an appropriate amplifier. As a result, both qualitative data (position on the M–C axis) and quantitative data (ion current at the specific M–C ratio point) can be obtained on the sample.

This method of analysis was applied to atmospheric and emission samples as early as 1949. In order to obtain sufficient concentrations of the constituents of interest, freeze-out techniques were used for organic samples, since their concentrations in the ambient were in the parts per million range, and on introducing the sample into the evacuated chamber, the number of molecules was too small to be detected at the mass ion collector. The freeze-out technique is especially suitable for measuring the total hydrocarbon content in the atmosphere.

It must be remembered that complex molecules break down into ions of different sizes and different charges in the ionization chamber, depending on the energy of the ionizing electrons and the nature of the molecule. Each molecule has a different "cracking" pattern, and in a mixture of complex molecules, the cracking patterns are used to determine the nature of the original molecules, since specific ions of a given charge ratio may be derived from different molecules.

An extension of the mass spectrometer described above has recently become important; it is the *quadrupole* mass spectrometer. In this instrument, the gas is admitted to a sampling chamber that reduces the gas sample from ambient to analyzer pressure without altering relative concentrations of components. The sample is ionized by electron bombardment in a manner similar to that described above and is then injected into a quadrupole mass filter section. Ions with a specific M–C ratio traverse the filter section under the influence of appropriate RF and DC voltages, placed on two pairs of metal rods forming the quadrupole. Species with different M–C ratios are filtered out, because they are not accelerated in the appropriate fashion by the specific RF and DC voltages. By continuously varying these voltages, the species can be sequentially stabilized so as to travel the length of the quadrupole filter and the entire range of M–C values is scanned. The ions sorted in this manner are detected by an ion collector or an independent electron multiplier. The output current is recorded on an oscilloscope or a chart recorder. As in mass spectroscopy, described above, the location of each peak on the x-axis indicates the M–C ratio, while the height of the peak indicates its quantity. A great many pollutants of interest can be analyzed in this manner, including carbon monoxide, carbon dioxide, sulfur dioxide, nitrogen oxide, ozone, hydrocarbons of various kinds, water, hydrogen sulfide, ammonia, and lead and other particulates that can be ionized in the electron bombardment chamber. Very great sensitivities are claimed for this technique with accuracies sufficient for most purposes. Molecular weights in the range from 1 to 300 can be handled, and extremely high sensitivities (in the parts per million range) are claimed.[127] Early instruments of this type were expensive and required a great deal of manpower input, hence were not suited for continuous monitoring in the field. However, they are well suited for research purposes with respect to qualitative and quantitative analysis of samples about which little information is available, and improved models are expected.

10.3.5. Thermal Conductivity Methods

A hot-wire thermal-conductivity gas-analysis cell consists of an electrically conductive wire mounted in a chamber containing the gas to be analyzed. The hot wire in the cell is maintained at an elevated temperature with respect to the temperature of the cell walls by passing an electric current through it. Heat is thus transferred from the hot wire to the cell walls by means of the gas contained in the cell. An equilibrium temperature is attained by the wire when the heat generated by the passage of the electric current is exactly equalized by thermal losses due to conduction by the gases in the cell and by thermal losses of the wire by other means. Aside from heat loss by gaseous conduction, there are losses due to convection, radiation, and conduction through the solid supports of the wire. With a suitable design, the other losses can be minimized and heat loss due to gaseous conduction maximized. If the thermal conductivity of the gas in contact with the hot wire changes, the heat loss from the wire will change. The equilibrium temperature will change and thus the resistance of the wire itself will change. If the wire is made part of the suitable Wheatstone bridge, then the changes in the temperature of the wire that depend on the thermal conductivity of the gas surrounding it can be followed by changes in the bridge balance. In order to obtain sufficiently high sensitivity to changes in gaseous thermal conductivity, the heater element is usually made of a material having a high temperature coefficient of resistance. Platinum is frequently used for hot wire because it has a sufficiently high coefficient and because it is chemically inert. Other hot-wire materials that have been used are tungsten and gold. In recent years thermistor beads have been used because their very high temperature coefficient permits more sensitive analyses.

As indicated above, the Wheatstone bridge is usually employed to follow changes in the resistance of the wire due to changes of the thermal conductivity of the gas surrounding it. The current used to energize the bridge also serves to heat the wire. In nearly all cases, at least two cells are used in the bridge: one serves as a reference, while the other serves to measure the conductivity of the sample of interest. The bridge is balanced with some reference gas in both cells usually air, but in the case of gas chromatography, the helium carrier gas is used to balance the bridge. Then, any change in the composition of the sample gas results in a bridge imbalance that is proportional to the change in composition in the sample cell. It is clear from this description that, in the first place, thermal conductivity analyses can only be obtained for binary mixtures of the reference gas and some other gas that changes its conductivity. Second, the nature of the sample gas must be known, since the analytical method is not specific but depends on a general property, thermal conductivity. Thus, in the case of the air pollutants, one can determine changes in the composition of the background air by noting changes in the conductivity of the sampled air in the measuring cell.

In order to increase the sensitivity more than two cells may be used. Four-cell bridges with the sample passing through sampling cells on opposite arms of the bridge are frequently used. However, this method is rarely if ever used to determine pollutants in the parts per million range, since thermal-conductivity methods are normally not sensitive enough.

Combustible materials in the atmosphere may be quantitatively determined by using a hot-wire platinum cell. Combustible materials striking the hot wire in the presence of air are catalytically oxidized and thus raise the temperature of the platinum wire. The bridge imbalance resulting therefrom is a measure of the amount of the combustibles. Here again, the sensitivity is normally not sufficient to measure materials in the parts per million range. However, thermal-conductivity methods are frequently of great use in measuring combustibles in stack and automobile emission gases and in industrial effluents.

In gas-chromatographic analyses using a thermal conductivity detector great sensitivites can be achieved by suitable cell design, but they depend very greatly on the difference in conductivity between the carrier gas, helium, and the component which is being diluted from the column. The ratio of conductivities is about 6:1. In the case of air, the ratio of thermal conductivities between background air and pollutants in it is very much closer to 1:1, so that sufficient sensitivity simply cannot be achieved. Therefore, thermal-conductivity methods are very rarely used in the analysis of air pollutants, except possibly for combustible materials as discussed above.

10.3.6. Electrochemical Measurements

Most instruments referred to as coulometric analyzers may more accurately be called amperometric instruments. In a coulometric technique, the current is integrated over time so as to obtain the total number of coulombs passing through the circuit during a measured time period. In the usual case, the number of coulombs is proportional to the quantity of COI present in the sample; if the sample volume is known, the concentration may be calculated. In amperometric instruments, the value of the instantaneous current is proportional to the concentration of the COI and it is the current that is measured. As will be seen below, most analyzers measure the current so that the instrument should rightfully be referred to as amperometric.

For example, the "Toxgard"[128] is a commercial amperometric instrument specifically designed for quantitating toxic gases in the parts per million range: HCN (0–50 ppm), H_2S (0–50 ppm), and Cl_2 (0–5 ppm). Its primary application is to provide an alarm in industrial environments. Two electrodes are immersed in an electrolyte that flows into a porous glass cell: the center electrode is the reference, and the outer electrode reacts with the gas that has diffused into the electrolyte via the porous glass. A current is generated proportional to the gas concentration.

A two-channel instrument has been described that quantitates SO_2 and O_3 in the 0.1 ppm range electrochemically.[129] This instrument is also commercially available. Filtered air is drawn into the instrument by a pump on the downstream end of the system. The sample stream is divided in two,

10.3. Instruments and Techniques That Measure the Sample Directly

and each portion travels down a channel for independent electrochemical determination of the COI. In one channel, SO_2 consumes I_2 in an NaI solution; the loss of I_2 is measured amperometrically. O_3 is removed from this channel by use of a granular ferrous sulfate heptahydrate filter, which does not affect the SO_2. In the second channel, O_3 reacts with NaI to liberate I_2, which is also measured amperometrically. The SO_2 in this channel is removed by a filter containing quartz chips soaked in a solution of equal weights of chromium trioxide, water, and phosphoric acid; this filter does not affect the O_3. The measurements in both channels depend on the observation that the current flow between a pair of polarized platinum electrodes in an iodide solution containing a trace of iodine is linearly related to the iodine concentration, provided the applied potential is below the decomposition potential of the iodide salt. In addition to the emf, the electrode area, the rate of flow of electrolyte past the electrodes, and the electrolyte concentration must be kept constant. Complete reaction of the COI in the air with the iodide solution is assured by employing a specially designed gas absorber in which the air stream is sprayed through the liquid reagent so that the latter forms a thin layer over the walls of a 1-liter spherical flask. The iodide used is sodium iodide because it has less tendency to crystallize than KI. The solution is kept close to pH 7 so that NO_2 does not interfere. The reagent is self-buffering, since NaOH produced in the electrolysis absorbs CO_2 from the air to form $NaHCO_3$. The reactions are

$$O_3 + 2I^- + H_2O \rightarrow O_2 + I_2 + 2OH^-, \tag{56}$$

$$SO_2 + I_2 + 2H_2O \rightarrow SO_4^= + 4H^+ + 2I^-. \tag{57}$$

Ozone and other oxidants are quantitated in an instrument operating amperometrically, although termed a coulometric analyzer, the Mast ozone recorder.[130] Potassium iodide solution is pumped down an annulus formed by a cylindrical chamber and a concentric electrode support member, while the air sample is drawn up through the same annulus. In this way, intimate contact between air and liquid is achieved. Many turns of a wire cathode are wound around the entire support member, while one turn of an anode wire is wound at the bottom. A small potential is supplied to the electrode by a battery, and the current flow in the circuit is measured by a microammeter in series. Any ozone or other sufficiently powerful oxidant in the air sample reacts with iodide to liberate iodine as in Eq. (56).

At a polarization voltage of 0.25 V, a layer of hydrogen builds up on the cathode and the current ceases to flow:

$$2H^+ + 2e \rightarrow H_2. \tag{58}$$

Iodine produced by the ozone reaction immediately depolarizes the cathode,

$$H_2 + I_2 \to 2H^+ + 2I^-, \tag{59}$$

so that the current in the circuit is directly proportional to the ozone in the sample. The basic electrochemistry is identical to that involved in the Schulze instrument discussed above. The annulus is made sufficiently long and the flowrates of air and liquid adjusted so that no O_3 escapes reaction, and the current is a measure of total O_3 in the air stream. Field experience with this commercially available instrument has been discussed in a paper by Potter and Duckworth.[131] Since SO_2 will react with I_2 produced by oxidants, low results will be obtained if appreciable amounts of SO_2 are also present in the sample. Saltzman and Wartburg[132] indicate that the SO_2 interference on the O_3 determination is on a 1:1 basis (i.e., 1 ppm SO_2 will reduce the measured O_3 concentration by 1 ppm). They recommend the use of a chromic oxide–sulfuric acid filter to remove the SO_2 and show that it does not affect the O_3. These authors also point out that NO_2 causes a positive interference in iodometric instruments. NO will interfere in the same manner when the CrO_3–H_2SO_4 filter is used, because it is converted to NO_2, causing a positive error of 10% of the NO present. They designed another absorber for NO_x: silica gel saturated with sodium dichromate and $0.72M$ sulfuric acid and dried at 120°C. While dry, this filter is effective in removing NO_x but fails when it becomes damp. The CrO_3–H_2SO_4 filter is more practical, since it is not sensitive to humidity. Since NO is commonly present to the extent of 0.1–0.2 ppm in the atmosphere and causes an error of only 10% of its amount, in the usual case, adequate oxidant results will be obtained with the CrO_3–H_2SO_4 filter to remove SO_2. If an NO correction is required it will be all or nothing (i.e., if enough NO is present, O_3 will be negligible). This filter would be disadvantageous in only a few areas, such as Los Angeles, where the NO is high and SO_2 low.

A similar device is marketed by Beckman Instruments. Countercurrent streaming of the air sample against buffered KI solution is used, as in the Mast, but the glass cell has a somewhat different design. Oxidants react with KI in solution to liberate iodine–ozone as in Eq. (56) and NO_2 as in

$$NO_2 + 2I^- + H_2O \to I_2 + NO + H_2O. \tag{60}$$

The iodine is reduced at the platinum cathode, as in Eq. (59), which may also be written

$$I_2 + 2e \to 2I^- \text{ (at the cathode)}. \tag{61}$$

At the carbon anode an equivalent oxidation of the carbon takes place completing the circuit. The current in the circuit is proportional to the

10.3. Instruments and Techniques That Measure the Sample Directly

concentration of oxidant COI. Specificity for an NO_2 instrument is achieved by a scrubber at the air intake that removes O_3, SO_2, H_2S, and mercaptans without affecting the NO_2. The same instrument is also used for total oxidants. In the latter case a CrO_3-sand scrubber is used to remove SO_2, H_2S, and mercaptans. Finally, the identical glass cell is used for SO_2 determination, using a special scrubber for the removal of O_3, H_2S, and mercaptans. In this case three electrodes are employed. The SO_2 destroys iodine in solution, as in Eq. (57). This reaction lowers the steady state I_2-I^- ratio maintained in the solution by the passage of a constant current through the solution via the platinum anode and cathode. Since the cell is driven by a constant voltage source and since there is less I_2 available to react at the cathode, a current equivalent to the loss in iodine is passed through the carbon reference electrode in parallel with the cathode by the reaction:

$$C(\text{oxidized}) + ne \rightarrow C(\text{reduced}). \tag{62}$$

The current in this branch is measured with an amplifier and is proportional to the SO_2 in the sample.

In the Philips SO_2 monitor[133-135] the sample is passed through a silver gauze filter at 120°C, which destroys O_3 and holds up H_2S but passes the SO_2. The measuring cell consists of an inner glass titration vessel surrounded by an outer vessel filled with electrolyte solution. The latter contains $0.1M$ KBr, Br_2, and $2N$ H_2SO_4. The electrolyte is freely exchanged between vessels via a hole in the bottom of the inner vessel. As the filtered air sample bubbles through the inner solution, SO_2 reacts with the bromine present:

$$SO_2 + Br_2 + 2H_2O \rightarrow SO_4^{2-} + 2Br^- + 4H^+. \tag{63}$$

Two electrodes in this solution monitor the Br_2-Br^- ratio potentiometrically. This voltage is compared to a reference voltage; the difference is zero when no SO_2 is in the sample. When Br_2 is used up by SO_2 in the sample, the difference between the two voltages is used to drive a current through a pair of generator electrodes, which restores the preset Br_2-Br^- ratio:

$$2Br^- \rightarrow Br_2 + 2e \quad \text{(at the anode)}. \tag{64}$$

It is this current that is a measure of the SO_2 content of the air sample. Thus, the difference feedback voltage drives the system to maintain a constant Br_2-Br^- ratio, while the current required to do this measures the SO_2. According to Stevens,[135] the Philips system is the most trouble-free, drift-free wet-chemical system tested in his laboratory. Trouble-free, maintenance-free operation of up to 3 months is claimed by the manufacturer. This SO_2 monitor has been used in the Rijnmond area of Holland on a regional basis; the SO_2 data from a large number of points is used as an indicator of general pollution (see Section 6.1.1).

A sensor using a membrane for selective permeation of the pollutant of interest and operating on the principle of electrooxidation or electroreduction of the species, has been described by Shaw[136] and is commercially available.[137] This device differs from the amperometric instruments described above in that the latter require other chemical reagents in order to quantitate the COI, while in the membrane device, direct oxidation or reduction of the COI takes place at an internal electrode. A schematic of the sensor cell is shown in Figure 21. The sensor is a galvanic cell in which the current flow is proportional to the COI reacting at the anode (electrode where oxidation occurs). This electrode contains a catalyst promoting the electrooxidation. Presumably, specificity to the particular COI is obtained by varying the properties of the membrane and of the catalyst in the sensing electrode. The reactions at the sensing electrode (all oxidations) are

$$SO_2 + 2H_2O \rightarrow SO_4^{2-} + 4H^+ + 2e, \qquad (65)$$

$$NO + H_2O \rightarrow NO_2^- + 2H^+ + e, \qquad (66)$$

$$NO + 2H_2O \rightarrow NO_3^- + 4H^+ + 3e, \qquad (67)$$

$$NO + H_2O \rightarrow NO_2 + 2H^+ + 2e, \qquad (68)$$

$$NO_2 + H_2O \rightarrow NO_3^- + 2H^+ + e. \qquad (69)$$

Linear response for NO in the range 0–5000 ppm and for SO_2 in the range 0–2.5 ppm has been obtained.[136] In addition to SO_2, NO, and NO_2, the principle has been applied to formaldehyde (HCHO) and CO.

The iodide–iodine reaction of the Mast oxidant instrument has been modified to quantitate pollutant gases that on absorption into water generate an acid.[138] The reagent is iodide–iodate (I^-/IO_3^-), which in the presence of acid generates iodine (I_2):

$$IO_3^- + 5I^- + 6H^+ + 6H_2O \rightarrow 3I_2 + 9H_2O. \qquad (70)$$

Gases such as SO_2 and NO_2 on solution in water form acids, and in principle, each proton (H^+) produced by the COI will liberate an equivalent amount of iodine. For strong acids, such as HCl, theoretical response is obtained. For a weak acid, such as formic acid, only 40% of the theoretical response was obtained, while CO_2 and H_2S gave responses well below 1% of theoretical. SO_2 gave greater than theoretical response, presumably because of a direct reduction of iodate by SO_2:

$$2IO_3^- + 5H_2SO_3 \rightarrow I_2 + H_2O + 5SO_4^= + 8H^+. \qquad (71)$$

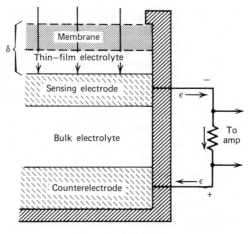

Figure 21 Selective membrane electrochemical sensor (courtesy of Dynasciences Corp.)

In the case of NO_2, a linear response in the range 0.03–1.3 ppm was obtained, but this represents only 10% of theory. The quantitation of NO_2 is interfered with by O_3 (positive interference) and SO_2 (negative interference). Specificity was achieved by using chemical filters to remove the interfering components:

1. Ascarite removes SO_2 and NO_2 but passes NO.
2. Ferrous sulfate heptahydrate removes oxidants but passes SO_2.
3. Dichromate paper removes SO_2, oxidizes NO to NO_2, and passes O_3.

Thus, by a judicious combination of filters, specificity may be obtained. For example, use of filters 2 and 3 permitted analysis of NO_x (NO + NO_2).

10.4. CALIBRATION AND STANDARDIZATION

To validate an analytical technique or instrument one must be prepared to answer in quantitative terms the question: To what extent is the experimental result obtained a "correct" or "true" one? More colloquially, the question may be phrased, How do you know the answer you have obtained is correct? Analytical chemists have always concerned themselves with this question, and a variety of approaches have been developed classically that permit one to estimate the validity of a method as well as provide the basis for instrument calibration. The most direct and generally useful technique is to submit for analysis a sample whose composition is known independently of the method to be tested. Samples of known composition may be prepared in a variety of ways, but they all depend on the availability of standard materials of high and known purity. A major activity of the National Bureau of Standards is devoted to preparation of certified standard reference materials of this type. Another approach is to apply a variety of proven analytical procedures to a material and, after applying appropriate statistical weights, to certify the composition. Materials of this type may be natural or manufactured in origin and are especially useful because they approximate closely the properties and composition of samples to be analyzed. The chemist may employ these techniques in his own laboratory, but his statements about analytical validity tend to carry more weight if NBS-traceable materials are used.

The need for suitable standards and techniques for validiting air-pollutant analyses and analyzers is aggravated by the fact that the concentrations dealt with are small, many of the species are chemically active, and the chemistry of some procedures is in doubt, so that assumptions about stoichiometry and/or reactivity are likely to lead to great error. The appearance of a text specifically devoted to the preparation of known gas mixtures[139] attests to the importance of the subject and is especially welcome in view of the great activity in the field of air-pollutant research and instrument development.

The creation of gas-calibration mixtures may be divided roughly into three broad classes: static systems, dynamic systems, and systems for special cases. In static systems, known weights, volumes, or pressures of the gases of interest are introduced into a container. Depending on the physical parameter measured, the dimensions of the container, the known amounts of the gases, and the gas laws are used to estimate the concentration of the COI. A number of techniques are used to bring the gases together: injection of COI with high-precision calibrated syringe into diluent contained in a known volume; dilution techniques with rigid containers of known volume; flow metering into nonrigid containers.[140] Cylinder mixtures prepared from high pressure tanks and a suitable manifold have been used for many years. In this technique, pure gases at high pressure (2000 lbs/in.^{2}) supply gas to an evacuated chamber fitted with a pressure gauge. Component 1 flows into the empty cylinder until the desired pressure is reached. Component 2 is admitted to the manifold at a higher pressure than exists in the cylinder and is permitted to enter it until the desired amount of 2 has been added. From the recorded partial pressures of all the components, a rough estimate of composition may be made. It is advisable to check this by analyzing the gas by a known procedure. Low parts per million concentrations can be prepared by evacuating the system and introducing the trace COI through the system. This is then swept into the mix cylinder and pressurized with diluent gas. The use of this technique for calibrating process analyzers is described by Villalobos and Gill.[141] Although mixtures from the parts per million to the percent range may be prepared by static methods, they frequently give poor results with air-pollutant COI's because of their reactivity. For this reason dynamic methods have warranted increasing attention in air-pollution work.

In dynamic systems for generating calibrating mixtures of a COI in a diluent, the two gases are blended in a continuous, uninterrupted manner in such a way that the amount of both components is known with sufficient precision to estimate the COI to the required accuracy. Dynamic systems permit the production of large quantities of known mixtures from the percent range to the parts per billion range over long periods of time and are especially useful in air-pollution monitors, since the calibration procedure can be made to match the sample quantitation procedure, thus giving increased accuracy to the instrument. Nelson[142] discusses a number of techniques for dynamic blending, but by far the most widely used and accepted procedure is the use of the permeation tube. Because of the importance of the technique and its endorsement by the EPA,[143] it will be briefly described.

If a liquid is confined in a container with walls permeable to the vapor of the liquid, then at constant temperature, the rate of permeation of the vapor through the walls is constant, provided the vapor pressure outside the container is kept constant. In the usual case the outer surface of the permeation tube is swept by the diluent gas, so that the vapor pressure outside the container is essentially zero. Since the vapor pressure of a pure liquid depends only on its temperature, the driving force through the container walls is constant under these conditions. The weight loss of the container per unit time is an accurate measure of the amount of vapor added to the diluent gas. If the flowrate of the diluent gas is constant, then the composition of the mixture can be known with a precision limited only by the precision of weighing and measuring flowrates. Furthermore, by varying the flowrate and the constant temperature the composition of the trace gas in the diluent may be varied over rather wide limits. First described and studied in detail by O'Keeffe and Ortman,[144] the permeation tube has achieved wide acceptance in air-pollution calibration work. Their paper should be consulted for a method of filling and sealing permeation tubes. A wide variety of COI's may be prepared in this manner: NO_2, propane, butane and other hydrocarbons, alkyl halides, olefins, and other gases. Detailed studies on preparation, behavior, and contitioning of permeation tubes have been made by other investigators.[149-153]

The National Bureau of Standards provides calibrated certified SO_2 permeation tubes,[145] and a number of commercial firms provide devices containing a calibrated permeation tube and accurate flow meters for air and scrubbers to purify air to provide a zero COI standard.[146-148]

Several materials must be considered as special cases when calibrating mixtures are prepared, because of unusual properties of these COI's. Ozone standards must be generated dynamically because the gas is extremely reactive. In addition, as noted above, extreme care in choice of materials and length of tubing must be exercized if accurate results are to be obtained with ozone. The most widely used method of ozone generation is UV irradiation of air or oxygen. The ozone level may be adjusted by dilution with nonirradiated air or by varying the area of the UV source that irradiates the air. All downstream apparatus should be glass or Teflon, since these have the least effect on O_3. Polyethylene or polyvinyl connections, if they must be used, should be as short as possible. Rubber tubing anywhere must be avoided. Although O_3 mixtures in air in parts per million can be generated in this manner, the composition is not known and the mixture must be analyzed by an independent, known procedure. A great many procedures are available to quantitate O_3 in air, but the most widely used and generally accepted method is the KI method previously discussed. When ozone is bubbled through a neutral KI solution, iodine is generated and in the presence of excess iodine, the triiodide ion is formed:

$$O_3 + H_2O + 3I^- \rightarrow I_3^- + 2OH^- + O_2. \tag{72}$$

The I_3^- may be measured spectrophotometrically at 352 nm or determined amperometrically as described previously (Section 10.3.6) or titrated.[154] Devices based on UV irradiation of air have been described by the EPA[154] and Hodgeson et al.[155] They are also available commercially.[156]

Nitrogen dioxide may give difficulties when test atmospheres are prepared, because it dimerizes:

$$2NO_2 \leftrightarrows N_2O_4. \tag{73}$$

The equilibrium is strongly influenced by temperature and pressure. Therefore, blending procedures using the pure gas at normal cylinder pressure at room temperature using single or double dilution will usually fail.[157] However, permeation tubes are effective because they are calibrated gravimetrically and the very small concentrations of the COI produced shift the equilibrium almost quantitatively to the left.

For other gases, such as HCN, hydrogen halides, the free halogens, and H_2S, Nelson's volume should be consulted.

REFERENCES

1. S. S. Butcher and R. E. Ruff. "Effect of Inlet Residence Time on Analysis of Atmospheric Nitrogen Oxides and Ozone." *Anal. Chem.*, **43**, 1890–1892 (1971).
2. E. B. Sandell. "Errors in Chemical Analysis." In I. M. Kolthoff and P. J. Elving, Eds., *Treatise on Analytical Chemistry*, Vol. I, Pt. 1. Wiley: New York, 1959, Chapter 2.
3. W. J. Youden. "Accuracy and Precision: Evaluation and Interpretation of Analytical Data." In Ref. 2, Chapter 3.
4. W. J. Youden. *Statistical Methods for Chemists*. Wiley: New York, 1951.
5. J. Mandel. *The Statistical Analysis of Experimental Data*, Wiley: New York, 1964.
6. W. J. Dixon and F. J. Massey. *Introduction to Statistical Analysis*. McGraw-Hill: New York, 1969.
7. Advisory Board of Analytical Chemistry. "Guide for Use of Terms in Reporting Data in Analytical Chemistry." *Anal. Chem.*, **45**, 630 (1973).
8. W. Leithe. *The Analysis of Air Pollutants*. Ann Arbor Science: Ann Arbor, Mich., 1970, p. 265.
9. J. P. Lodge. "When NOT to Use On-Line Air Pollution Instruments." *Inst. Tech.*, **18**(12), 28–29 (1971).
10. B. D. Tebbens. "Gaseous Pollutants in the Air." In A. C. Stern, Ed., *Air Pollution*, 2nd ed., Vol. I. Academic: New York, 1968, p. 24.
11. K. B. Schnelle and R. D. Neeley. "Air Pollution Analyzer Dynamics." *Inst. Tech.*, **19**(6), 53–55 (1972).
12. K. B. Schnelle and R. D. Neeley. "Transient and Frequency Response of Air Quality Monitors." *J. Air Poll. Control Assn.* **22**, 551–555 (1972).
13. G. D. Gill. *Data Validation, Environmental Measurements*, Public Health Service Publication No. 999-AP-15 or No. 999-WP-15 85-100 (July 1964).
14. D. R. Coughanour and L. K. Koppel. *Process Systems Analysis and Control*, McGraw-Hill: New York, 1965.
15. W. C. Clements, Jr., and K. B. Schnelle, Jr. *Mathematical Modeling of Dynamic Systems with Application to Non-ideal Flow Systems*, Tech. Report 24. Environmental and Water Resources Eng. Dept., Vanderbilt Univ.: August, 1969.
16. W. C. Clements, Jr., and K. B. Schnelle, Jr. "Pulse Testing for Dynamic Analysis." *IEC Proc. Des. Devel.*, **2**, 2, 94–102 (April 1963).
17. N. A. Anderson. "Step-Analysis Method of Finding Time Constant." *Inst. Cont. Syst.* **36**, 130–135 (November 1963).
18. D. T. Mage and J. Noghrey. "True Atmospheric Pollutant Levels by Use of Transfer Function for An Analyzer System." *J. Air Poll. Control Assn.* **22**, 115–118 (1972).

19. M. Adelman. *Air Pollution—What's It All About.* Technicon Instruments Corp.: Tarrytown, N.Y., 1971.
20. Technicon Industrial Systems. *Air Monitor IV.* Brochure 2281R-7-1-5M, Technicon Instruments Corp.. Tarrytown, N.Y., 1971.
21. EPA National Primary and Secondary Ambient Air Quality Standards. *Federal Register,* **36**(84), 8186–8201 (April 30, 1971).
22. Ref. 21, **36**(228), 22384–22397 (November 25, 1971).
23. E. R. Hendrickson. "Air Sampling and Quantity Measurements." In Ref. 10, Vol. II, Chapter 16.
24. M. Katz, *Measurement of Air Pollutants,* World Health Organization: Geneva, 1969.
25. American Society for Testing and Materials. *1972 Annual Book of ASTM Standards, Part 23: Water: Atmospheric Analysis.* Philadelphia.
26. Ref. 25, p. 396.
27. Horizon Ecology Co., Chicago, Ill.
28. Matheson Scientific Co., Chicago, Ill.
29. Research Appliance Co., Allison Park, Pa.
30. Gelman Instrument Co., Ann Arbor, Mich.
31. Tudor Scientific Glass Co., Belvedere, S.C.
32. G. W. Ewing. *Instrumental Methods of Chemical Analysis,* 2nd ed. McGraw-Hill: New York, 1960, Chapter 3.
33. P. W. West and C. C. Gaeke. "Fixation of Sulfur Dioxide as Sulfitomercurate III and Subsequent Colorimetric Determination." *Anal. Chem.,* **28**, 1816 (1956).
34. M. Katz. In Ref. 10, Vol. II, p. 95.
35. Ref. 34, p. 58
36. Ref. 34, p. 76 ff.
37. Ref. 34, p. 106
38. B. E. Saltzman. "Colorimetric Microdetermination of Nitrogen Dioxide in the Atmosphere." *Anal. Chem.,* **26**, 1949–1955 (1954).
39. N. A. Lyshkow. "A Rapid and Sensitive Colorimetric Reagent for Nitrogen Dioxide in Air." *J. Air Poll. Control Assn.,* **15**, 481–484 (1965).
40. M. B. Jacobs and S. Hochheiser. "Continuous Sampling and Ultramicro-Determination of Nitrogen Dioxide in Air." *Anal. Chem.,* **30**, 426–428 (1958).
41. L. J. Purdue et al. "Reinvestigation of the Jacobs–Hochheiser Procedure for Determining Nitrogen Dioxide in Air." *Environ. Sci. Technol.,* **6**, 152–154 (1972).
42. "Error in NO Measurements Discovered." *Chem. Eng. News,* June 26, 1972, p. 26.
43. Ref. 21, **38**, 15174–15180 (June 8, 1973).

44. C. Thorp. "Starch-Iodide Method of Ozone Analysis." *Ind. Eng. Chem., Anal. Ed.*, **12**, 209 (1940).
45. D. H. Byers and B. E. Saltzman. "Determination of Ozone in Air by Neutral and Alkaline Iodide Procedures." *Amer. Ind. Hyg. Assn. J.*, **19**, 251–257 (1958).
46. B. E. Saltzman and N. Gilbert. "Iodometric Microdetermination of Organic Oxidants and Ozone." *Anal. Chem.*, **31**, 1914–1920 (1959).
47. B. E. Saltzman and A. F. Wartburg. "Absorption Tube for Removal of Interfering SO_2 in Analysis of Atmospheric Oxidant." *Anal. Chem.*, **37**, 779–789 (1965).
48. M. D. Thomas et al. "Automatic Apparatus for Determination of Small Concentrations of Sulfur Dioxide in Air." *Ind. Eng. Chem., Anal. Ed.*, **4**, 253–256 (1932); **15**, 287–290 (1943).
49. M. D. Thomas et al. "Automatic Apparatus for Determination of Small Concentrations of Sulfur Dioxide in Air; New Countercurrent Absorber for Rapid Recording of Low and High Concentrations." *Ind. Eng. Chem., Anal. Ed.* **18**, 383–387 (1946).
50. E. R. Kuczynski. "Effects of Gaseous Air Pollutants on the Response of the Thomas SO_2 Autometer." *Environ. Sci. Technol.*, **1**, 68–73 (1967).
51. *SO_2 Ultraportable Analyzer*, Brochure U-2-D/745. Combustion Equipment Assn.: New York.
52. *Ultragas SO_2 Analyzer*, Brochure, Calibrated Instruments, Inc.: Palo Alto, Calif.
53. *Model 70 SO_2 Stack Monitoring System*, Brochure, Scientific Industries, Inc.: Springfield, Mass., (1969).
54. W. J. Moore. *Physical Chemistry*, 3rd ed. Prentice-Hall: Englewood Cliffs, N.J., 1962, p. 337.
55. Ref. 54, Chapter 10.
56. L. A. Elfers and C. E. Decker. "Determination of Fluoride in Air and Stack Gas Samples by Use of an Ion Specific Electrode." *Anal. Chem.*, **40**, 1658–1661 (1968).
57. *Analytical Methods Guide*, 5th ed. Orion Research, Inc.: Cambridge, Mass., 1973.
58. R. DiMartini. "Determination of NO_2 and NO in the ppm Range in Flowing Gaseous Mixtures by Means of the Nitrate-Specific-Ion Electrode." *Anal. Chem.*, **42**, 1102–1105 (1970).
59. J. N. Driscoll et al. "Determination of Oxides of Nitrogen in Combustion Effluents with a Nitrate Ion-Selective Electrode." *J. Air Poll. Control Assn.*, **22**, 119–122 (1972).
60. Honeywell Specification S509-1. *Series 4801 Hydrogen Sulfide Analyzer*, Brochure S509-1-5M. Honeywell, Inc.: Fort Washington, Pa., 1968.

61. H. P. Sanderson, R. Thomas and M. Katz. "Limitations of Lead Acetate Impregnated Paper Tape Method for Hydrogen Sulfide." *J. Air Poll. Control Assn.*, **16**, 328–330 (1966).
62. J. P. Pare. "New Tape Reagent for Determination of Hydrogen Sulfide in Air." *J. Air Poll. Control Assn.*, **16**, 325–327 (1966).
63. S. DalNogare and R. J. Juvet. *Gas-Liquid Chromatography.* Interscience: New York, 1962.
64. D. H. Desty and A. Goldup. *Gas Chromatography,* R. P. W. Scott, Ed. Butterworths: London, 1960, p. 162.
65. W. F. Wilhite. "Developments in Micro Gas Chromatography." *J. Gas Chromatog.*, 47–50 (1966).
66. *Model 1000 Process Gas Chromatograph,* Brochure D-394. Honeywell Process Analyzer Center: Houston, Tex.
67. R. K. Stevens and A. E. O'Keeffe. "Modern Aspects of Air Pollution Monitoring." *Anal. Chem.*, **42**, 143A–148A (February 1970).
68. R. K. Stevens et al. "Gas Chromatography of Reactive Sulfur Gases in Air at The Parts-per-Billion Level." *Anal. Chem.*, **4**, 827–831 (1971).
69. R. K. Stevens, A. E. O'Keeffe, and G. C. Ortman. "Current Trends in Continuous Air Pollution Monitoring Systems." *ISA Trans.*, **9**, 1–8 (1970).
70. F. Bruner et al. "Improved Gas Chromatographic Method for the Determination of Sulfur Compounds at the ppb Level in Air." *Anal. Chem.*, **44**, 2070–2074 (1972).
71. R. Villalobos and R. L. Chapman. "A Gas Chromatographic Method for Automatic Monitoring of Pollutants in Ambient Air." *ISA Trans.*, **10**, 356–362 (1971).
72. E. L. Szonntagh. "Selective Subtraction in Gas Chromatography: A Systematic Investiation of Its Theory and Application to Process and Environmental Analysis," Ph.D. thesis. Technical University of Budapest, June 1973.
73. G. W. Ewing. *Instrumental Methods of Chemical Analysis.* McGraw-Hill: New York, 1954, p. 137 ff.
74. H. A. Strobel. *Chemical Instrumentation.* Addison-Wesley: Reading, Mass., 1960, p. 151 ff.
75. H. H. Willard, L. L. Merritt, and J. A. Dean. *Instrumental Methods of Analysis,* 4th ed. Van Nostrand: New York, 1965, p. 74 ff.
76. H. G. Pfeiffer and H. A. Liebhafsky. "The Origin of Beer's Law." *J. Chem. Ed.*, **28**, 123–125 (1951).
77. H. A. Liebhafsky and H. G. Pfeiffer. "Beer's Law in Analytical Chemistry." *J. Chem. Ed.*, **30**, 450–452 (1953).
78. H. K. Hughes. "Suggested Nomenclature in Applied Spectroscopy." *Anal. Chem.*, **24**, 1349–1354 (1952).

79. N. L. Morrow, R. S. Brief, and R. R. Bertrand. "Sampling and Analyzing Air Pollution Sources." *Chem. Eng.*, **79**, 84–98 (Janaury 24, 1972).
80. F. L. Cross, Jr., and H. F. Schiff. "Continuous Source Monitoring." *Chem. Eng.*, **80**, 125–127 (June 18, 1973).
81. S. S. Ross. "Designing Your Plant for Easier Emission Testing." *Chem. Eng.*, **79**, 112–118 (June 26, 1972).
82. R. L. Chapman. "Instrumentation for Stack Monitoring." *Pollution Engineering*, September 1972, pp. 38–39.
83. D. P. Meffert et al. "Stack Testing and Monitoring." *Pollution Engineering*, June 1973, pp. 25–33.
84. Hewlett-Packard Advertisement. *Chem. Eng. News*, August 24, 1970, p. 12.
85. D. T. Williams and R. N. Hager, Jr. "The Derivative Spectrometer." *Appl. Optics*, **9**, 1597–1605 (1970).
86. *Model III d^2 Analyzer*, Technical Bulletin 1971. Spectrometrics of Florida: Pinellas Park, Fla.
87. C. F. Dewey, Jr. "Excitation of Gases Using Wavelength-Tunable Lasers." In D. S. Dosanjh, Ed., *Modern Optical Methods in Gas Dynamics Research*. Plenum: New York, 1971, pp. 221–270.
88. E. D. Hinkley and P. L. Kelley. "Detection of Air Pollutants with Tunable Diode Lasers." *Science*, **171**, 635–639 (1971).
89. I. Melngailis. "The Use of Lasers in Pollution Monitoring." *IEEE Trans. Geosci. Electron.*, **GE-10**(1), 7–17 (1972).
90. E. D. Hinkley, "Tunable Infrared Lasers and Their Applications to Air Pollution Measurements." *J. Opto-Electron.*, **4**(2), 69–82 (1972).
91. H. R. Schlossberg and P. L. Kelley: "Using Tunable Lasers." *Physics Today*, **25**(7), 36–42, 44 (1972).
92. H. Kildal and R. L. Byer. "Comparison of Laser Methods for the Remote Detection of Atmospheric Pollutants." *Proc. IEEE*, **59**, 1644–1663 (1971).
93. P. L. Hanst. In J. N. Pitts, Jr., and R. L. Metcalf, Eds., *Advances in Environmental Science and Technology*, Vol. II. Wiley: New York, 1971, Chapter 4.
94. L. B. Kreuzer and C. K. N. Patel. "Nitric Oxide Air Pollution: Detection by Optoacoustic Spectroscopy." *Science*, **173**, 45–47 (1971).
95. "Air Pollution to be Measured Quickly by Portable Equipment." *Machine Design*, August 19, 1971, p. 18.
96. "Lasers Analyze Air." *Chem. Eng. News*, July 12, 1971, p. 12.
97. E. J. Rosenbaum. *Physical Chemistry*. Meredith: New York, 1970, p. 305 ff.
98. T. Kobayasi and H. Inaba. "Lasar-Raman Radar for Air Pollution Probe." *Proc. IEEE*, **58**, 1568–1571 (1970).
99. J. H. Davies. "Correlation Spectroscopy." *Anal. Chem.*, **42**, 101A–112A (May 1970).

100. D. T. Williams and B. L. Kolitz. "Molecular Correlation Spectrometry." *Appl. Optics,* **7,** 607–616 (1968).
101. B. M. Harney, D. H. McCrea, and A. J. Forney. *The Application of Remote Sensing to Air Pollution Detection and Measurement* Bureau of Mines Information Circular 8577. U.S. Dept. of Interior: Washington, D.C., 1973.
102. M. Bottema, W. Plummer, and J. Strong. "Water Vapor in the Atmosphere of Vensus." *Astrophys. J.,* **139,** 1021–1022 (1964).
103. M. Bottema, W. Plummer, J. Strong, and R. Zander. "Composition of the Clouds of Venus." *Astrophys. J.,* **140,** 1640–1641 (1964).
104. M. Bottema, W. Plummer, and J. Strong. "A Quantitative Measure of Water-Vapor in the Atmosphere of Vensus." *Ann. Astrophys.,* **28,** 225–228 (1965).
105. J. Strong. "Infrared Astronomy by Balloon." *Scientific American,* January 1965, pp. 28–37.
106. J. Strong. "Balloon Telescope Optics." *Appl. Optics,* **6,** 179–189 (1967).
107. R. B. Kay. "Absorption Spectra Apparatus Using Optical Correlation for the Detection of Trace Amounts of SO_2." *Appl. Optics,* **6,** 776–778 (1967).
108. A. R. Barringer. "Chemical Analysis by Remote Sensing," 23rd Annual ISA Instrumentation Automation Conference, October 1968.
109. A. R. Barringer and J. D. McNeil. "Advances in Correlation Techniques Applied to Spectrometry," ISA-AID Symposium, New Orleans, May 5, 1969.
110. A. R. Barringer, B. C. Newbury, and J. Robbins. "Field Experience with the Ambient Correlation Spectrometer for Pollutant Gases." Air Poll. Control Assn., 62nd Annual Meeting, June 22, 1969.
111. A. Fontijn, D. Golomb, and J. A. Hodgeson. *A Review of Experimental Measurements Based on Gas-Phase Chemiluminescence,* Technical Report AC-12-PV prepared for Office of Naval Research, November 1972, NTIS Document AD-753, 222.
112. R. K. Stevens and J. A. Hodgeson. "Applications of Chemiluminescent Reactions to the Measurement of Air Pollutants." *Anal. Chem.,* **45,** 443A (April 1973).
113. G. W. Nederbragt, A. VanderHorst, and J. Van Duijn. "Rapid Ozone Determination Near an Accelerator." *Nature,* **206,** 87 (1965).
114. Ref. 21, p. 8195.
115. A. Fontijn, A. J. Sabadell, and R. J. Ronco. "Homogeneous Chemiluminescent Measurement of Nitric Oxide with Ozone." *Anal. Chem.,* **42,** 575–579 (1970).
116. J. A. Hodgeson et al. "Measurements for Atmospheric Oxides of Nitrogen and Ammonia by Chemiluminescence," Paper No. 72-12, 65th Annual Meeting of Air Poll. Control Assn, Miami, June 1972.

117. J. E. Sigsby et al. "Chemiluminescent Method for the Analysis of Nitrogen Compounds in Mobile Source Emissions (NO, NO_2 and NH_3)." *Environ. Sci. Technol.* **7,** 51–54 (1973).
118. V. H. Regener. "On a Sensitive Method for Recording of Atmospheric Ozone." *J. Geophys. Res.,* **65,** 3975–3977 (1960).
119. V. H. Regener. "Measurement of Atmospheric Ozone with Chemiluminescent Method." *J. Geophys. Res.,* **69,** 3975–3800 (1964).
120. J. A. Hodgeson et al. "Chemiluminescent Measurement of Atmospheric Ozone." *Anal. Chem.,* **42,** 1795 (1970).
121. B. Draeger. West German Patent 1,133,918 (7-6-62).
122. W. L. Crider. "Hydrogen Flame Emission Spectrophotometry in Monitoring Air for Sulfur Dioxide and Sulfuric Acid Aerosol." *Anal. Chem.,* **37,** 1770–1773 (1965).
123. S. S. Brody and E. Chaney. "Flame Photometric Detector." *J. Gas Chromatog.* **4,** 42–46 (1966).
124. R. K. Stevens and A. E. O'Keeffe. "Modern Aspects of Air Pollution Monitoring." *Anal. Chem.,* **42,** 143A–148A (February 1970).
125. R. K. Stevens et al. "Ratio of Sulfur Dioxide to Total Gaseous Sulfur Compounds and Ozone to Total Oxidants in the Los Angeles Atmosphere–An Instrument Evaluation Study." In G. Mamantov and W. D. Schults, Eds. *Determination of Air Quality.* Plenum: New York, 1972, p. 83 ff.
126. C. H. Hartmann. "Gas Chromatography Detectors." *Anal. Chem.,* **43,** 113A (February 1971).
127. *Pollution Analyzer,* Brochure 94025, Model EIC-A10, Environmental Instruments Co.: Menlo Park, Calif., 1970.
128. M-S-A Toxgard, Bulletin No. 0714-9; MSA Instrument Div., Mine Safety Appliances Co., Pittsburgh, Pa.
129. F. Schulze. "Versatile Combination Ozone and Sulfur Dioxide Analyzer." *Anal. Chem.,* **38,** 748–752 (1966).
130. G. M. Mast and H. E. Saunders. "Research and Development of the Instrumentation of Ozone Sensing." *ISA Trans.,* **1**(4), 325–328 (1962).
131. L. Potter and S. Duckworth. "Field Experience with the Mast Ozone Recorder." *J. Air Poll. Control. Assn.,* **15,** 207-209, (May 1965).
132. B. E. Saltzman and A. F. Wartburg. "Absorption Tube for Removal of Interfering SO_2 in Analysis of Atmospheric Oxidant." *Anal. Chem.,* **37,** 779–782 (1965).
133. F. Cabot. "So Goes SO_2." *Industrial Research,* September 1970, pp. 70–72.
134. *SO_2 Monitor PW 9700 for Air Pollution Monitoring,* Philips Bulletin WA4-C13, Philips Inc.: Eindhoven, The Netherlands.
135. R. K. Stevens. "Review of Analytical Methods for the Measurement of Sulfur Compounds in the Atmosphere," Paper presented at 11th Conf. on

Methods in Air Pollut. and Ind. Hygenic Studio, California Dept. of Health, March 30, April 1, 1970.

136. M. Shaw. "Electrochemical Transducers for Air Pollution Control," Proc. 1st National Symposium Heterogen. Catalysis for Control of Air Poll., Philadelphia, November 21–22, 1968.

137. *Dynasciences Air Pollution Monitor System,* Brochure 6/70-5M. Dynasciences Corp.: Chatsworth, Calif.

138. D. F. Miller et al. "A Versatile Electrochemical Monitor for Air Quality Measurements." *J. Air Poll. Control Assn.,* **21,** 414–417 (1971).

139. G. O. Nelson. *Controlled Test Atmospheres.* Ann Arbor Science: Ann Arbor, Mich., 1971.

140. Ref. 139, 59 ff.

141. R. Villalobos and H. G. Gill. "Use of Cylinder Mixtures of Volatile and Non-Volatile Hydorcarbons for Calibration of Process Analyzers." *ISA Trans.,* **9,** 51–60 (1970).

142. Ref. 139, p. 95 ff.

143. Ref. 21, pp. 8189–8191

144. A. E. O'Keeffe and G. C. Ortman. "Primary Standards for Trace Gas Analysis." *Anal. Chem.,* **38,** 760–763 (1966).

145. National Bureau of Standards, Standard Reference Material 1625, Sulfur Dioxide Permeation Tube.

146. Metronics Associates, Inc., Palo Alto, Calif.

147. TraCor Analytical Instruments Div., Austin, Tex.

148. Analytical Instrument Development, Inc., West Chester, Pa.

149. R. K. Stevens et al. "Absolute Calibration of a Flame Photometric Detector for Volatile Sulfur Compounds at Sub-part-per-million Levels." *Environ. Sci. Tech.,* **3,** 652–655 (1969).

150. B. E. Saltzman et al. "Performance of Permeation Tubes as Standard Gas Sources." *Environ. Sci. Tech.,* **5,** 1121–1128 (1971).

151. L. J. Purdue and R. J. Thompson. "A Rapid Sensitive Method for Calibration of Permeation Devices." *Anal. Chem.,* **44,** 1034–1036 (1972).

152. D. P. Lucero. "Performance Characteristics of Permeation Tubes." *Anal. Chem.,* **43,** 1744–1749 (1971).

153. F. P. Scaringelli et al. "Preparation of Known Concentrations of Gases and Vapors with Permeation Devices Calibrated Gravimetrically." *Anal. Chem.,* **42,** 871 (1970).

154. Ref. 21, p. 8196–8199.

155. J. A. Hodgeson et al. "A Stable Ozone Source Applicable as a Secondary Standard for Calibration of Atmospheric Monitors," Paper No. 71-560, AID-ISA Symposium, Houston, April 1971.

156. *Model 1020 Ozone Generator,* Bulletin 2721020. McMillan Electronics Corp.: Houston, Tex.

157. Ref. 139, p. 170.

11

Particulate-Monitoring Instruments

11.1. NOMENCLATURE AND DEFINITIONS

A great many terms are used to describe particulate matter suspended in the atmosphere. The operative concept is suspended: the particles do not go into solution and will ultimately settle out, although those within a certain size range may remain aloft for months and even years. Section 1.4.4 discusses this size effect and also lists a number of terms describing particles and their commonly accepted meanings. Although atmospheric particles are chemically diverse, they do have a number of physical properties in common and are thus discussed as a single class, sometimes called aerosols (i.e., particulate matter suspended in the air).

Usually the term *particle* refers to a solid of small size. We shall take particulate matter to mean any dispersed solid or liquid in which the individual aggregates are larger than single small molecules but smaller than about 500 μm.[1] Since small molecules are of the order of 0.0002 μm, we can take the range of airborne particulates to be 0.0005–500 μm. In the following discussion we shall use the terms *particle, particulate matter,* and *aerosol* interchangeably.

11.2. SOURCES AND DISTRIBUTION

From the point of view of geographic location, there are three main sources of particulates: natural continental aerosols, natural oceanic aerosols, and man-made aerosols.[2]

Natural continental aerosols are particulates that arise from natural sources over the continental landmasses and include the following:

• Particles from dust storms and desert areas, with sizes ranging above 0.3 μm in radius

• Particles from photochemical gas reactions between ozone and hydrocarbons that are derived from plants, resulting in particles less than 0.2 μm in radius

- Particles from photochemical reactions between trace gases such as SO_2, H_2S, NH_3, and O_3 or atomic oxygen. These reactions usually lead to sulphuric acid mists or ammonium sulphate salt dissolved in water droplets. Their formation is strongly influenced by humidity and the presence of other particulates, which serve as nucleation centers.
- Particles from volcanic eruptions which emit particles of all sizes in addition to trace gases, especially SO_2, which may become particulate in the atmosphere.

Natural oceanic aerosols are particles that arise from evaporation of ocean water thrown into the atmosphere in the form of spray. They have essentially the same composition as sea salt, with sizes ranging above 0.3 μm.

Man-made aerosols arise most usually from combustion processes, either from stationary sources or mobile sources, such as automobiles, trucks, railroad, and aircraft and may be characterized as follows:

- Particulates from smoke (i.e., solid particles resulting directly from combustion)
- Particulates from photochemical gas reactions between unburned or partially burned organic fuel and oxides of nitrogen in the atmosphere. Particles from this source usually are of small size, less than 0.2 μm in radius.
- Particles from photochemical reactions between SO_2 and ozone or atomic oxygen in the atmosphere, these being essentially the same as those from natural continental sources.

The number of small particles (less than 0.1 μm) per unit volume falls off with increasing altitude, while the concentration of large particles (0.1–1 μm) shows a maximum at about 18 km, indicating that these particles are formed in the stratosphere, as suggested by Junge.[3] It is believed that these particles originate from the oxidation of gaseous SO_2 and formation of sulphuric acid mists. The sulphate layer at 18 km appears to be markedly dependent on volcanic eruptions, undergoing marked increases after they occur. Typical concentrations of such stratospheric sulphate particles measured years after the Mount Agung eruption of 1963, average around 0.25 $\mu g/m^3$ at the 18-km level. At this level, 1 $\mu g/m^3$ represents about 20 ppb by weight of sulphate particles, so that on the average there are about 5 ppb of the sulphates in the stratosphere at 18 km. There is reason to suspect that a major portion of this stratospheric sulphate is caused by man-made sulphur injection into the atmosphere because the chlorine-bromine ratio for stratospheric particles is one-twentieth of its value in seawater.[4]

11.3. MECHANISMS OF AEROSOL FORMATION

In section 1.4.4 we stated that aerosols are formed by comminution (pulverizing, grinding, and atomization), by condensation from supersaturated vapors and gaseous chemical reactions, and by agglomeration of smaller liquid and/or solid aerosols to form larger ones.

Many industrial processes and the burning of carboniferous fuels produce particles. Combustion, a major source of particulates, is a complex process that may produce four distinct types of particles in the following ways:[5]

1. The heat of combustion may vaporize materials that subsequently condense to yield particles in the range 0.1–1 μm.
2. The high thermal energy available produces particles of very small size (< 0.1 μm) from the fuel. These may have a very short lifetime.
3. Mechanical processes may reduce fuel or ash to particles greater than 1 μm and may then entrain them.
4. If the fuel itself is an aerosol, a very fine ash may be produced that escapes directly to the atmosphere.
5. Partial combustion of the fuels may result in unburned fuel particulates and soot.

Examples of formation of aerosols are provided by the behavior of SO_2. As indicated previously (Chapter 2), in the presence of moisture and oxygen, SO_2 is oxidized to SO_3, which then forms a sulfuric acid mist by reaction with water vapor. SO_2 also takes part in the reactions forming photochemical smog. Organic gases and vapors also contribute to smog formation.

Particles larger than 10 μm frequently result from mechanical processes such as wind erosion, grinding, and spraying.

A detailed accounting of man-made sources of particulates and their amounts for four representative U.S. metropolitan areas has been compiled by NAPCA.[6] Over 20 distinct industrial, governmental, and residential sources are mentioned, including power generation, refuse disposal, motor vehicles, airplanes, asphalt roofing, cement plants, as well as metals, glass, and sulfuric acid plants. In addition, the suspended particle concentrations of 60 U.S. metropolitan areas are given. For the years 1961–1965, the geometric mean concentration of total suspended particles ranged from 58 to 180 μg/m^3 for these cities.

Data on man-made particulate emissions in the United States and on a global scale for broad classes of industrial and agricultural activities are given in Table I.[7] Table II[7] gives data on particulate emissions of selected U.S. industries for 1968. While the total tonnage of particulate-emissions

Table I
U.S. and Global Industrial Particulate Emissions, 1968 (10^6 Metric Tons)[7]

Industry	1968 global production[a]	U.S. 1968 percentage global production[b]	U.S. 1968 uncontrolled emissions[c]	U.S. 1968 controlled emissions[d]	U.S. percentage emissions controlled[e]	Foreign 1968 uncontrolled emissions[f]	Global 1968 uncontrolled emissions[g]
Iron and steel gray iron foundries	915[h]	22	12.0	1.890	84	42.6	54.6
Grain handling, storage flour, feed milling	112[i]	10	2.3	1.020	56	20.7	23.0
Cement	513	13	7.9	0.790	90	53.0	60.9
Pulp and paper	90[j]	38	3.6	0.660	82	5.9	9.5
Miscellaneous[k]	—	—	8.0	2.465	69	—	—
Total			33.8	6.825	81		148.0

[a] All data in this column from United Nations, *Statistical Yearbook*, 1969.
[b] Derived from United Nations, *Statistical Yearbook*, 1969, by taking the ratio of U.S. production to world production.
[c] NAPCA Division of Air Quality and Emissions Data, unpublished data.
[d] NAPCA, 1970.
[e] $\dfrac{\text{Column 3} - \text{Column 4}}{\text{Column 3}}$
[f] Column 3 $\dfrac{(100-\text{Column 2})}{(\text{Column 2})}$
[g] Column 3 + Column 6.
[h] Combination of figures for production of pig iron and crude steel.
[i] Includes only production of wheat flour.
[j] Combination of figures for chemical and mechanical production of wood pulp.
[k] See Table II.

Table II

U.S.[a] Particulate Emissions for Miscellaneous Sources, 1968 (10^6 Metric Tons)[4]

Industry	Uncontrolled emissions[b]	Controlled Emissions[c]
Sand, stone, rock, etc.	2.4	0.790
Asphalt batching	2.5	0.490
Lime	0.8	0.410
Phosphate	0.4	0.185
Coal cleaning	0.5	0.170
Other minerals	0.4	0.160
Oil refineries	0.2	0.090
Other chemical industries	0.3	0.085
Other primary and secondary metals	0.5	0.085
Total	8.0	2.465

[a] The industries for which there were global production figures are in Table I.
[b] NAPCA Division of Air Quality and Emissions Data, unpublished data.
[c] NAPCA, 1970.

due to activities of men is impressive, it is esimated that this consitutes only 10% of the particulate matter entering the atmosphere every year.[8] Nature accounts for most of the particles in the atmosphere—more than 50 million tons of dust in the northern hemisphere and between 10 and 20 billion tons of sea salts thrown up by the ocean. In comparison, the EPA estimates that in 1969 man-made sources accounted for only 35.2 million tons in the United States, a slight decline from 1966.[9] Other than forest fires, the main sources of particulate pollution are coal-burning incinerators, industrial furnaces and power plants, and some manufacturing and industrial processes. The transportation sector accounted for 800,000 tons in 1969, with automobiles accounting for about half, or 400,000 tons. Except for possible global effects (see Section 3.2), particulates are a pollution problem only in population and industrial centers, as we have seen for gaseous pollutants (Chapter 2).

In sum, particles from natural sources consist primarily of fine sand, soil particles, pollen, bacteria, dried salts from ocean spray, and water droplets. Particulates from human activities are derived mainly from combustion processes, although comminution processes (mining operations, cememt plants, metal working, abrasion of tires, etc.) contribute a significant fraction. Thus, although nature makes the preponderent contribution on a global basis, men frequently tip the scale the other way in local areas.

11.4. SIZE, SHAPE, AND MASS PARAMETERS

Aerosols may be divided broadly into three groups by size as follows:

Particle radius, μm	Name
<0.1	Aitken particles
0.1–1.0	Large particles
>1.0	Giant particles.

Large particles are involved in scattering visible radiation because their dimensions are of the same order as wavelengths in the visible region (see Section 1.4.4). Both large and Aitken particles are sufficiently small to be influenced by Brownian motion of air molecules and thus may remain suspended for long times and travel over large distances. Giant particles are sufficiently large that Brownian motion is unimportant and the atmosphere may be treated as a uniform fluid. From the point of view of numbers, the Aitken particles are overwhelmingly greater than the large and giant particles; from the point of view of mass, Aitken particles constitute about 10–20% of the total aerosol mass.[10]

The three broad classifications according to size many be broken down descriptively, as shown in Figure 1.[11]

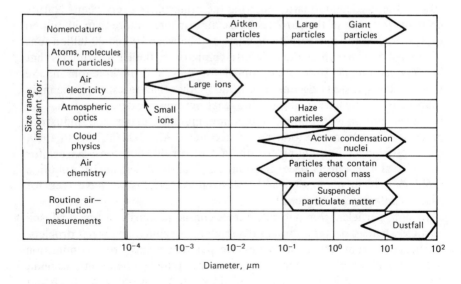

Figure 1 Ranges of particle diameters of various types of atmospheric particulate matter.[11]

The sizes shown range from about 10^{-4} to about 10^3 μm. Particles larger than 10 μm (termed *dustfall*) rapidly settle out of the air by sedimentation. The sedimentation of spherical particles above 1 μm in diameter is best described by Stokes's law (see Section 1.4.4). The balance of gravity settling, Brownian effects, and agglomeration were seen to lead to the formation of a mature aerosol with a limited "self-preserving" size distribution. While appealing from a theoretical viewpoint, it is found that many urban aerosols fail to conform to this model. Continental aerosols show one peak in the size distribution curve that may shift upward or downward with time, retaining the same qualitative shape. If local manmade sources are prevalent, several peaks may appear, but these are smoothed out in time. A striking feature of many continental aerosols is a regular decrease in number concentration of particles above 0.1 μm. Junge[3] has given an empirical formula describing the distribution of such particles:

$$\Delta n = C R^{-(\beta+1)} \Delta R, \tag{1}$$

where Δn is the number concentration of particles with radius between R and $R + \Delta R$, C is a constant, and β is a number in the exponent, ranging from 2.5 to 4. A plot of log $(\Delta n/\Delta R)$ versus log R is a straight line for many mixed aerosols in Europe and the United States.[12]

We noted in Chapter 1 that the average wavelength of daylight and the average size of mature urban aerosols are about the same (0.5 μm). This coincidence is the basis of pronounced optical effects that lead to methods of measuring particulate concentration in the atmosphere through visibility loss.

11.5. COMPOSITION, REACTIONS, FATE

Particles from different industrial and agricultural sources will, of course, have quite different chemical compositions. It is not possible to discuss all of them, but it is useful to indicate the diversity of materials which are emitted into the atmosphere. Table III[13] provides representative data on particulate composition from a number of sources of interest. Some idea of range and average composition of suspended particulate matter in the atmosphere is given in Table VI of Chapter 1.

11.5.1. Removal Mechanisms

Particulate matter in the atmosphere, both soluble and insoluble, is washed out primarily by rain or snow (see also Section 8.2.4). Very small sized particles are collected by small cloud droplets under the influence of Brownian motion and subsequently are precipitated out when the cloud droplets form rain. Larger particles are removed by the same two-stage

Table III

Representative Particulate Compositions from Some Sources[54]

A. Open-Hearth Furnaces[a]

Compound	Percent
Fe_2O_3	89.1
FeO	1.9
SiO_2	0.9
Al_2O_3	0.5
MnO	0.6
Alkalis	1.4
P_2O_5	0.5
S	0.4

C. Fly Ash Collected at 3 N.Y. Incinerators

	Weight percent	
	Collected	Emitted
Silicon as SiO_2	49.5	36.3
Aluminum as Al_2O_3	22.9	25.7
Iron as Fe_2O_3	6.3	7.1
Calcium as CaO	8.8	8.8
Magnesium as MgO	2.2	2.8
Sodium as Na_2O / Potassium as K_2O	6.0	10.4
Titanium as TiO_2	1.3	0.9
Sulfur as SO_3	3.0	8.0

B. Portland Cement Plant, Lehigh Valley, Pa.

	Weight percent	
Compound[b]	Raw kiln feed dust avg. for 3 types cement	Dust from precipitator outlet (avg. of 3 samples)
CaO		40.9
$CaCO_3$	75.9	
SiO_2	13.4	18.8
Al_2O_3	3.7	7.1
Fe_2O_3	2.1	9.6
MgO		2.5
Na_2O		1.1
K_2O		7.3
MnO		0.2
TiO_2		0.1
CuO		Trace
Ignition loss		12.7

D. Analysis of Fly Ash from a Variety of Coal Combustion Units

Compound	Percentage of fly ash
Carbon, C	0.37–36.2
Iron (as Fe_2O_3 or Fe_3O_4)	2.0 –26.8
Magnesium (as MgO)	0.06– 4.77
Calcium (as CaO)	0.12–14.73
Aluminum (as Al_2O_3)	9.81–58.4
Sulfur (as SO_3)	0.12–24.33
Titanium (as TiO_2)	0 – 2.8
Carbonate (as CO_3)	0 – 2.6
Silicon (as SiO_2)	17.3 –63.6
Phosphorus (as P_2O_5)	0.07–47.2
Potassium (as K_2O)	2.8 – 3.0
Sodium (as Na_2O)	0.2 – 0.9
Undetermined	0.08–18.9

[a] Fluorides may be present in open-hearth-furnace particulate emissions if fluorspar fluxes or fluoride-containing ores are used.
[b] No determination of sulfur made.

11.5. Composition, Reactions, Fate

process or by direct washout as the raindrops fall. The average lifetime of particles in the lower atmosphere depends on the pattern of rain- or snowfall in the particular region. Studies using radioactive tracers give lifetimes ranging from 6 to 14 or more days in the lower troposphere. In the upper troposphere, the residence time appears to be considerably greater, ranging from 2 to 4 wk. In the lower stratosphere, the residence time varies from 6 months at high latitudes to about a year just above the tropical tropopause. The residence time continues to increase with altitude and is about 3–5 yr in the upper stratosphere and 5–10 yr in the mesosphere.[14]

Because the residence time of particles are appreciable and are subject to the influence of moisture and active gaseous materials such as SO_2, NO_x, CO_2, NH_3, and other materials, it is unusual for a particle to retain its original chemical and physical identity for any length of time. In addition, it may change by accretion of other types of particles because of coagulation, condensation, and evaporation processes.

An important physical characteristic of small particles is their ability to serve as condensation nuclei for water vapor. Particles in the range 0.1–1 μm and perhaps larger are most effective as condensation nuclei, but ionic salts such as ammonium and sodium sulfate, as well as sulfuric acid droplets, also are very effective as condensation nuclei. Similarly, the particles serve as nuclei for the initiation of freezing of supercooled water droplets.

Gases and particles in the troposphere are fairly well mixed vertically within periods of a few days to about a month. At the same time, they are removed by scavenging mechanisms and direct contact with surfaces at ground level. In the lower part of the stratosphere, however, residence times are of the order of 2 yr.[15]

11.5.2. Chemical Reactions of Aerosols

Relatively few studies have been made of particulate reactions in the atmosphere compared to their great diversity of composition; some of these have been reported in Ref. 1. The role of particulates in photochemical reactions is of particular interest, but most of those studies (e.g., Section 8.4.1, Table V) take place in the gas phase. Particle–gas and particle–particle reactions must both take place, the latter particularly if the particles are small (0.1 μm or less), but these events are difficult to observe and collisions between larger particles in the atmosphere are rare (reactions between particles collected on filters must be regarded as atypical).

Reference 1 reports a few specific cases: sulfuric acid mist and ammonia gas reactions[16] were shown to be greatly influenced by humidity. The effect of both humidity and surface area of particles on photochemical reactions was shown in another study;[17] the reaction product was shown clearly proportional to the surface area, but the influence of humidity was complex. Other studies have shown the interactions between SO_2 and metal oxide aerosols.[18]

This paucity of data is being remedied in part, by current studies. These include the RAPS study discussed in Chapter 6, which, among other things, is expected to improve knowledge of particulate reaction roles in smog and pollution formation.

11.6. MEASUREMENT PROBLEMS OF PARTICULATE MATTER

The primary problem in particulate measurement, as in other measurements, is why we are making the measurement in the first place. What are we going to do with the information once we have obtained it? As pointed out in Chapter 10, we may seek qualitative and quantitative information for survey and research purposes—that is, to answer the question, *What* is out there, *how much,* and does it *change* with time? We may be interested in qualitative and quantitative information with respect to effects, having established tentatively from survey work that a relationship exists between a measured variable and some important effect. Finally, we may require quantitative measurements for surveillance, planning, and control purposes in order to satisfy legal requirements with respect to emission sources and ambient air quality.

For survey work, generalized laboratory techniques in the hands of trained and skillful analysts are probably the most efficient and cost-effective ways of obtaining the required information. The entire armamentarium of physical, chemical, and biological techniques may be brought to bear on the investigation in a flexible and effective manner. Many different particle parameters will be measured by a variety of instrumental and manual methods. The particle parameter to be measured depends most closely on the effect the particles are likely to have after emission into the atmosphere. Some effect–particle parameter relationships are listed in Table IV.

The most important single parameter of airborne particulate matter is mass concentration, expressed as micrograms per cubic meter. Effects on which air-quality criteria are based, and particularly health effects, are cited in these terms. In contrast, nearly all automatic instruments actually measure the volume concentration, cubic microns per cubic meter, which is converted to mass units by use of an assumed average density (the fact that this density and its variations are really not known in most cases might suggest a profitable field for future research).

11.6. Measurement Problems of Particulate Matter

Table IV
Effect of Particulates and Appropriate Measurement Parameter

Effect	Parameter	Remarks
Radiation from the sun (effect on local and global climate)	Spectral radiation transmission	Energy reaching the earth's surface; visibility, haze, etc.
	Spectral radiation scattering	Energy lost to earth; color effects; visibility
	Mass per unit volume, total mass	Relationship moot
	Particle size distribution	
Visibility, visual range	Mass per unit volume	Same
	Particle size distribution	Same
Soiling of materials	Dustfall	Same
Corrosion of materials	Dustfall	Particles accelerate or catalyze corrosion; synergism with SO_2, NO_2
	Composition	Particles themselves are corrosive; H_2SO_4, HF, HNO_3, chlorides
Mechanical effects	Dustfall, composition	Electrical contacts; abrasion of materials
Effects on vegetation	Dustfall	Cement plant dust
	Composition	Fluorides, H_2SO_4, soot; effect on animals ingesting fluorided vegetation
Effects on humans	Particle size distribution	Respirable particles (0.1–5 μm)
	Composition	Silicosis, toxic metals, H_2SO_4 mists, asbestos

Second in importance is the particle size distribution. The respirability of particulates—and hence their retention in the lungs—is a function of particle size, and so, therefore, is the most serious health hazard. The size statistical distribution is generally, but not universally,[55] assumed to be lognormal (see Section 1.4.4). Therefore, the mean radius (or diameter) is the geometric mean, which is equal to the particle count median, or the 50 percentile value (see Figure 2). But the most significant parameter is not the count median but the mass mean or mass median diameter (half the mass lies on either side of this diameter). This can be computed from the count median, given the geometric standard deviation of radius (Section 1.4.4). In most of the literature, the distinction between these mean values is not finely drawn, although they may differ by a factor of 4 or 5 or more.

11.6.1. Sampling

Particulates, like other pollutants, are measured for three basic reasons: (1) to determine whether hazardous levels exist in the ambient, (2) to evaluate the effectiveness of abatement procedures, and (3) to trace pollutants to a specific source. To accomplish these objectives, an effective sampling procedure matched to the measurement objective must be used. Because of the different objectives of a measurement program, as well as the wide variety of conditions encountered, it is probably not possible to predict the performance of a particular sampling technique in a specific situation. For this reason it is necessary to appreciate the principles and limitations of the several techniques available so that a suitable match between technique and application can be made.

The essential parameters of a sampling method are as follows:

1. Threshold of sample detection
2. Accuracy of sampling device; collector efficiency
3. Variations in pollutant concentration with time, location, and meteorological conditions
4. Total sample size with respect to representativeness of the sample, space- and/or time-averaging, and requirement not to miss significant peaks
5. Range of expected concentrations.

When sampling for particulate matter, values for collector efficiency may be misleading, since efficiency values may be calculated on the basis of weight of particles removed, the number of particles removed, or reduction in soiling capability. For example, if the aerosol being sampled contains a broad distribution of particles sizes, the sampling efficiency on a weight basis may be quite high, exceeding 99%, while efficiency on the basis of the number of particles removed may be relatively low, say 50%. As noted above, the purpose of the measurement will determine the sampling method to be used and the best way to calculate its efficiency. If we are concerned with visibility effects, a particle count sampler is used and the efficiency calculated from the number removed. If soiling characteristics are significant, a mass sampler is used and the efficiency calculated on a weight-removed basis. Although it is not necessary that a sample collection device be 100% efficient, it is important that the efficiency be known and repeatable; an efficiency of at least 75% is required to obtain meaningful data. However, there is relatively little reliable information on collector efficiencies as a function of sampling rates, pollutant concentrations, and reagents used in the collection device. It is possible that great variations in results and inconclusive or inaccurate data are caused by variation in collector efficiency with the factors mentioned. The effectiveness of a sampling device and/or procedure is a function of the factors mentioned as well as the accuracy of subsidiary apparatus such as flow meters. It is best to evaluate overall effectiveness by use of a calibrating standard or some known technique (see Section 11.10).

Since any sampling period is a time-averaging operation, the sampling interval should be carefully chosen to achieve the optimum compromise between conflicting objectives: short enough to detect significant peaks; long enough to collect a sufficient amount of pollutant consistent with the analytical technique to be used as well as to minimize "noisy" peaks. Factors influencing the minimum sampling time are the actual concentrations likely to be present, the rate of sampling imposed by the sampling technique or device itself, and the constraints imposed by the final analytical procedure.

Continuous sampling and quantitation provide the most useful record of pollutant concentration variations (i.e., real-time monitoring). Peak-concentration data obtained from real-time analysis are essential when monitoring the air to determine pollutants hazardous to health or unpleasant or to comply with legal standards (see Section 10.1.3).

11.6.2 Ambient and Source (Stack and Mobile) Measurements

Ambient particulate measurements are easier than those from sources, because of the many adverse analytical conditions associated with the latter: high temperature, high humidity, corrosive components, ongoing chemical reactions that may alter the sample composition. As with gaseous pollutants, the concentrations are much lower for ambient. This poses no special problems when optical or filtration methods are used, but point measurements in a small region are difficult to obtain. Usually space- or time-averaged data are obtained for ambients.

Aside from the problems already mentioned, there is a specific particulate sampling problem for sources—that of isokinetic sampling (see Section 9.2). There are two main forces acting on a particle in a gaseous laminar stream:[19]

1. Inertial (I): $F = ma$, K.E. $= 1/2\ mv^2$, where a and v are vectors
2. Drag (D): Frictional dissipation of kinetic energy (K.E.) by collision with air molecules and other particles depends, among other things, on surface area and shape.

For larger particles, the I–D ratio is greater than for smaller particles. Therefore, if the suction velocity of a sampling probe is smaller than that of the stream, the larger particles will tend to continue in their same path because of their high inertia; the smaller particles will follow the slip stream around the probe because of their small inertia. Thus, the sample withdrawn will be too high in large particles compared to the stream. If the suction velocity of the probe is greater than that of the stream, again, the larger particles will tend to continue in their previous path, while the smaller particles will follow the new path because of their low inertia. In this case, the sample withdrawn will be too low in large particles. For very small particles or for uniform particle size distribution, no problem arises. For a wide particle size distribution, an unrepresentative sample will be withdrawn unless care is taken to match probe and stream velocities (i.e., isokinetic sampling). The breakpoint between large and small particles, in this respect, is about 5 μm.

11.6.3. Form of Reporting

Mass

The EPA has specified a preference for reporting pollutants as micrograms per cubic meter. This unit is readily understood and relevant to particulate matter and has the advantage of being completely metric. It is also independent of temperature. However, it is not clear whether in expansion of a gas with suspended particles, the particle mass density decreases along with the gas density. For small particles this is probably the case.

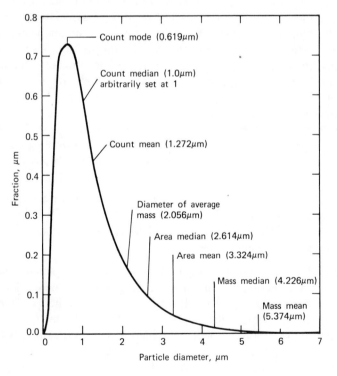

Figure 2 Log-normal distribution of particle diameters, assuming count median diameter of 1.0 μm.[1]

Particle Size Distribution

The preferable form would appear to be a frequency plot as in Figure 2 along with a table showing percent by mass in a given size range.

Number

Pollen counts are now reported as number per cubic meter. In many cases a similar reporting form for other particulates may be of advantage (haze, visibility, etc.).

Composition

The ideal form is percent by weight of the species as present in the sample. This is frequently difficult to obtain. A useful form, especially for toxic metals, is percent by weight of the element or the oxide. These provide a means to make a rough charge or acid–base balance. If the sample is considered to be the total volume of air from which the particulates were separated, the reporting form is micrograms per cubic meter. Hendrickson[20] recommends the following:

- Particle fallout: Count: number per square centimeter per time interval; Weight: milligrams per square centimeter per time interval.
- Airborne particulates: Count: number per cubic meter; Weight: micrograms per cubic meter.

11.6.4. Real-Time, Space-Averaged, and Time-Averaged Measurements

These topics have been discussed in Chapter 10 in relation to analysis of gaseous pollutions, and the considerations discussed there also apply to particulate analysis. It is only necessary to point out that in general, optical methods will be in real-time and space-averaged, while filtration methods are time-averaged with an element of space-averaging. Composition analysis of particulates will generally not be in real time, since a finite collection time and work-up in the laboratory are required. In the description of instruments below, their position in this classification will be mentioned.

11.7. INTEGRATED INFORMATION MEASUREMENTS

11.7.1. Particle Collection and Concentration

Filtration

Methods using filtration depend on the passage of relatively large volumes of gas through some device that separates and concentrates the particulate matter. After separation, a variety of techniques are brought to bear on the separated mass of particles to evaluate them qualitatively and/or quantitatively. In a few cases, a pump operating upstream of the device may be used, but since this will nearly always affect the sample adversely, the practice is rarely used and is to be discouraged. The preferred placement of the pump is downstream of the collection device (i.e., the pump operates in the vacuum mode). Motor-driven vacuum pumps, hand-operated pumps, or aspirators are employed.

In order to provide quantitative data on the extent of pollution, the flow-rate or cumulative volume of air sample through the device must be measured. The flow meter is usually downstream of the sampling device, before the vacuum pump. A manometer and thermometer should be used at the flow meter to correct flow conditions to standard conditions.

A filter consists of some porous material through which a fluid sample is forced. The fluid passes along tortuous paths in the filtration medium. Inertial forces on the particles or direct collision bring the particles into direct contact with a large surface area and stop particle motion. Particles collecting in the filter reduce the pore size, so that finer particles are caught and the resistance to flow increases. Filter media have many different structures:[21]

• Fiber filters: Papers or mats made of cellulose, glass fibers, mineral wool, plastic fibers

• Granular filters: Porous ceramics, fritted glass or metal, sand, and other solid particles (sugar, salicylic acid)

• Controlled pore filters: Known as membrane or molecular filters, commonly made from cellulose esters. The method of manufacture permits a uniform, controlled pore size.

The tape sampler is the most widely used filtration device. Successive portions of the filter medium, paper or cloth, from a 1–2-in.-wide tape roll, are positioned in an airtight manner between an intake tube and a vacuum pump. Air is drawn through the filter for selected periods (15 min to 24 h); at the end of each period a new portion of the filter is moved into sampling position. The spots resulting from the collection of particulate matter may be evaluated optically during the course of sampling, immediately afterwards, or in the laboratory after some time lapse. They may also be examined microscopically or chemically. All of these methods of evaluation of particulates collected on tape will be described in more detail below (Section 11.7.2).

Impingement

In wet impingement, the air sample is bubbled through liquid at the bottom of a vessel by means of an inlet tube whose tip is below the liquid surface.[22] The type of liquid selected depends on the analytical end in view. For gaseous pollutants special solutions are used to insure quantitative removal of the COI from the air; for example, a solution of sodium tetrachloromercurate for SO_2, in the West–Gaeke method. For particulates, however, pure water or an organic solvent is normally used.

In dry impingers, or impactors, a series of progressively smaller jets direct the air stream against a series of glass slides or other surfaces. This results in fractionating the particles according to size. The fractions may be weighed and subjected to other analytical procedures (see below). Thus, the cascade impactor permits sizing the particulates as well as composition analysis.

The Andersen impactor[23, 24] has a sampling head containing up to nine circular metal plates, each plate with several hundred holes (jet orifices). The size of the orifices is progressively smaller for each downstream plate, so that the velocity of the gas stream passing through increases for constant overall sample throughput. On the basis of the same I–D considerations discussed above (Section 11.6.2), large particles are caught on the first plates and smaller particles acquire increasing inertia as they pass through. A stage is reached where a smaller particle is also caught on a portion of the plate that does not contain a hole, since it cannot follow the slip stream. The particles are thus fractionated by size, the number of fractions corresponding to the number of plates. Fractionation in the size range 0.01–10 μm is possible, depending on the orifice plates used. Here again, in addition to fractionation by aerodynamic size, the fractions may be weighed and subjected to qualitative and quantitative physical and chemical analysis. An interesting example of the latter is the size distribution of trace-metal components in particulates.[25]

Sedimentation

Sedimentation depends on the natural gravitational settling of particulates. In still air, the rate of settling is best described by Stokes's law (Section 1.1.4) and depends mainly on the density and the square of the linear dimension ("effective" area). Sedimentation devices may be flat plates (microscope slides), but more frequently used are jars with a funnels at the open ends on the dustfall bucket. To prevent the loss of collected material, water or other liquid may be placed in the jar. The time for collection varies from days to months. Because of the influence and variability of meteorological factors (wind, rain, sunshine, etc.), it is difficult to relate mass collected in a scientific manner to other variables. The collected material is usually weighed and reported as mass per unit area. It may then be subjected to qualitative and quantitative chemical analysis as well as size-distribution studies. The technique is used principally to estimate dustfall or sootfall and thus gives an indication of "dirtiness" of a locality.

Multiplication of the mass per collection area (area of funnel or top of jar) to mass per community area probably gives results with great inherent error.

Electrostatic Precipitation

The basic principle of operation in electrostatic precipitators is the charging of suspended particulate matter in an electric field and the discharge and immobilization of the latter on a collecting electrode. A potential difference of 10,000–30,000 V imposed on two metal electrodes results in ionization of the air molecules between them. The air molecules become positively charged because of removal of electrons from the outer orbits. Particles in the air stream collide with the charged molecules, assuming the charge. The electric field results in a drift velocity of the particles toward the negative electrode where they are discharged and collected.

Electrostatic precipitators are nearly 100% effective for particles with diameters of 0.01–10 μm^3, which includes the respirable range. Because of their efficiency, they are not only used for investigative purposes but provide an effective means of particulate pollution abatement.

Thermal Precipitation

If a thermal gradient of 750°C/cm or greater is created in an air space, it is found that small particles migrate from the hot side to the cool side. A complete explanation for the drift mechanism has not yet been developed, but commercial precipitators based on the technique are available. High efficiency over a wide particles range is claimed (0.001–100 μm). An advantage is the gentle treatment of the particle in the course of immobilizing, so that subsequent examination relates more nearly to the original particle rather than a mechanically or chemically altered particle. A review of thermal precipitation, theory and practice, has been written by Gieseke,[26] which should be consulted for detailed background in this field. Thermal precipitators both for aerosol sampling and air cleaning are discussed.

Centrifugal Separation

In this technique, used in cyclone separators, an air stream at high velocity enters a cylindrical body tangentially at the top. A vortex or spiral flow downward is created between the wall of the cylinder and the wall of a smaller outlet cylinder placed concentrically in the top cover of the main cylinder. The bottom of the main cylinder is closed by a cone. While used extensively for abatement of particle emissions in industry, small versions are

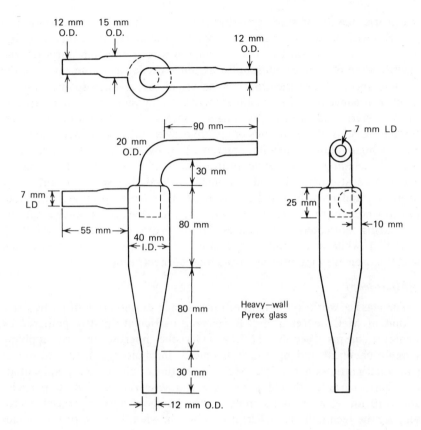

Figure 3 Glass cyclone that makes a cut at about 3 μm.[52] (Reprinted, with permission, from *J. Air Pollut. Control Assoc.,* **13,** No. 4.)

used for investigative purposes (Figure 3).[52] They are especially efficient for particles greater than 5 μm, but submicron particles are not captured.

Centrifugal forces are also employed in a collector-particle sizer that operates on the principle of spiral centrifugation. The theory, construction, and operation of the instrument are described by Stober and Flachsbart,[27] and the instrument is available commercially.

11.7.2. Analysis and Measurement of Collected Particulate Matter

Particulates in industrial plumes and effluents or in the atmosphere (aerosols) may be collected as in the foregoing and the samples measured to determine the total concentration or analyzed by chemical or physical means to identify their nature or origin. General practice in monitoring atmospheric quality or stack emission focuses on the total concentration of particles without regard to size distribution or analysis, since this is all that is required to conform to current regulations. For research purposes and identification of sources, more information is needed as to the nature of the particle species. Identification can be visual, as in microscopy, or by a number of conventional or exotic analytical techniques. Mass concentration can be determined directly, by weighing samples collected on tared filters through which large quantities of air are drawn (such as the ubiquitous Hi-Vol, or high-volume, filtration apparatus), or by a variety of indirect techniques, including photometric measurement of filter-paper discoloration and light scattering. The indirect methods lend themselves to automatic operation and recording and so are better adopted to routine, continuous monitoring. In this section we will describe means of analyzing and measuring collected particulates (apart from the obvious weighing methods), while in the following section we will discuss those means that lend themselves to monitoring of the total concentration.

Microscopy

The classic work in the field of microscopic examination of particulates is that of McCrone et al.[28, 29] By placing particulate matter gathered by whatever means (Section 11.7.1) under the microscope and applying suitable chemical and optical techniques, morphological identification of individual particles is possible. McCrone's atlas of photomicrographs of atmospheric particles should permit the characterization of most particles above 10 μm in an urban sample. In the hands of an experienced microscopist, the technique is powerful because the identification of the species present, not merely the elements, gives information not only as to the nature and quantity of possibly harmful materials but also a strong indication of their source.

Photometry

Photometry in association with tape samplers attempts quantitation of the concentration of particulates as a class, without differentiation as to chemical or physical nature. As such, it lacks the refinement of the microscopic visual or analytical methods described in this section and the convenience of remote and real-time methods discussed further on. Nevertheless, it is the most common "automatic" particulate monitoring method now in use.

The spots collected on a tape sampler may be evaluated as follows:

1. Visual comparison with a standard gray scale
2. Reflectance measurement photometrically using a clean portion of the tape as a reference

3. Transmittance measurement photometrically using a clean portion of the tape as a reference.

Transmission measurements are more or less related to visibility conditions of the atmosphere, while reflectance measurements are related to soiling. Visual comparison and reflectance measurements attempt to quantitate the blackness of the deposit; transmittance measurements are a function of all the particles, not merely the black ones. The two types of measurement may not correlate well. For example, a white material such as MgO smoke collected in a spot may not appear gray or may even reflect better than the filters, but in transmittance, the spot might show close to zero or no transmission at all. In fact, however, most urban particulates correlate fairly well on the three measures—visual, reflectance, and transmittance.

The problem with these methods of estimation is that they do not show a well-defined and repeatable relationship between the quantity measured and the number or mass of particles collected. For example, reflectance changes will depend on whether later deposits are on the surface or penetrate into previously deposited layers, while transmittance changes will depend in a like manner on whether additional deposits add to the thickness or fill pores in the previous deposit. In addition, the rate of deposition changes with time as pores in the filter medium grow smaller by immobilization of particles; hence, the structure of the filter medium changes as well as the structure of the deposit, which itself becomes part of the filter medium. Finally, beyond a certain density, reflectance measurements do not discriminate, because only the top surface, which may not change in character, contributes to reflectance, while the thickness or total mass of the deposit does not affect the reflectance.

Measurements from a photometer may be reported in a number of ways (see Section 1.4.4.). The first is percent transmittance, T, where

$$T = \frac{P(T)}{P(T)_0} \times 100 = \frac{\text{Radiant power transmitted by spot}}{\text{Radiant power transmitted by clean filter}} \times 100. \tag{2}$$

The second is optical density, OD, where

$$\text{OD} = \log \frac{100}{T} = \log \frac{P(T)_0}{P(T)} = A = \text{absorbance.} \tag{3}$$

A frequently used reporting form is the COH, or coefficient of haze. A COH unit is defined as that quantity of particulate matter on the filter that produces an OD = 0.01:

$$\text{COH} = \frac{\text{OD}}{0.01}. \tag{4}$$

Thus, if a transmittance measurement reduces to an OD = 0.01, COH = 1.00. More frequently, COH is reported as COH per 1000 linear feet of air passing through the filter. The entire volume of air sampled is visualized as a cylinder of equal volume whose cross-sectional area, A, is that of the spot and whose length is such that LA = volume sampled = V:

$$L = \frac{V}{A} = \frac{F \times t}{A}, \tag{5}$$

where V is the volume of sample in cubic feet, A is the area of spot in square feet, F is the flowrate in cubic feet per minute, and t is the time of sampling in minutes. The length in thousands of linear feet is then $L/1000$ and COH per 1000 linear feet is:

$$\frac{\text{COH}}{1000 \text{ linear ft}} = \frac{\text{OD}}{0.01(L/1000)} = \frac{(\text{OD})}{L} \times 10^5 \tag{6}$$

$$= \frac{(\text{OD}) \times A}{F \times t} \times 10^5.$$

The table in Section 1.4.4 indicates roughly the significance of the COH/1000 ft numbers.[30]

For reflectance, a frequently used soiling index is RUDS, or reflectance unit of dirt shade. It is defined as a deposit that produces an optical reflectance of 0.01 due to 10,000 linear feet of sample.[31] In analogy to the transmittance we may define the reflectance R as follows:

$$R = \frac{P(\text{R})}{P(\text{R})_o} \times 100 = \frac{\text{Radiant power reflected by spot}}{\text{Radiant power reflected by clean filter}} \times 100$$

$$\text{OR} = \log \frac{100}{R} = \text{optical reflectance}, \tag{7}$$

$$\text{RUDS} = \frac{(\text{OR})}{0.01(L/10,000)} = \frac{(\text{OR})}{L} \times 10^6$$

$$= \frac{(\text{OR}) \times A}{F \times t} \times 10^6. \tag{8}$$

Katz[32] defines the reflectance, R, in the same manner as we have defined OR in Eq. (7) with some confusion in the use of the symbol R. To avoid this confusion we define R and OR as above.

Wet Chemistry

Although many workers have described wet methods for the examination of collected particulate matter, the classical work in this field has been the product of P. W. West and his co-workers at Louisiana State University. A compact review of this work is given in Stern's compendium.[53] West discusses the determination of a wide variety of elements in collected particulate matter, using wet chemical techniques. The general approach will be briefly described below; the original literature should be consulted for experimental details.

Particulate matter is collected on a filter paper by an appropriate filtration technique (tape sampler or high-volume sampler). The total mass collected may be determined by weighing the filter before and after collection, making sure to equilibrate the filter under controlled standard conditions. Since the total amount of matter collected is quite small, classic gravimetric or titrimetric procedures are rarely used, although a titration for sulfuric acid mists is described.[33] The main technique used is to leach the desired elements from the central spot of collected particulates with suitable reagents. Using the ring-oven technique of Weisz,[34] the leaching solution, added to the center of the spot, dissolves the desired constituents and transports them by capillarity toward the edge of the paper. The edge of the paper is clamped on the top of a heated metal cylinder, which causes the water in the solution to evaporate (Figure 4). After leaching and drying, specific reagents are applied to the paper to form colored compounds with various materials leached into the ring. The formation of a characteristic color with a species is a qualitative indication of the presence of a specific species. The intensity of the color is a quantitative measure of the species. The intensity may be quantitated by running through the procedure with known amounts of the species of interest and comparing the unknown with standards. Table V is a partial listing of some species that may be determined in this manner, the range, and limit of detection. As indicated in Figure 5, a number of species may be determined from one spot by cutting the filter after leaching.

Other final quantitating techniques besides the ring-oven plus color-forming techniques may be used on the collected particulate sample. These involve various instrumental techniques. Digestion with mineral acids is the method of choice to bring the species into solution. Quantitation may then be achieved by using flame photometry, atomic absorption spectroscopy, and polarography. Of these, atomic absorption spectroscopy has been increasingly used in recent years because of the great sensitivity and selectivity and, with recent instrumentation, ease of manipulation and quantitation.[35, 36]

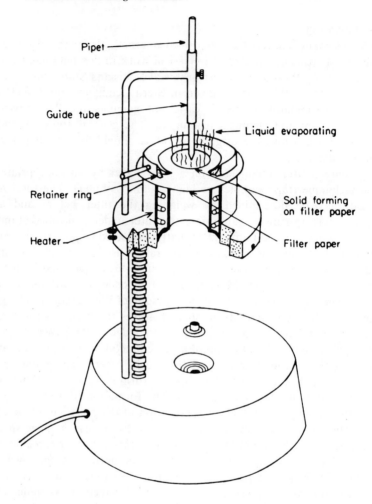

Figure 4 Weisz ring oven for analysis of particulates.[53]

Nonwet Quantitation

A number of instrumental methods may be applied directly to the collected particulates without the necessity for wet-chemical manipulations. Among these may be mentioned neutron activation analysis,[37] X-ray fluorescence,[38] and spark source mass spectrometry.[39] The literature is voluminous for each of these topics and should be consulted for details.

Table V
Colorimetric Reactions Used with Ring-Oven Methods

Species	Reagent; color	Limit of detection, µg	Range, µg
Aluminum	Morin; green-yellow under UV	0.01	0.03 – 0.5
Antimony	Phosphomolybdate; blue	0.08	0.1 – 1.0
Beryllium	Morin; yellow-green under UV	0.01	0.01 – 0.2
Chromium	Diphenylcarbazide; violet	0.15	0.3 – 1.0
Cobalt	1-nitroso-2-napthol; red brown	0.02	0.04 – 0.5
Copper	Dithiooxamide; dark green-black	0.04	0.01 – 0.5
Iron	$K_4Fe(CN)_6$; blue	0.01	0.01 – 0.5
Manganese	Malonic acid; brown	2.0	2.0 –10
Nickel	Dimethylglyoxime; red	0.08	0.10 – 1.0
Nitrite	Diazo reagent; pink-red	0.01	0.025– 0.25
Phosphate	Orthodianisidine molybdate; brown	0.02	0.05 – 2.0
Selenium	Diaminobenzidine; citron yellow	0.08	0.1 – 0.5
Sulfate	Barium chloride-$KMnO_4$; pink	0.1	0.1 – 1.0
Vanadium	Anthranilic acid-salicylaldehyde; gray-yellow	0.01	0.01 – 3
Zinc	KCN-chloralhydrate; orange	0.04	0.05 – 1.0

Source. Adapted from Ref. 53.

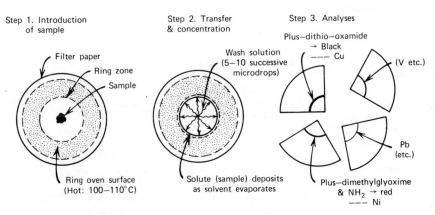

Figure 5 Essential steps of ring-oven analysis.[53]

Organic Materials in Particulates

The techniques for collection and analysis of inorganic materials in particulates described above may be applied to organic materials. However, the quantitation problem is complicated by the fact that elemental analysis gives little information of value. Characterization of the organic compounds present is required to assess health hazards as well as meteorological effects (smog formation). The general approach is to collect the particles by an appropriate technique (see Section 11.7.1) and then separate the organic matter from the inorganic material by extraction with appropriate solvents. This step may be included in a total fractionation scheme the object of which is to separate the organic materials into broad classes of compounds, which permits easier identification and quantitation. Such a broad fractionation scheme might divide the organics into a neutral, an acidic, and a basic fraction, depending on the nature of the solvent used for separating the class of materials from the other classes. Each fraction is then subjected to further fractionation and final quantitation.

Table VI

The Neutral Fraction in Organic Particulate Material[40]

Class of material	Analytical approach	Remarks
Saturated hydrocarbons	Subfractionate, then chromatography (gas column)	Indicate combustion products of petroleum, gasoline, diesel fuel, wood, leaves
Unsaturated hydrocarbons	Same	Indicate plant products; tumorigenic
Benzene, naphthalene, and their derivatives	Same	Indicates gasoline, exhaust of internal combustion engines, organic combustion; carcinogenic
Polynuclear aromatic hydrocarbons (PAH)	Fractionation, column chromatography, thin layer and paper chrom; UV and fluorescence quantitation	Carcinogenic; PAH results from high temperature pyrolysis in reducing atmosphere
Oxygenated neutral components	IR, GC	Irritants, tumorigenic; industrial sources, photochemical reactions
Nitrosamines	Thin-layer chromatography, color reactions.	Mutagenic, carcinogenic, tumorigenic
Pesticides	Microcoulometric GC	Agricultural operation; toxic, carcinogenic

The chemical analysis and carcinogenic assay of organic particulates has been described in an extensive review of Hoffman and Wynder.[40] Those authors give a number of fractionation schemes starting with the collected particle sample and means for identification and quantitation of materials. The original may be consulted for experimental details. At this point it is useful to give a brief summary of the scope of the analytical schemes and the importance of some of the determinations.

The neutral fraction includes a wide variety of materials. Their importance in air-pollution work is indicated in Table VI,[40] where it is seen that many of the materials have health effects. In addition, their presence in the particulate is a clue to their source. The other fractions have not been as well studied as the neutral fraction. It is probable that many of the components in these fractions will be found, on further study, to have significant health effects.

11.8. METHODS ADAPTED TO CONTINUOUS MONITORING OF PARTICULATES

The following techniques are those that attempt to measure concentration of particulates as a class without regard to their chemical nature and, as such, are employed as monitors of total aerosol concentration or of a particular range of particle sizes. Monitors are generally required to be automatic or capable of being modified for unattended operation and to measure the current concentration with little or no time lag. The types of instruments described below have or can be utilized as ambient or source monitors or both.

Some of these techniques permit discrimination of aerosols by particle size (diameter, surface area, volume, or mass). Current research on the effects of particulate matter on health emphasizes the role of smaller particles capable of being taken into the human lung. It can be anticipated that particle-size discrimination will be important in future automatic monitoring instruments.

11.8.1. Counters and Classifiers

Most of these instruments depend on light scatter caused by individual particles. The latter are led, one by one, into an illuminated space monitored by a photocell. The observed scattering angle may range from 70° down to 20° or even 10°. In general, the smaller the angle, the less the dependence on particle color or refractive index.[41, 42] A typical forward-scattering instrument has a 30–71° half-angle optical system that maximizes signal per particle with minimal effects of particle color.

A problem common to all optical particle counters is the coincidence error, which is proportional to the square of the concentration and the observed zone volume. The effect is to shift upward the apparent mean size. The problem is minimized by insuring sufficient dilution to reduce the probability of two particles appearing in the zone. Sensing by light scattering is possible with particles down to 0.2 μm in gas and to about 3 μm in liquid. The largest measurable particle size in gases is about 70 μm, since large particles tend to settle out despite agitation of the sample.[43]

Another optical counting technique, not dependent on scattering, is the obscuration method carried out in a liquid medium. This is limited to particles 2 μm or greater. A variation is the focused laser technique, which resolves particles down to 3 μm (in liquid), using a HeNe laser. Diffraction effects in particles below 0.5 μm render this technique useless in this range.

The conductivity technique developed by Coulter is a nonoptical method that depends on the increase in resistance of an electrolyte in a circuit that includes a small orifice. When a particle passes through and partially blocks the hole, a change in resistance occurs that is directly proportional to particle volume if its diameter does not exceed 30% of the orifice size. Sophisticated electronics have been developed to give a complete particle size distribution within the range of the orifice used. The working range of the Coulter counter is on the order of 0.1–5 μm for a single orifice. A modified instrument drives the particles through the orifice electrophoretically and counts particles down to 0.05 μm.

A system developed by Whitby[44] counts and classifies particles in the range 0.007–0.6 μm by charging them and measuring their mobility in an electric field. The accuracy ascribed to this instrument depends on the assumptions made for charging the particles. Condensation of particles in supersaturated vapor to form countable droplets is another technique that can enumerate particles as small as 0.003 μm, although size discrimination within this range is not practical. Commercial counters using this principle are available.

11.8.2. Light Scattering

An aerosol photometer, as distinguished from a particle counter, measures the light scattered simultaneously from many particles illuminated in the sensing zone. When the physical characteristics of the particle remain relatively constant, the intensity of the scattered light is directly proportional to aerosol concentration, so long as the concentration is sufficiently low and secondary scattering does not occur.[45]

The advantages of scattering photometers include the following:

1. In situ measurements. It is not necessary to separate the particles from the air by sampling means.

11.8. Methods Adapted to Continuous Monitoring of Particulates

2. Instantaneous response.
3. A high degree of sensitivity, stability, and accuracy.
4. The quality measured (the scattering coefficient) correlates directly with atmospheric visibility, an important effect of urban aerosols.
5. Instruments can be relatively low in cost.

The disadvantages are as follows:

1. Assumptions must be made as to particle size distribution, density, refractive index, and absorption (or color) before the readings can be correlated with mass concentration. In general, such a relationship holds only for "well-aged" haze—that is, the particulate population from combustion, for example, that must be at least 1 hour old.
2. Multiple scattering and absorption effects would prohibit direct use in high-concentration atmospheres, as stacks.

In Section 3.5.3 it was pointed out that the visible range in hazy atmospheres is a simple function of particle scattering, the relationship described by Middleton,[45]

$$L_V = \frac{3.9}{b_{scat}}, \quad (9)$$

where L_V is the visual range in meters and b_{scat} is the scattering coefficient of the particles. The light absorption by gases and aerosols and the molecular scattering are assumed to be negligible. Further cautions in using these relations are given in Section 3.5.3.

Also, over a wide range of observations,

$$\text{Mass concentration } (\mu g/m^3) \simeq 3 \times 10^5 \, b_{scat}, \quad (10)$$

and hence,

$$L_V \times \text{Concentration} \simeq 1.2 \, (g/m^2). \quad (11)$$

In view of the caveats discussed in Section 3.5.3, the accuracy of Eqs. (10) and (11) is only a factor of 2 either way; that is, the coefficients are 3 $(+3, -1.5) \times 10^5$ and 1.2 $(+1.2, -0.6)$, respectively.

The relationships expressed in Eqs. (10) and (11) apply especially to the integrating nephelometer designed by Charlson.[46] Physically, this instrument is a 2 m by 15 cm pipe, fitted with a Xenon flash tube illuminating the volume between 8 and 170°. A phototube monitors the scattered light. No lenses or mirrors are used in the system, only collimating disks and diffusing (opal) glass. The sensitivity is sufficient to calibrate with molecules of pure gases, such as CO_2, Freon 12, or dust-free air. Various filters and photocathodes can be used for modifying spectral response. Overall error is about $+10\%$.

The integrating nephelometer is designed to measure Middleton's b_{scat}, which is the integral over the incident angle range $(0-\pi)$ of the volume scattering function times the sine of the incident angle. This is the quantity (in the absence of absorption) used in the Lambert–Beer law $I/I_o = \exp(-b_{scat}x)$. Other instruments measure the scatter at a fixed angle or range. The right-angle scatter photometer of Mendenhall is one such device that is commercially available. Single and double-beam instruments have been built. Absorption effects are essentially canceled out in a 90°-scatter instrument, but the readings are affected by the size distribution and refractive index (m) in accordance with Mie theory. Forward scatter instruments have been designed to reduce the dependence on m. In an extreme case where collection angles are as small as 0.5°, through use of a laser illuminator, the scattering may be controlled by Fraunhofer diffraction and is independent of m. It can be shown that the ratio of scattering at 2° to that at 0.5° is essentially a function of diameter, thus yielding the possibility of obtaining size or volume distribution information. Unfortunately, this relationship only holds for particles larger than 3 or 4 times the wavelength, so that it would be difficult to apply to those less than 1–1.5 μm.

The turbidity of the atmosphere may be estimated by the Volz "sun photometer," which measures scattering loss in the atmosphere in line with the sun. Corrections are made for sun angle and the depth of intervening atmosphere. The corrected "turbidity" value B, which is related to b_{scat}, can be used to estimate mass concentration by the formula.

$$m'_0 = 10^3 B_0 \quad \mu g/m^3,$$

where B_o is the value of B at the ground. This rough approximation only holds for the conditions of light winds, moderate city pollution, and clear sky; hence, the Volz sun photometer is difficult to calibrate.

Backscatter, using a powerful laser source as a radar or lidar can provide a measure of the concentration of particulate matter in a cloud or plume at a remote distance (the instrument also has ability to identify certain gaseous emissions as discussed in Section 6.2.2). With the assistance of a computer and specialized detectors, lidar reflections can plot stack effluent concentrations from a remote location. Commercial instruments are available in the United Kingdom, Sweden, and Japan.

11.8.3. Electric Charge Measurements

Electric charge current is also a basic technique for continuous monitoring of particulates. This is another bulk method, depending on the ability of fine particles to accept an electric charge, which is later measured with an electrometer. The greater the electrometer current, the greater the particle concentration. A fundamental paper by Whitby et al.[47] showed that under certain conditions the current is proportional to the volume concentration of the particles—that is, the mass concentration, given the average density. The conditions require that the size distribution of particulates must follow the "self-preserving" form; that is, the distribution is proportional to r^{-4}. This is approximately true over the range $0.05 > r > 5$ μm or from 0.1 to 10 μm diameter. As we have noted, this range contains the bulk of particles in a mature aerosol. The other requirement has to do with the means of charging the particles. Unipolar diffusion charging, rather than field charging, is recommended. Specifically, it is necessary to remove all particles below 0.1 μm, which can be done with a mobility analyzer (see Section 11.8.1); that is, the particles are unipolarly charged, passed through a weak electric field, which removes those most mobile below 0.10 μm, before collecting the remaining current. Experimental results cited by Whitby show fair correlation when these conditions are fulfilled, indicating that the volume fraction or mass density can be estimated within a factor of 2, perhaps. The best correlation coefficient was for a bivariate (nonlinear) model, 0.92 for particles over 0.1 μm diameter.

11.8.4. Radiant Energy Absorption

A fourth basic principle of particulate monitoring is energy absorption. This is the method most common in densitometer-type smoke monitors. Both optical photometers and IR (bolometer) types are commercially available. A source, a sealed-beam automobile lamp, for example, is placed at one end of a sample tube or in the stack, and the detector is installed a fixed distance from it. The readings of the latter can be calibrated in terms of "smoke points" (color density), grains per cubic foot, or Ringelmann numbers (see below). The calibration must be fairly unique for each installation and is dependent on the effluent color as well as particle size.

The energy absorption technique has also been utilized in the Transmissometer to measure visibility of the atmosphere, more or less directly. This instrument, used commonly at airports, might be described as a "long-path" photometer. A typical base line is 500 ft. The instrument is calibrated in miles of visibility. As discussed above (scattering principle), the relationship between visibility and mass concentration is linear for the urban areas tested but is accurate only to a factor of 2. Apart from any possible advantages of the averaging characteristic, the integrating nephelometer is probably a more convenient way to get the same information.

11.8.5. Light Reflection

The well-known Ringelmann method (see Section 4.3.8 and Figure 1, Chapter 4), in use since the turn of the century, is a manual application of this method.[48] Since light reflection is influenced by many factors other than mass concentration, such as position of the light source, particle size, and color, this method (manual or automatic) can hardly be related quantitatively to any property of the plume other than an aesthetic one. Even though the technique lends itself conveniently to remote observation with little or no equipment, it has little else to commend it. Its entrenched position in the statute books of many communities may not survive the development of practical instruments measuring more meaningful parameters, such as mass concentration and particle-size distribution.

11.8.6. Methods Related to Mass of Particles

If beta particles (electrons) are passed through a medium, some will be absorbed and some reflected. The loss in beam intensity is called *beta-radiation attenuation* and is a measure of the mass of material through which the beam has passed.[49] The attenuation depends statistically on the number of electrons in the matter encountered by the beta-particle beam. Correlation of attenuuation with particulate mass thus depends on the relationship between the number of electrons per atom (atomic number) and the mass of atom (atomic weight). This ratio is between 0.4 and 0.5 for all elements except hydrogen and is found to be between 0.45 and 0.5 for nearly all elements found in coal-combustion particles, for example. Over a wide variety of materials found in airborne particles (soot, fly ash, cement,

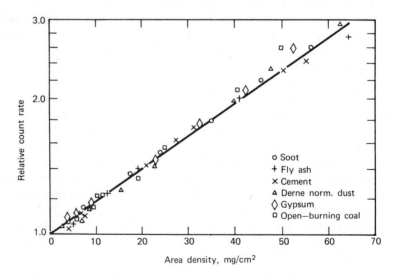

Figure 6 Calibration of a beta-radiation instrument for various dusts.[49] (courtesy of Chemical Engineering Progress)

11.8. Methods Adapted to Continuous Monitoring of Particulates

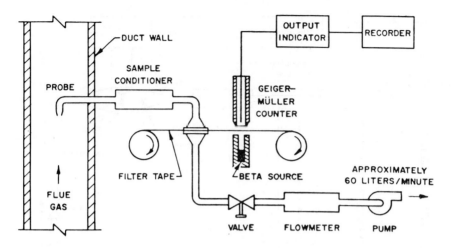

Figure 7 Beta-radiation attenuation instrument, schematic.[49] (courtesy of Chemical Engineering Progress)

gypsum, coal-combustion particles), the calibration curve is linear (Figure 6).[49] Instruments using this technique (Figure 7) have been utilized to measure the mass concentration of airborne particulates in ambient and stack effluents. Carbon-14 with a half-life of 5568 years is a typical beta-radiation source: ^{204}Tl, ^{137}Cs, and ^{147}Pm have also been used. Geiger–Miller counters are most frequently employed for detectors but proportional and scintillation counters have also been used.[49] As shown in Figure 7, the instrument is only quasicontinuous, since a sample must be collected on tape, but the sampling times can be made as low as 1 min if high detector sensitivities are employed. The main shortcomings are expense and complexity; there are also the sampling problems mentioned earlier.[24]

The piezoelectric microbalance measures particle mass directly without any theoretical factors being involved. Particles in the sample to be measured are deposited on a piezoelectric quartz crystal by means of electrostatic precipitation. Since the crystal's natural frequency of oscillation depends on its mass, the additional mass due to deposition of particulates causes a decrease of the oscillation frequency that is linear for small changes in mass.[24, 49] The frequency is monitored by placing the crystal in a resonant electronic oscillating circuit. Most particles smaller than 20 μm in ambient air adhere to the crystal well enough to be weighed. A schematic of a monitoring system utilizing a piezoelectric microbalance is shown in Figure 8. There are a number of problems: a sampling system is required to bring a representative sample to the crystal; crystals must be

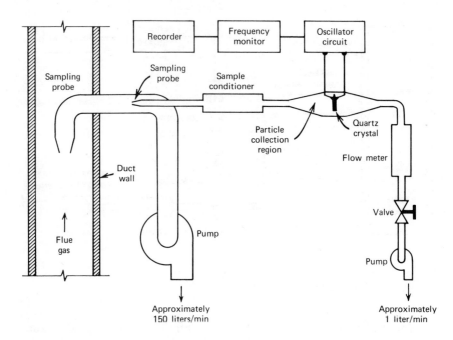

Figure 8 Piezoelectric microbalance instrument, schematic.[49] (courtesy of Chemical Engineering Progress)

cleaned periodically because they go out of the linear range and ultimately stop oscillating when overloaded; the oscillation is temperature-dependent. The great advantage of the device is its ability to sense true mass and its high sensitivity.

11.9. EVALUATION OF PARTICULATE MONITORS FOR SYSTEM PURPOSES

11.9.1. Instrument Evaluation Criteria

It is assumed that survey work and cost–benefit analysis have led to the conclusion that continuous monitoring of particulates and particulate components is required. The question is, What are the criteria by which choice of instrument or technique can be made? A set of such criteria relevant to the systems approach of this work is given in Table VII. These criteria cover most of the possibilities for system monitoring that we have been discussing. For each application, a scale of 0–4 has been used to rate the importance of the criterion for the particular application, with 4 representing great importance and 0 representing little or no importance.

Table VII

Aerosol Measurement Criteria

	Ambient air	Stationary source	Mobile source
Composition	(1) Unimportant (currently); active or inert	(1) Unimportant (except special cases); includes smoke-fumes.	(1)
Mass concentration	(4) Important, 10–300 $\mu g/m^3$ (mean); daily max. 3 × or more.	(4) Most important 80–5000 $\mu g/m^3$.	(4)
Size distribution	(4) Important (esp. <1 μm), 0.1–10 μm mass median.	(2) Not yet important 0.1–200 μm, depending on source	(2)
Detects particles <1 μm	(4) Respirable particles	(3)	(2)
Optical effects correlation	(2) Some importance; haze or low visibility via scattering	(3) Currently important; plume opacity and reflection	(3)
Unattended–automatic operation	(4) Most important	(3) Important; daily maintenance available	(1)
Continuous operation	(4) Most important (for alerts)	(4) Most important (control)	(3)
Real-time response	(4) Most important (for alerts)	(4) Most important (control)	(4)
Sensitivity	(4) Important	(2) Unimportant	(2)
Stability	(4) Important (for cumulative count)	(3) Some importance; daily check possible	(2)
Absolute accuracy and reproducibility	(3) Some importance	(4) Important; legal standards	(3)
Remote measurement	(1) Unimportant (see text)	(3) Some importance (see text)	(4)
In situ	(3)	(4)	(4)
Average measurement	(0) Unimportant (see text)	(4) Most important	(0)
Auxiliary measurements	(NA) Pollen detector; automated dustfall	(NA) Mass flow in stack	(NA)

485

Table VIII
Real-Time Particulate Instruments

Class	Atmosphere			Source			Measures			Remarks
	Point	Avg.	Profile	Sampled	Avg.	Remote	Conc.	Distr.	Vis./Scat.	
Counter-classifiers										
Optical-scatter	X			X			X	X		0.4–6 μm (typical)
Obscuration	X			X			X	X		In liquid; 2–60 μm
(Laser)	X			X			X	X		In liquid; HeNe laser; over 0.5 μm
Conductivity	X			X			X	X		Dispersed in electrolyte
Mobility	X			X			X	X		Whitby design for submicron
Sonic							X			N.G. under 20–40 μm
Light scatter										
90° scatter	X			X			X			
Fwd. scatter (10°)	X			X			X			
Integrating nephelometer (8–170°)	X			X					X	Charlson design

Method						Notes
Small-angle (0.5°)				?		Size ind. of refr. index; uses HeNe laser
Sun "turbidity"	X				X	Averaged through atmosphere
Plus absorption	X				X	Conc. plus average size
Laser backscatter (Lidar)	X	X				
Electric charge						
Mobility classified	X				X	Smallest part removed
Strong field chg.			X		X	Total current
Ionization (diffuse chg.)			X		X	Commercial (Germany)
Energy absorption						
Transmissometer	X			X	X	"Long-path" 500-ft base
Stack photometer				X		Bolometer
IR			X	X		
Beta ray	X		X	X		Dust on filter tape
X-ray (gamma)				X		10 kV, 1 ma
Sonic	X			X		
Light reflection				X Ringelmann		
Condensation	X			X		Total particle count down to 0.003 μm

487

11.9.2. Ambient Versus Source Requirements

An examination of Table VII shows that both ambient and stationary-source surveillance require a continuous automatic mass concentration instrument with a real-time response. The ambient monitor, however, must be 10^4 times as sensitive as a source monitor. Furthermore, it must provide a particle size distribution or at least a median particle size value. The source monitor need not presently provide size information, although it is probable that this will be required in the future. Absolute accuracy is only significant for source instruments because of the need to provide legal evidence of compliance with emission standards. The precision for ambient instruments should be within 5%.

Average (long-path) measurements are not needed for the atmospheric instrument; in fact, some evidence indicates that the point measure is better, since there are apparently small-scale inhomogeneities even over short distances in atmospheric aerosol concentrations. On the other hand, average readings across a stack or plume are essential, since there is no adequate theory to relate total emission from a stack to eddies and turbulence within the stack and in the plume.

A remote instrument to measure the particulate profile of the lower atmosphere would be useful, since meteorological data (inversion height) can be deduced thereby. However, this is a specialized requirement (met adequately by lidar) and would not affect the design of a general instrument. On the other hand, a remote stack or plume reading instrument would be more than a convenience for industry, since it would obviate stack installation and many operation and maintenance problems.

11.9.3. Real-Time Versus Integrating Techniques

Table VIII summarizes the results of a survey of instrument principles operating in real-time and capable of measuring the particulate parameters of interest. Table IX lists, for completeness, those instruments and techniques that provide time-averaged information (i.e., are inherently incapable of providing information in real-time). They either integrate an effect (such as soiling) over long time periods or collect and/or sort particulates by size for later examination.

In the authors' opinion, it appears that the instrumental techniques with the greatest current capability for mass monitoring of emission sources on a real-time in situ basis are beta attenuation and the piezoelectric microbalance. For remote real-time monitoring of sources, the lidar appears to be the technique of choice. For air quality over a region, transmission measurements are adequate. Since the costs of these and other instruments vary widely, other types, including long-term discontinuous

Table IX

Time-Averaged Particulate Instruments and Techniques[a]

Class	Atmosphere (Point)	Source Sampled	Measures Conc.	Dist.	Remarks
Filter/Hi-Vol	C	C	X		Manually weighed
Tape-spot	X	X	X		Measures COH and RUD
Impinger (liquid)	C	C	X		General purpose collector
Cascade impactor	X	X	X	X	Classifier
Thermal gradient	C		X		
Electrostatic	C	C	X		Portable sampler
Cyclone (midget)		C	X		
Centrifuge	C	C	X	X	
Elutriation	X	X		X	
Sedimentation		X	X	X	Automatic weighing, X-rays
Precipitation	C		X	X	Cloud chamber droplet count
Auto-Scan Microscope and Scanning Electron Microscope (SEM)	X	X	X	X	Speed 1000× manual

[a] X = measuring instrument; C = collector (fractionator) only.

methods such as simple dustfall collectors, have a valuable place in the total monitoring system.

11.10. CALIBRATION OF MEASUREMENTS ON PARTICULATES

In physical measurements of partculates, as well as other measurements, the basic question that must be answered is: What does the measurement mean? The measurements of particle size, mass, or concentration can only be given meaning by reference to known and accepted particle parameters (a calibration standard) or an instrument or technique that is known on the basis of theory or experience to give correct results. In particulate measurements, parameters can be given this meaning directly only for collected particles: for mass, when they are weighed, and for size, when they are examined microscopically. While the collection process may lead to error, the measurement processes mentioned can be made as accurate as one wishes. It is for this reason, for example, that the piezoelectric microbalance is an absolute instrument that measures mass without equivocation. While there are problems and errors associated with the technique as a whole, there is no question about the measurement per se. Light scatter methods for particle size distribution and/or mass, as another example, will always be subject to doubt, since a great many assumptions (density, color, refractive index, shape) must be made to relate the experimentally measured quantity (radiation scattered at fixed angle or integrated over a sphere) to mass and/or size distribution. This also holds true for sedimentation and impact methods where we wind up with an "aerodynamic" size distribution (see Section 11.3).

There are two approaches presently available to give size measurements more significance in this sense. One is the use of fine particle spheres of known composition and with well-defined particle size distribution. Commercially available (Dow Chemical Company) latex spheres in the range 0.09 µm to 3.5 µm have been used to calibrate the spiral centrifuge particle sizer of Stöber and Flachsbart.[27] They are also available from commercial standards houses[50] in the range 0.09–2 µm (sized by electron microscope) and 3.5–20 µm (sized by optical microscope). The materials from which the particles are made are polystyrene, polyvinyltoluene, polystyrene divinylbenzene, and polyethylene. In addition to the mean size, a standard deviation figure on a size range is provided. Coarse spheres (5 to several hundred µm) in various metals and glass are also available. Fluorescent particles (zinc sulfide) are also available for checking efficiency of filters.

The second approach is to use an aerosol generator that disperses a material into a known particle size distribution of uniform or known shape and of known concentration. Two types of such generators are commercially available.[51] The spinning-disk aerosol generator uses a centrifugal atomizer and air classifier to produce aerosols from both solutions and suspensions. An electrically neutral, practically monodisperse aerosol is produced in the 1–10 µm range with a geometric standard deviation as low as 1.1. It is used where precise control of particle size is required. The fluid atomization aerosol generator utilizes air-blast atomization and inertial impaction to produce aerosols. An electrically neutral aerosol in the range 0.03–3 µm is produced. For details of operation and method of sizing, the original literature should be consulted.

REFERENCES

1. HEW. *Air Quality Criteria for Particulate Matter,* NAPCA Publication No. AP-49. Washington, D.C., January 1969, p. 5.
2. Report of the Study of Critical Environmental Problems. *Man's Impact on the Global Environment.* MIT Press: Cambridge, Mass., 1970, pp. 56–58.

3. C. E. Junge. *Air Chemistry and Radioactivity.* Academic: New York, 1963.
4. Ref. 2, p. 58.
5. Ref. 1, p. 11.
6. Ref. 1, pp. 12-13.
7. Ref. 2, pp. 264-265.
8. S. J. Williamson. *Fundamentals of Air Pollution.* Addison-Wesley: Reading, Mass., 1973, p. 345.
9. C. H. Connolly. *Air Pollution and Public Health.* Dryden: New York, 1972, p. 219.
10. Ref. 8, p. 351.
11. Ref. 1, p. 6.
12. Ref. 8, pp. 355-356.
13. HEW. *Air Quality Data 1964-1965.* Div. of Air Pollution: Cincinnati, 1972, pp. 1-2.
14. Ref. 2, pp. 58-59.
15. Ref. 2, p. 44.
16. R. C. Robbins and R. D. Cadle. "Kinetics of the Reaction Between Gaseous Ammonia and Sulfuric Acid Droplets in an Aerosol." *J. Phys. Chem.* **62,** 469-471 (1958).
17. A. Goetz and R. Pueschl. "Basic Mechanisms of Photochemical Aerosol Formation." *Atmos. Environ.* **1,** 287-306 (1967).
18. B. M. Smith, J. Wagman, and B. R. Fish. "Interaction of Airborne Particles with Gases," 156th Nat. Meeting, Amer. Chem. Soc. Atlantic City, N.J., September 11, 1968.
19. M. R. Jackson and E. Hynsley. "A Discussion of the Need for Isokinetic Sampling in the Real World." In *Proc. First Nat'l Symp. "Forum on Air Pollution Regulations."* Amer. Soc. Mech. Engrs., Air Poll. Cont. Div., Pittsburgh, May 4-5, 1971.
20. E. R. Hendrickson. "Air Sampling and Quality Measurement." In A. C. Stern, Ed., *Air Pollution,* 2nd ed., Vol. II. Academic: New York, 1968, p. 14.
21. Ref. 20, p. 24.
22. Ref. 20, p. 28.
23. (a) A. A. Andersen. "A Sampler for Respiratory Health Hazard Assessment." *Amer. Ind. Hyg. Assn. J.,* **27,** 160-165 (1966); (b) Anderson 2000, Inc. *Particle Sizing Deadline Nears,* Brochure. Atlanta, Ga., 1972.
24. N. R. Iammartino. "Fine Particles Start Coming Under Scrutiny." *Chem. Eng.,* July 10, 1972, p. 32.
25. R. E. Lee et al. "National Air Surveillance Cascade Impactor Network, II, Size Distribution of Trace Metal Components." *Environ. Sci. Technology,* **6,** 1025-1030 (1972).

26. J. A. Gieseke. "Thermal Precipitation of Aerosols." In W. Strauss, Ed., *Air Pollution Control,* Pt. II. Wiley: New York, 1972, p. 211ff.
27. W. Stober and H. Flachsbart. (a) "Size-Separating Precipitation of Aerosols in a Spinning Spiral Duct." *Environ. Sci. Technol.* **3,** 1280–1296 (1969); (b) "High Resolution Aerodynamic Size Spectrometry of Quasi-Monodisperse Latex Spheres with a Spiral Centrifuge." *Aerosol Sci.* **2,** 103–116 (1971).
28. (a) W. C. McCrone. "Morphological Analysis of Particulate Pollutants." In Ref. 26, p. 281ff. (b) W. C. McCrone, *The Particle Analyst-Compiled.* Ann Arbor Science: Ann Arbor, Mich. 1968.
29. (a) W. C. McCrone, R. G. Draftz, G. Ronald, and J. G. Delly. *The Particle Atlas.* Ann Arbor Science: Ann Arbor, Mich., 1967. (b) Ibid., 2nd ed., 1973.
30. M. Corn. "Nonviable Particles in the Air." In Ref. 20, Vol I, p. 89.
31. ASTM Standard D1704-61. "Particulate Matter in the Atmosphere." In *1972 Annual Book of ASTM Standards,* Pt. 23. American Society for Testing and Materials: Philadelphia, Pa., 1972, pp. 427–433.
32. M. Katz. *Measurement of Air Pollutants.* World Health Organization: Geneva, 1969, p. 65.
33. W. Leithe. *The Analysis of Air Pollutants.* Ann Arbor Science: Ann Arbor, Mich., 1970, p. 168.
34. H. Weisz. *Mikrochim. Acta* 140, 376, 460, 785 (1954); *Microanalysis by the Ring Oven Technique.* Pergamon: Oxford, 1961.
35. S. L. Sachder and P. W. West. "Concentration of Trace Metals by Solvent Extraction and Their Determination by Atomic Absorption Spectrophotometry." *Environ. Sci. Technol.,* **4,** 749–751 (1970).
36. J. Y. Hwang. "Trace Elements in Atmospheric Particulates and Atomic Absorption Spectroscopy." *Anal. Chem.* **44,** 21A–27A (December 1972).
37. W. H. Zoller and G. E. Gordon. "Instrumental Neutron Activation Analysis of Atmospheric Pollutants Using Ge(Li)γ-Ray Ray Detectors." *Anal. Chem.* **42,** 257–265 (1970).
38. C. L. Luke et al. "X-Ray Spectrometric Analysis of Air Pollution Dust." *Environ. Sci. Technol.* **6,** 1105–1109 (1972).
39. R. Brown and P. G. T. Vossen. "Spark Source Mass Spectrometric Survey Analysis of Air Pollution Particulates." *Anal. Chem.,* **42,** 1820–1822 (1970).
40. D. Hoffman and E. L. Wynder. "Chemical Analysis and Carcinogenic Bioassays of Organic Particulate Pollutants." In Ref. 26, p. 187ff.
41. W. T. Ingram et al. "Size Differential Particle Counter Related to Particulate Density." In *Analysis Instrumentation,* Vol. VIII. Instrumentation Society of America: Pittsburgh, 1970.
42. P. F. Mullaney and P. N. Dean. "Cell Sizing: A Small-Angle Light-Scattering Method for Sizing Particles of Low Refractive Index." *Appl. Optics,* **8,** 2361–2362 (1969).

43. P. M. Giever. "Analysis of Number and Size of Particulate Pollutants." In Ref. 20, pp. 276–277.
44. K. T. Whitby. "Apparatus and Method for Determining Aerosol Particle Concentration and Particle Size Distribution," U.S. Patent No. 3,413,545, June 23, 1965.
45. W. E. K. Middleton. *Vision Through the Atmosphere.* Univ. of Toronto Press: Toronto, 1952.
46. R. J. Charlson et al. *Atmospheric Environment,* Vol. I. Pergamon. London, 1967, pp. 469–478.
47. K. Whitby et al. "Electrical Measurement of the Mass Concentration of a Self-Preserving Aerosol Size Distribution." *J. Air Poll. Control Assn.*, **18**, 760–764 (1968).
48. P. W. Purdom. "Source Monitoring." In Ref. 20, pp. 542–543.
49. G. J. Sem, J. A. Borgos, and J. G. Olin. "Monitoring Particulate Emissions." *Chem. Eng. Prog.*, **67** (10), 83–89 (1971).
50. Duke Standards. *Bulletin 171R.* Duke Standards Co.: Palo Alto, Calif., (September 15, 1971).
51. Environmental Research Corp. *1970-71 Catalog.* Environmental Research Corp.: St. Paul, Minn.
52. W. F. Patton and J. A. Brink. *J. Air Poll. Control Assn.*, **13**, 162–166 (1963).
53. P. W. West, "Chemical Analysis of Inorganic Pollutants." In Ref. 20, pp. 147–185.
54. Ref 1, pp. 23–27.
55. K. T. Whitby, R. E. Charlson, W. E. Wilson, and R. K. Stevens. *Science,* **183**, 1098–1099 (March 15, 1974). Also R. L. Lee *Ibid.,* 1099–1100.

12

The Future of Air-Pollution Control Systems

Following an era of almost complete neglect by the technical and financial communities and by the public at large, air-pollution control has, in the present decade, made a resounding impact. It is now recognized as a major factor in urban and industrial evolution, as testified by the most recent U.S. estimate of $48 billion of annual expenditures in 1980 for capital investment in air-pollution abatement equipment.[1]

Having achieved this recognition, it is unlikely (barring catastrophe) that the subject will fall back into neglect; however, it is equally improbable that present generations will see such a complete societal solution to the problem of evaluating pollution externality costs as the creation of a new "environmental currency" suggested by some authors.[2]

In the United States, at least, the evolution of air environmental controls appears to be shaped by four factors, that will be operative in the near term. We will discuss in this concluding chapter the effect of new data; interface with other societal problems; an expanded goal, preservation of "clean" areas; broadened cost-effective and cost–benefit strategies.

12.1. NEW DATA AND REVISED OBJECTIVES

In the normal course of setting up large systems programs such as air-pollution control, the logical procedure would be (1) to gather the necessary data, (2) thorough examination of the data to determine the quantitative goals and improvement objectives, and (3) to establish the necessary control strategy and models and verify them, using the data.

We saw that this process was interrupted by the 1970 Clean Air Act because of the slow pace in implementation. Goals, including ambient standards and time scales, were established by law, using the best data where available as to the current status of air pollution, its effects, and the national cost. It is inevitable that this data base would be found lacking or in error in many respects. Hence, to this extent, the normal procedure has been inverted: first, goals; then, strategy; and, finally, data! The inadequacy of the air-data base has been clearly stated by one EPA official:[3] "We simply do not have adequate data as to the quality of the air... a limited number of monitoring stations tell us we have a problem but we don't know what the scope of the problem is." Resolution of this inadequacy has been mandated by this agency,[4] and as a first step, the five-year RAPS program in St. Louis, described in Section 6.2.1, has been initiated at an annual funding level of $5 million. The RAPS experiment now underway utilizes 30 fixed monitoring sites in concentric circles about the St. Louis downtown center to collect air-quality data over a distance of 100 km. The data, telemetered to a central computer for immediate evaluation and storage, will evaluate the ability of mathematical simulation models to predict transport, diffusion, and concentration of both inert and reactive pollutants over the area and will be used to develop improved control strategies, including those for selective source reduction, rather than uniform cutbacks.[5] The RAPS data will include the concentration of six pollutants, wind, temperature, humidity, and four bands of solar radiation. Models tested will include photochemical and aerosol dynamics types, the latter attempting to predict the gas–solid equilibria and the agglomeration of airborne particles.[6]

Following the RAPS program, it is hoped that the data-gathering and model-verification effort will be extended to more complicated terrains, including coastal and mountainous regions, deliberately excluded as a means of simplifying the initial experiment.

Other programs for gathering physical data under way include the Large Power Plant Effluent Study (LAPPES) conducted by the EPA in western Pennsylvania, which attempts to describe the dispersion and physicochemical changes of effluents from large single sources, and the washout, ground contamination, and ultimate disposal of pollutants emitted from tall (700–1000 ft) stacks of large power plants.[7] In a related EPA program,[8] a fluid model facility (FMF) using water and wind tunnels will permit data to be obtained on the complex flow patterns of pollutants around buildings and topographical features (see Section 8.2.5).

Further data is being collected during implementation by the states of the 1970 federal act. These include quarterly reports of air quality from each surveillance network that is part of each state's plan. These reports are made available in computer-compatible form and standard format (SAROAD, or Storage and Retrieval of Aerometric Data). Source emission data is required semiannually for new or altered sources emitting more than 100 tons of any controlled pollutant and for stacks or points emitting more than 25 tons.[9] These data also are placed in computer-com-

12.1. New Data and Revised Objectives

patible format (NEDS point source coding) for ease of retrieval. New types of stationary sources now coming under federal emission standards, (see Tables IX and X of Chapter 4) are required to submit results of emission tests. These include[10] asphalt concrete plants, petroleum refineries and storage tanks, secondary brass and bronze ingot plants, secondary lead smelters, sewage treatment plants and sludge incinerators, and iron and steel plants (basic oxygen process). This regulation will lead to new emission inventory data on particulates, CO, and sulfur compounds.

The physical data are only part of the story and must be supplemented by adequate health information in order to reestablish the ambient quality goals. The Community Health and Environmental Surveillance System (CHESS) has been established to fill this gap. CHESS studies have simultaneously measured environmental quality and sensitive health indicators in communities that represent a gradation of exposure to particulates, sulfur oxides, nitrogen oxides, and photochemical oxidants. These communities (in 1972) totaled 33 neighborhoods in 6 areas (one is St. Louis, supplementing the RAPS program). Health indicators monitored include respiratory disease in adults and children, asthma frequency, acute irritation symptoms during air-pollution episodes, pulmonary function of school children, and cumulative pollutant tissue residues in humans.[11]

A major finding already evolved from this study is the fact that suspended sulfates are more consistently associated with adverse human health effects than are either sulfur dioxide or suspended particulates (TSP). This finding is considered highly significant in that sulfates are more widely dispersed from urban sources than is gaseous SO_2. Research of this nature can lead to a qualitative or quantitative change in ambient and emission standards and methods of control.

Modification of annual and 24-h ambient-concentration secondary standards for SO_2 (60 and 260 $\mu g/m^3$), based on studies of damage to white-pine growth, has already been accomplished. On reexamination, these data show that the damage was the result of short-term, rather than long-term, exposure. Ironically, increased emphasis on short-term standards (3-h average) may make control more difficult for isolated operations like copper smelters who initiated the reexamination in the first place,[12] because of the greater (geometric) standard deviation of ambient concentration surrounding point sources.

The data on nitrogen oxides is also currently in need of review. The 1970 act mandated a 90% reduction in NO_x emissions from automobiles, originally by the 1976 model year. Nitrogen oxides were considered a serious urban problem based on results of a study of respiratory illness among school children in Chattanooga, Tennessee, during 1968. Using the Jacobs–Hochheiser method of measuring ambient NO_2, it was determined that 47 of the 247 air-quality regions in the United States were in violation of the 100 $\mu g/m^3$ annual average air-quality standard, justifying the reduction in automobile emissions. However, it has been found that the reference NO_2 measurement technique is in error because of variable collection efficiencies and positive nitric oxide interferences. Remeasurement of 200 sites, including the 47 suspected regions, using other NO_2 methods; the arsenite bubbler, chemiluminescent, and Greiss–Saltzman; have resulted in a finding that only 4 regions need to be classified Priority I with respect to NO_2 ambient standard violation.[13]

Difficulties in implementing the NO_x reduction in conventional automobiles (which apparently will require not only an exhaust catalyst but also a feedback control on air–fuel ratios) have further impelled the EPA to postpone the 90% reduction to 1977, the longest period permitted by law.[14] Meanwhile, new data is being sought to confirm the 100 ppm annual standard (which may be found either too high or too low), and the agency may be given more regulatory leeway relative to automobile emission standards as a result.

12.2. INTERFACE WITH OTHER SOCIETAL PROBLEMS

Air-pollution reduction, even in its health aspects, cannot stand alone as a societal goal, and it is inevitable that conflicts would arise, even beyond those based on economic costs and self-interest. Typical of these interfaces are transportation, energy, and materials use.

To the extent that air-quality goals and timing cause real and apparent cost–benefit shifts in these other areas, pressures will exist to offset either or both equilibria.

The conflicts with the private automobile as a means of public transportation have been the most visible to date. The requirements for increasingly stringent emission controls, including the NO_x problem mentioned above, have involved difficult technical problems, costs to user and producer, and claimed increase in fuel consumption in the face of apparent shortages, all of which have caused bitter attack on the air-pollution control agency and its supporters.

These constraints on the individual car buyer, however, are miniscule compared to those needed to meet the clean-air ambient standards in Los Angeles and many cities by the mandatory Congressional time limit (1975, extended to 1977). Complete control of new cars and stationary sources cannot meet photochemical oxidant and nitrogen oxide standards, according to current predictive models, because of the prevalence of older cars on the streets and the projected increase in car population. Hence, the control agency (by law) is required to impose other "strategic" constraints

12.2. Interface with Other Societal Problems

on the offending communities. These include a wide variety of measures, such as high-speed-bus and car-pool lanes, parking restrictions and limitations, mandatory in-use vehicle inspection and maintenance, limitations on gasoline consumption, and retrofit emission controls on older cars.

Many of these transportation strategies have been termed "apparently impractical and unworkable" by the EPA itself,[15] although a 20% reduction in city traffic is considered feasible without economic disruption.[16] It appears obvious that the most attractive and permanent solution to these problems is the provision of attractive mass transport, but currently only one city (Washington) is building a system which will cut traffic by as much as 5%. It has been estimated that the number of buses needed to meet the mass-transport requirements in 1977 (45,000) is twice the projected production.

In European cities where auto traffic has been restricted, the results have been favorable (to merchants and shoppers). Hence, it is possible that this apparent imposition will have beneficial results. Without going into all the details of these strategies,[15] it is obvious that an immense area for modeling and tradeoff exists for each affected urban area.

The situation in Denver, Colorado, however, is worthy of special mention. This region is an average of 6800 ft, above sea level and is surrounded by mountains, rising to 8000 ft, which subject the area to downslope or drainage winds (see Chapters 7 and 8). The result of this topographical anomaly is to concentrate motor vehicle-generated CO and oxidants resulting from hydrocarbons during stable atmospheric conditions. Average 1-h oxidant concentrations measured in Denver exceed 0.18 ppm, mandating a 60% reduction in hydrocarbon emissions using a "proportional model."[17]

The unique problem of Denver is the high altitude, which precludes the use of catalysts or emission controls suitable for low-altitude use. Among strategies proposed for reduction of vehicle miles traveled (VMT) is the identification and restriction of one-fifth of the automobile population from the city area each working day.

Automobile restrictions have both positive and negative results with respect to the energy utilization interface. Engine pollution controls (on 1973 models) tend to increase fuel consumption, but greater utilization of the bus and other forms of mass transport will reduce it. The other facets of impact with energy use are even more obvious. Restrictions on SO_2 emission from power plants require low-sulfur fuel, which at any given time and place may be in short supply. Reliance on greater numbers of nuclear energy plants raises another set of environmental problems and risks that must be evaluated.

Still another interface is that of air pollution versus materials use. The most prominent of these overlaps is that of waste disposal. Solid waste may be utilized for landfill or dumped into the seas, raising one set of problems. If incinerated, the waste burner is a potential air polluter. Sewage-treatment sludge raises similar problems. A cost-effective solution may be incineration and recovery of waste heat, utilizing the economic credits to insure adequate treatment of the air effluent.

Recycling of all wastes is an ultimate solution, discussed under "Grand Strategies" (Chapter 5). Recent studies by the SRI predict that economic recycling incentives, such as lower freight rates for secondary materials, tax credits for consumption of post-consumer waste, and public education programs, will be in effect before 1980. The trend will be toward reduced incineration of solid wastes and recycling of both materials and energy. The impact on air-pollution control is seen to be both direct and indirect.

12.3. PRESERVATION OF "CLEAN" AREAS

The assessment of changes in air-quality goals due to the interactions discussed above are subject to some discussion; however, the court decision that now prohibits "significant deterioration" of air quality in geographic areas now exceeding the national standards, is a clear-cut interpretation and demands an entirely new look at the implementation of the 1970 Clean Air Act. If dirty urban areas were allowed to be cleansed by removal of pollution sources, such as power plants, to places with substantially background level of air quality (such as the well-publicized Four Corners power plants feeding Los Angeles), the result might well be cost-effective within the criteria of national ambient standards, but the "benefit" of a uniform countrywide level of pollution just under health and welfare limits is certainly negative. The opposite policy, now defined in the United States, will have "substantial impact on... future industrial, commercial, and residential development" and "affect utilization of... mineral resources, ... availability of employment and housing, ... and the costs of... electricity and manufactured goals."[18] Carried to the reductio ad absurdum, a "no-deterioration" policy would prevent any incursion of man's works in presently wild areas. To offset the implication of "no growth," which is rejected as national policy, it is necessary to define *significant deterioration* and to specify how it will be prevented.

In effect, the problem is to define some standard that lies between the background or zero degradation and the upper bound established by the national secondary standards.

In addition to requiring the best available control technology for all new sources of controlled pollutants (SO_2, CO, NO_x, hydrocarbons, and particulates) listed in Table I and for any other stationary source emitting more than 4000 tons/yr of any of these pollutants, and in addition to

12.3. Preservation of "Clean" Areas

Table I

New or Modified Sources Proposed for Preconstruction Review For Prevention of Air-Quality Degradation

---

A. Fossil-fuel-fired steam electric plants of more than 1000 million Btu/h heat input
B. Coal cleaning plants (thermal dryers)
C. Kraft pulp mill recovery furnaces
D. Portland cement plants
E. Primary zinc smelters
F. Iron and steel mill metallurgical furnaces
G. Primary aluminum-ore-reduction plants
H. Primary copper smelters
I. Municipal incinerators capable of charging more than 250 tons of refuse per day
J. Sulfuric acid plants
K. Petroleum refineries
L. Lime plants
M. Phosphate-rock-processing plants
N. By-product coke-oven batteries
O. Sulfur recovery plants
P. Carbon-black plants (furnace process)

---

requiring a preconstruction review of each of these sources to determine if it has, in itself, the capacity to cause "significant deterioration," the EPA has proposed four alternate methods of control, each of which defines *deterioration* in a different way.

12.3.1. Air-Quality Increment Plan

The first method permits a maximum allowable increase in SO_2 and particulate above the 1972 amounts in areas where these were below federal ambient standards. These increments are as follows:

> Particulate:
> 10 $\mu g/m^3$ annual average
> 20 $\mu g/m^3$ 24-h average
>
> SO_2:
> 15 $\mu g/m^3$ annual average
> 100 $\mu g/m^3$ 24-h average
> 300 $\mu g/m^3$ 3-h average.

These amounts are deemed "nonsignificant" on a purely arbitrary basis and offer a good opportunity for economic trade-off. For example, a large (15,000-unit) apartment complex increasing the 3-h SO_2 average by 70 $\mu g/m^3$ would be allowed under these criteria, as would a well-controlled 1000–1500 MW coal-fired power plant (which would be expected to raise 24-h SO_2 between 50 and 200 $\mu g/m^3$ depending on stack height, terrain, and meterological conditions).

This plan also requires new sources to install continuous air-quality and meteorological monitoring instruments in areas of expected maximum concentration both for data-gathering purposes and protection of the deterioration ceiling. One reason for this requirement is the lack of data in most very clean areas. Initially, it will be necessary to estimate the background by diffusion modeling and similar techniques.

It is considered that the socioeconomic effects of this procedure would be favorable to clean areas, since it would force large sources to use increasingly effective control techniques and conduct strong control research and development programs but would not prevent the economic development entirely. It would prevent construction in difficult terrain areas with poor meteorological conditions, and the clustering of large sources to badly pollute one locale.

In areas already built up, it would curb new development until the current level of pollution was reduced below standards but would then permit more development. The general trend could be to protect undeveloped areas and maintain those presently developed at a constant acceptable level. Since no other land-use criteria are involved in the decisions, it is not clear that the protection automatically given to waste lands, as well as valuable park and scenic areas, is deserved, especially at the expense of continued marginal levels in cities.

12.3.2. Emission Limitation Plan

This proposal states another philosophy—to prevent deterioration by limiting significant increases in emissions. The intent is to control those effects that are presumed more responsive to total atmospheric loading than to specific ambient concentration—that is, visibility and solar radiation reduction, acidification of rain and bodies of water, sulfates and nitrates formation, and background increases. These effects are significant even when ambient concentration levels are below those affecting health and may correlate poorly with ground-level ambient measurements. However, emission density (the regional emissions divided by the regional area) is an excellent indicator of atmospheric loading. Thus, the average emissions in a central region (AQCR), a number easily acquired, and the known AQCR area can be utilized to control the emission density.

Some difficulty arises because of the varying sizes of AQCR's and the consequent lack of emission-averaging flexibility imposed on small areas. With an emission density limit, this may lead to development opportunity inequities in the smaller units. Some procedure to adjust AQCR sizes would be incorporated in this plan.

The proposed plan would limit the average emission density of each control region to 10 tons/(yr)(mile^2) for SO$_2$ and 3 tons/(yr)(mile^2) for particulate or 120% of the 1972 baseline emissions would be allowed, whichever is larger. No development that would exceed these criteria would be allowed, but the allocation of the available emissions below the ceiling is at the choice of the state.

12.3.3. Local Definition Plan

A third proposal would allow each locality to determine in each instance whether a proposed new source is considered to cause significant deterioration. Every proposed new source, after incorporating the best available control technology in its plans, would be required to submit complete and detailed data to the state as to the amount and type of emissions expected and their effect on air quality in the surrounding area. It is presumed that the state would analyze this data in relation to other sources in the area by means including a multiple-source modeling procedure. (This would identify existing sources subject to further emission control, which might thereby offset the additional pollution attributed to the proposed new source.)

This data, then, would be subject to public disclosure and perhaps hearings of "town-hall" character in which community goals and attitudes would be explored, relative to the air data. Following, the state would determine the significance of the new source's deterioration of air quality and would have authority to prevent construction if the findings were adverse. The federal authority would be permitted to review the decision from the standpoint of what constitutes "best available control" technology or the state's procedures but could not upset the local decision on deterioration.

This proposal clearly has the advantage of local rule and would permit the community to consider its own goals and interests without arbitrary restriction to national views. Such local options are already in effect in some states. The plan's disadvantages include the "sliding baseline," in which further deterioration is always measured from existing levels in the direction of worse quality. Furthermore, pressure by industry desirous of expanding in the area might influence the population to approve short-range advantage to its ultimate long-range disadvantage. Furthermore, the effects of creeping deterioration cannot be completely localized. It is probable that a "clean-air-oriented" community surrounded by "growth-oriented" neighbors will lose the control over its own air quality that this plan is supposed to provide.

12.3.4. Area Classification Plan

This alternative is very similar to the Air Quality Increment Plan discussed in Section 12.3.1 but permits more flexibility in the choice of increments. Two kinds of zones are established by and within each state, one

Table II

Incremental Air-Quality Deterioration Allowable in Proposed "Area Classification Plan"

	Zone I, $\mu g/m^3$	Zone II, $\mu g/m^3$
Particulate matter:		
Annual	5	10
24-h	15	30
SO_2:		
Annual	2	15
24-h	5	100
3-h	25	300

limiting the incremental air-quality deterioration from the 1972 base to the same limits previously listed and the other zone being considerably more restrictive (see Table II). The Zone I classification would prevent even one small oil or coal-fired electric power plant, or municipal incinerator, or medium-sized, oil-heated apartment complex to be built, unless unusually good emission-control techniques were used. Such a restriction would normally be applied to national and state park or forest areas or to urban or suburban areas where growth limitation is desired.

This plan would also allow an escape hatch—in essence, a third zone—in places where growth is encouraged for exploitation of valuable raw materials or to support a long-range development plan. This "zone" would be allowed to increase pollution levels up to the federal ambient air-quality standards.

The flexibility allowed in this plan would relieve many of the local inequities that may be imagined to result from the basic increment scheme. All zones may be permitted some growth, even Class I, providing active development programs find better ways to limit emissions. Rational, long-range growth patterns can be planned by states with less inhibition from ceilings.

The difficulties previously mentioned in connection with incremental limitation still exist in some form. The needed data on baseline air quality for most clean areas is not available and must be approximated by diffusion modeling. This plan, as well as the basic version, places the burden of additional data collection on the source. Furthermore, the boundaries between zones must be cleverly drawn so that the integrity of Zone I areas is not threatened by the looser regulation of surrounding zones. Interstate cooperation is required in this regard.

12.3.5. Emission Charges and Other Schemes

Numerous permutations and gradations of these basic plans may be envisioned. The parameter most obviously subject to optimization is the area on which an average emission density increment is based. If this area is too small, numerous small sources will be permitted but no large source, such as a power plant. If too large an area is used, the peak concentration at some point becomes excessive, even though total atmospheric loading remains low.

Emission charges have also been considered in this context. This scheme permits air-quality deterioration up to the federal standards if the polluter pays for the amount emitted. The result is to make real the "external" costs of pollution and require the polluter to consider them in the normal alternatives of plant design and location. An economic incentive toward better pollution control also results. The emission charge plan fails to achieve environmental preservation mainly because there is no known relationship between the charge rate and air quality. Even if the charge is adjusted from place to place to offset the natural topographical and meteorological variations in effluent dispersion capability, the magnitude of the economic utility of air emissions is not known. Consequently, there is no guarantee that some significant deterioration would not occur by a polluter willing to pay for the "privilege".

Regardless of which type of plan is selected to maintain air quality, certain problems remain:

1. Jurisdictional ambiguities—the problems arising when a source in one state pollutes the air in another, "using up" the latter's growth potential
2. Lack of long-range land use plans; encouraging rapid use of any region's growth potential through de facto decisions without regard to long-range goals and plans
3. Urban sprawl—the creeping growth of built-up areas between cities which, accompanied by numerous residential heating emissions, will rapidly use up any area's capacity for growth
4. Fuel switching inpact—shortage of low-sulfur fuel needed to prevent serious health hazards in urban areas may require "reverse switching" to higher sulfur content in clean areas. The sulfur oxide emissions being proportionally increased, a degradation in air quality that is not insignificant, albeit temporary, may result.
5. The problem of electric power growth. Since most large urban areas cannot provide enough electric power for their own needs, in view of the existing threat to national air-quality standards, the power must come from relatively clean rural areas. But these may need to supply

more power for their own use in the future. For example, Iowa, an agricultural state, will need 1700 MW of additional power per year by 1980.[19] Using the best emission-control techniques and available fuel, this will produce 160,000 tons of additional SO_2 per year or a 50% increase in emissions over the 1970 level. These factors must be considered during implementation of any plan.

12.4. COST-EFFECTIVE AND COST-BENEFIT ANALYSIS

The need to optimize efforts to achieve air quality has been emphasized throughout this volume. Both cost-effective and cost-benefit analyses are required to achieve this objective, and it is important to understand the distinction between them. *Cost-effective analysis* is a means of selecting from all feasible alternatives a least-cost strategy of obtaining a predetermined goal, such as the ambient air-quality standards or, on a plant scale, a particular emission standard. *Cost-benefit analysis,* on the other hand, attempts to rationalize the goals themselves by equalizing the marginal cost of both benefits and control and relating the optimal benefit level (called, by economists, Pareto-optimality) to an air-quality goal. The concept of cost-benefit was introduced in Chapter 5, but it is much less developed than cost-effective techniques, which are in fairly routine use.[20] The tools of cost-effective analysis are the diffusion model, emission factors and cost data-base, and the computer. Cost-benefit, however, requires seemingly impossible inputs such as the dollar value of sunshine or aesthetic scenery and has thereby been the subject of much more controversy and skepticism. Nevertheless, both types of analysis will be increasingly developed and utilized in the future—cost-effective models to determine the strategy for an abatement program and cost-benefit to fix the long-range goals.

Cost models were first applied in the Model Cities program, sponsored by HEW, in which the accounting firm of Ernst and Ernst constructed automated models to determine the cost and effectiveness of various strategies meeting present levels of SO_2 and particulate from fossil-fuel and industrial sources.[21, 22] The computer programs utilized as inputs a data base the firm had developed, giving the costs of particulate-control equipment as a function of capacity and effectiveness, the costs of fuel switching for each source, a source-emissions inventory, and meteorological parameters. A transport and diffusion model in conjunction with the source emissions computes the concentration of pollutants at receptor points. Control strategies are then applied that reduce emissions in controlled amounts, under constraints of equal reduction per source or of least cost, in order to meet the chosen goal. The total cost of the strategy is then computed. The goal may be reduction of the highest concentration in the area, reduction of integrated human exposure, reduction of average annual concentration, or merely cutback of source emissions. In all these aspects, the techniques were applied successfully during 1968 and 1969 to the New York–New Jersey area, Kansas City, and Washington, D.C., in support of abatement conferences held under the then existing (1967) federal air-pollution control act.[23]

At about the same time, the TRW company contracted with HEW to develop two computer programs, one of which, the Air Quality Display Model (AQDM), produces a printout of the average spatial distribution of SO_2 and particulates, given the meteorology, emission inventory data, card control technology, and assumed strategy for a given area. The effectiveness of given strategies can be studied, and the program has been used for evaluation of state implementation plans under the 1967 and 1970 Clean Air Acts.[24, 25] The other program, Implementation Plans Program (IPP), extends AQDM by introducing the cost–control data for alternate abatement strategies.[26, 27] It assumes that control costs are piecewise linear and achieves a least-cost strategy by means of a linear programming model (see Figure 1). It is thus a complete cost-effectiveness model, permitting the user to determine least-cost control strategies for point sources and to relate regional air quality levels of cost–control.[28]

In a typical application for this type of model, IPP has been used by Platkin and Lewis to study the cost-effectiveness of three alternate emission control strategies in the St. Louis AQCR.[29]
These included the following:

• A "conventional" source-category strategy that achieves an ambient air quality standard with one uniform emission standard for every source in each of three categories: fuel combustion, industrial processes, and solid waste incinerators

• A rollback strategy based on similar source-category standards, that achieves a net reduction in regional emissions

$$R = \frac{X_{\max} - X_{\text{standard}}}{X_{\max} - X_{\text{background}}},$$

where X is the ground level concentration of a given pollutant, $X_{\max}$ is the highest measured regional concentration, and the other variables represent the background and the standard to be achieved

• A least-cost strategy that minimizes the cost of achieving an ambient standard by tailoring the emission standard for each source, based on its location in the region and its cost–control

Rollback is an allowable means of "demonstrating" adequacy of a state control plan according to current law and is attractive to regions that do not have access to the computers and models otherwise required. It is,

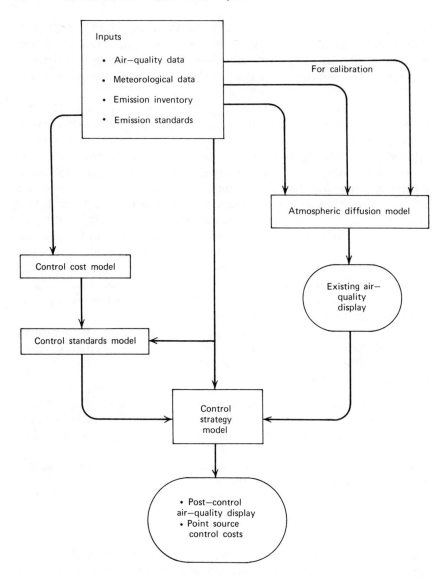

Figure 1 IPP microflow chart (after Ref. 29).

further, politically attractive, because it is considered, by some, inequitable to apply emission standards that vary with plant location. But it is a dangerous strategy to apply, since it does not in any way assure reaching air-quality goals in every part of the area, dealing as it does only with the average volume concentration. Thus, a gross failure of rollback will

12.4. Cost-Effective and Cost–Benefit Analysis

necessitate reinstallation of expensive control equipment. On the other hand, a least-cost method allows plants with high marginal-control costs to bear a smaller burden of emission control, which can have the effect of preserving old or inefficient plants.

The application of either strategy is not simple, and the reader is referred to the references for details. It is interesting to note the results, however, since they indicate that this type of study will find increasing utilization.

In this case, the rollback strategy was very successful in achieving the particulate goals (as established by the conventional strategy) and at an equivalent annual cost-effectiveness ($0.509 $\times$ 10^6/μg-m^{-3} reduction compared to $0.551 $\times$ 10^6/μg-m^{-3}). It was much less cost-effective for SO_2 reduction. Here, the comparable figures were $2.7 $\times$ 10^6/μg-m^{-3} for rollback and 1.88 $\times$ 10^6/μg-m^{-3} for the conventional strategy.

With respect to the least-cost strategy, a saving of 42% of the annual cost of controlling particulate was achieved ($6 million compared to $10.4 million for conventional strategy). These savings indicate a fast payback for investment in further studies of this nature.

The least-cost linear programming model used by Platkin has practical constraints on size: approximately 30 sources and 45 source-plus-receptors. More advanced types of optimization models have been used, but none has yet provided a complete solution. Consequently, there exists a choice between the heuristic, microlevel model such as IPP, which treats each point source individually, and the mathematical, macrolevel model, which measures air quality at a single point but treats sources in groups, regardless of their geographic location. Kohn has used a linear programming model of the latter type.[30] This was also applied to the St. Louis airshed and allowed 200 control measures (such as fuel switching, leaf collection, automobile and process controls, etc.) to be applied against the constraints of availability, maximum effectiveness of each measure, and total requirements for pollution reduction. The model accounted for the five major pollutants and a large number of sources. The output of the computation was a set of control methods that eliminated the desired weight of pollutants from the airshed at the least total cost.

Although this model is useful as a tool for air-pollution control and is simple enough to be applicable generally to other airsheds, it does not contain enough detail to determine localized or neighborhood effects or to solve short-term problems (to be fair, neither did Platkin's IPP study, because of the small number of receptor points considered).

Burton and his associates at the EPA have attempted to improve on both Kohn's model and IPP[31] by combining the latter with another model, the Direct Cost of Implementation Model (DCIM).[32] The resulting structure (Figure 2) will optimize the cost-effectiveness according to built-in decision criteria. These include least-cost for source standards and least-cost for air-

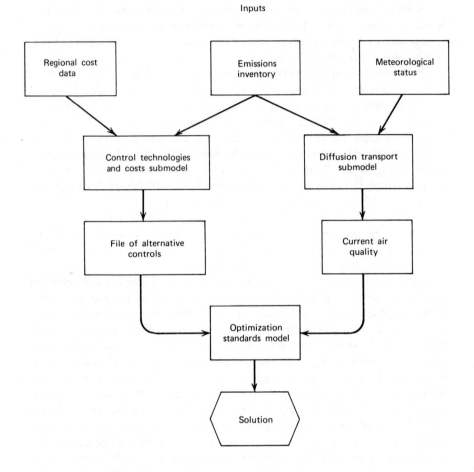

Figure 2 IPP–DCIM microstructure (after Ref. 31).

quality and fuel restraints. The latter approaches an interface with the energy and fuel problem, encouraging visions of even more universal models. At this stage, however, even microscale models such as IPP–DCIM, representing the state-of-the-art in application to real cities, fail to take into account such realities as emissions growth and other economic impact and do not quantitatively account for benefits. This work remains for the future.

REFERENCES

1. Creative Strategies, cited in *Prod. Eng.,* August 1973, p. 9.
2. W. E. Westman and R. M. Gifford. *Science,* **181,** 819–824 (August 31, 1973).
3. J. R. Quarles, Jr., EPA deputy administrator, cited in *Government Executive,* July 1973, p. 40.
4. R. E. Train. Hearings before the Senate Public Works Committee, August 1, 1973.
5. *National Environmental Research Center. Annual Report.* Research Triangle Park, N.C., 1972, p. 22–23.
6. *Environ. Sci. Technol.,* **7,** 598–599 (1973).
7. Ref. 5, pp. 63–64.
8. Ref. 5, p. 62.
9. *Federal Register,* **38,** 20832–20834 (August 3, 1973).
10. Ref. 9, 15406–15413 (June 11, 1973).
11. Ref. 5, pp. 3, 20–21, 33.
12. Ref. 9, **38,** 11355–11356 (May 7, 1973). Also *Chemical Week,* May 16, 1973, p. 23.
13. Ref. 9, **38** 15180 (June 8, 1973).
14. R. Fri. Statement by EPA Acting Administrator, July 30, 1973 (EPA).
15. Ref. 9, **38,** 17683ff. (July 2, 1973).
16. *Business Week,* August 4, 1973, p. 21.
17. Ref. 9, **38** 20752–20754 (August 2, 1973).
18. Ref. 9, **38** 18986ff. (July 16, 1973).
19. Ref. 9.
20. *Mathematical Models for Air Pollution Control Policy Decision-Making.* National Industrial Pollution Control Council: Washington, D.C., 1971.
21. E. S. Burton. *A Cost Effectiveness Approach to Urban Air Pollution Abatement,* 1968 Joint National Meeting, Operation Res. Soc., AIMS, San Francisco, California (May 3, 1968).
22. Ernst and Ernst. *Application of Cost-Effectiveness Analysis to Air Pollution Control,* Contract CPA-22-69-17. HEW, PHS: Washington, D.C., 1970.
23. *A Cost-Effectiveness Study of Air Pollution Abatement in the Greater Kansas City Area.* HEW: Washington, D.C., 1968.
24. *Air Quality Display Model (AQDM).* HEW: Washington, D.C., 1969.
25. EPA. "Regulation on Preparation of Implementation Plans." In Ref. 9, **36,** 22398, 24002, 25233 (1971); **37,** 26310; **38,** 6279, 15194, 15834 (1973).
26. *Control Techniques for Particulate Air Pollutants,* Pub. AP-51, NAPCA. HEW: Washington, D.C., 1969.

27. N. C. Edmisten and F. L. Bunyard. *J. Air Poll. Control Assn.*, **20,** 446–456. (1970).
28. TRW Systems Corp. *Air Quality Implementation Planning Program,* Vol. I, EPA: Washington, D.C., 1970.
29. S. E. Platkin and D. H. Lewis. "Control Strategy Evaluation Using Models," AICLE 72nd National Meeting, May 21–24, 1972, St. Louis, Mo.
30. R. E. Kohn. "Linear Programming Model for Air Pollution Control: A Pilot Study of the St. Louis Air Shed." *J. Air Poll. Control Assn.*, **20,** 78–82 (February 1970).
31. E. S. Burton, E. H. Pachan, and W. Sanjour. *Environ. Sci. Technol.* **7,** 412–415 (1973).
32. CONSAD Research Corp. *The Direct Cost of Implementation Model,* Vol. I. EPA: Washington, D.C., 1972.

Author Index

This list includes only those cited by name in the text. Authors of other works cited are listed in the references following each chapter.

Anderson, A. A., 467
Arrhenius, S., 58

Babcock, L. R., 271, 273, 276
Badgley, F. I., 339
Barrett, L. B., 107, 110, 114, 117, 118, 120
Barringer, A. R., 238, 427
Benedict, H. M., 117
Black, G., 189, 190
Bosanquet, C. H., 324
Bottema, M., 427
Boulding, K., 13
Boussinesq, J., 307
Briggs, G. A., 321, 324, 326, 327
Brody, S. S., 430
Broecker, W. S., 88
Burton, E. S., 509
Byer, R. L., 416, 420, 421

Cadle, R. D., 67, 79
Callendar, G. G., 57, 78
Cassell, E. J., 95, 96
Chamberlain, T. C., 58
Chan, S. H., 240
Chaney, E., 430
Chang, P. C., 329
Chapman, R. L., 400
Charlson, R. J., 479
Clarke, J. F., 340
Collis, N., 243
Cramer, H. E., 298
Crider, W. L., 430
Croke, E. M., 351

Dabberdt, W. B., 329
Dale, E., 12, 126
Davies, J. H., 427
Deacon, E. L., 288
DeMartini, R., 393

Dinman, B. D., 103
Djuric, D., 329
Doyle, G. J., 67
Draeger, B., 430
Driscoll, J. N., 393
Duckworth, S., 436
Duprey, R. L., 169

Eriksson, E., 49, 50

Fairhill, A. W., 70
Fiocco, G., 243
Flachsbart, H., 469, 490
Fontijn, A., 427
Forrester, J. W., 6

Gaeke, C. C., 389
Gieseke, J. A., 468
Gifford, F. A., 321
Gill, H. G., 441
Golomb, D., 427

Haagen-Smit, A. J., 5, 69
Hager, R. N., 410
Hanst, P. L., 239
Hamilton, P. M., 243
Hardin, G., 13
Harney, B., 427
Hecht, T. A., 332
Heist, H. E., 125, 126
Hendrickson, E. R., 383, 388, 465
Hidalgo, J., 239
Hilst, G. R., 339
Hinkley, E. D., 415, 423
Hochleiser, S., 390
Hodgeson, J. A., 427, 430, 442
Hoffman, D., 477
Holzworth, G. C., 340

513

Inaba, H., 419

Jacobs, M. B., 390
Johnson, F. S., 43
Johnson, W. B., 243, 329
Junge, C. E., 49, 50, 51, 68, 85, 452, 457

Katz, M., 67, 388, 444, 472
Kay, R. B., 427
Kellogg, W. W., 45, 69, 85, 86
Kelley, P. L., 415, 423
Kildal, H., 416, 420, 421
Kobayasi, T., 419
Koenig, H., 187, 188
Kohn, R. E., 511
Kolitz, B. L., 427
Koogler, J. B., 340
Kreuzer, L. B., 417

Lamb, R. G., 341
Landsberg, H. E., 124
Larsen, R. I., 227, 269
Lave, L. B., 111
Lawyer, R. E., 121
Leahey, D. M., 339
Lederman, S., 239
Ledford, M., 67
Ledingham, G. A., 67
Leithe, W., 21, 377, 388
Lettau, H. H., 339
Lewis, D. H., 507
Little, G. C., 243
Lodge, J. P., 378
Ludwig, F. L., 329

McAllister, L. G., 243
McCaldin, R., 203
Mc Crone, W. C., 470
Machol, R. E., 7, 8, 9, 11
Mage, D. T., 386
Maley, S. W., 246
Middleton, W. E. K., 122, 479
Miller, M. E., 340
Morreal, J. A., 239
Moses, H., 301, 339
Mulvaney, B. D., 125, 126

Nederbragt, G. W., 428
Neeley, R. D., 384, 385
Nelson, G. O., 441

Noghrey, J., 386

O'Keeffe, A. E., 399, 441
Ortman, G. C., 441

Pare, J. P., 394
Pasquill, F., 319, 320, 321
Patel, C. K. N., 417
Peterson, J. T., 124
Pickwell, G. V., 60
Platkin, S. E., 507, 509
Pooler, F., Jr., 319, 340
Portman, D. J., 288
Potter, L., 436
Prandtl, L., 288
Prengle, H. W., 240

Rasool, S. T., 58
Ridker, R. G., 12, 104, 106, 107, 118, 132, 133, 203
Regener, V. H., 429
Reiquam, H., 269
Robbins, R. C., 43
Roberts, J. J., 351
Robinson, E., 43, 46, 49, 58, 61, 62, 74, 76
Robinson, G. D., 82, 83, 87
Roth, P. M., 334
Ryznar, E., 288

Saltzman, B. E., 390, 436
Salvin, V. S., 116
Sandell, E. B., 372
Savas, E. S., 194
Schlossberg, H. R., 447
Schneider, S. H., 58
Schnelle, K. B., 384, 385
Schulze, F., 436
Schwartz, S., 209
Scorer, R., 290
Seinfeld, H., 336, 339
Seskin, E. P., 111
Shaw, M., 438
Shults, W. D., 276
Siegel, G., 209
Sisler, F. D., 56
Slade, D. H., 305, 312
Smullen, L. D., 243
Snyder, L. E., 239
Somers, E. V., 320
Stern, A. C., 145, 339, 473

Stevens, R. K., 399, 427, 430, 437
Stober, W., 469, 490
Strong, J., 427
Sutton, O. G., 314
Szonntagh, E. L., 400

Taylor, G. I., 313, 314
Tebbens, B. D., 382
Thomas, J. C., 329
Thomas, W. A., 276
Turner, D. B., 319, 340

Villalobos, R., 400, 441

Waddell, T. E., 107, 110, 114, 117, 118, 120

Walther, E. G., 272, 273
Wanta, R. C., 339
Wartburg, A. F., 436
Watt, A. D., 82, 83
Weisz, H., 473
Went, F. W., 61
West, P. W., 389, 473
Widhopf, J., 239
Whitby, K. T., 478, 480
Williams, D. T., 410, 427
Wynder, E. L., 477

Yankelovich, D., 120
Yocum, J., 203
Youden, W. J., 443
Young, J. A., 70

Subject Index

Absenteeism, 97, 98
Absorbance, 403
Absorption measurements, 401-407
Absorption spectrophotometry, 401, 405, 416, 418
Accuracy, 370, 372-376
 of sampling networks, 230
Actinometer, 298
Adiabatic lapse, 284-285, 308-309
Aerial and satellite photograph, 221
Aerodynamic diameter, 32. *See also* Stokes's law
Aerometric data, 250. *See also* Air quality data
Aerosol, 16, 31, 451-452. *See also* Particles; Particulate matter
 chemical reactions of, 459-460
 formation of, 31, 453-455
 lead, 16, 98-100, 234
 parameters: size, shape, mass, 456-457
 self-preserving, 32, 84
 removal and fate of, 457-459
 urban, 31
Aesthetic degradation, 120-126
Agung-Bali, 84-85, 452
Air, normal, 21-22
Air basin (airshed), 139-140, 191, 283
Airborne health hazards, 101
Aircraft emissions, 158-159
Air parcel, 354
Air pollutant, primary, 16. *See also* Emissions
 secondary, 16
Air pollution, 1, 42
 concentration levels, 24, 26, 366
  maximum, 24
 control, constraints on, 199-202
  costs, 13, 202, 203, 350, 495, 509
  at high altitudes, 499
  versus solid waste disposal, 499-500
  transportation interface, 498-499
 control system (U.S.), 190-199, 204-207

 second-generation, 207-209
 control system objectives, 204-207
 definitions, 1, 15-16, 42
 in state regulations, 165
 economic basis, 13
 reports, 268, 270
 societal problem of, 6, 11-15
 systems problem, 6
Air Pollution Control Association, 117
Airports, emissions, 158-159
Air quality, significant deterioration, 500
Air Quality Act, 141-145
Air Quality Control Techniques, 144
Air Quality Control Regions (AQCR), 143, 282. *See also* Air basin
 priority class, 149-150
Air Quality Criteria, 15, 141-142, 144
 as standards basis, 215-217
Air Quality data, collection, 213-250
 from RAPS program, 496
 summary, 266
Air-quality increment plan, 501
Air-quality standards, 23, 146-148, 216-217
 ambient, 146-148
 index basis, 269-271
 primary, 23, 147-148
 modification of SO_2, 497
 secondary, 147-148, 274, 275
 state, 166, 167
Airshed, 191. *See also* Air basin
Aitken particles, 122, 456. *See also* Particulate matter
Albedo, 84
Alert criteria, episode, 181, 184
Ambient air monitoring, 150-152
 systems, *see* Monitoring
Ambient standards, state, 166, 167
American Chemical Society, 276
Amino acids, 1
Ammonia, 1, 29, 43, 53-54, 69-70
 measurement of, 429
Amperometric instruments, 434-437

517

Anabatic wind, 289
Analysis, *see* Measurement of specific species
Anemometer, 297
Antarctica, 58
AQCR, *see* Air Quality Control Regions
AQDM (Air Quality Display Model), 507
Area classification plan, 503-504
Area source, 304, 319
  data collection from, 255-256
Arsenite method, 391. *See also* Nitrogen oxides
Asbestos, 102
Asbestos emissions control, 163
ASTM (American Society for Testing and Materials), 338
Atmosphere, composition, 20, 41, 42
  of background, 41, 43, 276-277
  changes, 2
    affecting climate, 11-12
  concentration of gases, 20
  of normal dry, 21-22. *See also* specific gases
  primitive, 1
Atmospheric reactions, 18-20, 101, 336
  of carbon compounds, 70-73. *See also* Photochemical reactions
Atomic absorption, 453
Austauch, 307
Autoanalyzer, 388
Automobile and smog, 5
Automobile emissions control, 144-145, 146, 155, 158, 179-180. *See also* Motor-vehicle
Averaging time, 226-227
Awareness, public, 121
Aztec, 2

BAAPCD (Bay Area Air Pollution Control District), 384-385
Background concentration, 41, 43
  ACS values, 276-277
Bacterial action, 51, 53, 60, 63
Bahia de los Humas, 4
Beaver commission (Great Britain), 3
  report, 119
Beer's law, 389, 401, 412, 424, 480
  plot, 403
Bernoulli distribution, 313
Bernoulli's law, 290

Beryllium, 102
Beryllium emissions control, 163
Beta-ray attenuation, 482
Binomial distribution, *see* Bernoulli
Biological action, 43, 50, 51, 65, 72
Boulding, Kenneth, 13
Box model, 304-305
  urban, 338-339
Buckingham π Theorem, 307
Buoyancy, in plume rise, 324-328
Buoyancy forces, 285, 308-309
Burning, open, 153-155, 169, 173, 256
Butanol, 62

Calcium carbonate, as CO_2 sink, 55
Calibration, 376, 381, 439-442
  particulate monitors, 489
  by uniform spheres, 490
California Agriculture Department study, 117-118
California Institute of Technology, 5
CAMP (Continuous Air-monitoring Program), 226
  measurements from, 23, 28
  of hydrocarbons, 27
Cape Kennedy-Vandenberg model test, 333-334
Carbon compounds, 55-56
  inorganic, 17
  reactions and sinks, 70-73. *See also* Hydrocarbons and under specific compounds
Carbon dioxide, 2, 42, 55, 56-59, 70-71
  analysis by conductimetry, 392
  concentration in air, 25
  global temperature, 55, 58-59, 83-84
  sinks, 70-71
Carbon monoxide, 43, 55, 59-61, 71-73, 74
  absorption in bloodstream (COHb), 92-95
  concentration in air, 25
  determination of, 394
  sinks, 71-73
  standard, air quality, 148, 275
Carboxyhemoglobin, 92-95
Carcinogens, 35
Carotene, 63
Catalytic converter (automotive), 155, 159
Cement plant, emission standards, 160-161
Charcoal, 2
Chemical reactions, atmospheric, 101-336

Eschenroeder's model, 336
  kinetic models, 335-338
  see also, photochemical reactions
Chemiluminescence, 427-430
Centrifugal separation of particles, 468-469
CHESS (Community health and environmental surveillance system), 497
Chloride electrode, 393
Chlorine, 30
  analysis of, 434, 442
Chlorophyll, 1
Chromatography, gas, see Gas chromatography
Chronic disease, 95-96
City size model, 340-341
City-street model, 329-331
Cities, as climatic changes cause, 124. See also Urban air pollution
Clean areas preservation, 500-506
  by area classification, 503-504
  air-quality increment plan, 501
  by emission limitation, 502
  local definition plan, 503
Clean Air Act, Great Britain, 3
  of 1963 (U.S.), 5, 141
  of 1970, amended, (U.S.), 5, 13, 145-163
Climate, air pollution effects, 123-125
  of cities, 124
  effect of atmospheric changes, 11-12
  effect of carbon dioxide, 55, 58-59, 83-84
  statistics, 87
Cloud, 31
Coal, 2, 91, 243, 501-502. See also Combustion; Smoke
Coefficient of haze (COH), 36, 471-472. See also Particulate measurements
COHb (Carboxyhemoglobin), 92-95
COI (component of interest), 385
Coincidence error, of counter, 478
Colorimetry, 388-392
Combustion, carbon dioxide from, 56-58, 77, 83, 222
  carbon monoxide from, 59-61, 77, 222
  of coal, 2, 91, 243, 501-502
  efficiency of, 3, 4, 63-65
  of fossil fuels, 2, 176, 190, 221, 253-260, 274, 304
  of gaseous fuels, 176
  of gasoline and petroleum, 2, 4-5, 60-61, 129-131, 293

global effects of, 83, 89
  hydrocarbons from, 63-65, 77, 222
  hydrogen flame (FPD, FID), 430-431
  incineration of wastes, 61, 63-65, 153-155, 160, 162, 172, 173, 175, 221-222, 261, 500
  industrial, see Stationary sources
  internal (engine), 59, 144, 155-158, 179, 281 282, 499. See also Motor vehicle, Automobile
  nitrogen oxides from, 17, 52, 77, 130, 152, 176, 222
  open burning, 153-155, 169, 173, 256
  particulates from, 31, 152, 222, 453
  stationary sources, 109, 110, 129, 131, 160-162, 172, 175, 264, 281-282
  sulfur oxides from fossil fuel, 44-48, 77, 152, 169, 222, 241, 506
Combustible gases, analysis by thermal conductivity, 433-434
Comminution, 31
Commons, tragedy of, 13
Community health, 95-103
Community Health and Environmental Surveillance System (CHESS), 497
Community management of air quality, 349-362
Computer requirements for math models, 340-345
Concentration, atmospheric gases, 20-21
  background, 41, 43, 276-277
  of pollutants, 21, 24, 26, 366
Concentration coefficient, 328
Condensation, 31
Condensation nuclei, 459
Conductimetry, 392
Confidence limits, 374
Coning, 293, 323. See also Plumes
Constraints on control systems, 199-202
Consumer protection, 13
Contaminants, atmospheric, 15, 36
Continuity model, 334. See also Diffusion equations
Control Regions, see Air Quality Control Regions
Control technology, "best available", 503
Coriolis force, 288
Cornell Aero Labs (Calspan), 125
Cornell Family Illness Study, 95-96
Corner reflector, see Retroreflector

Subject Index

Cost-benefit, 43
   analysis, 506
   relationships, 104-105, 202-204
Cost-effective analysis, 506-510
   control of air quality, 14
Cost models, see Models
Costs of air pollution, 3, 81
   economic, 12-13, 104-120
   measure of disbenefit, 189
   in Pittsburgh (Mellon Institute), 107
   on residential property loss, 107-110, 118-119
   in U.S., 12, 106-119
      in 1968 (tables), 107, 108
      in 1977 (tables), 110, 112
   of materials loss, 113-116
Coulometric analyzers, 435-438. *See also* Amperometric
Coulter counter, 478
Council on Environmental Quality (CEQ), 145
Counter, *see* Particle measurement
Criteria, *see* Air quality criteria
Cyanide, hydrogen, analysis, 434, 442

Damage function, 104
Data, acquisition of, 213-266
   characteristics of, 214-233
      density, 224
         in time, 224-229
         in space, 229-233
         function of population, 232
      real time, 225-226
      transmission, 245-248, 249
Data-processing requirements, math models, 342-345
Dead time, 384
Denver, Colorado, 499
Deposition, particle, 327-328
   dry velocity, 327
   Hanford model factors, 333
Description, 400
Detector, flame, ionization (FID), 398, 400, 430
   optoacoustic, 417
   photometric (FPD), 398, 399, 430
   thermal conductivity, 398, 432
Diabatic lapse, 309
Diffusion coefficients, 311, 316, 320-321, 322, 332

Diffusion equation, vector, 305, 334-335, 341
Diffusion model, urban, 329-331
Diffusion theory, 305, 306
   Fickian, 310
   relative, 315-316
   Sutton's model, 314-315
"Dirtiness" measure, 467
Display, 244, 358-359. *See also* Instrument readout
Dry deposition velocity, 327
Dust, 31
Dustfall, 35. *See also* Particulate matter
Dyes, fading of, 116
Dynamic response of instruments, 382-388

Earth Day, 13
Echo sounder, 243
Ecological effects, 88
Economic costs of air pollution, 104-120. *See also* Costs of air pollution
Economics of Clear Air (EPA report), 106
Eddies, 289, 290
   wake, 290
Eddy-diffusion coefficient, 307, 310, 311, 314
   horizontal, 315
      heat conductivity coefficient, 309
   see Eddy-diffusion
   see also Turbulence; Viscosity
Effects, of air pollution, 81-126
   global, 82-89
   irreversible, 82
   urban, 89-103
Elastomer losses, 114-116
Electric power growth, 505-506
Electrochemical measurement of sulfur dioxide, 434, 437, 438
Electrostatic precipitation, 468
Emission charges, 505
Emission control, aircraft and airports, 158-160
   hazardous pollutants, 162-163
   hydrocarbons, 177-178
   motor vehicles, 144-145, 155-158
      New Jersey Inspection standards, 179
      state regulations, 179-180
   sources, 151-153
      new stationary, 160-162
   see also Emissions standards

Emission data, new plants, 497
Emission factors, 220, 222
  ammonia, 54
  carbon dioxide, 56
  carbon monoxide, 60
  hydrocarbons, 64
  nitrogen oxides, 52
  sulfur dioxide, 47
Emission limitation plan, 502
Emission spectroscopy, IR, 240
Emission standards, area-based, 351
  automotive vehicle, California, 156-157
    federal, 144, 146, 155-158
  federal, 144, 146, 351-352
    for hazardous pollutants (NESHAP), 162-163, 220
    for motor vehicles, 155-158
    for new stationary sources (NSPS), 160-162, 220
  gases, 353-354
  particulate, 171-173, 352-353
  state, for hydrocarbons, 178
    for motor vehicles, 156-157, 179
    for particulates, 171-173
  sulfur oxides, 169-171
  visible emissions, 173-175
  worldwide, 182-183
Emissions, "non-significant," 501
  inventory, 220, 253-265
  potential, 353
  primary, 16
  traffic, 330
  U.S., 23, 109
    ammonia, 53-54
    carbon compounds, 55
      dioxide, 56-58
      monoxide, 60
    hydrocarbons, 63-65
    hydrogen sulfide, 49-50
    methane, 55, 63
    nitrogen oxides, 52
    sulfur, 45
    sulfur oxides, 44, 46-48
    terpenes, 63, 65
Empirical function, oxidant prediction, 218-219
  model, see Model
Energized atoms, 19
Energy crisis, 13, 14, 505-506
Entropy of waste materials, 188

Environmental insults, 81
Environmental Protection Agency (EPA), 15, 129, 145, 146-148, 152, 155, 158, 160, 163, 164, 176, 351, 377, 388, 390, 429, 441, 442, 444, 455, 464
Environmental Science Services Administration (ESSA), 305
EPA, see Environmental Protection Agency
Epidemiology, 95-103
Episodes, acute air pollution, 90-95
  carbon dioxide, 92-95
    Birmingham (Alabama), 91
    Donora (Pennsylvania), 3, 66, 90
    London, 2, 3, 66, 91
    Muese Valley (Belgium), 3, 90
    New York-New Jersey, 91-92
    oxidant level during, 92
    statistics, of, 191, 192
Episode control, 360-362
  alert criteria for, 181, 184
  quick-reaction system, 361-362
  significant levels for, 94
  state plans for, 153-155
  state regulations, 180-181, 184
Equation of state (gas law), 20, 285, 308
Ernst and Ernst, 506
Errors, 372
ESSA, 305
Eulerian frame, 314
  model coordinates, 334
Evelyn, John, 2
EVI (Extreme Value Index), 279-281
"Excess" deaths, 12
Exchange coefficient (Austauch), 307
Excitation analysis methods, 437-431
  chemiluminescence, 427-430
  flame ionization detector, 430-431
  flame photometric detector, 430
  high temperature, 430
  low temperature, 428
Expected value, 9

Fanning, 293, 321. *See also* Plumes
Federal Register, 148, 158, 160, 162
Federal Regulations, Code of, 164
Feedback control, 267
Feedforward control, 192
Fermi, E., power plant, 298
Fickian theory, 310

Fick's law, 290. *See also* Diffusion theory
Filters, 466
Filtration, 394
Fires, 129, 131
Fires, forest, 107, 109
Fires, open, 153-155, 169, 173, 222, 256
Flame ionization detector (FID), 430-431
   photometric detector (FPD), 430
Flow, around structures, 328-331
   over flat plate, 328
   see also Street model
Fluid Model Facility (FMF), 496
Fluorescence, resonance backscatter, 421-423
   X-ray, 474
Fluoride determination, specification, 393
Forcing functions, 235
Forecasting, air pollution levels, 208, 281-283
   air quality model, 267-268
   short-term, 283
   trends, 281-297
Forrester, Jay W., 6
Formaldehyde detection, 438
FORTRAN, 245, 345
Four Corners power plant, 121-122, 500
Fuel, additives, 146, 158
   consumption, 499, 510
   fossil, *see* Coal, Petroleum
   low-sulfur, 505
   state regulations for, 169-171
   sulfur-bearing, 23
   switching, 505
   see also Combustion
Free radicals, 85
Frictional velocity, 307, 308
Froude number, 328
Fumigation, 286-288, 321

Gas chromatography, 394-400
   application, 399-400
   carrier gas, 397
   columns, 397-399
      capillary, 398-399
   detectors, 398. *See also* Detectors and specific types
   gas-liquid, 399
   gas-solid, 399
   injection port, 397
   readout, 398

   solid support, 397
   subtraction technique in, 400
   trapping in, 400
Gas law, 20, 285, 308
Gaussian (normal) distribution, 32, 230, 311, 313
   of plumes, 292
      formulas for, 321-324
      Hanford, 332-333
      models, 316-324, 342-344
      urban, 339-341
Gasoline, *see* Petroleum
Glaciation, 12
Global buildup of pollution, 281-282
Global temperature change, 55, 58-59, 83-84
GNP, 126-127
Gradient transport theory, 309-311
Greenhouse coefficient, 84
   effect, 58, 83
Greenland, 59
Grey, Thomas, 6
Grid model, 334
Griess-Ilosvay method, 390
Griess-Saltzman, 498
Growth of air pollution, 126-132, 193
   of electric power, 505-506
   of hydrocarbon production, 129-131
   of nitrogen oxides, 129-131
   of population, 132-134

Haagen-Smit, A. J., 5
Halides, halogen compounds, 18, 30
Hanford model, 332-333
HAPPA (High Air-Pollution Potential Advisory), 296-297, 299
Hardin, Garrett, 13
Hazardous pollutants, 101-102
   federal emission standards (NESHAP), 162-163, 220
Health, costs, 104-106, 107, 108, 109, 110-113
   effects, function of concentration-duration, 96-98
   of suspended sulfates, 497
   public, 89-103
Health Education and Welfare Department (HEW), 141-145, 490, 491
Height-of-rise formulas, 324-327
High altitude emission control, 499

Subject Index 523

Holland-Stürmke formula, 324-326
Hydrocarbons, concentration in air, 25, 27
　emissions, 5, 63-65
　　control of, 177-178
　　growth, 129-131
　　reactive, 55, 63, 64, 178
　　standards, air quality, 148, 275
Hydrofluoric acid, 18
Hydrogen, 18
　flame, 430-431
Hydrogen cyanide, 434, 442
Hydrogen sulfide, 1, 17, 43, 44, 66, 113, 152, 394, 434, 442

Identification of air-pollutants, 198
ILAMS (Infrared laser monitoring system), 241
ILLIAC IV, 88, 345
Impactor, Anderson, 467
Incinerator emissions standards, 160-161.
　　See also Combustion, incineration
Index of air pollution, 214, 267, 268-281
　ambient standards basis, 269-271
　application, 281
　BAAPCD, 269-280
　comparative values, 280
　Extreme Value (EVI), 279-281
　media reports, 268-270
　Mitre (MAQI), 273-276
　Oak Ridge (ORAQI), 276-279
　Pindex, 271-272
　sulfur dioxide-based, 268
　Walther rankings, 272-273
　weighting of effects, 269-271
Infrared (IR), 87, 239, 240, 241-242, 356, 408, 411-412, 414-417, 419, 422-423. See also Optical measurements, Laser
Inspection, of motor vehicles (New Jersey), 179-180
　by state and local agencies, 168-169
　of stationary sources, 220
Instruments, automatic, 366-367, 378-379
　characteristics of, 378-388
　cost, 382
　drift-free, 381
　field, 379-382
　general considerations, 365
　ideal, 379
　integrating, 382

interferometer, 240
　long-path, 238-242, 244, 367, 408-409
　manual, 378
　optical, 401-427
　particulate, 484, 488
　　averaging, 488-489
　　criteria, 485
　　real-time, 486-487
　readout, 377
　real-time, 382, 486-487
　separation, 388
　short-path, 367, 408. See also Measurements, Monitoring, specific instruments
Interfacing of monitoring instruments, 378
Interferometer spectrometer, 240
Internal combustion engine, see Combustion
Inversion, 90, 180, 285-286
　anticyclonic, 286
　ground-based, 286-287
　lid, 292, 338
　nocturnal, 285-287
　penetration of, 292-295, 325-326. See also Height-of-rise
　subsidence, 286
　urban elevated, 286-287
Impinger, 388
IPP (Implementation Plans Program), 507
Irreversible effects, 82
Isokinetic sampling, 161-162, 355, 463-464
Isopleth, 323

Jacobs-Hochheiser method, 390
Japan, 127

von Karman's constant, 295
Katabatic wind, 289
Kelp, 61
Kentucky, ambient standards, 167
Kew Observatory, 83
K-theory, 311-312
Kraktoa, 84
Kinetics, atmospheric reactions, 335-338

LAAPCD (Los Angeles Air Pollution Control District), 140
Lagrangian frame, 315
　one-point velocity correlation, 313
　model coordinates, 334
　statistical diffusion, 315

Subject Index

Lambert-Beer law, see Beer's Law
Laminar flow, 306
Laminar layer of atmosphere, 307
Land-use strategy, 188
LAPPES (Large power-plant effluent study), 496
Lapse rates, 284-285, 308-309
   adiabatic, 284, 293, 308
      wet, 285
   diabatic, 309
   environmental, 284
   measurement of, 297
   plume behavior, 293
   process, 284
   stability, 309
   superadiabatic, 284
Lasers, 411-422
   tunable, 412, 413
Laser spectrometer, 239, 241-242, 414-418
   Raman backscatter, 418-421
   resonance fluorescence, 412-423
   see also Lidar
Lead, 16, 98-100, 158, 234
   airborne, health effects of, 98-100
   in gasoline, 146-158
Least-cost strategy, 190. See also Cost-benefit
LeChatlier's principle, 69
Legislation, air pollution control, 139-185
   federal, 140-163
   state and local, 163-181
   worldwide, 182-184
Lid, see Inversion
Lidar, 241-242, 419
   mixing layer height from, 298
Linear programming models, 351, 509
Line source, 318-319
   infinite, 311-312
   with wind, 319
Livestock-losses, 119
Local definition plan, 503
Lofting, 293, 323. See also Plumes
Log normal frequency distribution, 21, 32-33, 464
   of gaseous pollutant concentration, 227
   of particles, 464
   of sampling, site data, 230
London smog, 2, 4
Long-path sensors, 233, 238-242, 367, 408-409
   effectiveness of, 240-241, 242
   prospects for, 244
Long-period average concentration, 323
Los Angeles smog, 4-5, 128-132
Looping, 293, 322. See also Plumes
Low-sulfur fuel shortage, 505

Macroviscosity, 314
Management air resource, 349-362
MAQI (Mitre Air Quality Index), 273-276
Mass concentration and visibility, 123
Mass conservation model, 341-345
Mass spectrometer, 431-432
   quadrupole, 432
   spark source, 474
Mass transportation, 499
Massachusetts Institute of Technology (MIT), 6, 45, 56, 58, 78, 490
Mast monitor, 435, 438
Materials deterioration, costs, 113-116
Mathematical models, see Models
Mauna Loa, 57-59
Maximum Allowable Concentration (MAC), 96
Maximum concentration from point source, 323
Mean, 33, 374-376, 464
Measurements, 367. See also Instruments
   absorption, 401-407
   continuous, 367-380
   electrochemical, 434-439
   in situ, 368
   long-path, see Long-path sensors
   particulates, see Particulate measurements
   point, 238, 367
   range of, 372
   remote, 367-368
   sampled, 368
   taxonomy of, 369
Median, 33, 464
Mellon Institute, 106
Membrane sensor, 438-439
Mercaptans, 28
Mercury, emissions, 102-103, 163
Mesoscale model, 330
Meteorology, air pollution, 283-297
Meteorology and Atomic Energy - 1968, 305
Meteorological forcing functions, 235
   control strategy, 14

Subject Index 525

measurements, 297-299
submodel, 190
Methane, 1, 41, 55, 61, 63
Mexico, 2
Michelson interferometer, 240
Microbalance, piezoelectric, 483-484
Microscopic analysis of particulates, 470
Microwave spectrometry, 239-240
  rotational resonance, 409
Midwest Research Institute (MRI), 114-116
Mie scattering, 32, 84, 122, 417
Mist, 31
Mitre Corp., 225, 273-276
Mixing layer, depth of, 243, 284, 298. *See also* Inversion
Model, city size, 340-341
  city street, 329-331
  proportional, 176, 499
  scale, 234
  mesoscale, 330
Model Cities program, 506
Model input-output, 303
Models, cost, 506-510
  AQDM-IPP, 507-508
  IPP-DCIM, 509-510
  least cost, 507-509
  linear programming, 351, 509
  macro-and microlevel, 509
  rollback, 507-508, 509
Model smoke ordinance, 173, 352
Models, climate, 87-88
  empirical, 301, 304, 333-334
  statistics of, 87
Model, mathematical, 6, 9, 301-345
  applications, 316-341
  Gaussian, 316-324, 339-340, 343-344
  objectives of, 338
  photochemical, 334-338, 341-342
Monitoring, 349
  ambient atmosphere, 150-152
    CAMP measurements (U.S.), 23, 26
    global, 89
  chemical species, 215-216
  legal requirements, 384
  source, 354-356
Monitoring instruments, 151-152, 365-367
  application, 365
  see also Instruments
Monitoring (ambient) systems, air quality
  in U.S., 194-195

networks, 9, 233-250, 356-360
  cost, 231-232
  network design, grid, 229-231
  European, 229
  number of stations, 229-232
  optimum, 247-248
  RAPS (St. Louis), 235-238
  second generation, 208
  U.S.S.R., 230
Mortality, correlation with air pollution, 12
  premature, cost of, 106
Motor-vehicle emissions control, 144-145, 146, 155-158, 179-180
  state regulations, 179-180
  New Jersey standards, 179
Multivariable control, 267

NAPCA (National Air Pollution Control Administration), 15, 141-145, 453-455, 490
National Academy of Sciences, 98
National Ambient Air Quality Standards (NAAQS), *see* Air quality standards
National Air Surveillance Network (NASN), 195
National Bureau of Standards, 440, 442, 450
National Center for Atmospheric Research, 243
National Research Council, 98
NEEDS (Neighborhood Environmental Evaluation and Decision System), 223-224
Nephelometer, 479-480
Nernst equation, 392
NESHAP, 162-163, 220
Network design for data acquisition, 233-250
  optimum size, 247-248
Neutron activation, 474
New York Times, 126
Nitric acid, 50, 53
  new plant emissions, 160-161
Nitrogen, 41, 42
Nitrogen compounds, 17, 50, 68
  ammonia, 1, 29, 43, 53-54, 69-70
    measurement of, 429
  ammonium nitrate, 50, 53
  gaseous, 43, 50, 68
  nitric acid, *see above*
  nitric oxide, 43, 50, 51-52

526 Subject Index

analysis, 429
  electrochemical, 438
  optoacoustic, 418
  potentiometric, 393
  Saltzman, 390
nitrogen dioxide, see below
nitrogen oxides, see below
nitrous acid, 50, 53
nitrous oxide, 41, 43, 50, 51, 68
organic nitrates, 29
PAN and PPZN, 29-30
Nitrogen dioxide, 43, 50, 51-52
  analysis methods, 429, 442
    correlation spectroscope, 426
    electrochemical, 438
    Griess-Ilosvay, 390
    Griess-Saltzman, 498
    Lyshkow, 390
    potentiometric, 393
    Saltzman, 390-391
    see also Nitrogen oxides, analysis
  global effects of, 88
  standards, air-quality, 148, 275
Nitrogen oxides (NO_x), 28-29
  analysis of, 429
    arsenite method, 391, 498
    chemiluminescence, 429, 498
    Jacobs-Hochheiser, 390-391
    phenoldisulphonic acid, 391
    see also Nitrogen dioxide
  control in auto emissions, 498
  growth, 129-131
  revised data, 497, 498
  standards, state, 175-177
    see also Air quality standards
  see also specific compounds
NOAA, 243
Normal air composition, 21-22
Normal air-movement patterns, 191
Normal distribution, see Gaussian, Log-normal
North Atlantic, CO_2 concentration, 57
NSPS, 160-162, 220

Objectives of control systems, 204-207
Obsolescence, 188-189
Oceans, 60, 61, 70-72
  1 and 2 layer models, 71
Odor, 125-126
  dilutions (measurement), 167

Olefins, 67, 127, 128
Optical attenuation, 401
Optical density, 471-472
Optical measurements, 401-409
Optimum systems, 8, 11
ORAQI, 276-278
  nomograph, 278
Opacity, 161, 162. See also Ringelmann; Smoke, visible
Organic compounds, 17. See also Hydrocarbons; PAN and PBZN
Organoleptic method, 125
Oxidants, 5, 30. See also Nitrogen compounds; Ozone
  photochemical, 148
  precursors, 25, 217-219
  prediction of maximum, 218-219
  standards, air quality, 148, 275
Oxygen, 42
  depletion, 88
  early atmosphere, 1
Ozone, 30, 41, 67
  analysis of, chemiluminescence, 428
    electrochemical, 434-439
    Haagen-Smit and Brunelle, 392
    Mast, 435, 438
    phenolphthalein, 392
    potassium iodide (neutral, alkaline), 391-392
  effect on materials, 114
  standards for, 442

Paint, drying, SO_2 inhibited, 113
  solvent emissions, 178
PAN and PBZN, 29-30, 85
Pareto-optimality, 506
Particles, Aitken, 122
  analysis of, 469-477
    organic, 476-477
  atmospheric, 31, 85
  coarse, 16
  concentration, 34-35
  composition, 34, 457-458
  definitions, 451
  distribution, 451-452
  dustfall, 456-457
  effects of, 461
  fire, 16
    radioactive, 36
  formation mechanisms, 31, 453-455

Subject Index 527

giant, 122
lead, 16
parameters, size, shape, mass, 31-34, 456, 464-465
reactions of, 459-460
removal mechanisms, 457-459
residence time, 457-458
respirable, 219, 461, 468
sampling of, 461-463
sources of, 451-452
  combustion, 31, 152, 222, 453
  man-made, 453-455
  volcanic, 84, 85
standards for emission, 171-173
statistics of, 32
stratosphere, 85-86, 452
  chlorine-bromine ratio, 452
sulfate, health effects, 497
  stratosphere, 85-86, 452
see also Aerosols
Particulate matter, air quality standards for, 148, 275
Particulate measurements, 460-490
  ambient, 463, 488
  averaged, 463, 465, 488-489
  calibration, 489-490
  centrifugal spearator, 468
  COH, 471-472
  composition, 465
  counter-classifier, 477-478
  electric charge, 480-481
  energy absorption, 481, 482
  evaluation criteria for, 485
  filtration, 463, 465-466
  impingement, 466-467
  light-reflection, 481-482
  light-scattering, 478-480
  mass concentration, 460, 464
  mass related, 482-484
  microscopy, 470
  mobile source, 463
  number of particles, 465
  optical density, 471
  photometry, 470-472
  precipitation electrostatic, 468
    thermal, 468
  real-time, 465, 486-488
  RUDS, 472
  size distribution, 465
  sources, 463, 488

space-averaged, 465
stacks, 463
transmittance, 471
Partitioning, 395-397
  coefficient, 395
Pasquill stability conditions, 319-321
Penetration through inversion, 293-295
Pennsylvania, ambient standards, 166
  losses of vegetation, 117
Permeation tube, 385, 441-442
Permits and registrations, 165, 168
Petroleum and gasoline combustion, 2, 4-5, 60-61, 129-131, 293
Phillips monitor for SO_2, 437
Photochemical models, 334-338, 341-342
Photochemical reactions, 18-20, 27, 44, 335-338
Photochemical smog, 4-5, 128-132
Photometer, right angle scatter, 480
Photometry, 470-472
Photo-oxidant, see Oxidant
Photosynthesis, 1, 70
Pindex, 271-272, 273, 280
PL-1 language, 245, 343
Plume, 317
  bent-over, 294-295
  conical, 316
  fluctuating, 318
  Gaussian, 292
  model, verification of, 331-334
  reflection, 317-318
  rise, 324-327
    combined with Gaussian model, 326-327
  shape of, 290-292
  stability, effect on, 293, 321-323
Point Barrow, 59, 62
Point source, 221, 223
  continuous, 317
  data, 253-255
  instantaneous, 316. See also Puff
Polarizability, molecular, 419
Pollen, 36
Pollutants, see Air pollution
Population, and air pollution growth, 132-134
  data density function, 232
Porphyrins, 72
Posterior controls, 349, 351, 352. See also Emission standards

Potential temperature gradient, 325
Potentiometry, 392-393
Precision, 370, 373-376
Prediction of air pollution levels, 198-199
Primary air pollutant, 16. *See also* Emissions
Primary ambient air-quality standards, 23, 147-148
Primitive equations, 87
Prior controls, 349, 350, 351
Process-weight control basis, 171, 352-353
Property losses, residential, 118-119
Proportional model, 176, 499
Psychic cost, 121
Public awareness, 121
Public health, 89-103
Puff, 310, 316
  integrated, model, 341-342
  nonisotropic, 317

Quartz crystal microbalance, 483-484
Quick-reaction control, 194, 209, 225, 361-362

Radiation, radiant energy, 1, 5, 12, 18, 19, 58, 83-84, 87, 124, 218, 236, 336-338, 481
  infrared, *see* Infrared
  ultraviolet, 1, 5, 18, 19, 123, 124, 218, 239
Radiation-matter interaction, 402
Radioactivity, 4, 18, 36
Raman backscattering, 239, 416, 418
Raman effect, 419
RAPS (Regional Air Pollution Study), 233-235
  data collection, 235, 496-497
  network configuration, 235-238
Rare gases, 41
Rayleigh scattering, 32, 417
Reaction rates in atmosphere, 336-337
Reactions in atmosphere, 18-20, 53, 85, 70-73, 101, 336. *See also* Photochemical reactions
Reactive hydrocarbons, 55, 63, 64, 178
Readout, 377
Real-time data, 225-226. *See also* Instruments, Particulate measurements
Reasonable technology, 152
Recycling of resources, viii, 500
  control strategy, 187, 188

Reflection of plume, 317-318
Regions, *see* Air Quality Control Regions
Registration, 165, 168
Relative diffusion, 315-316
Reporting form for analysis, 377
  of particulates, 464
  by telemetry, 378
Representativeness, 370
Reproducibility, *see* Precision
Residence time, 41, 68, 70, 73
Residential property losses, 107-110, 118-119
Respirable particles, 219, 461, 468
Retroreflector, 241, 242, 416
Reynolds number, 306
  in stack plumes, 328
Rhodamine-B, 429
Richardson number, 309
Rijnmond (Netherlands), monitoring network, 214-215, 357
  sulfur dioxide monitor, 438
Ringelmann chart, 174-175
Ringelmann number, 481-482
Ring oven analysis, 473-475
ROM, 249
Rossby number, 328
Roughness length, 308
Rousseau, Jean-Jacques, 6
Rubber deterioration, cost, 114-116
RUDS, 36, 472

Saltzman method, 390-391
Sample handling, 370-371
Sampling, 368-370
  isokinetic, 161-162, 355, 463-464
  number of stations, 229-232
  state regulations for, 168-169
SAROAD (Storage and retrieval of aerometric data), 357, 359, 496-497
Scattering, Mie, 32, 84, 122, 417
  Raman, 239, 416, 418
  Rayleigh, 32, 417
  resonance fluorescent, 417-421
Scattering coefficient, 122
Scattering layer, 84-85
Scrubbers, aqueous, 366-367
Sea coal, 2
Seaweed, 61
Second derivative spectrometer, 410-411
Secondary pollutants, 16, 18-20

Secondary reactions, 5. *See also* Photochemical
Secondary ambient air-quality standards, 147-148, 274-275
Second-generation control system, 207-209
Sedimentation diameter (reduced), 32
Self-preserving aerosol, 32, 84
Sensor cost, monitoring network, 231-232. *See also* Instruments, cost
Sensors, long-path, 238-242, 244, 367, 408-409
  remote, 238, 242-243
  third generation, 232-233
  transfer function, 227-229
  see also Instruments; Monitoring
Separation of wake, 290
  around buildings, 328
Shear, 306
Significant deterioration of air quality, 500
Sigma computer, 297-298
Similarity theory, 328-329
Sinks, air pollutants, 41, 43, 65-73, 132
Siphonophores, 60-61
Smaze, 31
Smog, 2, 4-5, 31,
  growth, 128-132
  report, 270
  see also London; Los Angeles
Smoke, 2, 3, 4, 121-122, 179
Smoke, visible, 121-122, 139, 175, 182-183. *See also* Ringelmann
Smoke abatement ordinances, ASME model, 173, 352
  Chicago and Pittsburgh, 3
  London (1306 Act), 2
  worldwide, 281-282
Smokers, 95
SMA, 117, 118-119
Societal viewpoints of air pollution, 6, 11-15
Soil, 72-73
Soiling, 471
Soiling and cleaning costs, 119-120
Soiling index, 472
Solids, fine, *see* Particles
Solid waste disposal, 499-500
Soot, 4
Source control, 140
  federal, 151-153
  new stationary, 160-162

see also Emissions
Source data, 220-223
Source emission factors, *see* Emission factors
Source inventory, 220-221, 253-265
Source monitoring, 354-356
  by ambient measurements, 354
Sources, area, 304, 319
  line, 318-319
  point, 221, 223. *See also* Plumes
  superposition of, 319
  see also Models, mathematical
  see also Box model
Sources of air pollution, 43
  major, 154
  see also Emissions
Source testing, 354-355
Source types, 220-221
Specific ion electrode, 393
Spectrophotometry, 401-408
Spectroscopy, absorption, 401, 405, 416, 418
  correlation, 238-239, 423-427
  emission, 240
  interferometer, 240
  laser, 239, 241-242, 414-418
  mass, 431-432
  microwave, 409
  optoacoustic, 417
  quadrupole, 432
  Raman, 418-421
  resonance fluorescence, 421-423
  second derivative, 410-411
Spreading coefficients, 321. *See also* Diffusion coefficients
SST, 86-87
Stability, atmospheric, 309. *See also* Lapse
Stability conditions, Pasquill, 319-321
Stack height, 295, 350. *See also* Inversion, penetration
Stack monitoring, 355-356
Stagnation, 90, 91. *See also* Inversion
Standard deviation, 313, 374
Standards, *see* Air Quality; Emissions; and Motor Vehicle
Standards, analytical, 439-442
Stanford Research Institute (SRI), 62, 117
State ambient standards, Pennsylvania, 166
  Kentucky, 167

Subject Index

State emission standards, *see* Emission standards
State implementation plans, 147-155
State legislation and regulations, 163-184
Stationary source, *see* Emission; Source control
Statistical diffusion theory, 312-315
Statistics, of analytical measurements, 373-374
 climatic, 87
 plumes, 292
 pollution concentration, 230
 turbulent diffusion, 312-313
Steam plants, fossil fueled, emission standards, 160-161
Stochastic system, 8
Stockholm Conference (1972), 89, 184
Stokes diameter, 32
Stokes's law, particle deposition, 30-31, 68, 122, 327, 457, 467
Storage, volatile hydrocarbons, 177-178
Strategies of air pollution control, 14, 187-189
 emission control, 153
 land use, 188
 meteorological, 189-190
 optimum, 192-193
Stratosphere, sulfate particle layer, 85-86, 452
Stratman method for sulfur dioxide, 389, 390
Street model, 329-331
Study of critical environmental problems (SCEP), 45, 86, 89
Submodels, 189-190
Suboptimum, 10
Subsystems, control, federal, 205
 source, 207
 state and local, 207
Subtraction techniques in gas chromatography, 400
Sulfuric acid, 85
Sulfuric acid plant emission standards, 160-161
Sulfurous acid, 85
Sulfur compounds, 17, 43, 44, 66
 ammonium sulfate, 85
 hydrogen sulfide, 1, 28, 49, 67, 394, 434, 442
 mercaptans, 28

 oxides, 28
 state regulation of, 169-171
 sulfate, 45, 60, 66
 sulfate particles, health hazard, 497
 in stratosphere, 85-86, 452
 sulfites, 66
 see also Sulfur dioxide
Sulfur dioxide, 43, 46, 66, 74
 analysis, by barium chloranilate, 389
  by conductimetry, 392
  by correlation spectroscopy, 426, 427
  electrochemical methods, for, 434, 437, 438
  by fuchsin-formaldehyde, 389-390
  by starch-iodine, 389
  Strathman method for, 389-390
  by titrimetry, 393-394
  West-Gaeke method for, 389
 and death rate, 90, 96
 Phillips monitor for, 437
 particulates, source of, 452, 453
 reactions with, 460
Sun photometer, Volz, 480
Superadiabatic, 284. *See also* Lapse
Superposition of sources, 319
Sutton coefficients, 316, 332
Sutton diffusion equations, 314-315
 with particle deposition, 327-328
Synergy, 105, 269
 of SO_2 and particulates, 96
Synoptic meteorology, 297
System, analysis of, 189
 criteria for, 7-9
 definition of, 7
 design principles, 9-11
 functions, in air pollution control, 195-198, 199
 in second generation system, 207-209
 goals for air pollution control, 196-198, 204-207
 organization in U.S. air pollution control, 197-198

Tape samplers, 394, 466, 470, 473
 analysis of spots, 394
 photometry in, 470-472
Technology and pollution, vii, 7
Telemetry, 378
Temperatures, global changes, *see* Carbon dioxide

Subject Index   531

Terminal velocity of particles, 31
Terpenes, 61, 63, 65
Tetroon, 297
Textiles, deterioration costs, 114, 116
Thermal precipitation, 468
Thermal radiation, 418, 422
Thermal radiometer, 298
Thermal reactions, 18
Threshold, of biological effects, 103-104
Threshold limit value (TLV), 15, 83, 96, 97
Time constant, sensor, 227-229
Titrimetry, 393-394
TLV, 15, 83, 96, 97
Topographical and geographical data, 221, 235
Topography, effect on wind, 289
Towers, television, 298
Toxic limits, see TLV
Tracers, SO_2, 357
    ZnS, 331
Traffic, auto, 25
Transfer function, 228, 385
Transient response of sensors, 228
Transmission, data, 245-248, 249
    capacity, 246
Transmissometer, 481
Transmittance, optical, 481
Transportation, mass, 499
Trapping, 400
Trendex, Inc., 120
Troposphere, 41, 43
TSP (total suspended particulates), see Particulates
Turbidity, 11
Turbulence, 31, 306
    in flow around structures, 328
    see also Eddies
Turbulence computer (sigma), 297-298
Turbulent diffusion, 312-313
Turbulent viscosity, 307

Ultraviolet, 1, 5, 18, 123, 218, 239
"Unacceptable days," 268
UN Stockholm Conference, 89, 184

Urban air pollution, 15
    effects of, 89-103
Urban envelope, 286-287
Urban models, 338-345

Vegetation losses, 117-118
Ventilation rate, 304
Viscosity, dynamic, 306
    kinematic, 306
    macro-, 314
    turbulent, 307
Visible emission, 173-175
Visibility, 121-123
Volcanic activity, 84-85, 452
Volz sun photometer, 480
von Karman's constant, 307-308
Vortex generator, 295

Walther's rankings, 272-274
Waste gases, hydrocarbon, 178
Wave Propogation Laboratory (WPL-NOAA), 243
Weather, air pollution effects, 123-125
    in cities, 124
    models, 87-88
    records, 83
    see also Climate
Weighting of effects (Index), 269-271
West-Gaeke procedure, 389
Wet chemical analysis, 388-394, 473-475
Willingness-to-pay, 120-121
Wind, anabatic, 289
    katabatic, 289
    power law of, 312
    troposphere, 288-290
    wind rose, 323
Worldwide air-pollution regulations, 182-184

X-ray Fluorescence, 474

Zero emissions, 9, 14
Zinc sulfates, 91, 126